CONTEMPORARY HUMAN GEOGRAPHY

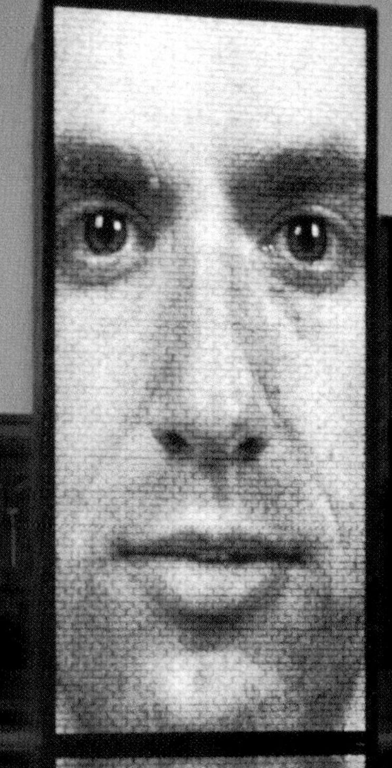

CONTEMPORARY HUMAN GEOGRAPHY

James M. Rubenstein

MIAMI UNIVERSITY; OXFORD, OHIO

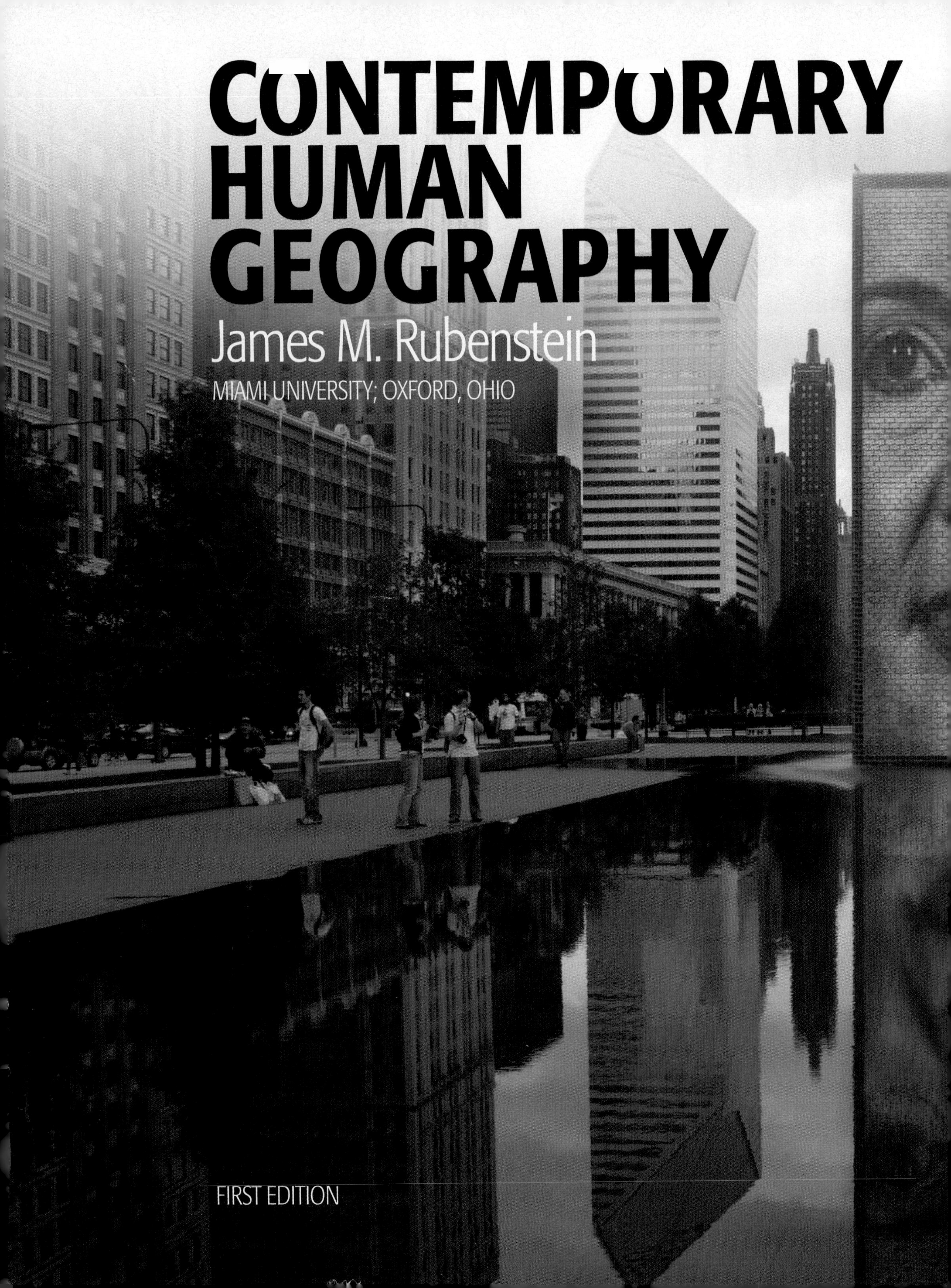

FIRST EDITION

Prentice Hall

New York Boston San Francisco

London Toronto Sydney Tokyo Singapore Madrid

Mexico City Munich Paris Cape Town Hong Kong Montreal

Library of Congress Cataloging-in-Publication Data
Rubenstein, James M.
Contemporary human geography/James M. Rubenstein.—1st ed.
p. cm.
Includes bibliographical references and index.
ISBN-13: 978-0-321-59003-9
ISBN-10: 0-321-59003-1
1. Human geography. I. Title.
GF41.R8 2010
304.2—dc22
2009000300

DORLING KINDERSLEY

Page Design, Development, and Layout: Stuart Jackman
DTP: David McDonald and Julie Thompson
Project Art Editor: Clive Savage
Maps: Advanced Illustration, Congleton, UK; and Iorwerth Watkins
Cover Design: Stuart Jackman

PRENTICE HALL

Acquisitions Editor, Geography: Christian Botting
Project Manager, Geography: Tim Flem
Editor in Chief, Geosciences and Chemistry: Nicole Folchetti
Development Editor: Melissa Parkin
Managing Editor, Geosciences and Chemistry: Gina M. Cheselka
Production Editor: Edward Thomas
Senior Media Producer: Angela Bernhardt
Media Project Manager: Natasha Wolfe
Marketing Manager: Amy Porubsky
Marketing Assistant: Keri Parcells
Senior Operations Supervisor: Alan Fischer
Art Director: Maureen Eide
Copy Editor: Jane Loftus
Indexer: Bob Richardson
Editorial Assistant: Haifa Mahabir
Photo Research Manager: Elaine Soares
Image Permission Coordinator: Debbie Hewitson
U.S. Composition: GGS Composition
Cover Photo: Millennium Park, Chicago. Don Nowak
Half Title and Title Page Photos: Crown Fountain at Millennium Park, Chicago. Jaume Plensa, © Art on File/Corbis

Printed in the United States of America
10 9 8 7 6 5 4 3 2 1

ISBN 10: 0-32-159003-1
ISBN 13: 978-0-32-159003-9

Prentice Hall is an imprint of

www.pearsonhighered.com

BRIEF CONTENTS

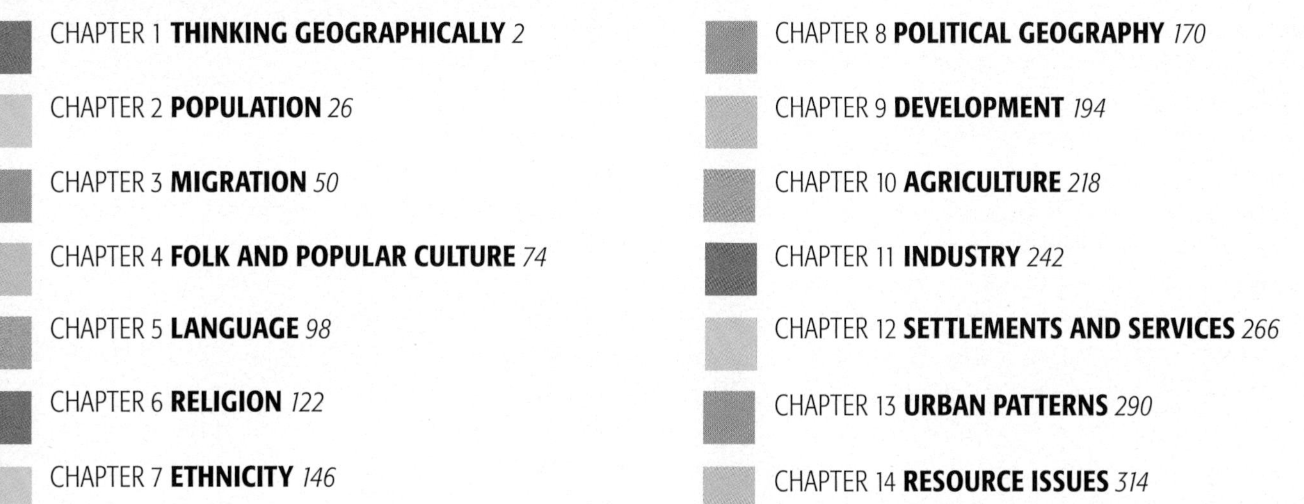

CONTENTS

CONTENTS

STARBUCKS COFFEE

CORRIER
EL P
ПОЛИТИКА
EON

Chapter 7 **ETHNICITY** *146*

Chapter 8 **POLITICAL GEOGRAPHY** *170*

Chapter 9 **DEVELOPMENT** *194*

Chapter 10 **AGRICULTURE** *218*

Chapter 11 **INDUSTRY** *242*

Chapter 12 **SETTLEMENTS AND SERVICES** *266*

PREFACE

Welcome to a new kind of geography textbook! We live in a visual age, and geography is a highly visual discipline, so Pearson Prentice Hall—the world's foremost publisher of geography textbooks—invites you to study human geography as a visual subject.

HUMAN GEOGRAPHY IS CONTEMPORARY

The main purpose of this book is to introduce you to the study of geography as a social science by emphasizing the relevance of geographic concepts to human problems. It is intended for use in college-level introductory human or cultural geography courses. The book is written for students who have not previously taken a college-level geography course.

A central theme in this book is a tension between two important themes—globalization and cultural diversity. In many respects we are living in a more unified world economically, culturally, and environmentally. The actions of a particular corporation or country affect people around the world. This book argues that, after a period when globalization of the economy and culture has been a paramount concern in geographic analysis, local diversity now demands equal time. People are taking deliberate steps to retain distinctive cultural identities. They are preserving little-used languages, fighting fiercely to protect their religions, and carving out distinctive economic roles.

Recent world events lend a sense of urgency to geographic inquiry. A decade into the twenty-first century, we face wars in unfamiliar places and economic struggles unprecedented in the lifetimes of students or teachers. Geography's spatial perspectives help to relate economic change to the distributions of cultural features such as languages and religions, demographic patterns such as population growth and migration, and natural resources such as energy and food supply.

For example, geographers examine the prospects for an energy crisis by relating the distributions of energy sources and consumption. Geographers find that the users of energy are located in places with different social, economic, and political institutions than the producers of energy. Geographers seek first to describe the distribution of features such as the production and consumption of energy, and then to explain the relationships between these distributions and other human and physical phenomena.

CHAPTER ORGANIZATION

Each chapter is organized into 12 two-page modular "spreads" that follow a consistent pattern:

Introductory Spread. The first spread includes a short introduction to the chapter, as well as an outline of nine key issues that will be addressed in the chapter. The nine key issues are grouped into three overall chapter themes.

Nine Key Issues Spreads. Nine spreads cover the principal issues of the chapter. Each of these two-page spreads is self-contained, making it easier for an instructor to shuffle the order of presentation. A numbering system also facilitates finding material on a particular spread.

Summary Spread. Following the nine key issues spreads are two concluding spreads. First is a summary spread, in which the three overall chapter themes are repeated with a brief review of the important concepts covered in detail in the text.

Geographic Consequences of Change. The summary spread includes an essay titled "Geographic Consequences of Change" that returns to the principal theme of the chapter with an eye towards the future.

Share Your Voice. The Summary spread includes an essay by a college student chosen from submissions from around the country. These essays represent the views of the students and have not been edited for content. Students are encouraged to submit essays for the next edition of the book.

Resources Spread. The final spread includes a detailed look at a topic related to the theme of the chapter that could not be fully covered in one of the key issue spreads.

Key Terms. The key terms in each chapter are indicated in bold type when they are introduced. These terms are defined both at the end of the chapter and at the end of the book.

On the Internet. The Resources spread includes information on useful Internet sites related to the theme of the chapter.

Further Readings. A list of books and articles is provided for students who wish to study the subject further.

OUTLINE OF TOPICS

This book discusses the following main topics:

What basic concepts do geographers use? Chapter 1 provides an introduction to ways that geographers think about the world. Geographers employ several concepts to describe the distribution of people and activities across Earth, to explain reasons underlying the observed distribution, and to understand the significance of the arrangements.

Where are people located in the world? Chapters 2 and 3 examine the distribution and growth of the world's population, as well as the movement of people from one place to another. Why do some places on Earth contain large numbers of people or attract newcomers whereas other places are sparsely inhabited?

How are different cultural groups distributed? Chapters 4 through 8 analyze the distribution of different cultural traits and beliefs and the problems that result from those spatial patterns. Important cultural traits discussed in Chapter 4 include food, clothing, shelter, and leisure activities. Chapters 5 through 7 examine three main elements of cultural identity: language, religion, and ethnicity. Chapter 8 looks at political problems that arise from cultural diversity. Geographers look for similarities and differences in the cultural features at different places, the reasons for their distribution, and the importance of these differences for world peace.

How do people earn a living in different parts of the world? Human survival depends on acquiring an adequate food supply. One of the most significant distinctions in the world is whether people produce their food directly from the land or buy it with money earned by performing other types of work. Chapters 9 through 12 look at the three main ways of earning a living: agriculture, manufacturing, and services. Chapter 13 discusses cities, the centers for economic and cultural activities.

What issues result from using Earth's resources? The final chapter is devoted to a study of issues related to the use of Earth's natural resources. Geographers recognize that cultural problems result from the depletion, destruction, and inefficient use of the world's natural resources.

INSTRUCTIONAL PACKAGE

In addition to the text itself, the author and publisher have been pleased to work with a number of talented people to produce an excellent instructional package. This package includes the traditional supplements that students and professors have come to expect from authors and publishers, as well as electronic media.

Online Study Guide. This website gives students the opportunity to further explore topics presented in the book using the Internet. The site contains human geography videos, numerous review exercises (from which students get immediate feedback), projects to expand understanding of human geography, and resources for further exploration. Access is provided with each new textbook. Please visit **www.mygeoscienceplace.com.**

Goode's World Atlas. The 22nd edition of the *Goode's World Atlas* includes 160 pages of new, digitally-produced reference maps, as well as new thematic maps on global climate change, sea level rise, CO_2 emissions, polar ice fluctuations, deforestation, extreme weather events, infectious diseases, water resources, and energy production. Pearson Prentice Hall offers the atlas at a substantially reduced price with *Contemporary Human Geography*.

Instructor Resource Center DVD and on-line. The Pearson Prentice Hall *Instructor Resource Center* provides instructors with everything they need in one, well-organized, easy-to-access place. Instructor resources include:

Figures: JPEGs of all maps, graphs, and charts, and select photographs from the text.

PowerPoint™: Pre-authored slides outlining the concepts of each chapter with embedded art can be used as is for lecture, or customized to fit individual lecture needs.

Test Bank: Multiple choice, short answer, essay, and true/false questions for each chapter available in TestGen or Microsoft Word format.

THE TEAM

At this point in the preface, an author usually goes through the motions of perfunctorily thanking many people who performed jobs that resulted in the book's production. In this case, collaborative partnership is a better way to describe the process.

Let's face it, some textbooks have been slow to adapt to our visual age. This is because the steps involved in producing most textbooks haven't changed much. The book passes from one to another like a baton in a relay race; those responsible for producing the book typically start their work only after the author's words have been written, reviewed, and approved.

In contrast, *Contemporary Human Geography* started as a genuine partnership among the key editorial and production teams. Each two-page spread was assembled in the reverse order of traditional textbooks. Instead of beginning with an author's complete manuscript, this book started with a sketch of a visual concept for each two-page spread in the book. What would be the most important geographic idea presented on the spread, and what would be the most effective visual way to portray that idea? The maps and images were placed on the page first, and then the text was written around the graphics.

The traditional separation of editorial and production personnel did not occur, and in fact the lines between the two were deliberately blurred. Key members of the team included Tim Flem, Stuart Jackman, Jane Loftus, Melissa Parkin, and Ed Thomas.

Tim Flem, Geography Project Manager at Pearson Education, was the ringmaster. He kept track of what was where and who was doing what.

Stuart Jackman, Design Director at DK Education, is the creative genius responsible for the spectacular graphics. I consider Stuart a "co-author" of this book.

Copyeditor Jane Loftus and Development Editor Melissa Parkin did their essential work "on the fly" seamlessly, despite often being outside the typical workflow.

Ed Thomas, Science Production Project Manager at Pearson Education, juggled the unconventional production schedule.

Many others at DK and Pearson have contributed to the success of this project. At DK, Clive Savage, Project Art Editor, created the book's distinctive maps and graphics. At Pearson Education, Managing Editor Gina Cheselka was a major organizing force in the nonstandard production workflow. Editorial Assistant Haifa Mahabir organized the substantial reviewing process for the project. Marketing Manager Amy Porubsky expertly created the marketing package for this unique book. Media Producers Angela Bernhardt and Ziki Dekel managed the production of the *Online Study Guide Companion Website* that accompanies this book.

Essential to the success of this innovative project was the support of senior leaders at DK and Pearson. Sophie Mitchell, Publisher at DK Education, provided the strategic vision for the design team. Pearson has been the dominant publisher of geography textbooks for a quarter century, and it is no coincidence that for most of that time Pearson's geography program has been led by only two editors: Paul Corey, now President of Pearson Science, and Dan Kaveney, now Publisher for Pearson Chemistry. These two individuals have played such a central role in my life as a geography author that I will always be grateful to them. During the development of this book, Christian Botting took over as Pearson's Geography Editor. I look forward to working closely with Christian for many years to come.

REVIEWERS

I'd like to extend a special thanks to my colleagues who served as reviewers as we developed this new textbook:

Roger Balm, Rutgers University
Joby Bass, University of Southern Mississippi
Craig S. Campbell, Youngstown State University
Edward Carr, University of South Carolina
Owen Dwyer, Indiana University-Purdue University, Indianapolis
Barbara E. Fredrich, San Diego State University
Piper Gaubatz, University of Massachusetts, Amherst
Daniel Hammel, University of Toledo
Leila Harris, University of Wisconsin
Susan Hartley, Lake Superior College
Marc Healy, Elgin Community College
Scot Hoiland, Butte College
Wilbur Hugli, University of West Florida
Anthony Ijomah, Harrisburg Area Community College
Karen Johnson-Webb, Bowling Green State University
Oren Katz, California State University, Los Angeles
Olaf Kuhlke, University of Minnesota, Duluth
Claudia Lowe, Fullerton College
Ken Lowrey, Wright State University
Eric C. Neubauer, Columbus State
Ray Oman, University of the District of Columbia
Tim Scharks, Green River Community College
Debra Sharkey, Cosumnes River College
Wendy Shaw, Southern Illinois University, Edwardsville
Daniel Vara, College Board Advanced Placement Human Geography Consultant
Anne Will, Skagit Valley College
Lei Xu, California State University, Fullerton
Daisaku Yamamoto, Central Michigan University
Robert C. Ziegenfus, Kutztown University of Pennsylvania

ABOUT THE AUTHOR

Dr. James M. Rubenstein received his Ph.D. from Johns Hopkins University in 1975. His dissertation on French urban planning was later developed into a book entitled *The French New Towns* (The Johns Hopkins University Press, 1978). In 1976 he joined the faculty at Miami University, where he is currently Professor of Geography. Besides teaching courses on Urban and Human Geography and writing textbooks, Dr. Rubenstein also conducts research in the automotive industry and has published three books on the subject, entitled *The Changing U.S. Auto Industry: A Geographical Analysis* (Routledge, 1992), *Making and Selling Cars: Innovation and Change in the U.S. Auto Industry* (The Johns Hopkins University Press, 2001), and *Who Really Made Your Car? Restructuring and Geographic Change in the Auto Industry*, co-authored with Thomas Klier (W. E. Upjohn, 2008). Originally from Baltimore, he is an avid Orioles fan (despite their mediocrity). He also spends much of the year in Chicago, and is an active painter. Winston, a lab–husky (?) mix with one brown eye and one blue eye, takes Dr. Rubenstein for a long walk in the woods every day.

This book is dedicated to Bernadette Unger, Dr. Rubenstein's wife, who has been by his side through many books. Dr. Rubenstein also gratefully thanks the rest of his family for their love and support.

Note to Students

A NEW APPROACH TO HUMAN GEOGRAPHY TEXTBOOKS

CONTEMPORARY HUMAN GEOGRAPHY BY JAMES M. RUBENSTEIN

This book takes a dramatic new approach to Human Geography. Instead of supporting words with graphics—or supporting graphics with words—it integrates words and graphics in the bold visual style that distinguishes Dorling Kindersley publications.

James Rubenstein's best-selling textbook *The Cultural Landscape: An Introduction to Human Geography* has always been known for the author's sharp focus on key geographic issues and concepts. In his new ***Contemporary Human Geography,*** Dr. Rubenstein applies this same focus, and teams up with the information architects at Dorling Kindersley to produce an exciting, clear-cut treatment of the topic. The result is a dramatic view of Human Geography that is at once inviting, informative, and challenging.

THE BOOK'S ORGANIZATION CONSISTS OF A SERIES OF TWO-PAGE MODULES. Each spread is devoted to one important topic.

INTEGRATED MAPS highlight the spatial dimensions of the topics presented, without sacrificing the structure so important to building an understanding of Human Geography.

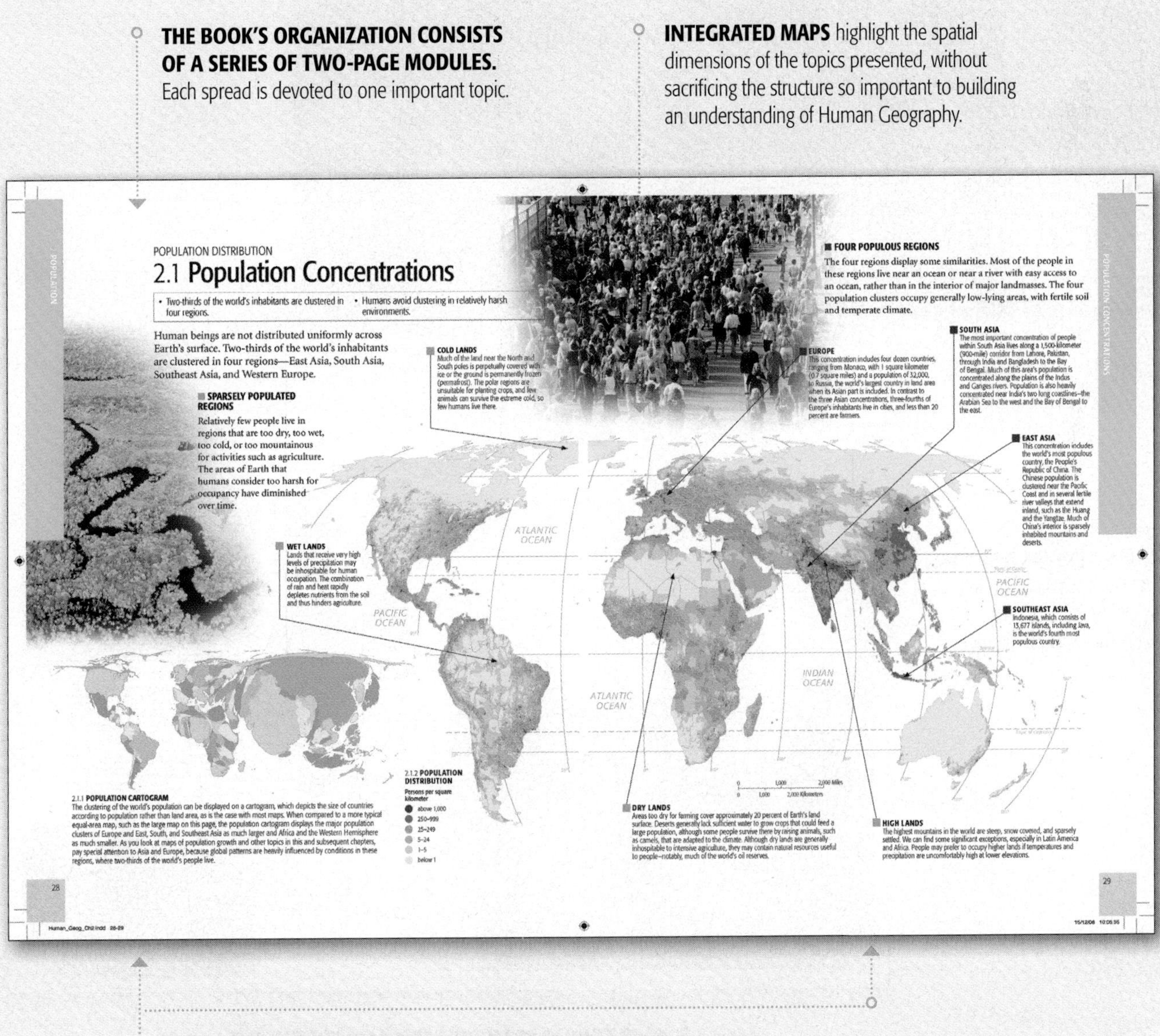

POPULATION DISTRIBUTION

2.1 Population Concentrations

- Two-thirds of the world's inhabitants are clustered in four regions.
- Humans avoid clustering in relatively harsh environments.

Human beings are not distributed uniformly across Earth's surface. Two-thirds of the world's inhabitants are clustered in four regions—East Asia, South Asia, Southeast Asia, and Western Europe.

SPARSELY POPULATED REGIONS

Relatively few people live in regions that are too dry, too wet, too cold, or too mountainous for activities such as agriculture. The areas of Earth that humans consider too harsh for occupancy have diminished over time.

COLD LANDS

Much of the land near the North and South poles is perpetually covered with ice or the ground is permanently frozen (permafrost). The polar regions are unsuitable for planting crops, and few animals can survive the extreme cold, so few humans live there.

WET LANDS

Lands that receive very high levels of precipitation may be inhospitable for human occupation. The combination of rain and heat rapidly depletes nutrients from the soil and thus hinders agriculture.

FOUR POPULOUS REGIONS

The four regions display some similarities. Most of the people in these regions live near an ocean or near a river with easy access to an ocean, rather than in the interior of major landmasses. The four population clusters occupy generally low-lying areas, with fertile soil and temperate climate.

EUROPE

This concentration includes four dozen countries, ranging from Monaco, with 1 square kilometer (0.7 square miles) and a population of 32,000, to Russia, the world's largest country in land area when its Asian part is included. In contrast to the three Asian concentrations, three-fourths of Europe's inhabitants live in cities, and less than 20 percent are farmers.

SOUTH ASIA

The most important concentration of people within South Asia lives along a 1,500-kilometer (900-mile) corridor from Lahore, Pakistan, through India and Bangladesh to the Bay of Bengal. Much of this area's population is concentrated along the plains of the Indus and Ganges rivers. Population is also heavily concentrated near India's two long coastlines—the Arabian Sea to the west and the Bay of Bengal to the east.

EAST ASIA

This concentration includes the world's most populous country, the People's Republic of China. The Chinese population is clustered near the Pacific Coast and in several fertile river valleys that extend inland, such as the Huang and the Yangtze. Much of China's interior is sparsely inhabited mountains and deserts.

SOUTHEAST ASIA

Indonesia, which consists of 13,677 islands, including Java, is the world's fourth most populous country.

2.1.1 POPULATION CARTOGRAM

The clustering of the world's population can be displayed on a cartogram, which depicts the size of countries according to population rather than land area, as is the case with most maps. When compared to a more typical equal-area map, such as the large map on this page, the population cartogram displays the major population clusters of Europe and East, South, and Southeast Asia as much larger and Africa and the Western Hemisphere as much smaller. As you look at maps of population growth and other topics in this and subsequent chapters, pay special attention to Asia and Europe, because global patterns are heavily influenced by conditions in these regions, where two-thirds of the world's people live.

2.1.2 POPULATION DISTRIBUTION

Persons per square kilometer

- above 1,000
- 250–999
- 25–249
- 5–24
- 1–5
- below 1

DRY LANDS

Areas too dry for farming cover approximately 20 percent of Earth's land surface. Deserts generally lack sufficient water to grow crops that could feed a large population, although some people survive there by raising animals, such as camels, that are adapted to the climate. Although dry lands are generally inhospitable to intensive agriculture, they may contain natural resources useful to people—notably, much of the world's oil reserves.

HIGH LANDS

The highest mountains in the world are steep, snow covered, and sparsely settled. We can find some significant exceptions, especially in Latin America and Africa. People may prefer to occupy higher lands if temperatures and precipitation are uncomfortably high at lower elevations.

28

29

Human_Geog_Ch2.indd 28-29

15/12/08 10:05:55

THE INTEGRATION OF TEXT, PHOTOGRAPHS, DIAGRAMS, AND MAPS encourages a complete and compelling view of human geography—in a brief format.

A Bold New Architecture from James M. Rubenstein

In this brief format, Dr. Rubenstein presents carefully selected topics that are a springboard to each professor's Human Geography course.

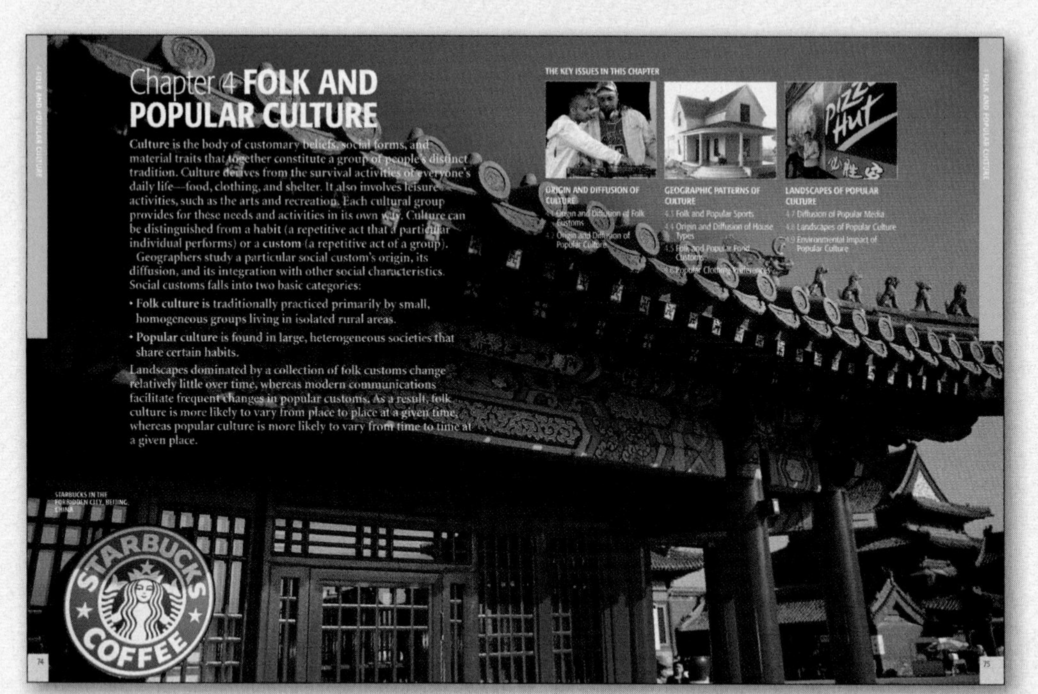

Chapter 4 **FOLK AND POPULAR CULTURE**

Culture is the body of customary beliefs, social forms, and material traits that together constitute a group of people's distinct tradition. Culture derives from the survival activities of everyone's daily life—food, clothing, and shelter. It also involves leisure activities, such as the arts and recreation. Each cultural group provides for these needs and activities in its own way. Culture can be distinguished from a habit (a repetitive act that a particular individual performs) or a custom (a repetitive act of a group).

Geographers study a particular social custom's origin, its diffusion, and its integration with other social characteristics. Social customs falls into two basic categories:

- **Folk culture** is traditionally practiced primarily by small, homogeneous groups living in isolated rural areas.
- **Popular culture** is found in large, heterogeneous societies that share certain habits.

Landscapes dominated by a collection of folk customs change relatively little over time, whereas modern communications facilitate frequent changes in popular customs. As a result, folk culture is more likely to vary from place to place at a given time, whereas popular culture is more likely to vary from time to time at a given place.

STARBUCKS IN THE FORBIDDEN CITY, BEIJING, CHINA

THE KEY ISSUES IN THIS CHAPTER

ORIGIN AND DIFFUSION OF CULTURE

GEOGRAPHIC PATTERNS OF CULTURE
4.3 Folk and Popular Sports
4.4 Origin and Diffusion of House Types

LANDSCAPES OF POPULAR CULTURE
4.7 Diffusion of Popular Media
4.8 Landscapes of Popular Culture

CHAPTER OPENERS OUTLINE THE MODULES students are about to explore with a visually engaging spread and an introduction that inspires curiosity.

VISUALS, TEXT, AND SPATIAL GRAPHICS are designed to highlight multiple dimensions of the content within the module.

ETHNIC CONFLICTS

7.8 Ethnic Cleansing

- Ethnic cleansing is the forcible removal of an ethnic group by a more powerful one.
- Ethnic cleansing has been practiced in several African countries.

Ethnic cleansing is a process in which a more powerful ethnic group forcibly removes a less powerful one in order to create an ethnically homogeneous region.

Ethnic cleansing is undertaken to remove an entire ethnicity from an area so that the surviving ethnic group can be the sole inhabitants. Rather than a clash between armies of male soldiers, ethnic cleansing involves the removal of every member of the less powerful ethnicity—women as well as men, children as well as adults, the frail elderly as well as the strong youth.

Sub-Saharan Africa has been a region especially plagued by conflicts among ethnic groups competing to become dominant within the various countries. Conflict is widespread in Africa largely because the present-day boundaries of states were drawn by European colonial powers about a hundred years ago without regard for the traditional distribution of ethnicities. When the European colonies in Africa became independent states, especially during the 1950s and 1960s, the boundaries of the new states typically matched the colonial administrative units imposed by the Europeans. As a result, many African states contained large numbers of ethnicities.

EACH TWO-PAGE MODULE VISUALLY COVERS ONE CONCEPT, incorporating **CAREFULLY CRAFTED TEXT** that stimulates classroom discussion.

END-OF-CHAPTER RESOURCES CONCLUDE EACH CHAPTER

CHAPTER SUMMARIES reinforce **KEY ISSUES** raised at the beginning of each chapter, listing overarching themes from each section and referencing familiar photographs that are consistently utilized to represent and help classify diversified information.

GEOGRAPHIC CONSEQUENCES OF CHANGE extends student understanding by drawing on relevant, contemporary issues in Human Geography. Located within each chapter's concluding two-page spread, this feature equips students with the ability to appreciate information presented in the preceding chapter as it relates to greater schemes in current events.

POPULATION

Chapter Summary

POPULATION DISTRIBUTION

- Global population is concentrated in a few places. Two-thirds of the world's people live in four clusters: East Asia, South Asia, Europe, and Southeast Asia.
- Human beings tend to avoid those parts of Earth's surface that they consider to be too wet, too dry, too cold, or too mountainous.

POPULATION GROWTH

- A population increases because of fertility and decreases because of mortality. The natural increase rate measures the difference between fertility and mortality.
- Virtually all the world's natural increase is concentrated in the LDCs of Africa, Asia, and Latin America, whereas the population is growing slowly or actually declining in MDCs.
- The demographic transition is a process of change in a country's population from a condition of high birth and death rates, with little population growth, to a condition of low birth and death rates, with low population growth.
- MDCs in Europe and North America have reached or neared the end of the demographic transition, whereas LDCs in Africa, Asia, and Latin America are at the stages of rapid population growth.

OVERPOPULATION THREATS

- More than 200 years ago, Thomas Malthus argued that population was increasing more rapidly than the food supply. With the rate of world population growth during the second half of the twentieth century at the highest level in history, some contemporary analysts believe that Malthus's prediction was accurate.
- Accounting for the record high natural increase rate during the twentieth century was a dramatic decline in death rates, especially from infectious diseases.
- Since the 1990s, birth rates have declined sharply, slowing world population growth and reducing fear of overpopulation in most regions. Meanwhile, death rates have increased in MDCs because of chronic disorders associated with aging and in some LDCs because of new infectious diseases, especially AIDS.

SHIBUYA, A DISTRICT OF TOKYO, JAPAN

46

CHAPTER SUMMARY

Geographic Consequences of Change

Worldwide population increased rapidly during the second half of the twentieth century because few countries were in the two stages of the demographic transition that have low population growth—no country remains in stage 1, and few have reached stage 4.

The four-stage demographic transition is characterized by two big breaks with the past. The first break—the sudden drop in the death rate that comes from technological innovation—has been accomplished everywhere. The second break—the sudden drop in the birth rate that comes from changing social customs—has yet to be achieved in many countries.

If most countries in Europe and North America have reached—or at least are approaching—stage 4 of the demographic transition, why aren't other countries elsewhere in the world? LDCs moved into stage 2 of the demographic transition in the twentieth century for different reasons than did MDCs in the nineteenth century. In the past, the CDR declined in MDCs as part of the Industrial Revolution. In contrast, the sudden drop in the CDR in LDCs occurred because of injection of medical technology invented in MDCs.

Having caused the first break with the past through diffusion of medical technology, MDCs now urge LDCs to complete the second break with the past—the reduction in the CBR. However, reducing the CBR is difficult. A decline in the CDR can be induced through introduction of new technology by outsiders, but the CBR will drop only when people decide for themselves to have fewer children.

Geographers fear that as a result, LDCs, especially in Sub-Saharan Africa, are threatened by overpopulation—too many people for the available resources. Geographers caution that the number of people living in a region is not by itself an indication of overpopulation. Some densely populated regions are not overpopulated, whereas some sparsely inhabited areas are. The capacity of the land to support life derives partly from characteristics of the natural environment and partly from human actions to modify the environment through agriculture, industry, and exploitation of raw materials.

SHARE YOUR VOICE Student Essay

ONE TOO MANY
By Julieta Gomez de Mello
Florida International University

JULIETA GOMEZ DE MELLO

I had a history teacher once who joked that South Florida should be proclaimed as an autonomous state, separate from the rest of Florida. While this may sound a bit far-fetched, people who live or have lived down here know how isolated we are from the rest of the country in terms of population and culture.

Simply put, we are one too many.

Thousands upon thousands of people fill Miami's streets under the scorching sun either in cars, riding bicycles or simply walking by foot. Classrooms that are equipped for about 20 students are sometimes replete with 30, even 40 students.

Though most people originate from a Spanish culture, it is meshed with English as well as other languages to create an entirely new dialect that is particular to the South Florida niche. So, how does one distinguish themselves amidst such a large crowd? The answer might not be so simple, but with an open mind, anything is possible.

As a native Spanish speaker myself, I'm always eager to take in a new culture and new experiences. When I was in fourth grade I met a Muslim girl. She was polite, funny and inteligent. We became friends and soon were inseparable. She came to my birthday parties and helped me with homework assignments and was the best friend I had always wanted. Aside from any cultural differences we had, what I can recall the most is how good it felt to have someone who cared and was there for me everyday. I admired her for that.

When I was in middle school and 9/11 came about, I understood the difference between people like my friend and terrorists who, as she herself told me once, manipulated the Islamic religion, molded it to their liking, until it twisted like breadsticks. Living in Miami and associating with people from different countries and cultures has really shaped who I am today and allowed me to understand others. Therefore, the best solution for finding one's identity is to perhaps mesh with the crowd and learn from different cultures, languages and religions to then decide for yourself who you are among the many.

47

Human_Geog_Ch2.indd 46-47 15/12/08 10:10:07

STUDENT ESSAYS AT THE END OF EACH CHAPTER provide local insights and stories that connect the content of the chapter to students on the ground.

SUPPLEMENTS

INSTRUCTOR RESOURCE CENTER ON DVD & ONLINE

Teaching your Human Geography course just got easier! For your convenience, many of our instructor supplements are available on DVD or for download from your textbook's catalog page. Available resources include PowerPoint® presentations, Image Banks, and *Test Bank* files.

TESTGEN

This test-generating program allows instructors to add, edit, or delete questions from the *Test Bank*; analyze test results; and organize a database of exams and student results. The *Test Item File* is also available in Microsoft Word format.

COMPANION WEBSITE

www.mygeoscienceplace.com

The *Online Study Guide* guide is designed to accompany *Contemporary Human Geography.* This website provides both drill-based review exercises and web-based critical thinking exercises that expand upon the content of the text.

Available for the first time online, Human Geography videos cover a wide array of issues affecting people and places in the contemporary world, including international immigration, the HIV/AIDS epidemic, urbanization, homelessness, poverty, and environmental destruction. Students can test their understanding of the videos with self study or assessable quizzes.

Innovative Book Development Process

In this unique new text, James Rubenstein's geographic expertise meets the bold, graphic style that characterizes Dorling Kindersley books.

STAGE 1: CONCEPT DEVELOPMENT

The author chooses the most important concepts from *The Cultural Landscape* and builds the modules that become ***Contemporary Human Geography,*** creating an innovative approach to content delivery.

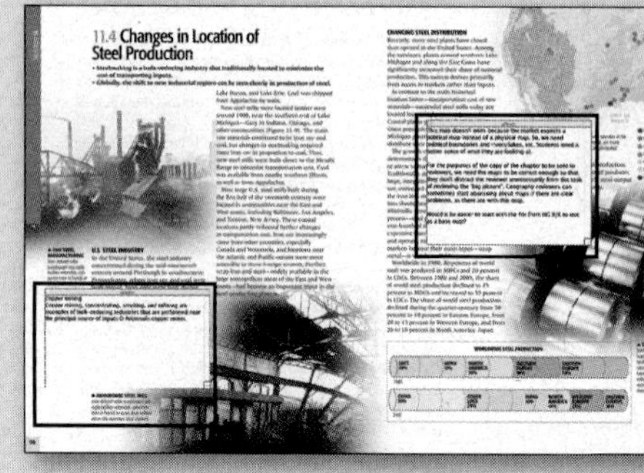

Author and publisher comment on the modules. This begins to build an information architecture that will give instructors a springboard for taking the course in any direction they choose.

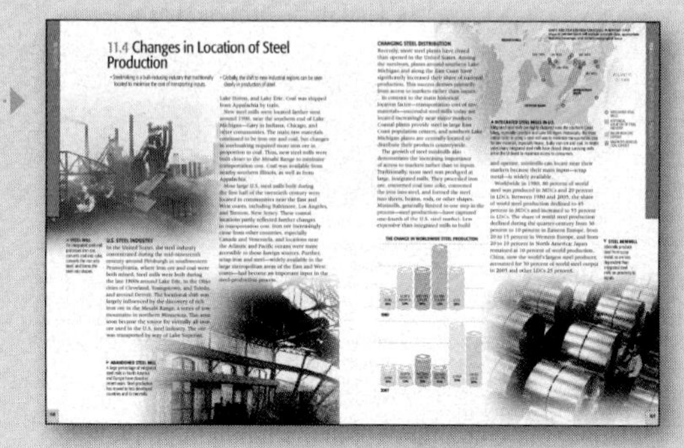

The new module is prepared for market testing.

STAGE 2: BOOK DEVELOPMENT

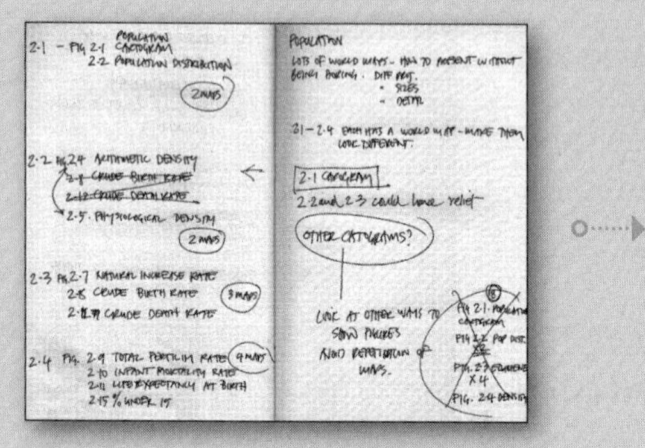

Reviewer comments are considered. Analyses of content notes are discussed at face-to-face meetings with the author.

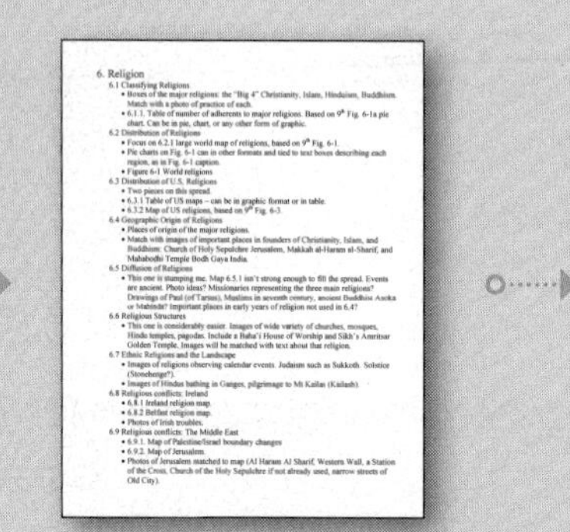

Working closely with the designer on each page, the author prepares brief contents for the complete chapter with image, map, and chart requirements.

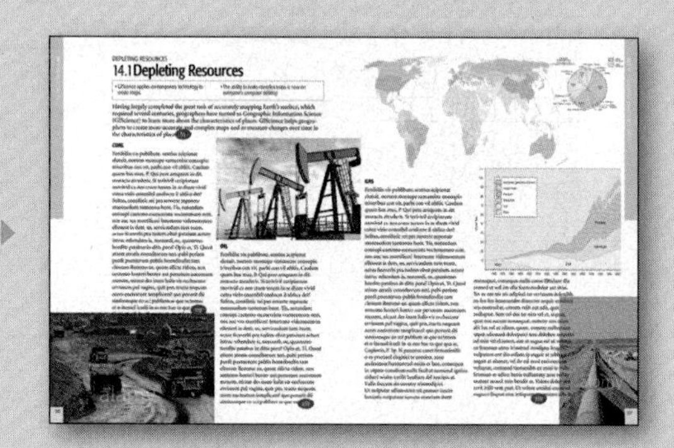

A first-pass design is created using image research, placeholder maps and charts, and placeholder text with word counts.

STAGE 3: BOOK CREATION

Author writes the text to length based on the designed pages.

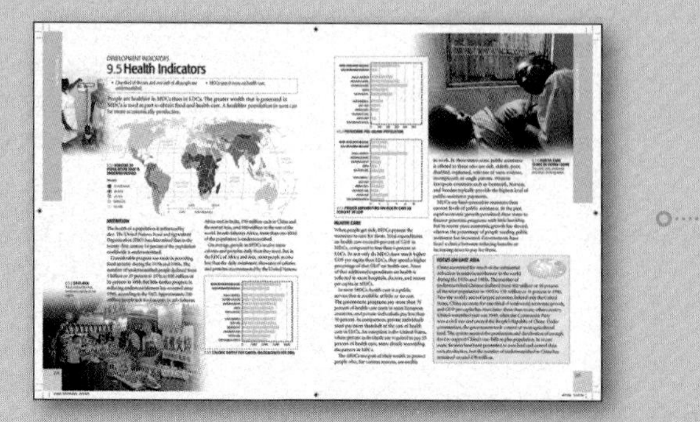

Pages are ready for reviewing and further class testing, where student and instructor response is overwhelmingly positive.

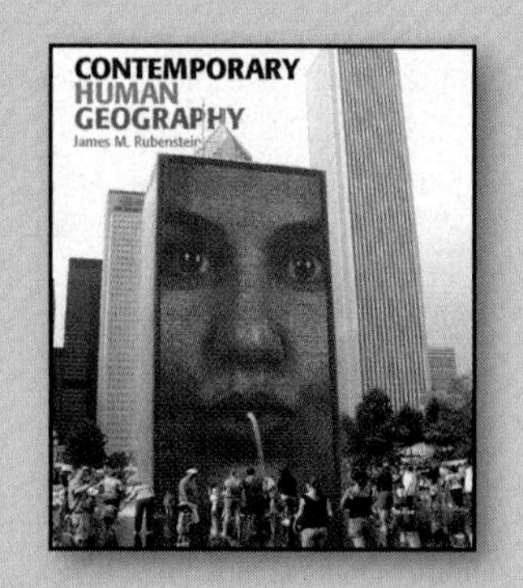

CONTEMPORARY HUMAN GEOGRAPHY

Chapter 1 THINKING GEOGRAPHICALLY

Thinking geographically is one of the oldest human activities. Perhaps the first geographer was a prehistoric human who crossed a river or climbed a hill, observed what was on the other side, returned home to tell about it, and scratched the route in the dirt. Perhaps the second geographer was a friend or relative who followed the dirt map to reach the other side.

Today, geographers are still trying to reach the other side, to understand more about the world in which we live. Geography is the study of where things are found on Earth's surface and the reasons for the location. Human geographers ask two simple questions: Where are people and activities found on Earth? Why are they found there?

HOUSING IN LEEUWARDEN, THE NETHERLANDS

THE KEY ISSUES IN THIS CHAPTER

DEVELOPMENT OF GEOGRAPHY

1.1 Geography Through the Ages
1.2 Geography Becomes a Science
1.3 Reading Maps
1.4 Contemporary Mapping Tools

THE UNIQUENESS OF EVERY PLACE

1.5 Places: Unique Locations
1.6 Regions: Unique Areas

SIMILARITIES AMONG PLACES

1.7 Scale: Global Patterns
1.8 Space: Distribution of Features
1.9 Connections Between Places

DEVELOPMENT OF GEOGRAPHY

1.1 Geography through the Ages

• Geography in the Ancient World • Geography in the Middle Ages

The word ***geography*** was invented by the ancient Greek scholar Eratosthenes (ca 276-ca 194 B.C.); it is based on two Greek words, ***geo*** meaning "Earth," and ***graphy***, meaning "to write." Eratosthenes accepted that Earth was round, as few did in his day; he also calculated its circumference within an amazing 0.5 percent accuracy. In one of the first geography books, he described the known world and correctly divided Earth into five climatic regions.

GEOGRAPHY IN THE ANCIENT WORLD

In ancient Greece, geography was associated with navigation.

- **Thales of Miletus** (ca 624–ca 546 B.C.) applied principles of geometry to measuring land area.
- **Anaximander** (610–ca 546 B.C.), a student of Thales, made a world map based on information from sailors, though he argued that the world was shaped like a cylinder.
- **Hecateus** (ca 550– ca 476 B.C.) may have produced the first geography book, called *Ges Periodos* (*Travels Around the Earth*).
- **Aristotle** (384–322 B.C.) was the first to demonstrate that Earth was spherical.
- **Strabo** (ca 63 B.C.– A.D. ca 24) described the known world in a 17-volume work entitled *Geography*.

1.1.1 **THE OLDEST KNOWN MAPS**
(above) A 7th century B.C. map of a plan for the town of Çatalhöyük, located in present-day Turkey. Archaeologists found the map on the wall of a house that was excavated in the 1960s. **(top right)** A color version of the Catalhoyuk map. A volcano rises above the buildings of the city.
(lower right) A world map from the 6th century B.C. depicts a circular land area surrounded by a ring of water. The ancient city of Babylon is thought to be shown in the center of the land area and other cities are shown as circles. Extending out from the water ring are seven islands that together form a star shape.

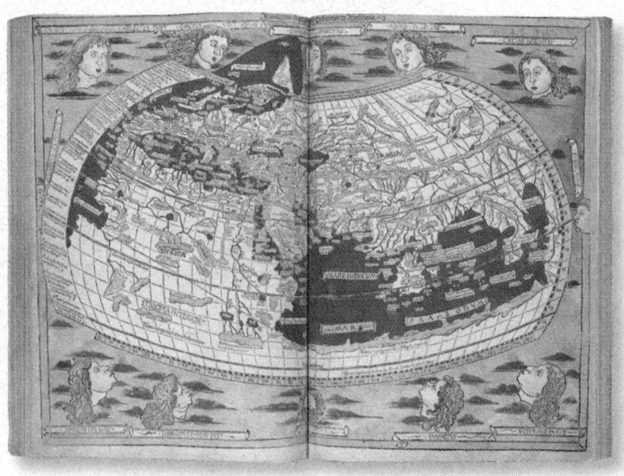

1.1.2 **A.D. 150. PTOLEMY'S MAP OF THE WORLD**
Two thousand years ago, the Roman Empire controlled an extensive area of the known world, including much of Europe, northern Africa, and western Asia. Taking advantage of information collected by merchants and soldiers who traveled through the Roman Empire, the Greek Ptolemy (ca A.D. 100–170) wrote an eight-volume *Guide to Geography*. He codified basic principles of mapmaking and prepared numerous maps, which were not improved upon for more than a thousand years.

GEOGRAPHY IN THE MIDDLE AGES

After Ptolemy, little progress in mapmaking or geographic thought was made in Europe for several hundred years. Maps became less mathematical and more fanciful, showing Earth as a flat disk surrounded by fierce animals and monsters. A revival of geography and mapmaking occurred in Europe during the Age of Exploration and Discovery. Geographic inquiry continued outside of Europe:

- The earliest surviving Chinese geographical writing, *Yu Gong* (*Tribute of Yu*), from the fifth century B.C., describes the economic resources of the country's different provinces.
- Pei Xiu, the "father of Chinese cartography," produced an elaborate map of the country in A.D. 267.
- Ibn-Battuta (1304–ca 1368) wrote *Rihlah* (*Journeys*) based on three decades of journeys covering more than 120,000 kilometers (75,000 miles) through the Muslim world of northern Africa, southern Europe, and much of Asia.

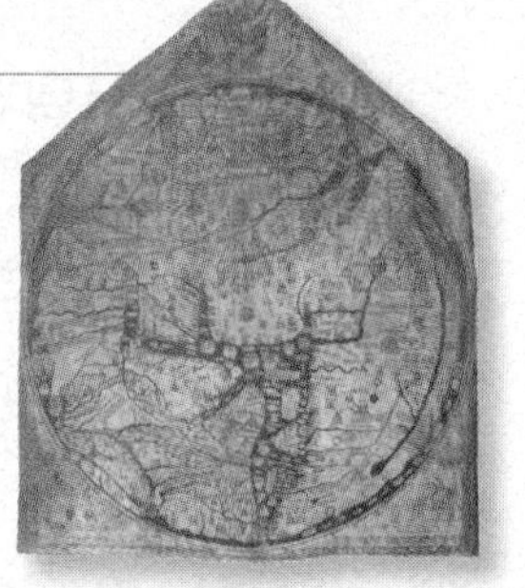

1.1.3. **1100. HEREFORD MAPPA MUNDI**
Jerusalem is at the center of this map made in England. The Garden of Eden is inside the circle near the top.

1.1.4. **1154. WORLD MAP BY AL-IDRISI**
Muslim geographer Muhammad al-Idrisi (1100–ca 1165) prepared a world map and geography text, building on Ptolemy's long-neglected work.

1.1.5. **1507. WORLD MAP BY WALDSEEMULLER**
German cartographer Martin Waldseemuller (ca 1470–ca 1521) is credited with the first map to use the label "America," which he wrote on the map derived "from Amerigo the discoverer."

1.1.6. **1508. WORLD MAP BY ROSSELLI**
Italian painter and engraver Francesco Rosselli (1445–ca 1513) was the first to draw a map on an oval projection and was the owner of the first known shop to sell maps.

1.1.7. **1530. WORLD MAP BY APIAN**
German cartographer Peter Apian (1495–1552) owned a print shop that produced high-quality maps; the Western Hemisphere coastline was still largely uncharted in the early sixteenth century.

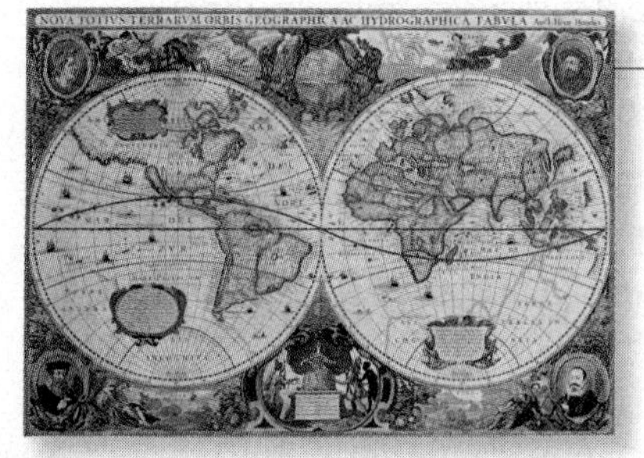

1.1.8. **1630. WORLD MAP BY HONDIUS**
Flemish engraver Jodocus Hondius (1563-1612) showed the world in two hemispheres. Along the borders, he engraved portraits of Caesar, Ptolemy, Mercator, and himself, as well as depictions of fire, air, water, and land.

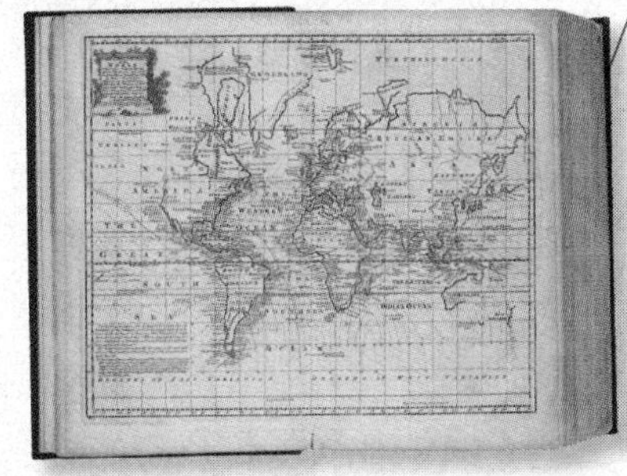

1.1.9. **1720. WORLD MAP BY VAN KEULEN**
Dutch cartographer Johannes van Keulen (1654–1711) published maps that made use of the Mercator projection, which was especially useful for mariners.

1.1.10. **SATELLITE IMAGE OF NEW YORK**

DEVELOPMENT OF GEOGRAPHY

1.2 Geography Becomes a Science

- The science of geography is based on human-environment interaction.
- Humans have the ability to modify the environment.

To explain relationships between human activities and the physical environment, geographers embrace the concept of **possibilism**. According to possibilism, the physical environment may limit some human actions, but people have the ability to adjust to their environment. People can choose a course of action from many alternatives in the physical environment.

ALEXANDER VON HUMBOLDT

CARL RITTER

FRIEDRICH RATZEL

ELLEN CHURCHILL SEMPLE

ELLSWORTH HUNTINGTON

PIONEERING SCIENTIFIC GEOGRAPHERS

A century ago, early scientific geographers had other views on the relationship between people and the environment:

- **Alexander Von Humboldt** (1769–1859), a German geographer, urged human geographers to adopt the methods of scientific inquiry used by natural scientists. He argued that the scientific study of social and natural processes is fundamentally the same.
- **Carl Ritter** (1779–1859), a German geographer, with Humboldt, argued that human geographers should apply laws from the natural sciences to understanding relationships between the physical environment and human actions. Humboldt and Ritter concentrated on how the physical environment caused social development, an approach called **environmental determinism.**
- **Friedrich Ratzel** (1844–1904), a German geographer, adopted environmental determinism. Environmental determinists argued that natural scientists had made more progress in formulating general laws than had social scientists, so an important goal of human geography would be to discover general scientific laws.
- **Ellen Churchill Semple** (1863–1932), a student of Ratzel, continued the environmental determinism approach. Contemporary geographers recognize that although the physical environment influences human behavior, people can adjust to the environment and act according to cultural values that may not be rooted in the environment.
- **Ellsworth Huntington** (1876–1947), an American geographer, argued that climate was a major determinant of civilization. For example, the temperate climate of maritime northwestern Europe produced greater human efficiency as measured by better health conditions, lower death rates, and higher standards of living.

Modern geographers reject the views of Huntington and other environmental determinists in favor of possibilism.

ENVIRONMENTAL SENSITIVITY

Modern technology has altered the historic relationship between people and the environment. Geographers are concerned that people sometimes use modern technology to modify the environment insensitively. Human actions can deplete scarce environmental resources, destroy irreplaceable resources, and use resources inefficiently.

For example, air-conditioning has increased the attractiveness of living in regions with warmer climates. But the refrigerants in the air conditioners have also increased the amount of chlorofluorocarbons in the atmosphere, damaging the ozone layer that protects living things from UV rays and contributing to global warming.

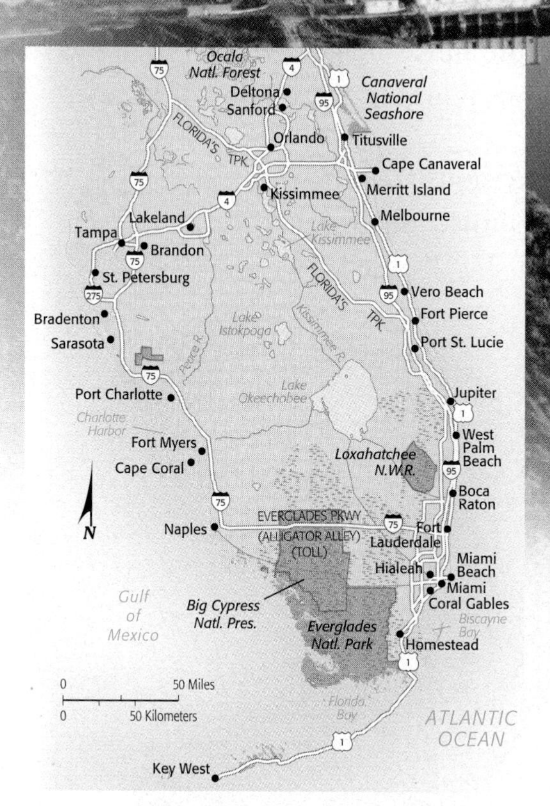

KISSIMMEE RIVER

1.2.1 FLORIDA: NOT-SO-SENSITIVE ENVIRONMENTAL MODIFICATION

What was once a very wide and shallow freshwater river is now the Everglades. A sensitive ecosystem of plants and animals once thrived in this distinctive landscape, but much of it has been destroyed by human actions. The U.S. Army Corps of Engineers built a levee around Lake Okeechobee during the 1930s, drained the northern one-third of the Everglades during the 1940s, diverted the Kissimmee River into canals during the 1950s, and constructed dikes and levees near Miami and Fort Lauderdale during the 1960s.

These modifications created farmland for growing sugarcane and protected cities from flooding. But they also killed fish in Lake Okeechobee and threatened rare vegetation and animal species in the Everglades. The 2000 Comprehensive Everglades Restoration Plan called for restoring the historic flow of water through South Florida while improving flood control and water quality.

1.2.2 THE NETHERLANDS: SENSITIVE ENVIRONMENTAL MODIFICATION

The Dutch have modified their environment with two distinctive types of construction projects:

- **Polders.** The Netherlands has 6,500 square kilometers (2,600 square miles) of polders, which are areas created by draining water. **Polders** are made by building a wall around an underwater area, then pumping out the water from inside the walled area. The dried out area can be used for housing, agriculture, or business.
- **Dikes.** Two sets of massive dikes have been constructed to prevent the North Sea from flooding much of the country.
    - The Zuider Zee dike completed in 1932 has created a large freshwater lake called the IJsselmeer.
    - The Delta Plan completed during the 1980s has prevented recurrence of a flood like the one that killed nearly 2,000 people in 1953.

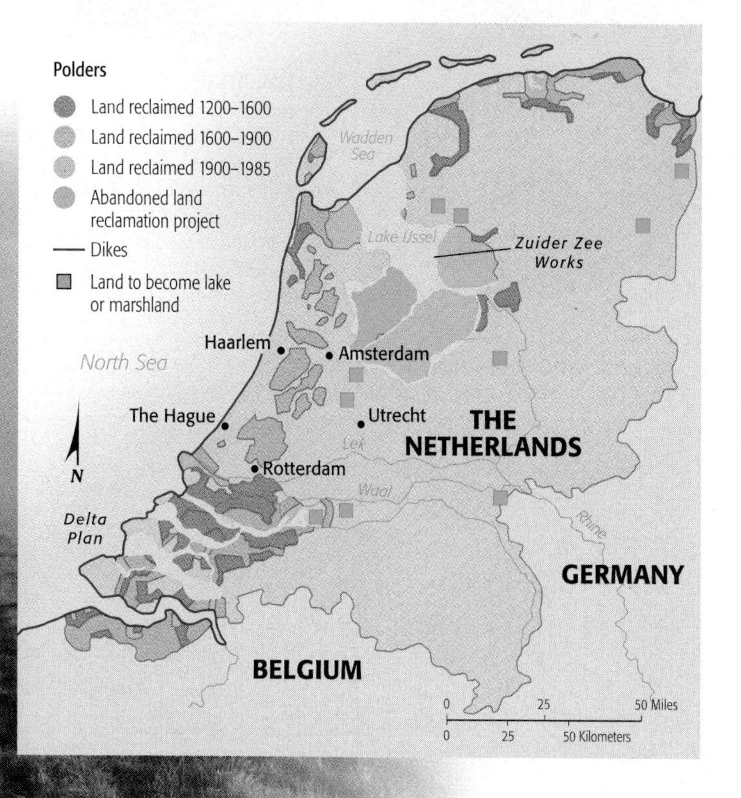

DEVELOPMENT OF GEOGRAPHY

1.3 Reading Maps

- The science of mapmaking is **cartography.**
- A map has a projection and a scale.

A **map** is a two-dimentional or flat representation of Earth's surface or a portion of it. Maps serve two purposes:

- Reference sources to help us find the shortest route and avoid getting lost.
- Tools for communicating geographic information about the distribution of features and reasons underlying the distribution.

To make a map, a cartographer must make two decisions:

- How to transfer a spherical Earth to a flat page (projection).
- How much of Earth's surface to depict (scale).

0° the prime meridian passes through Greenwich, England

0° 10° 20° 30° 40° 50° 60° 70° 80° 90° 100°

Lines of latitude measured from the center of Earth.

90° 80° 70° 60° 50° 40° 30° 20° 10° 0°

0° is the equator

Lines of longitude measured from the center of the Earth are wider near the Equator.

1.3.1 LONGITUDE

A **meridian** is an arc drawn between the North and South poles. The location of each meridian has a number known as **longitude**. The meridian that passes through the Royal Observatory at Greenwich, England, is 0° longitude, also called the **prime meridian**. The meridian on the opposite side of the globe is 180° longitude. All other meridians have numbers between 0° and 180° east or west.

1.3.2 LATITUDE

A **parallel** is a circle drawn around the globe parallel to the equator and at right angles to the meridians. The numbering system to indicate the location of a parallel is called **latitude**. The equator is 0° latitude, the North Pole 90° north latitude, and the South Pole 90° south latitude.

One degree of latitude is approximately 111 kilometers (69 miles).

1.3.3 HOW LONGITUDE AND LATITUDE WORK

Miami, Florida, is located near 26° north latitude and 80° west longitude.

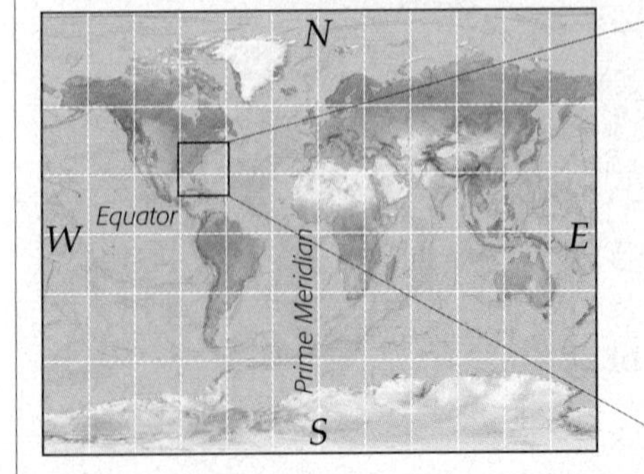

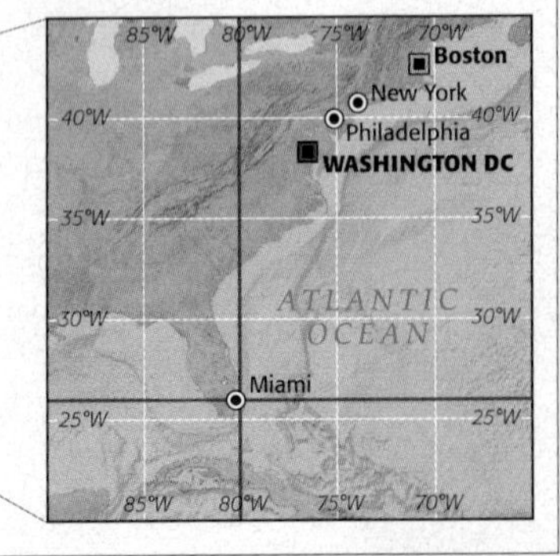

1.3.4 **PROJECTIONS**

Earth's spherical shape poses a challenge for cartographers because drawing Earth on a flat piece of paper unavoidably produces some distortion. The scientific process of transferring locations on Earth's surface to a flat map is called **projection**. Four types of distortion can result in the projection.

- The ***shape*** of an area can be distorted, so that it appears more elongated or squat than in reality.
- The ***distance*** between two points may increase or decrease.
- The ***relative size*** of different areas may be altered so that one area may appear larger than another on a map, but is in reality smaller.
- The ***direction*** from one place to another can be distorted.

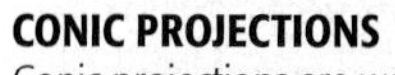

CONIC PROJECTIONS

Conic projections are well-suited for small area maps, and are used in this book for most of the maps of countries.

AZIMUTHAL PROJECTIONS

Azimuthal projections are well-suited for larger areas and are used for most of the world maps.

CYLINDRICAL PROJECTIONS

Cylindrical projections are used for specialized maps. The most widely used cylindrical world map, created by Gerardus Mercator in 1569, was widely used by mariners.

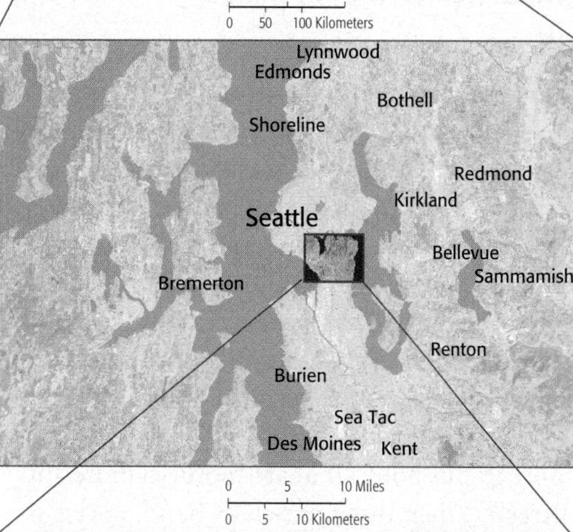

1.3.5 **SCALE**

The relationship of a feature's size on a map to its actual size on Earth is **scale**. Map scale is presented in three ways:

- A fraction (such as 1/24,000) or ratio (such as 1:24,000), where 1 refers to the map and 24,000 refers to Earth's surface.
- A written statement (such as "1 inch on the map equals 1 mile on Earth").
- A graphic bar scale (such as

0 50 100 Miles
0 50 100 Kilometers

Satellite Imagery Provided By Globexplorer.com

PROJECTIONS USED IN THIS BOOK

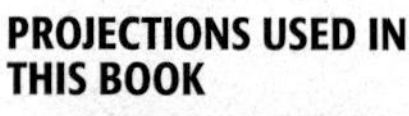

WORLD MAPS

The Eckert VII equal area projection.

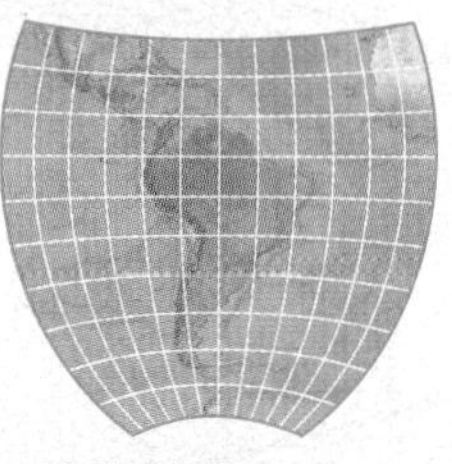

CONTINENTS

The Lambert azimuthal equal area.

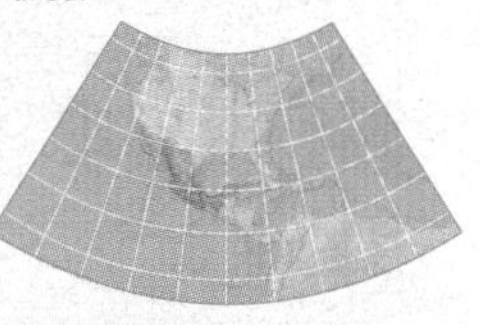

COUNTRIES

The Lambert conformal conic projection.

DEVELOPMENT OF GEOGRAPHY

1.4 Contemporary Mapping Tools

- GIScience applies contemporary technology to create maps.
- The ability to make complex maps is now on everyone's computer.

Having largely completed the great task of accurately mapping Earth's surface, which required several centuries, geographers have turned to Geographic Information Science (GIScience) to learn more about the characteristics of places. GIScience helps geographers to create more accurate and complex maps and to measure changes over time in the characteristics of places.

GIS

GIS (**geographic information system**) is a computer system that can capture, store, query, analyze, and display geographic data. The key to GIS is geocoding: the position of any object on Earth can be measured and recorded with mathematical precision and then stored in a computer. A map can be created by asking the computer to retrieve a number of stored objects and combine them to form an image.

GIS can be used to produce maps (including those in this book) that are more accurate and attractive than those drawn by hand. In the past, when cartographers drew maps with pen and paper, a careless moment could result in an object being placed in the wrong location, and a slip of the hand could ruin hours of work. GIS is more efficient for making a map than pen and ink: objects can be added or removed, colors brightened or toned down, and mistakes corrected (as long as humans find them!) without having to tear up the paper and start from scratch.

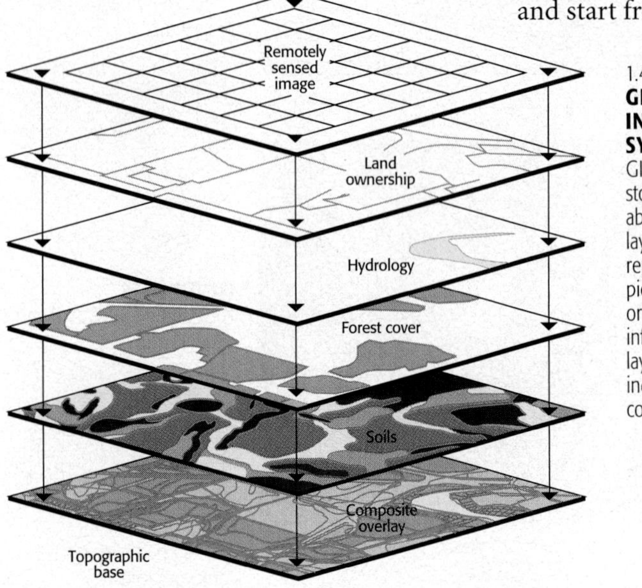

1.4.1 **A GEOGRAPHIC INFORMATION SYSTEM**
GIS involves storing information about a location in layers. Each layer represents a different piece of human or environmental information. The layers can be viewed individually or in combination.

Each type of information can be stored in a layer. For example, separate layers could be created for boundaries of countries, bodies of water, roads, and names of places. Depending on the desired purpose, a wide variety of maps can be created by turning on and off the various layers. A simple map might display only a single layer by itself, but most maps combine several layers, and GIS permits construction of much more complex maps than can be drawn by hand.

GIS enables geographers to calculate whether relationships between objects on a map are significant or merely coincidental. Layers can be compared to show relationships among different kinds of information. To protect hillsides from environmentally damaging development, for example, a human geographer may wish to compare a layer of recently built houses with a layer of steep slopes.

REMOTE SENSING

The acquisition of data about Earth's surface from a satellite orbiting Earth or from other long-distance methods is known as **remote sensing**. Remote-sensing satellites scan Earth's surface and transmit images in digital form to a receiving station on Earth.

At any moment a satellite sensor records the image of a tiny area called a picture element or pixel. Scanners are detecting the radiation being reflected from that tiny area. A map created by remote sensing is essentially a grid containing many rows of pixels. The smallest feature on Earth's surface that can be detected by a sensor depends on the resolution of the scanner. Human geographers are interested in remote sensing to map the distribution of urban sprawl and agricultural practices.

GPS

GPS (Global Positioning System) is a system that accurately determines the precise position of something on Earth. The GPS system in use in the United States includes two dozen satellites placed in predetermined orbits, a series of tracking stations to monitor and control the satellites, and receivers that compute position, velocity, and time from the satellite signals. GPS is most commonly used in the navigation of aircraft and ships. GPS is also being built into motor vehicles to provide the driver with directions. The GPS detects the vehicle's current position, the motorist programs the desired destination, and instructions are provided to tell the driver how to reach the destination.

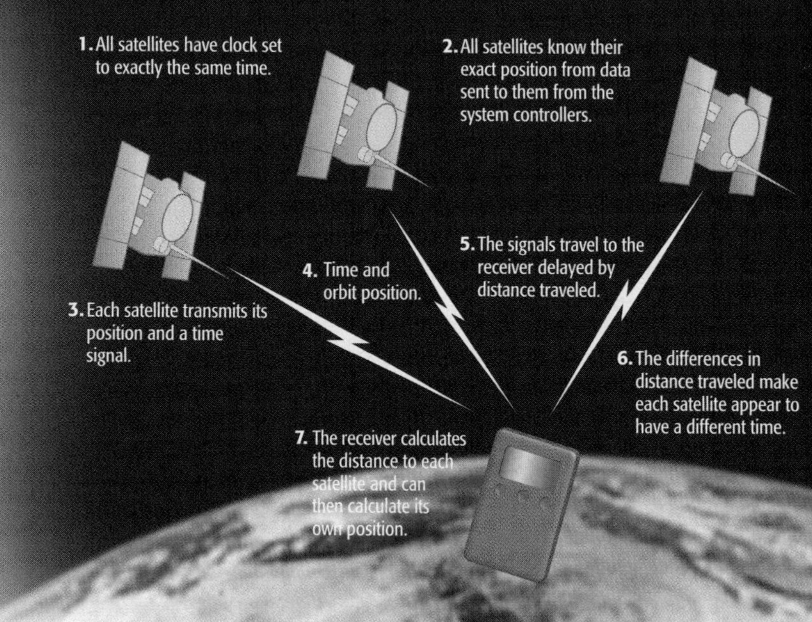

1.4.2 **HOW A GPS WORKS**

MASH-UPS

Mapping services, such as Google Maps, have given computer programmers access to the application programming interface (API), which is the language that links a database such as an address list with software such as mapping. The API for mapping software, available at such sites as **www.google.com/apis/maps**, enables a computer programmer to create a mash-up that places data on a map. The term *mash-up* refers to the practice of overlaying data from one source on top of one of the mapping services. The word comes from the popular practice of mixing two or more songs.

Mash-up maps can show the locations of businesses and activities near a particular street or within a neighborhood in a city. The requested information could be all restaurants within one-half mile of an address, or to be even more specific, all pizza parlors. Mash-ups assist in finding apartments, bars, hotels, sports facilities, transit stops, and crime scenes. Mapping software can show the precise location of commercial airplanes currently in the air, the gas stations with the cheapest prices, and current traffic tie-ups on highways and bridges.

1.4.3 **MASH-UP**
Google map shows pizza restaurants in Chicago's 60614 zip code.

THE UNIQUENESS OF EVERY PLACE

1.5 Places: Unique Locations

- Location is the position that something occupies on Earth.
- Location can be described by site, situation, and place names.

Humans possess a strong sense of **place**—that is, a feeling for the features that contribute to the distinctiveness of a particular spot on Earth. Geographers think about where particular places are located and the combination of features that make each place on Earth distinct. Geographers describe a feature's place on Earth by identifying its **location**, the position that something occupies on Earth's surface.

SITE

Geographers can describe the location of a place by **site**, which is the physical character of a place. Important site characteristics include climate, water sources, topography, soil, vegetation, latitude, and elevation. The combination of physical features gives each place a distinctive character.

Site factors have always been essential in selecting locations for settlements, although people have disagreed on the attributes of a good site. Some have preferred a hilltop site for easy defense from attack. Others located settlements near convenient river-crossing points to facilitate communication with people in other places.

An island combines the attributes of both hilltop and riverside locations, because the site provides good defense and transportation links. The site of the country of Singapore, for example, is a small, swampy island approximately 1 kilometer off the southern tip of the Malay Peninsula at the eastern end of the Strait of Malacca. The city of Singapore covers nearly 20 percent of the island.

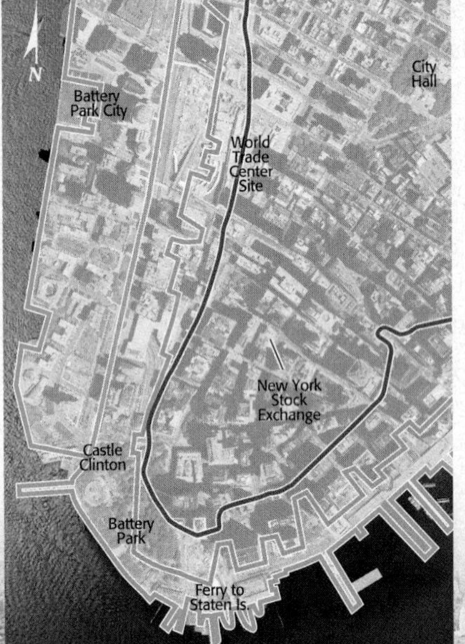

1.5.1 **SITE OF NEW YORK CITY**

Much of the southern part of New York City's Manhattan Island is built on landfill. The waterfront has been extended into the Hudson and East rivers to provide more land for offices, homes, parks, warehouses, and docks. In the eighteenth century, landfills were created by sinking old ships and dumping refuse on top of them. During the late 1960s and early 1970s, the World Trade Center was built partially on landfill in the Hudson River dating from the colonial era. Battery Park City was built on landfill removed from the World Trade Center construction site.

As a result of frequent site changes, the southern portion of Manhattan Island is twice as large today as it was in 1626, when Peter Minuit bought the island from its native inhabitants for the equivalent of $23.75 worth of Dutch gold and silver coins.

Satellite Imagery Provided By Globexplorer.com

1.5.2 **SITUATION OF SINGAPORE**
The small country of Singapore, less than one-fifth the size of Rhode Island, has an important situation for international trade. The country is situated at the confluence of several straits that serve as major passageways for shipping between the South China Sea and the Indian Ocean.

SITUATION

Situation is the location of a place relative to other places. Situation is a valuable way to indicate location, for two reasons:

- **Situation helps us find an unfamiliar place by comparing its location with a familiar one.**
  We give directions to people by referring to the situation of a place: "It's down past the courthouse, on Locust Street, after the third traffic light, beside the yellow-brick bank." We identify important buildings, streets, and other landmarks to direct people to the desired location.
- **Situation helps us understand the importance of a location.**
  Many locations are important because they are accessible to other places. For example, because of its location, Singapore has become a center for the trading and distribution of goods for much of Southeast Asia. Singapore is situated near the Strait of Malacca, which is the major passageway for ships traveling between the South China Sea and the Indian Ocean. Some 50,000 vessels, one-fourth of the world's maritime trade, pass through the strait each year.

PLACE NAMES

Because all inhabited places on Earth's surface—and many uninhabited places—have been named, the most straightforward way to describe a particular location is often by referring to its place name. A **toponym** is the name given to a place on Earth.

A place may be named for a person, perhaps its founder or a famous person with no connection to the community, such as George Washington. Some settlers select place names associated with religion, such as St. Louis and St. Paul, whereas other names derive from ancient history, such as Athens, Attica, and Rome.

A place name may also indicate the origin of its settlers. Place names commonly have British origins in North America and Australia, Portuguese origins in Brazil, Spanish origins elsewhere in Latin America, and Dutch origins in South Africa.

Some place names derive from features of the physical environment. Trees, valleys, bodies of water, and other natural features appear in the place names of most languages.

The Board of Geographical Names, operated by the U.S. Geological Survey, was established in the late nineteenth century to be the final arbiter of names on U.S. maps. In recent years the board has been especially concerned with removing offensive place names, such as those with racial or ethnic connotations.

1.5.3 **A PLACE NAME IN WALES**
The 58-letter name means "the Church of St. Mary's in the grove of the white hazelnut tree near the rapid whirlpool and the Church of St. Tisilio near the red cave." The town's name originally encompassed only the first 20 letters, but when the railway was built in the nineteenth century, the townspeople lengthened it. They decided that signs with the longer name in the railway station would attract attention and bring more business and visitors to the town.

THE UNIQUENESS OF EVERY PLACE

1.6 Regions: Unique Areas

- A region is an area of Earth with a unique combination of features.
- Three types of regions are functional, formal, and vernacular.

An area of Earth defined by one or more distinctive characteristics is a **region**. A region gains uniqueness from possessing not a single human or environmental characteristic, but a combination of them.

A region derives its unified character through the **cultural landscape**—a combination of cultural features such as language and religion, economic features such as agriculture and industry, and physical features such as climate and vegetation.

FUNCTIONAL REGION

A **functional region**, also called a nodal region, is an area organized around a node or focal point. The characteristic chosen to define a functional region dominates at a central focus or node and diminishes in importance outward. The region is tied to the central point by transportation or communications systems or by economic or functional associations.

Geographers often use functional regions to display information about economic areas. The region's node may be a shop or service, and the boundaries of the region mark the limits of the trading area of the activity. People and activities may be attracted to the node, and information may flow from the node to the surrounding area.

Examples of functional regions include the reception area of a television station, the circulation area of a newspaper, and the trading area of a department store. A television station's signal is strongest at the center of its service area, becomes weaker at the edge, and eventually can no longer be distinguished. A department store attracts fewer customers and a newspaper fewer readers from the edge of a trading area.

New technology is breaking down traditional functional regions. Newspapers such as *USA Today, The Wall Street Journal*, and *The New York Times* are composed in one place, transmitted by satellite to printing machines in other places, and delivered by airplane and truck to yet other places. Television stations are broadcast to distant places by cable or satellite. Customers can shop at distant stores by mail or Internet.

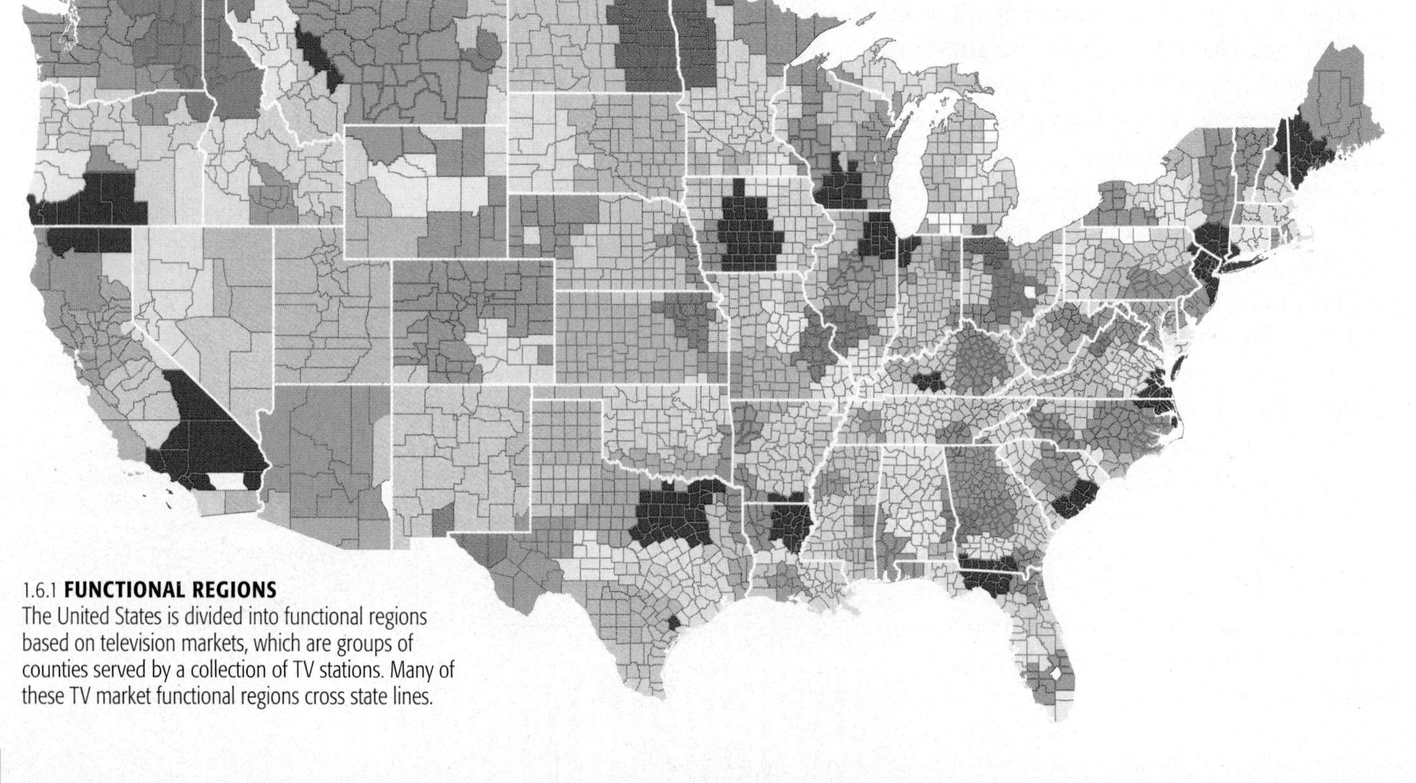

1.6.1 **FUNCTIONAL REGIONS**
The United States is divided into functional regions based on television markets, which are groups of counties served by a collection of TV stations. Many of these TV market functional regions cross state lines.

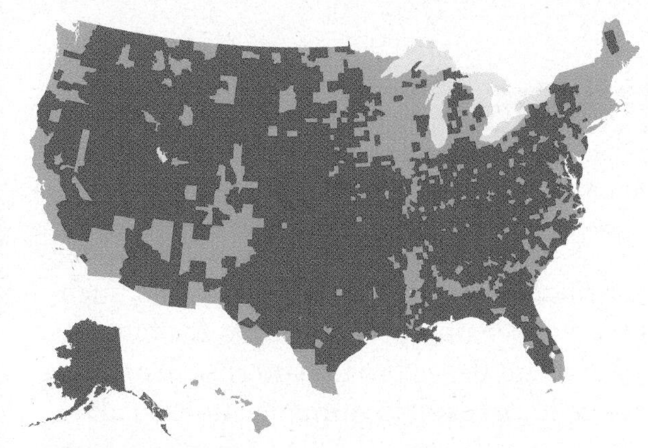

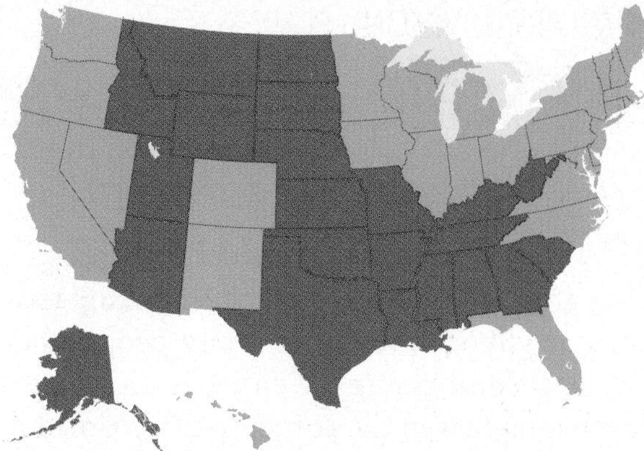

1.6.2 **FORMAL REGIONS** The two maps show the winner by county (left) and state (right) in the 2008 Presidential election. Counties and states are examples of formal regions. The extensive areas of support for Obama (blue) and McCain (red) are also examples of formal regions.

McCain
Obama

FORMAL REGION

A **formal region**, also called a uniform region or a homogeneous region, is an area within which everyone shares in common one or more distinctive characteristics. The shared feature could be a cultural value such as a common language, an economic activity such as production of a particular crop, or an environmental property such as climate. In a formal region, the selected characteristic is present throughout.

Geographers typically identify formal regions to help explain broad global or national patterns, such as variations in religion and levels of economic development. The characteristic selected to distinguish a formal region often illustrates a general concept rather than a precise mathematical distribution.

Some formal regions are easy to identify, such as countries or local government units. Montana is an example of a formal region, characterized by a government that passes laws, collects taxes, and issues license plates equally throughout the state.

In other kinds of formal regions, a characteristic may be predominant rather than universal. For example, we can distinguish formal regions within the United States characterized by a predominant voting for Republican candidates, although Republicans do not get 100 percent of the votes in these regions—nor in fact do they always win. However, in a presidential election, the candidate with the largest number of votes receives all of the electoral votes of a state (or Congressional District in Maine and Nebraska), regardless of the margin of victory.

A cautionary step in identifying formal regions is the need to recognize the diversity of cultural, economic, and environmental factors, even while making a generalization. A minority of people in a region may speak a language, practice a religion, or possess resources different from those of the majority. People in a region may play distinctive roles in the economy and hold different positions in society based on their gender or ethnicity.

VERNACULAR REGION

A **vernacular region**, or perceptual region, is a place that people believe exists as part of their cultural identity. Such regions emerge from people's informal sense of place rather than from scientific models developed through geographic thought.

As an example of a vernacular region, Americans frequently refer to the South as a place with environmental, cultural, and economic features perceived to be quite distinct from other places in the United States. Many of these features can be measured. Southerners and other Americans alike share a strong sense of the American South as a distinctive place that transcends geographic measurement. The perceptual region known as the South is a source of pride to many Americans.

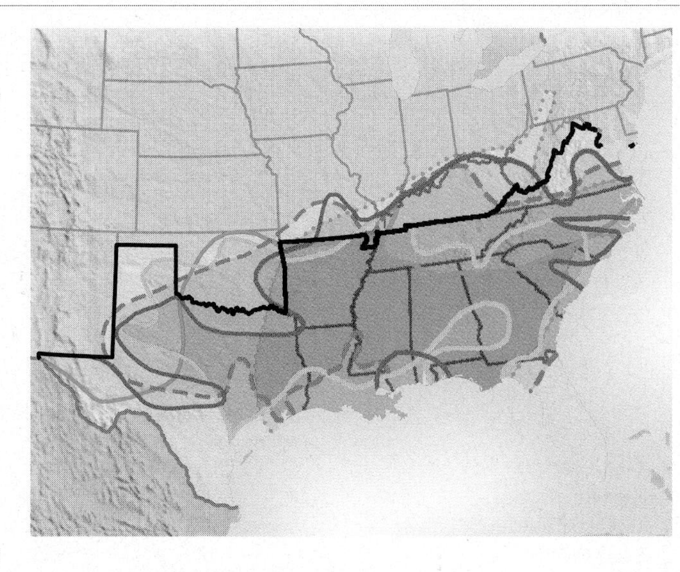

1.6.3 **VERNACULAR REGIONS** The South is popularly distinguished as a distinct vernacular region according to a number of factors.

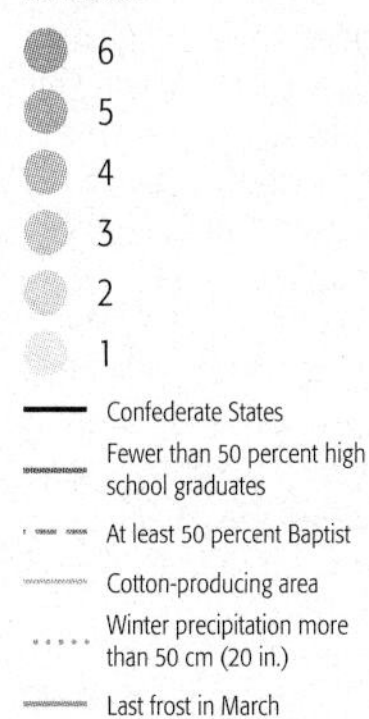

SIMILARITIES AMONG PLACES

1.7 Scale: Global Patterns

- People are plugged into a global economy and culture.
- Despite globalization, people play specialized economic roles and preserve cultural diversity.

Geographers think about scale at many levels, from local to global. At a local scale, such as a neighborhood within a city, geographers tend to see unique features. At the global scale, encompassing the entire world, geographers tend to see broad patterns. Geography matters in the contemporary world because it can explain human actions at all scales, from local to global.

GLOBALIZATION OF ECONOMY

Scale is an increasingly important concept in geography because of **globalization**, which is a force or process that involves the entire world and results in making something worldwide in scope. Globalization means that the scale of the world is shrinking—not literally in size, of course, but in the ability of a person, object, or idea to interact with a person, object, or idea in another place. People are plugged into a global economy and culture, producing a world that is more uniform, integrated, and interdependent.

A few people living in very remote regions of the world may be able to provide all of their daily necessities. But most economic activities undertaken in one region are influenced by interaction with decision makers located elsewhere. The choice of crop is influenced by demand and prices set in markets elsewhere. The factory is located to facilitate bringing in raw materials and shipping out products to the markets.

Globalization of the economy has been led primarily by transnational corporations, sometimes called multinational corporations. A **transnational corporation** conducts research, operates factories, and sells products in many countries, not just where its headquarters and principal shareholders are located.

Every place in the world is part of the global economy, but globalization has led to more specialization at the local level. Each place plays a distinctive role, based on its local assets. A place may be near valuable minerals, or it may be inhabited by especially well-educated workers. Transnational corporations assess the particular economic assets of each place.

Modern technology provides the means to easily move money, materials, products, technology, and other economic assets around the world. Thanks to the electronic superhighway, companies can now organize economic activities at a global scale.

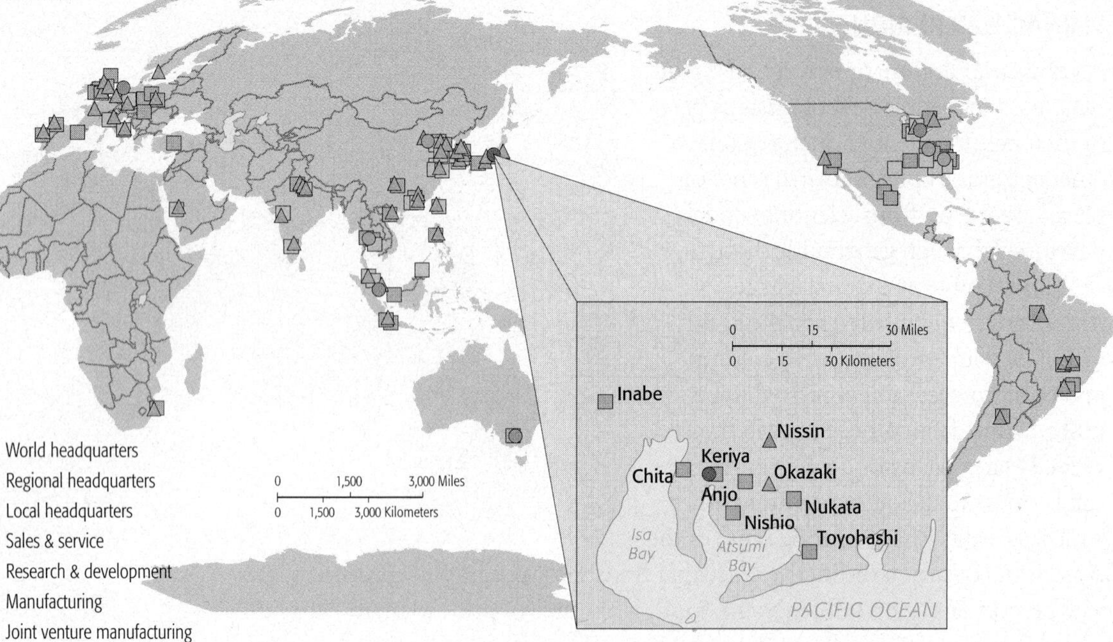

1.7.1 **GLOBALIZATION OF ECONOMY**
Denso, a transnational corporation that makes parts for cars, such as heaters and air conditioners, has its world headquarters, research labs, and eight factories in its "hometown" of Nagoya, Japan. Regional headquarters are located in the world's two other core regions—North America and Western Europe—the company's main overseas markets. A financial center is located in the Netherlands. Factories and sales centers are located in a number of more developed and less developed countries.

GLOBALIZATION OF CULTURE

Geographers observe that increasingly uniform cultural preferences produce uniform "global" landscapes of material artifacts and of cultural values. Fast-food restaurants, service stations, and retail chains deliberately create a visual appearance that varies among locations as little as possible so that customers know what to expect regardless of where in the world they happen to be.

The survival of a local culture's distinctive beliefs, forms, and traits is threatened by interaction with such social customs as wearing jeans and Nike shoes, consuming Coca-Cola and McDonald's hamburgers, and displaying other preferences in food, clothing, shelter, and leisure activities.

Underlying the uniform cultural landscape is globalization of cultural beliefs and forms, especially religion and language. Africans, in particular, have moved away from traditional religions and have adopted Christianity or Islam, religions shared with hundreds of millions of people throughout the world. Globalization requires a form of common communication, and the English language is increasingly playing that role.

As more people become aware of elements of global culture and aspire to possess them, local cultural beliefs, forms, and traits are threatened with extinction. Yet despite globalization, cultural differences among places not only persist but actually flourish in many places.

Global standardization of products does not mean that everyone wants the same cultural products. The communications revolution that promotes globalization of culture also permits preservation of cultural diversity.

Television, for example, is no longer restricted to a handful of channels displaying one set of cultural values. With the distribution of programming through cable and satellite systems, people may choose from hundreds of programs. With the globalization of communications, people in two distant places can watch the same television program. At the same time, with the fragmentation of the broadcasting market, two people in the same house can watch different programs.

Although consumers in different places express increasingly similar cultural preferences, they do not share the same access to them. And the desire of some people to retain their traditional cultural elements in the face of increased globalization of cultural preferences, has led to political conflict and market fragmentation in some regions.

Globalization has not destroyed the uniqueness of an individual place's culture and economy. Human geographers understand that many contemporary social problems result from a tension between forces promoting global culture and economy on the one hand and preservation of local economic autonomy and cultural traditions on the other hand.

CHINA

DUBAI

RUSSIA

THAILAND

JORDAN

SPAIN

SIMILARITIES AMONG PLACES

1.8 Space: Distribution of Features

- Distribution is the arrangement of a feature in space.
- Three properties of distribution are density, concentration, and pattern.

In his framework of all scientific knowledge, the German philosopher Immanuel Kant (1724–1804) compared geography's concern for space to history's concern for time:

- *Historians* identify the dates of important events and explain why human activities follow one another chronologically; *geographers* identify the location of important places and explain why human activities are located beside one another in space.
- *Historians* ask when and why; *geographers* ask where and why.
- *Historians* organize material chronologically because they understand that an action at one point in time can result from past actions that can in turn affect future ones; *geographers* organize material spatially because they understand that an action at one point in space can result from something happening at another point, which can consequently affect conditions elsewhere.

History and geography differ in one especially important manner: A historian cannot enter a time machine to study other eras firsthand; however, a geographer can enter an automobile or airplane to study other spaces. This ability to reach other spaces lends excitement to the discipline of geography—and geographic training raises the understanding of other spaces to a level above that of casual sightseeing.

Chess and computer games, where pieces are placed on a grid-shaped playing surface, require thinking about space. Pieces are arranged on the game board or screen in order to outmaneuver an opponent or form a geometric pattern. To excel at these games, a player needs spatial skills, the ability to perceive the future arrangement of pieces.

Similarly, spatial thinking is the most fundamental skill that geographers use to understand the arrangement of objects across surfaces considerably larger than a game board. Geographers think about the arrangement of people and activities found in space and try to understand why those people and activities are distributed across space as they are.

Each human and natural object occupies a unique **space** on Earth, and geographers explain how these features are arranged across Earth. On Earth as a whole, or within an area of Earth, features may be numerous or scarce, close together or far apart. The arrangement of a feature in space is known as its **distribution**.

0 200 400 Feet
0 60 120 Meters

1.8.1 **DISTRIBUTION**
The top plan for a residential area has a lower density than the middle plan (24 houses compared to 32 houses on the same 82-acre piece of land), but both have dispersed concentrations. The middle and lower plans have the same density (32 houses on 82 acres), but the distribution of houses is more clustered in the lower plan. The lower plan has shared open space, whereas the middle plan provides a larger, private yard surrounding each house.

PROPERTIES OF DISTRIBUTION

Geographers identify three main properties of distribution across Earth:

- **Density.** The frequency with which something occurs in space is its **density.** The feature being measured could be people, houses, cars, volcanoes, or anything. The area could be measured in square kilometers, square miles, hectares, acres, or any other unit of area.

  Remember that a large *population* does not necessarily lead to a high density. Density involves two measures—the number of objects and the land area.

- **Concentration.** The extent of a feature's spread over space is its **concentration.** If the objects in an area are close together, they are *clustered*; if relatively far apart, they are *dispersed*. To compare the level of concentration most clearly, two areas need to have the same number of objects and the same size area.

  Concentration is not the same as density. Two neighborhoods could have the same density of housing but different concentrations. In a dispersed neighborhood each house has a large private yard, whereas in a clustered neighborhood the houses are close together and the open space is shared as a community park.

- **Pattern.** The geometric arrangement of objects in space is **pattern.** Some features are organized in a pattern, whereas others are distributed irregularly. Geographers observe that many objects form a linear distribution, such as the arrangement of houses along a street or stations along a subway line.

  Objects are frequently arranged in a square or rectangular pattern. Many American cities contain a regular pattern of streets, known as a grid pattern, which intersect at right angles at uniform intervals to form square or rectangular blocks.

1.8.2 **DISTRIBUTION OF BASEBALL TEAMS.**
The changing distribution of North American baseball teams illustrates the difference between density and concentration.

These 6 teams moved to other cities during the 1950s and 1960s:

Braves–*Boston to Milwaukee in 1953, then to Atlanta in 1966*

Browns–*St. Louis to Baltimore (Orioles) in 1954*

Athletics–*Philadelphia to Kansas City in 1955, then to Oakland in 1968*

Dodgers–*Brooklyn to Los Angeles in 1958*

Giants–*New York to San Francisco in 1958*

Senators–*Washington to Minneapolis (Minnesota Twins) in 1961*

These 14 teams were added between the 1960s and 1990s:

Angels–*Los Angeles in 1961, then to Anaheim (California) in 1965*

Senators–*Washington in 1961, then to Dallas (Texas Rangers) in 1971*

Mets–*New York in 1962*

Astros–*Houston (originally Colt .45s) in 1962*

Royals–*Kansas City in 1969*

Padres–*San Diego in 1969*

Expos–*Montreal in 1969, then to Washington (Nationals) in 2005*

Pilots–*Seattle in 1969, then to Milwaukee (Brewers) in 1970*

Blue Jays–*Toronto in 1977*

Mariners–*Seattle in 1977*

Marlins–*Miami in 1993*

Rockies–*Denver in 1993*

Rays–*Tampa Bay in 1998*

Diamondbacks–*Phoenix in 1998*

As a result of these relocations and additions, the density of teams increased, and the distribution became more dispersed.

SIMILARITIES AMONG PLACES

1.9 Connections Between Places

- Connection between places results in spatial interaction.
- A characteristic spreads from one place to another through diffusion.

Geographers increasingly think about connections among places and regions. In the past, most forms of interaction among cultural groups required the physical movement of settlers, explorers, and plunderers from one location to another. Today, travel by motor vehicle or airplane is much quicker.

But we do not even need to travel to know about another place. We can communicate instantly with people in distant places through computers and telecommunications, and we can instantly see people in distant places on television. These and other forms of communication have made it possible for people in different places to be aware of the same cultural beliefs, forms, and traits.

SPATIAL INTERACTION

When places are connected to each other through a network, geographers say there is **spatial interaction** between them. Typically, the farther away one group is from another, the less likely the two groups are to interact. Contact diminishes with increasing distance and eventually disappears. This trailing-off phenomenon is called **distance decay**.

More rapid connections have reduced the distance across space between places, not literally in miles, of course, but in time. Geographers apply the term **space-time compression** to describe the reduction in the time it takes for something to reach another place. Distant places seem less remote and more accessible to us. We know more about what is happening elsewhere in the world, and we know sooner.

Space-time compression promotes rapid change, as the culture and economy of one place reach other places much more quickly than in the past. With better connections between places, people in one region are now exposed to a constant barrage of cultural traits and economic initiatives from people in other regions, and they may adopt some of these cultural and economic elements.

Electronic communications, such as the Internet and e-mail, have removed barriers to interaction between people who are far from each other. The birth of these electronic communications was initially viewed as the "death" of geography, because they made it cheap and easy to stay in touch with someone on the other side of the planet. Regardless of its location, a business could maintain instantaneous communications among employees and with customers.

In reality, geography matters even more than before. Internet access depends upon availability of electricity to power the computer and a service provider. Broadband service requires proximity to a digital subscriber line (DSL) or cable line. The Internet has also magnified the importance of geography, because when an individual is online, the specific place in the world where the individual is located is known. This knowledge is valuable information for businesses that target advertisements and products to specific tastes and preferences of particular places.

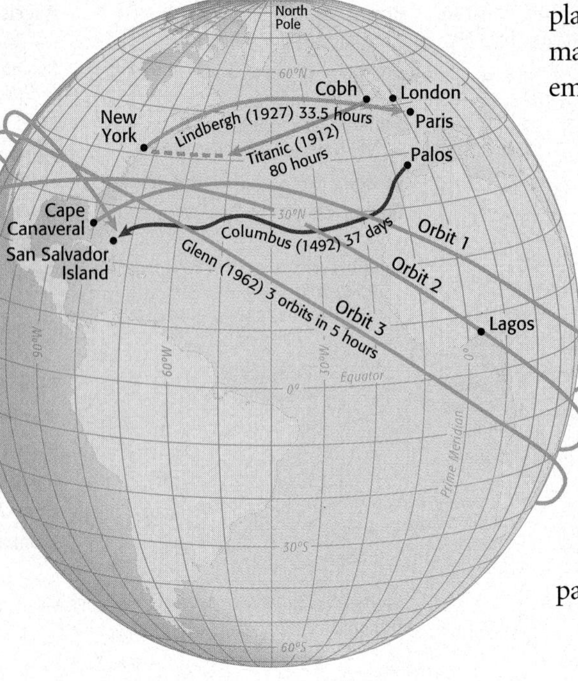

1.9.1 SPACE-TIME COMPRESSION Transportation improvements have shrunk the world. In 1492 Christopher Columbus took 37 days (nearly 900 hours) to sail across the Atlantic Ocean from the Canary Islands to San Salvador Island. In 1912 the Titanic was scheduled to sail from Queenstown (now Cobh), Ireland, to New York in about 5 days, although two-thirds of the way across, after 80 hours at sea, it hit an iceberg and sank. In 1927 Charles Lindbergh was the first person to fly nonstop across the Atlantic, taking 33.5 hours to go from New York to Paris. In 1962 John Glenn, the first American to orbit in space, crossed above the Atlantic in about a quarter-hour and circled the globe three times in 5 hours.

DIFFUSION

The process by which a characteristic spreads across space from one place to another over time is **diffusion.** Something originates at a **hearth** or node and diffuses from there to other places. Geographers document the location of nodes and the processes by which diffusion carries things elsewhere over time.

For a person, object, or idea to have interaction with persons, objects, or ideas in other regions, diffusion must occur. Geographers observe two basic types of diffusion:

- **Relocation diffusion** is the spread of an idea through physical movement of people from one place to another When people move, they carry with them their culture, including language, religion, and ethnicity.
- **Expansion diffusion** is the spread of a feature from one place to another in a snowballing process. This expansion may result from one of three processes:
- **Hierarchical diffusion** is the spread of an idea from persons or nodes of authority or power to other persons or places. Hierarchical diffusion may result from the spread of ideas from political leaders, socially elite people, or other important persons to others in the community.
- **Contagious diffusion** is the rapid, widespread diffusion of a characteristic throughout the population. As the term implies, this form of diffusion is analogous to the spread of a contagious disease, such as influenza. Ideas placed on the World Wide Web spread through contagious diffusion, because Web surfers throughout the world have access to the same material simultaneously—and quickly.
- **Stimulus diffusion** is the spread of an underlying principle, even though a characteristic itself apparently fails to diffuse. For example, early desktop computer sales in the United States were divided about evenly between Macintosh Apple and IBM-compatible DOS systems. By the 1990s Apple sales had fallen far behind IBM-compatibles, but principles pioneered by Apple, notably making selections by pointing a mouse at an icon rather than typing a string of words, diffused through a succession of IBM-compatible Windows systems.

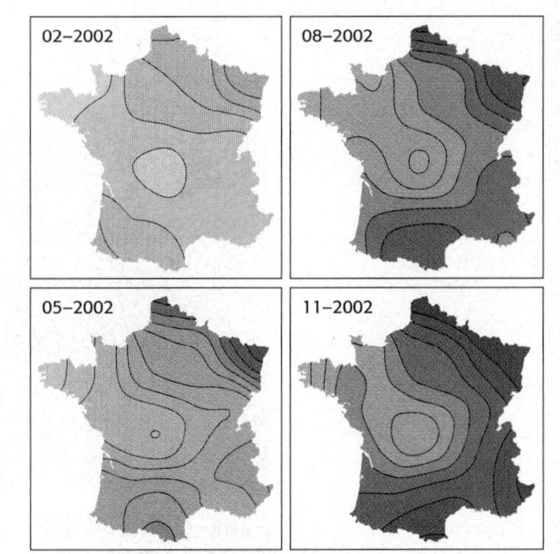

1.9.2 **RELOCATION DIFFUSION**
Introduction of a common currency, the euro, in 12 Western European countries on January 1, 2002, gave scientists an unusual opportunity to measure relocation diffusion. Although a single set of paper money was issued, each of the 12 countries minted its own coins in proportion to its share of the region's economy. A country's coins were initially distributed only inside its borders, although the coins could also be used in the other 11 countries. French scientists took month-to-month samples to monitor the proportion of coins from the other 11 countries. The percentage of "foreign" euro coins is a measure of the level of relocation diffusion into France. Not surprisingly, diffusion is highest near the borders with other countries.

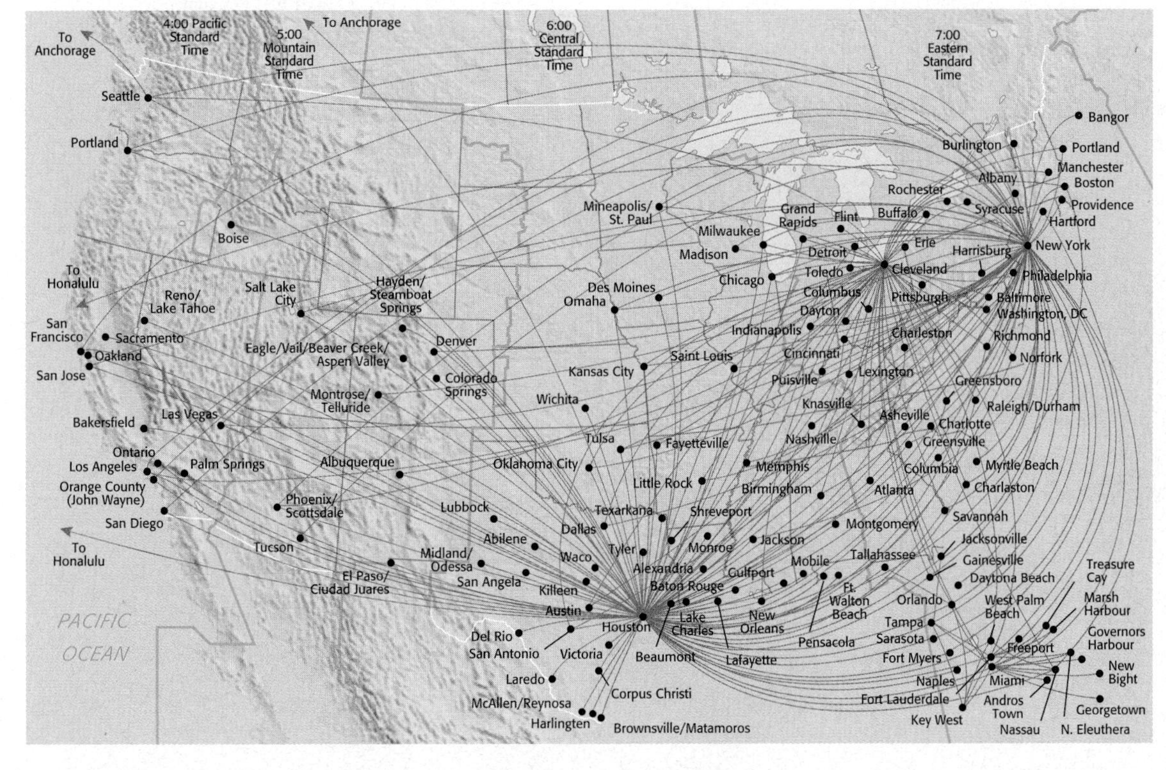

1.9.3 **CONNECTIONS BETWEEN PLACES: CONTINENTAL AIRLINES' NETWORK**
Continental, like other major U.S. airlines, has configured its route network in a system known as "hub and spokes." Lines connect each airport to the city to which it sends the most nonstop flights. Most flights originate or end at one of the company's hubs, especially at Houston, Newark, and Cleveland.

Chapter **Summary**

DEVELOPMENT OF GEOGRAPHY

- The word *geography* is based on two Greek words meaning "Earth" and "to write."
- Early geographic studies developed in large part to assist with exploration and discovery of Earth.
- Scientific geography began in the late nineteenth century as a search for natural laws and causal relationships.
- Geography's most fundamental tool is the map, which has become more accurate and complex over the centuries through exploration and electronic aids.

THE UNIQUENESS OF EVERY PLACE

- Every place on Earth is in some ways unique, because every place has a unique location that can be determined through site, situation, and name.
- Every area or region on Earth is also unique, because it contains a unique combination of features.

SIMILARITIES AMONG PLACES

- Because of the process of globalization, two places and regions can display similar economic and cultural features.
- Geographers document the distinctive distribution of particular economic and cultural features.
- Distribution has three properties: density, concentration, and pattern.
- Features spread from one place to another through a process of diffusion.
- Thanks to modern communications, distant places are more closely connected to each other.

GOOGLE EARTH IMAGE OF WHERE LAS VEGAS MEETS THE DESERT
Visible on the south is the V-shaped runways of McCarran International Airport.

Geographic Consequences of Change

A central theme in the contemporary world is the tension between two important geographic themes: globalization and cultural diversity. Geographers observe that people are being pulled in opposite directions by these two factors.

- Modern communications and technology have fostered globalization, pulling people into greater cultural and economic interaction with others.
- At the same time, people are searching for more ways to express their unique cultural traditions and economic practices.

Tensions between the simultaneous geographic trends of globalization and local diversity underlie many of the world's problems that geographers study, such as political conflicts, economic uncertainty, and pollution of the environment.

Geographers want to know why each place is in some ways unique. Geographers use two basic concepts to explain why every place is unique:

- A place is a specific point on Earth distinguished by a particular characteristic. Every place occupies a unique location, or position, on Earth's surface, and geographers have many ways to identify location.
- A region is an area of Earth distinguished by a distinctive combination of cultural and physical features. Human geographers are especially concerned with the cultural features of a group of people in a region—their body of beliefs and traditions, as well as their political and economic practices.

Geographers also want to know why different places on Earth have similar features. Three basic concepts help geographers explain why these similarities do not result from coincidence:

- Scale is the relationship between the portion of Earth being studied and Earth as a whole. Although geographers study every scale from the individual to the entire Earth, increasingly they are concerned with global-scale patterns and processes.
- Space refers to the physical gap or interval between two objects; geographers observe that many objects are distributed across space in a regular manner, for discernable reasons.
- Connections are relationships among people and objects across the barrier of space; geographers are concerned with the various means by which connections occur.

SHARE YOUR VOICE Student Essay

ANAHI CORTADA

Anahi Cortada
Miami Dade Honors College

A TRICK OF NATURE

Four years of taking Geography courses had not been enough for me to fully understand the role climate zones play in nature. I had learned the theory well, but I could not prove that knowledge since I had never been outside of that little island in the Caribbean where I was born.

In Cuba, exotic and spicy island of the torrid zone, I grew up enjoying the heat of an eternal summer. The sun, the beaches, and fresh heat were available the whole year. According to my professor there were only two annual seasons: dry and wet. I did not know what a real winter was. For me it was always summer from January to December.

It was not until I was 14 that I left my island and moved to the United States. New York was my destination and December 20th was the day I was traveling. I was moving to a temperate zone. This was the time when I had the pleasure (although I did not think this at that time) of encountering the winter. My pretty days of December were gone. My green trees with their rich and voluminous dresses of leaves had disappeared. The bright colors had died. In their place I only had gray days. The skeletal trees were an image I had never seen before; however, this was not too pleasant to observe when the cold even pierced my bones. The absence of sunshine made me live in an eternal nostalgia for a warm, sunny, and bright day.

I had no choice but to learn how to love the winter, although I never stopped enjoying and loving summer days. However, it is still amazing for me to contemplate the tricks that nature plays to each one of us in different zones of the world.

Chapter **Resources**

GENDER AND SPACE

Patterns in space vary according to gender.

Consider the daily patterns of what used to be considered an "all-American" family of mother, father, son, and daughter. Leave aside for the moment that this type of family constitutes less than one-fourth of American households. In the morning Dad gets in his car and drives from home to work, where he parks the car and spends the day; then, in the late afternoon, he collects the car and drives home. The location of the home was traditionally selected in part to ease Dad's daily commute to work.

The mother's travel patterns are likely to be far more complex than the father's. Mom takes the children to school and returns home. She also drives to the supermarket, visits Grandmother, and walks the dog. In between she organizes the several thousand square feet of space that the family calls home. In the afternoon she picks up the youngsters at school and takes them to Little League or ballet lessons. Later she brings them home, just in time for her to resume her responsibility for organizing the home.

Most American women are now employed at work outside the home, adding a substantial complication to an already complex pattern of moving across urban space. Where is her job located? The family house was already selected largely for access to Dad's place of employment, so Mom may need to travel across town. Who leaves work early to drive a child to a doctor's office? Who takes a day off work when a child is at home sick?

The importance of gender in space is learned as a child. Which child—the boy or girl—went to Little League, and which went to ballet lessons? To which activity is substantially more land allocated in a city—ballfields or dance studios?

All academic disciplines and workplaces have proclaimed sensitivity to issues of cultural diversity. For geographers, concern and deep respect for the dignity of all cultural groups is not merely a politically correct expediency; it lies at the heart of geography's explanation of why each place on Earth is unique.

TIME ZONES

The Mercator projection is typically used to display the world's 24 standard time zones. Earth as a sphere is divided into 360° of longitude. The international agreement designated the time at the prime meridian (0° longitude) as **Greenwich Mean Time (GMT)** or Universal Time (UT). As you travel eastward from the prime meridian, you must turn your clock ahead from GMT by 1 hour for each 15°. If you travel westward from the prime meridian, you turn your clock back from GMT by 1 hour for each 15°. The eastern United States, which is near 75° west longitude, is therefore 5 hours earlier than GMT. Thus when the time is 11 A.M. GMT, the time in the eastern United States is 5 hours earlier, or 6 A.M.

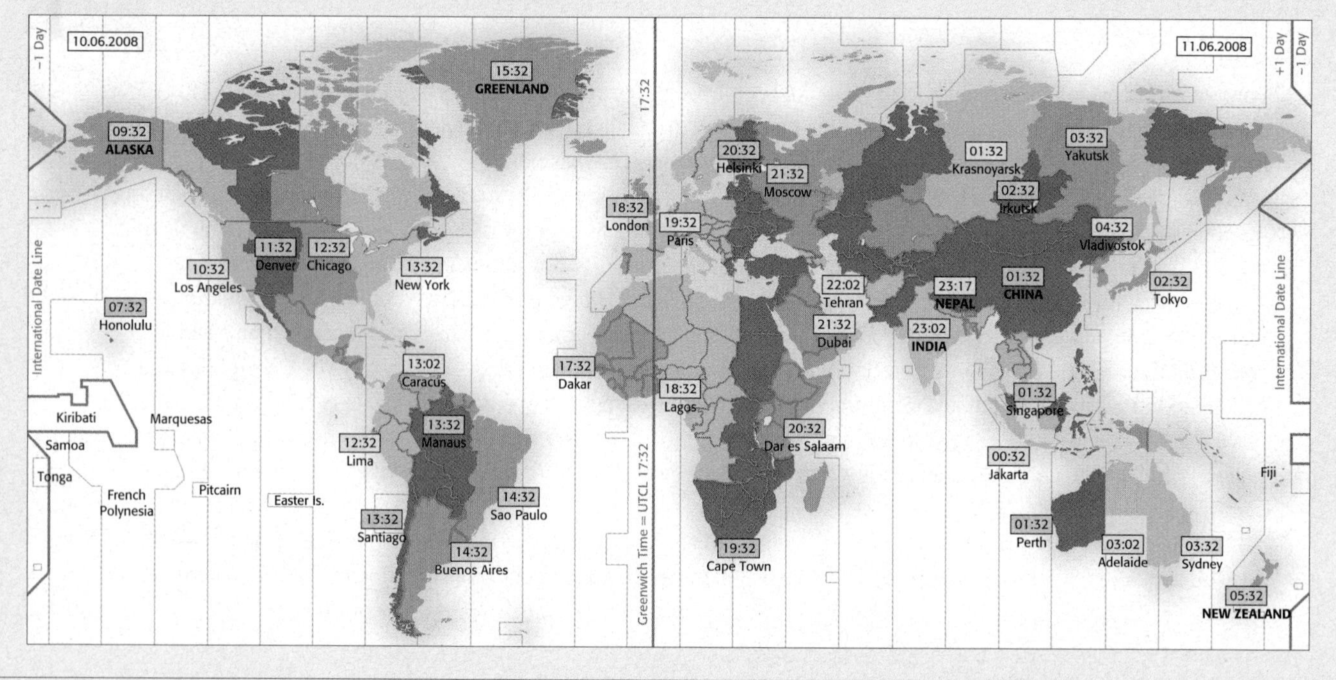

ON THE INTERNET

Useful places for learning about geography on the Internet are the Web sites maintained by the major professional organizations. In the United States, the Association of American Geographers maintains a Web site **(www.aag.org)**, as does the American Geographical Society (**www.amergeog.org**). The National Geographic Society offers access to material from its magazine and television programs, as well as online mapping, at its Web site **(www.nationalgeographic).**

FURTHER READINGS

Cutter, Susan L., Reginald Golledge, and William L. Graf. The Big Questions in Geography. *Professional Geographer* 54 (2002): 235–55. .

Geography Education Standards Project. *Geography for Life: National Geography Standards.* Washington, DC: National Geographic Research and Exploration, 1994.

Hanson, Susan, ed. *Ten Geographic Ideas That Changed the World.* New Brunswick, NJ: Rutgers University Press, 1997.

Hartshorne, Richard. *The Nature of Geography.* Lancaster, PA: Association of American Geographers, 1939.

James, Preston E. *All Possible Worlds: A History of Geographical Ideas.* New York: Bobbs-Merrill, 1972.

Johnston, Ron, Derek Gregory, Geraldine Pratt, and Michael Watts, eds. *The Dictionary of Human Geography*, 4th ed. Malden, MA: Blackwell, 2000.

Massey, Doreen B. *Space, Place, and Gender.* Minneapolis: University of Minnesota Press, 1994.

Sauer, Carl O. "Morphology of Landscape." *University of California Publications in Geography* 2 (1925): 19–54.

Wallach, Bret. *Understanding the Cultural Landscape.* New York: Guilford Press, 2005.

KEY TERMS

Concentration
The spread of something over a given area.

Contagious diffusion
The rapid, widespread diffusion of a feature or trend throughout a population.

Cultural landscape
Fashioning of a natural landscape by a cultural group.

Density
The frequency with which something exists within a given unit of area.

Diffusion
The process of spread of a feature or trend from one place to another over time.

Distance decay
The diminishing in importance and eventual disappearance of a phenomenon with increasing distance from its origin.

Distribution
The arrangement of something across Earth's surface.

Environmental determinism
A nineteenth- and early twentieth-century approach to the study of geography that argued that the general laws sought by human geographers could be found in the physical sciences. Geography was therefore the study of how the physical environment caused human activities.

Expansion diffusion
The spread of a feature or trend among people from one area to another in a snowballing process.

Formal region (or uniform or homogeneous region)
An area in which everyone shares in one or more distinctive characteristics.

Functional region (or nodal region)
An area organized around a node or focal point.

Geographic information system (GIS)
A computer system that stores, organizes, analyzes, and displays geographic data.

Global Positioning System (GPS)
A system that determines the precise position of something on Earth through a series of satellites, tracking stations, and receivers.

Globalization
Actions or processes that involve the entire world and result in making something worldwide in scope.

Greenwich Mean Time (GMT)
The time in that time zone encompassing the prime meridian, or 0° longitude.

Hearth
The region from which innovative ideas originate.

Hierarchical diffusion
The spread of a feature or trend from one key person or node of authority or power to other persons or places.

Latitude
The numbering system used to indicate the location of parallels drawn on a globe and measuring distance north and south of the equator (0°).

Location
The position of anything on Earth's surface.

Longitude
The numbering system used to indicate the location of meridians drawn on a globe and measuring distance east and west of the prime meridian (0°).

Map
A two-dimensional, or flat, representation of Earth's surface or a portion of it.

Meridian
An arc drawn on a map between the North and South poles.

Parallel
A circle drawn around the globe parallel to the equator and at right angles to the meridians.

Pattern
The geometric or regular arrangement of something in a study area.

Place
A specific point on Earth distinguished by a particular character.

Polder
Land created by the Dutch by draining water from an area.

Possibilism
The theory that the physical environment may set limits on human actions, but people have the ability to adjust to the physical environment and choose a course of action from many alternatives.

Prime meridian
The meridian, designated as 0° longitude, that passes through the Royal Observatory at Greenwich, England.

Projection
The system used to transfer locations from Earth's surface to a flat map.

Region
The system used to transfer locations from Earth's surface to a flat map.

Relocation diffusion
The spread of a feature or trend through bodily movement of people from one place to another.

Remote sensing
The acquisition of data about Earth's surface from a satellite orbiting the planet or other long-distance methods.

Scale
Generally, the relationship between the portion of Earth being studied and Earth as a whole, specifically the relationship between the size of an object on a map and the size of the actual feature on Earth's surface.

Site
The physical character of a place.

Situation
The location of a place relative to other places.

Space
The physical gap or interval between two objects.

Spatial interaction
The movement of physical processes, human activities, and ideas within and among regions.

Space-time compression
The reduction in the time it takes to diffuse something to a distant place, as a result of improved communications and transportation systems.

Stimulus diffusion
The spread of an underlying principle, even though a specific characteristic is rejected.

Toponym
The name given to a portion of Earth's surface.

Transnational corporation
A company that conducts research, operates factories, and sells products in many countries, not just where its headquarters or shareholders are located.

Vernacular region (or perceptual region)
An area that people believe exists as part of their cultural identity.

Chapter 2 POPULATION

More humans are alive at this time—about 6¾ billion—than at any point in Earth's long history. Most of these people live in less developed countries (LDCs), and nearly all of the world's population growth is concentrated in LDCs. More developed countries (MDCs), such as the United States, are growing slowly if at all.

Is the world overpopulated? Will it become so in the years ahead? Geographic approaches are well suited to addressing these fears. Geographers argue that overpopulation is not simply a matter of the total number of people on Earth, rather it depends on the relationship between the number of people and the availability of resources.

Overpopulation is a threat where an area's population exceeds the capacity of the environment to support it at an acceptable standard of living. The capacity of Earth as a whole to support human life may be high, but some regions have a favorable balance between people and available resources, whereas others do not. Further, the regions with the most people are not necessarily the same as the regions with an unfavorable balance between population and resources.

RUSH HOUR IN CHENGDU, CHINA

THE KEY ISSUES IN THIS CHAPTER

POPULATION DISTRIBUTION

2.1 Population Concentrations

2.2 Population Density

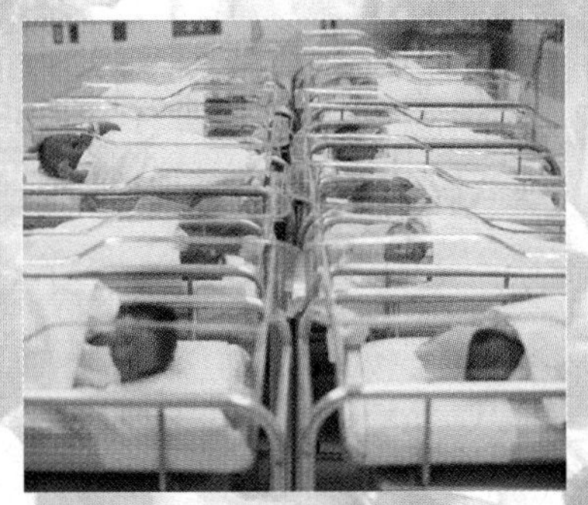

POPULATION GROWTH

2.3 Components of Change

2.4 Population Structure

2.5 The Demographic Transition

OVERPOPULATION THREATS

2.6 Malthus's Grim Forecast

2.7 Declining Birth Rates

2.8 The Epidemiologic Transition

2.9 Global Health Threats

POPULATION DISTRIBUTION

2.1 Population Concentrations

- Two-thirds of the world's inhabitants are clustered in four regions.
- Humans avoid clustering in relatively harsh environments.

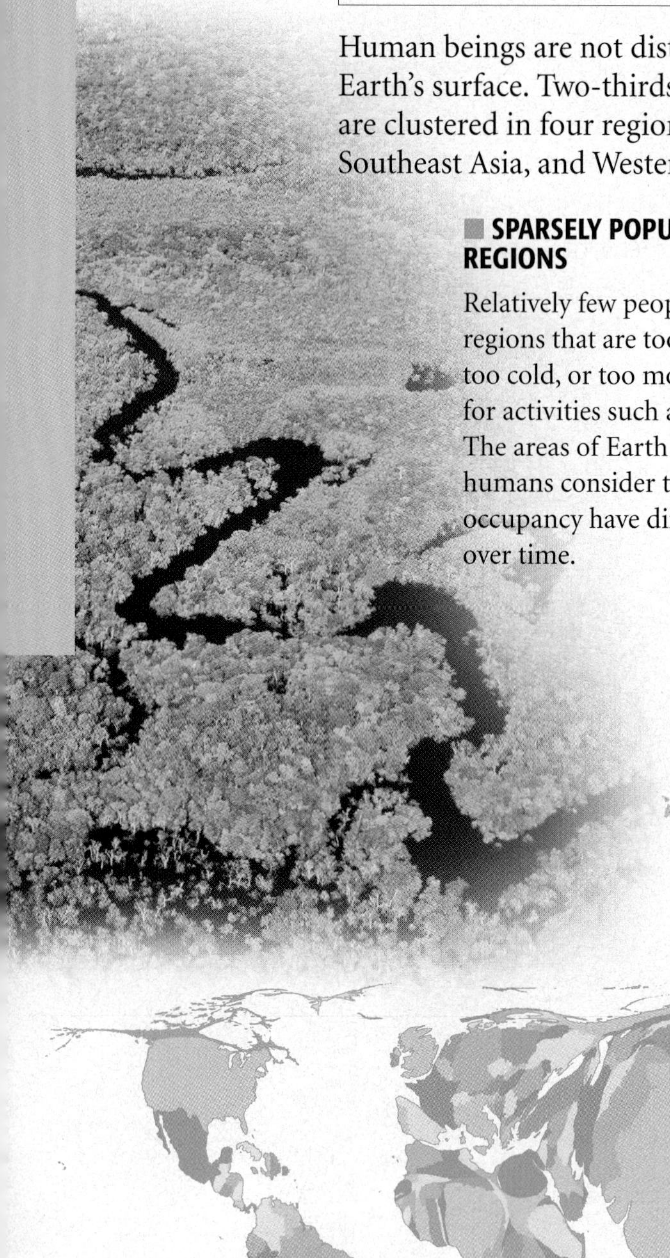

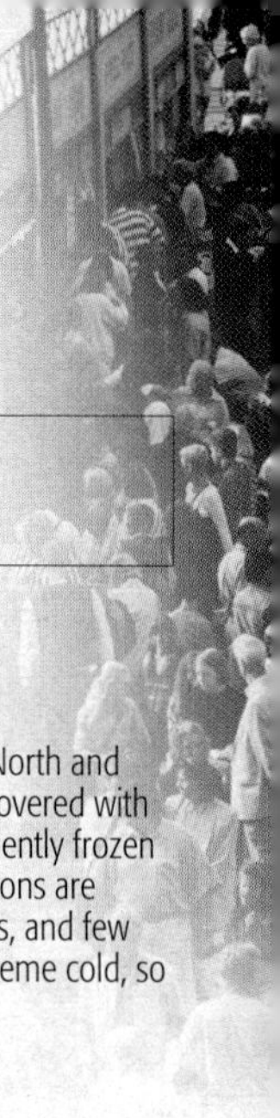

Human beings are not distributed uniformly across Earth's surface. Two-thirds of the world's inhabitants are clustered in four regions—East Asia, South Asia, Southeast Asia, and Western Europe.

SPARSELY POPULATED REGIONS

Relatively few people live in regions that are too dry, too wet, too cold, or too mountainous for activities such as agriculture. The areas of Earth that humans consider too harsh for occupancy have diminished over time.

COLD LANDS

Much of the land near the North and South poles is perpetually covered with ice or the ground is permanently frozen (permafrost). The polar regions are unsuitable for planting crops, and few animals can survive the extreme cold, so few humans live there.

WET LANDS

Lands that receive very high levels of precipitation may be inhospitable for human occupation. The combination of rain and heat rapidly depletes nutrients from the soil and thus hinders agriculture.

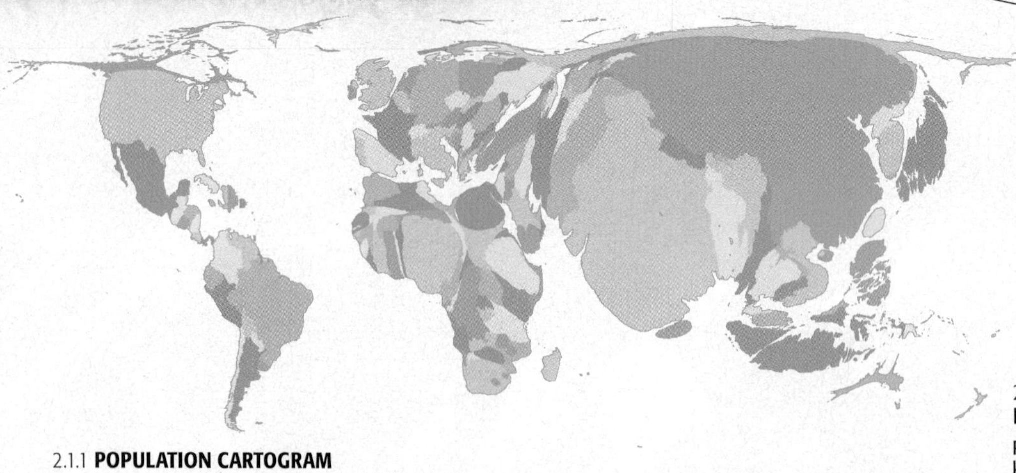

2.1.1 **POPULATION CARTOGRAM**

The clustering of the world's population can be displayed on a cartogram, which depicts the size of countries according to population rather than land area, as is the case with most maps. When compared to a more typical equal-area map, such as the large map on this page, the population cartogram displays the major population clusters of Europe and East, South, and Southeast Asia as much larger and Africa and the Western Hemisphere as much smaller. As you look at maps of population growth and other topics in this and subsequent chapters, pay special attention to Asia and Europe, because global patterns are heavily influenced by conditions in these regions, where two-thirds of the world's people live.

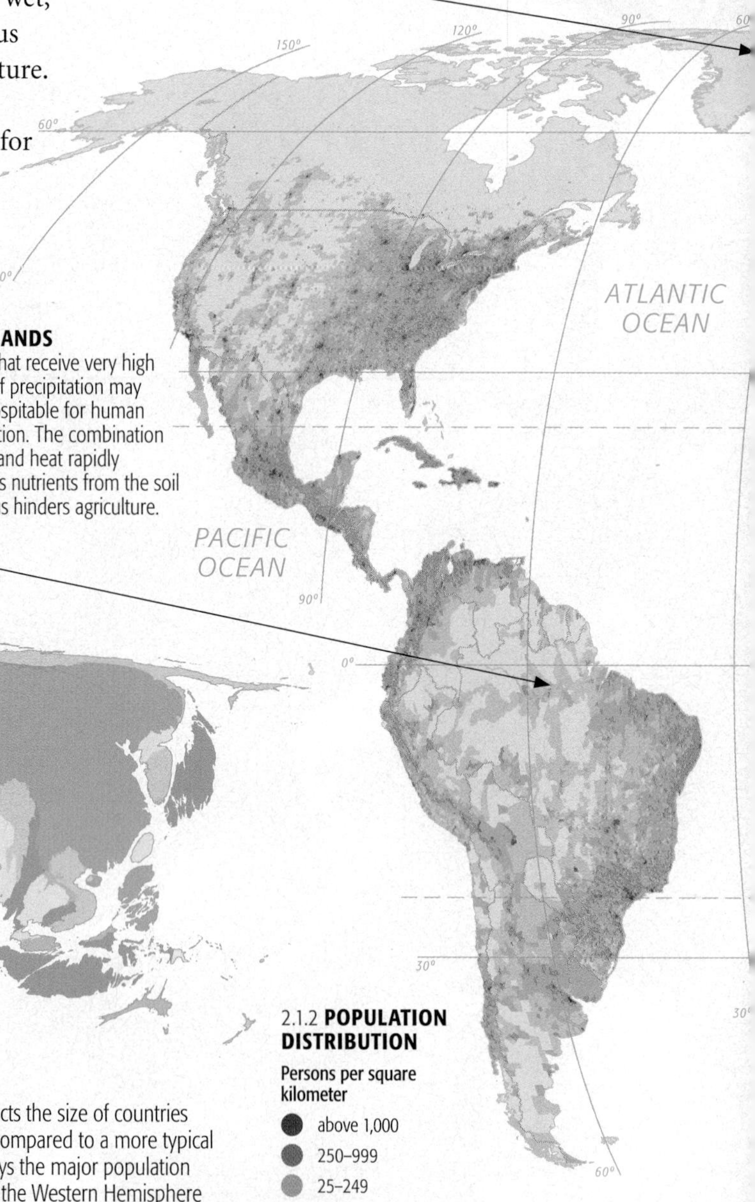

2.1.2 **POPULATION DISTRIBUTION**

Persons per square kilometer

- above 1,000
- 250–999
- 25–249
- 5–24
- 1–5
- below 1

FOUR POPULOUS REGIONS

The four regions display some similarities. Most of the people in these regions live near an ocean or near a river with easy access to an ocean, rather than in the interior of major landmasses. The four population clusters occupy generally low-lying areas, with fertile soil and temperate climate.

EUROPE
This concentration includes four dozen countries, ranging from Monaco, with 1 square kilometer (0.7 square miles) and a population of 32,000, to Russia, the world's largest country in land area when its Asian part is included. In contrast to the three Asian concentrations, three-fourths of Europe's inhabitants live in cities, and less than 20 percent are farmers.

SOUTH ASIA
The most important concentration of people within South Asia lives along a 1,500-kilometer (900-mile) corridor from Lahore, Pakistan, through India and Bangladesh to the Bay of Bengal. Much of this area's population is concentrated along the plains of the Indus and Ganges rivers. Population is also heavily concentrated near India's two long coastlines—the Arabian Sea to the west and the Bay of Bengal to the east.

EAST ASIA
This concentration includes the world's most populous country, the People's Republic of China. The Chinese population is clustered near the Pacific Coast and in several fertile river valleys that extend inland, such as the Huang and the Yangtze. Much of China's interior is sparsely inhabited mountains and deserts.

SOUTHEAST ASIA
Indonesia, which consists of 13,677 islands, including Java, is the world's fourth most populous country.

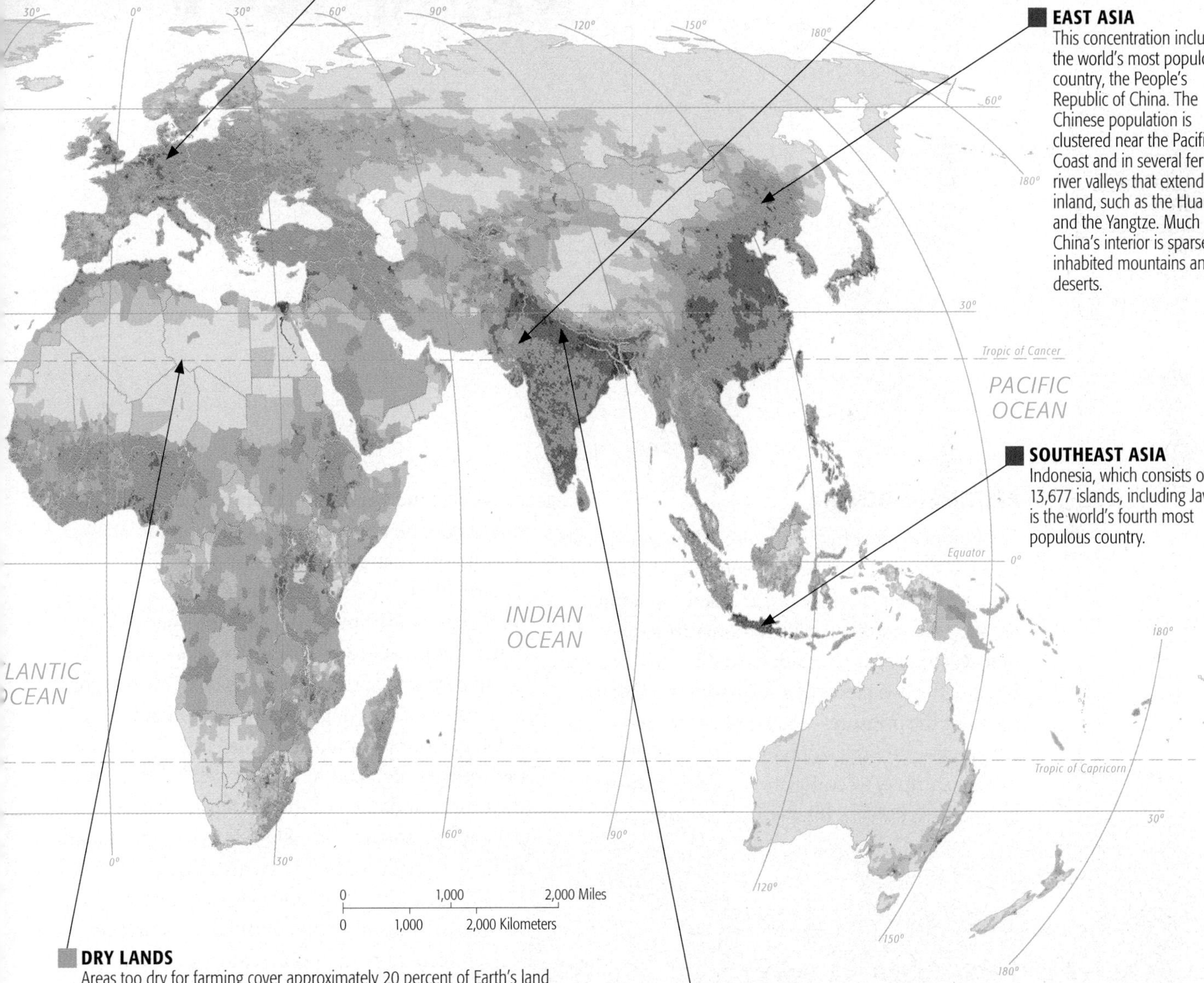

DRY LANDS
Areas too dry for farming cover approximately 20 percent of Earth's land surface. Deserts generally lack sufficient water to grow crops that could feed a large population, although some people survive there by raising animals, such as camels, that are adapted to the climate. Although dry lands are generally inhospitable to intensive agriculture, they may contain natural resources useful to people—notably, much of the world's oil reserves.

HIGH LANDS
The highest mountains in the world are steep, snow covered, and sparsely settled. We can find some significant exceptions, especially in Latin America and Africa. People may prefer to occupy higher lands if temperatures and precipitation are uncomfortably high at lower elevations.

POPULATION DISTRIBUTION

2.2 Population Density

- Arithmetic density measures the total number of people living in an area.
- Physiological density and agricultural density show spatial relationships between people and resources.

Density, defined in Chapter 1 as the number of people occupying an area of land, can be computed in several ways, including arithmetic density, physiological density, and agricultural density. These measures of density help geographers to describe the distribution of people in comparison to available resources.

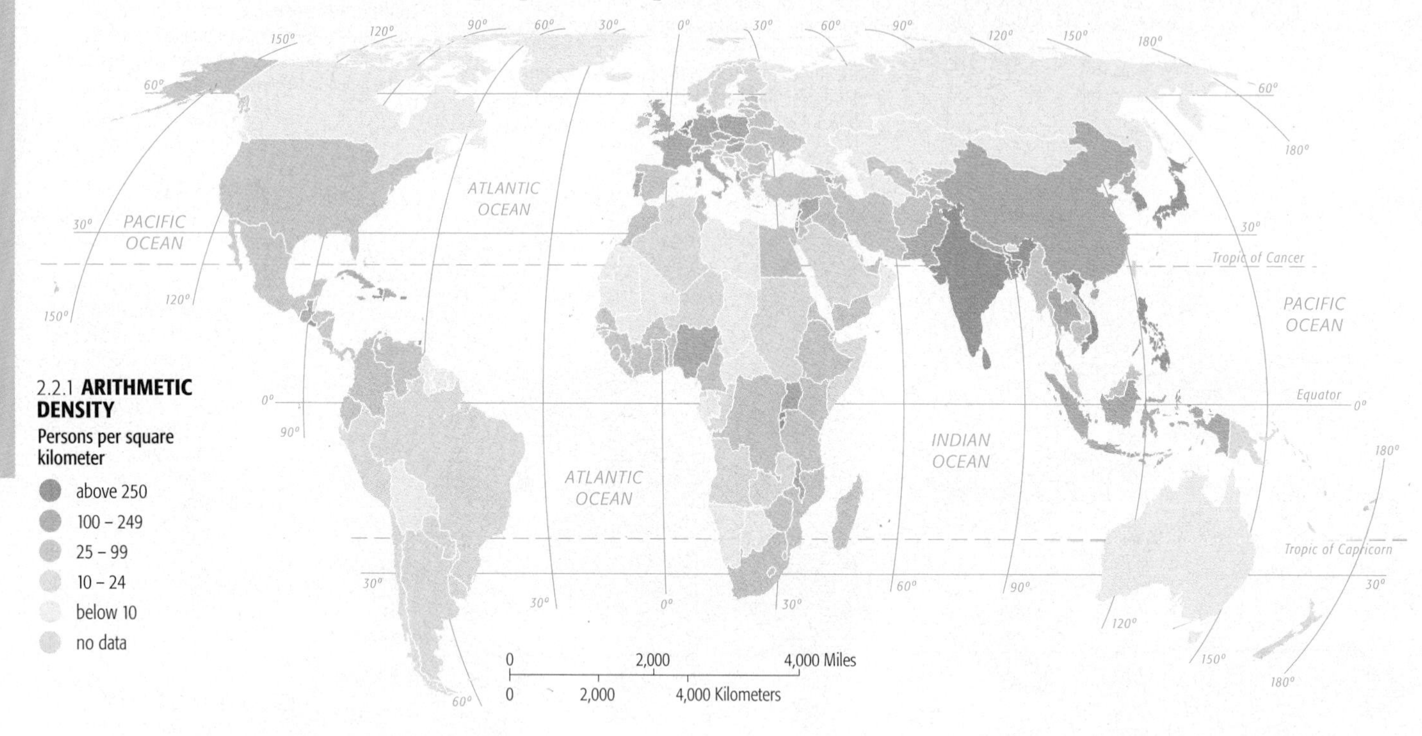

2.2.1 **ARITHMETIC DENSITY**
Persons per square kilometer

ARITHMETIC DENSITY

Geographers most frequently use arithmetic density, which is the total number of people divided by total land area. (This measure is also called *population density*.) Geographers rely on the **arithmetic density** to compare conditions in different countries because the two pieces of information needed to calculate the measure—total population and total land area—are easy to obtain.

For example, to compute the arithmetic or population density for the United States, we can divide the population (approximately 300 million people) by the land area (approximately 9.6 million square kilometers, or 3.7 million square miles). The result shows that the United States has an arithmetic density of 31 persons per square kilometer (80 persons per square mile). By comparison, the arithmetic density is much higher in South Asia. In Bangladesh, it is approximately 1,050 persons per square kilometer (2,700 persons per square mile), and in India it is 350 (900). On the other hand, the arithmetic density is only 3 persons per square kilometer (8 persons per square mile) in Canada and 3 (7) in Australia.

Arithmetic density enables geographers to compare the number of people trying to live on a given piece of land in different regions of the world. However, to explain why people are not uniformly distributed across Earth's surface, other density measures are more useful.

2.2.2 **EGYPT'S DENSLY POPULATED NILE DELTA**
Weekly market at Qutur

PHYSIOLOGICAL DENSITY

A more meaningful population measure is afforded by looking at the number of people per area of a certain type of land in a region. Land suited for agriculture is called **arable land**. In a region, the number of people supported by a unit area of arable land is called the **physiological density.** For example, in the United States the physiological density is 172 persons per square kilometer (445 per square mile) of arable land. This contrasts sharply with Egypt, which has 2,580 persons per square kilometer (6,682 per square mile) of arable land. This large difference in physiological densities demonstrates that crops grown on a hectare of land in Egypt must feed far more people than in the United States.

The higher the physiological density, the greater the pressure that people may place on the land to produce enough food. Physiological density provides insights into the relationship between the size of a population and the availability of resources in a region.

Comparing physiological and arithmetic densities helps geographers to understand the capacity of the land to yield enough food for the needs of the people. In Egypt, for example, the large difference between the physiological density (2,580 people per square kilometer of arable land) and arithmetic density (75 persons per square kilometer over the entire country) indicates that most of the country's land is unsuitable for intensive agriculture. In fact, all but 5 percent of the Egyptian people live in the Nile River valley and delta, because it is the only area in the country that receives enough moisture (by irrigation from the river) to allow intensive cultivation of crops.

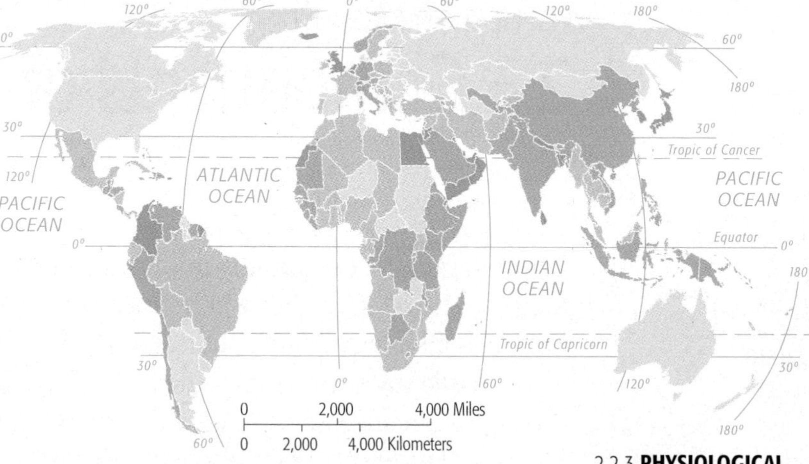

2.2.3 **PHYSIOLOGICAL DENSITY**
Persons per square kilometer of arable land
- above 1000
- 500 – 999
- 250 – 499
- below 250
- no data

AGRICULTURAL DENSITY

Two countries can have similar physiological densities, but they may produce significantly different amounts of food because of different economic conditions. **Agricultural density** is the ratio of the number of farmers to the amount of arable land. This density measure helps account for economic differences.

For example, the United States has an extremely low agricultural density (1 farmer per square kilometer of arable land), whereas Egypt has a very high density (826 farmers per square kilometer of arable land). MDCs such as the United States have lower agricultural densities because technology and finance allow a few people to farm extensive land areas and feed many people. This frees most of the MDC population to work in factories, offices, or shops rather than in the fields.

To understand the relationship between population and resources in a country, geographers examine a country's physiological and agricultural densities together. For example, the physiological densities of both Bangladesh and the Netherlands are high, but the Dutch have a much lower agricultural density than the Bangladeshis. Geographers conclude that both the Dutch and Bangladeshis put heavy pressure on the land to produce food, but the more efficient Dutch agricultural system requires fewer farmers than does the Bangladeshi system.

2.2.4 **AGRICULTRAL DENSITY**
Farmers per square kilometer of arable land
- above100
- 50 – 99
- 25 – 49
- below 25
- no data

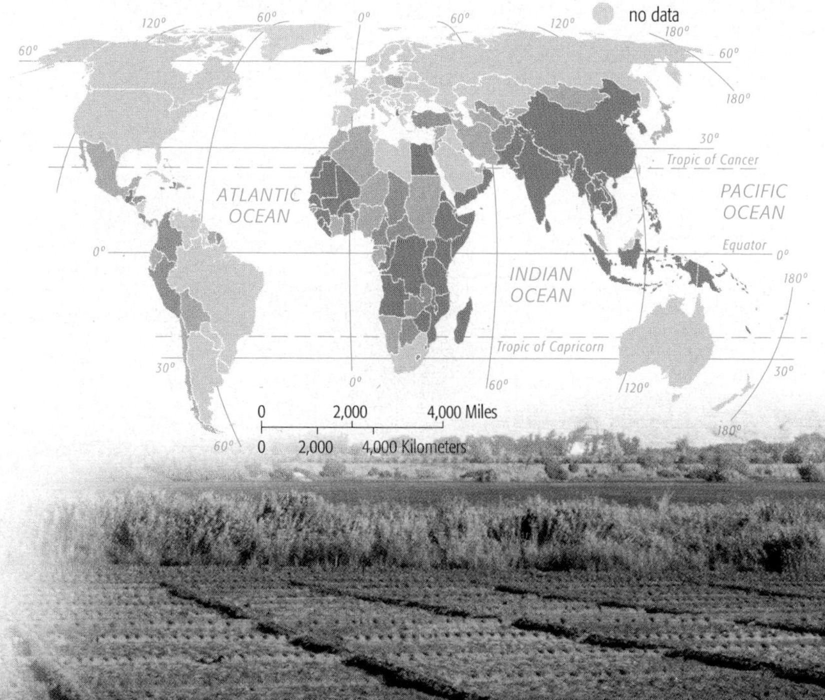

POPULATION GROWTH

2.3 Components of Change

- Geographers most frequently measure population change through three indicators.
- The three measures of population change vary widely among regions.

Population increases rapidly in places where many more people are born than die, increases slowly in places where the number of births exceeds the number of deaths by only a small margin, and declines in places where deaths outnumber births. Geographers most frequently measure population change in a country or the world as a whole through three measures—natural increase rate, crude birth rate, and crude death rate. The population of a place also increases when people move in and decreases when people move out. This element of population change—migration—is discussed in Chapter 3.

NATURAL INCREASE RATE

The **natural increase rate (NIR)** is the percentage by which a population grows in a year. The term *natural* means that a country's growth rate excludes migration.

The world NIR during the first decade of the twenty-first century has been 1.2, meaning that the population of the world has been growing each year by 1.2 percent. The world NIR is lower today than its all-time peak of 2.2 percent in 1963.

Virtually 100 percent of the natural increase is clustered in LDCs. The NIR exceeds 2.0 percent in most countries of Africa, Asia, Latin America, and the Middle East, whereas it is negative in Europe, meaning that in the absence of immigrants, population actually is declining. About two-thirds of the world's population growth during the past decade has been in Asia, with the remaining one-third divided about equally among sub-Saharan Africa, Latin America, and the Middle East. Regional differences in NIRs mean that most of the world's additional people live in the countries that are least able to maintain them.

The rate of natural increase affects the **doubling time**, which is the number of years needed to double a population, assuming a constant rate of natural increase.

- At the early twenty-first-century rate of 1.2 percent per year, world population would double in about 54 years.
- Should the same NIR continue through the twenty-first century, global population in the year 2100 would reach ***24 billion***.
- Should the NIR immediately decline to 1.0, doubling time would stretch out to 70 years, and world population in 2100 would be only ***15 billion***.

About 80 million people are being added to the population of the world annually. That number represents a decline from the historic high of 87 million in 1989. The number of people added each year has dropped much more slowly than the NIR because the population base is much higher now than in the past. World population increased from 3 to 4 billion in 14 years, from 4 to 5 billion in 13 years, and from 5 to 6 billion in 12 years. As the base continues to grow in the twenty-first century, a change of only one-tenth of 1 percent would produce very large swings in population growth.

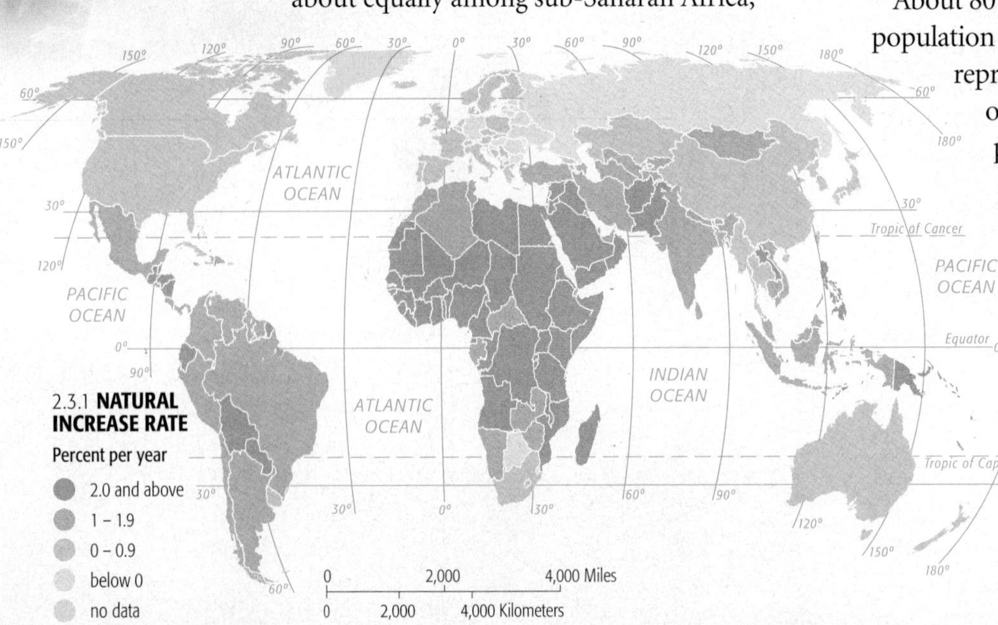

2.3.1 **NATURAL INCREASE RATE**

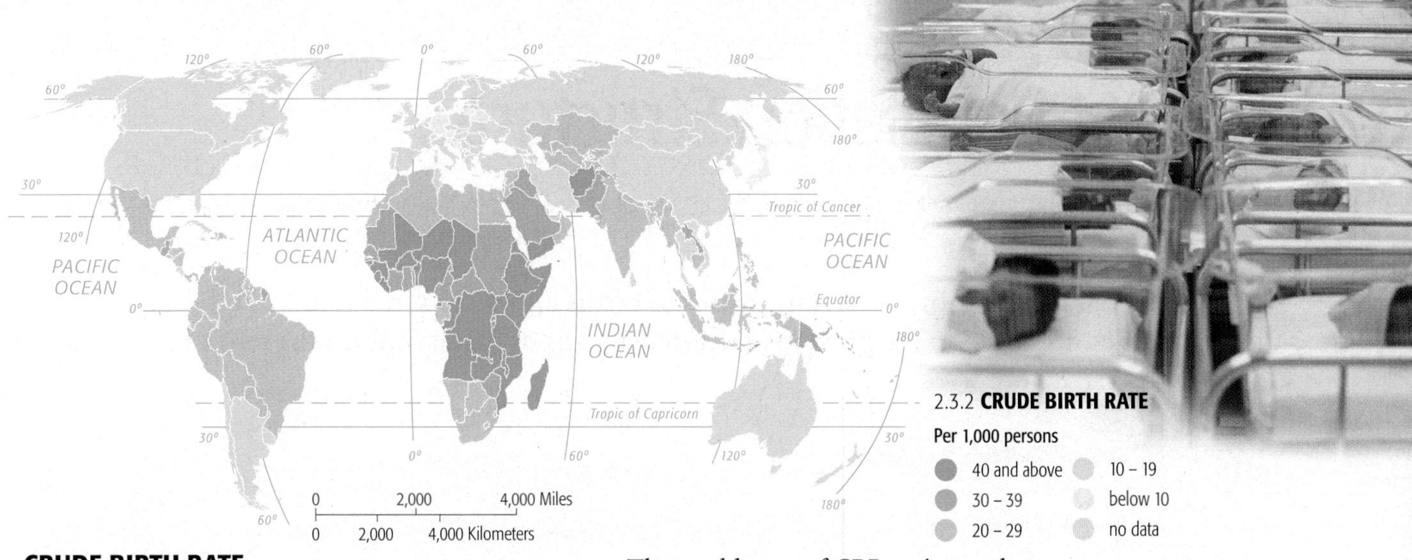

2.3.2 **CRUDE BIRTH RATE**

Per 1,000 persons

- 40 and above
- 30 – 39
- 20 – 29
- 10 – 19
- below 10
- no data

CRUDE BIRTH RATE

The **crude birth rate (CBR)** is the total number of live births in a year for every 1,000 people alive in the society. A CBR of 20 means that for every 1,000 people in a country, 20 babies are born over a 1-year period.

The world map of CBRs mirrors the distribution of NIRs. As was the case with NIRs, the highest CBRs are in sub-Saharan Africa, and the lowest are in Europe. Many sub-Saharan African countries have a CBR over 40, whereas many European countries have a CBR below 10.

CRUDE DEATH RATE

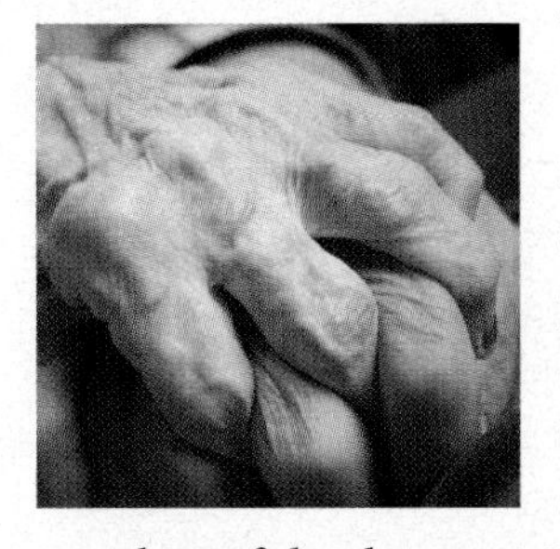

The **crude death rate (CDR)** is the total number of deaths in a year for every 1,000 people alive in the society. Comparable to the CBR, the CDR is expressed as the annual number of deaths per 1,000 population.

The NIR is computed by subtracting CDR from CBR, after first converting the two measures from numbers per 1,000 to percentages (numbers per 100). Thus if the CBR is 20 and the CDR is 5 (both per 1,000), then the NIR is 15 per 1,000, or 1.5 percent.

The CDR does not display the same regional pattern as the NIR and CBR. The combined CDR for all LDCs is actually lower than the combined rate for all MDCs. Furthermore, the variation between the world's highest and lowest CDRs is much less extreme than the variation in CBRs. The highest CDR in the world is 23 per 1,000 and the lowest is 2—a difference of 21—whereas CBRs for individual countries range from 8 per 1,000 to 50, a spread of 42.

Why does Denmark, one of the world's wealthiest countries, have a higher CDR than Mongolia, one of the poorest? Why does the United States, with its extensive system of hospitals and physicians, have a higher CDR than Mexico and every country in Central America? The answer is that the populations of different countries are at various stages in an important process known as the demographic transition (see Section 2.5).

2.3.3 **CRUDE DEATH RATE**

Per 1,000 persons

- above 20
- 15 – 19
- 10 – 14
- 5 – 9
- below 5
- no data

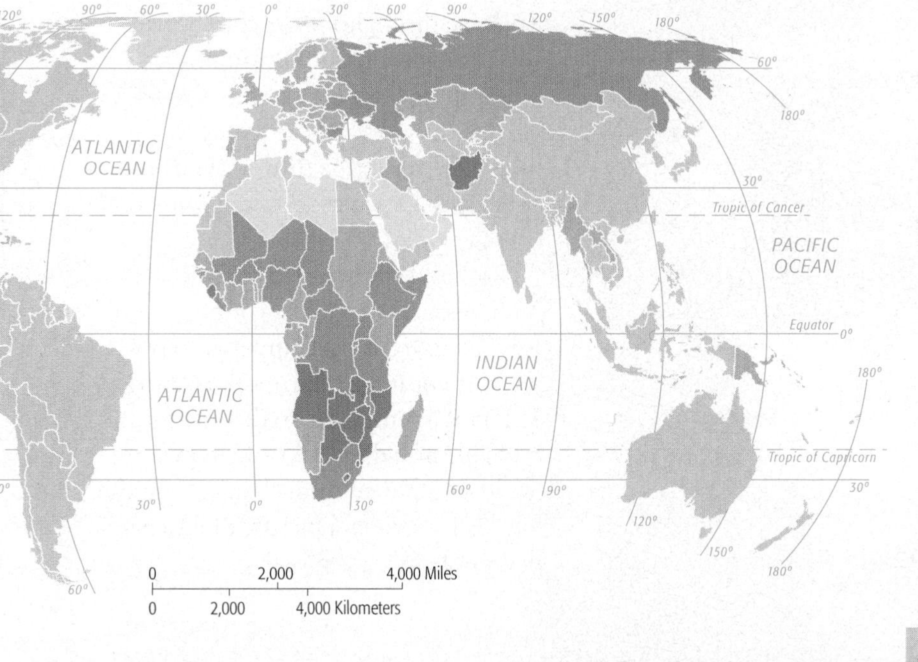

POPULATION GROWTH

2.4 Population Structure

- A country's birth rate is influenced by the actions of individual women.
- A country's death rate is influenced by the survival of infants and of older people.

Geographers utilize several measures of births and deaths in addition to the CBR and CDR discussed in the previous section. Patterns of births and deaths result in distinctive age structures.

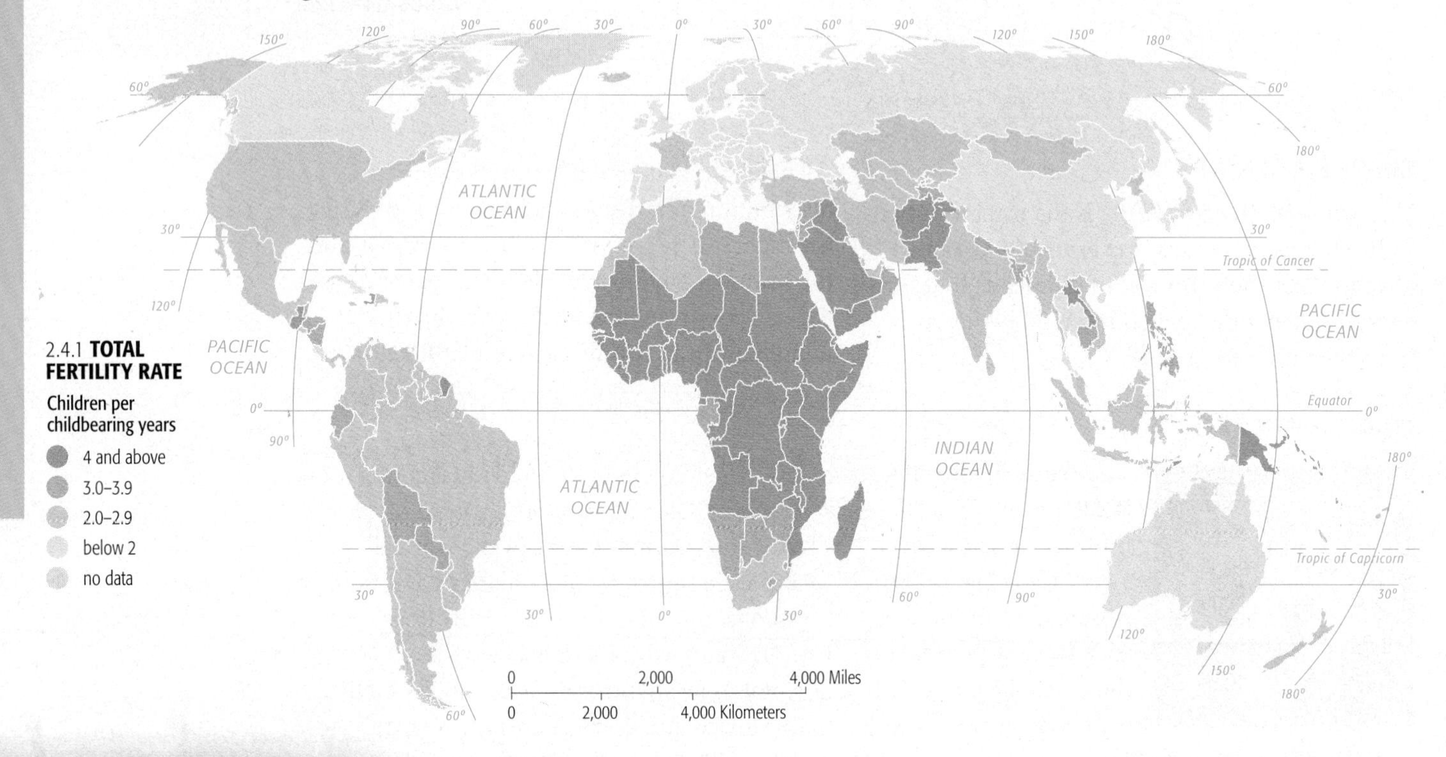

2.4.1 **TOTAL FERTILITY RATE**

TOTAL FERTILITY RATE

The **total fertility rate (TFR)** is the average number of children a woman will have throughout her childbearing years (roughly ages 15 through 49). To compute the TFR, scientists must assume that a woman reaching a particular age in the future will be just as likely to have a child as are women of that age today.

The TFR for the world as a whole is 2.7; it exceeds 6 in many countries of sub-Saharan Africa, compared to less than 2 in nearly all European countries. The TFR attempts to predict the future behavior of individual women in a world of rapid cultural change, whereas the CBR provides a picture of a society as a whole in a given year.

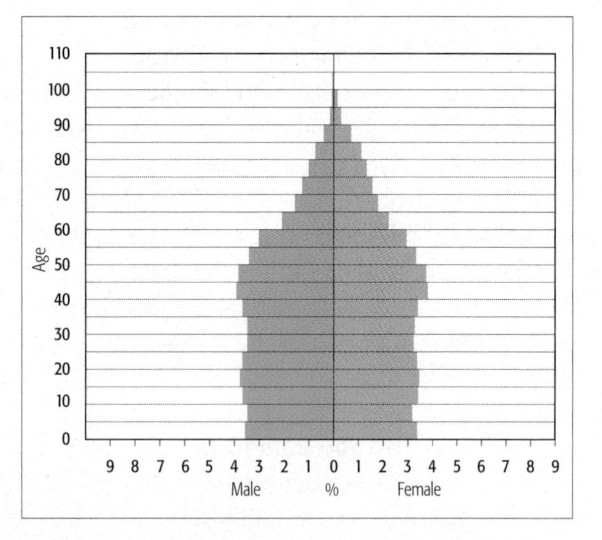

2.4.2 **POPULATION PYRAMID OF THE UNITED STATES**
A country's population can be displayed by age and gender groups on a bar graph called a population pyramid.

INFANT MORTALITY RATE

The **infant mortality rate (IMR)** is the annual number of deaths of infants under 1 year of age, compared with total live births. As was the case with the CBR and CDR, the IMR is usually expressed as the number of deaths among infants per 1,000 births rather than as a percentage (per 100).

The highest rates are in the poorer countries of sub-Saharan Africa, whereas the lowest rates are in Western Europe. IMRs exceed 100 in much of Africa, meaning that more than 10 percent of all babies die before reaching their first birthday. The IMR is less than 5 percent throughout Western Europe.

In general, the IMR reflects a country's health-care system. Lower IMRs are found in countries with well-trained doctors and nurses, modern hospitals, and large supplies of medicine.

Although the United States is well endowed with medical facilities, it suffers from a higher IMR than Canada and every country in Western Europe. African Americans and other minorities in the United States have IMRs that are twice as high as the national average, comparable to levels in Latin America and Asia. Some health experts attribute this to the fact that many poor people in the United States, especially minorities, cannot afford good health care for their infants.

2.4.3 **INFANT MORTALITY**

Per 1,000 live births

- 100 and above
- 50–99
- 25–49
- 10–24
- below 10
- no data

LIFE EXPECTANCY

Life expectancy at birth measures the average number of years a newborn infant can expect to live at current infant mortality rates. Like every other mortality and fertility rate discussed thus far, life expectancy is most favorable in the wealthy countries of Western Europe and least favorable in the poor countries of sub-Saharan Africa. Babies born today can expect to live into their late 70s in Western Europe but only into their 40s in most sub-Saharan African countries.

2.4.4 **LIFE EXPECTANCY AT BIRTH**

Per 1,000 live births

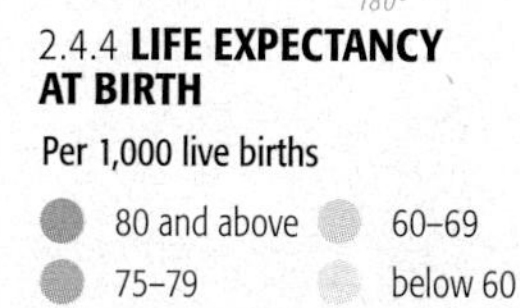

YOUNG AND OLD

The age structure of a population is extremely important in understanding similarities and differences among countries. The most important factor is the **dependency ratio**, which is the number of people who are too young or too old to work compared to the number of people in their productive years. The larger the percentage of dependents, the greater the financial pressure on those who are working to support those who cannot. People who are 0–14 years of age and 65-plus are normally classified as dependents.

One-third of the people are under age 15 in the LDCs, compared to only one-sixth in MDCs. The large percentage of children in LDCs strains ability to provide needed services such as schools, hospitals, and day-care centers. When children reach the age of leaving school, they need jobs, but the government must continue to allocate scarce resources to meet the needs of the still growing number of young people.

In contrast, MDCs face increasing percentages of older people, who must receive adequate levels of income and medical care after they retire from their jobs. The "graying" of the population places a burden on European and North American governments to meet these needs. More than one-fourth of all government expenditures in the United States, Canada, Japan, and many European countries go to Social Security, health care, and other programs for the older population.

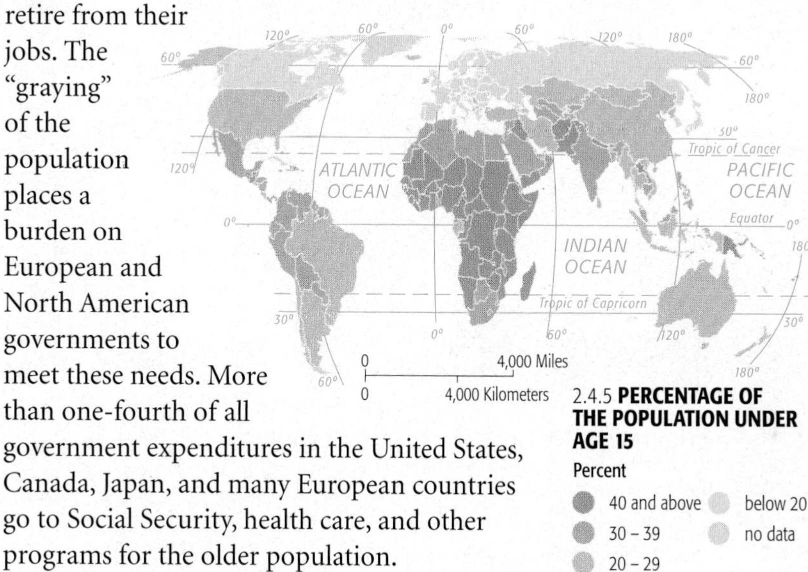

2.4.5 **PERCENTAGE OF THE POPULATION UNDER AGE 15**

Percent

- 40 and above
- 30 – 39
- 20 – 29
- below 20
- no data

POPULATION GROWTH

2.5 The Demographic Transition

- The demographic transition is the process of change of a country's population structure.
- Every country is in one of four stages of the demographic transition.

All countries have experienced some changes in natural increase, fertility, and mortality rates, but at different times and at different rates. Although rates vary among countries, a similar process of change in a society's population, known as the **demographic transition**, is operating.

FOUR STAGES OF DEMOGRAPHIC TRANSITION

The demographic transition is a process with several stages. Countries move from one stage to the next. At a given moment, we can identify the stage that each country is in.

STAGE 1

- Very high CBR
- Very high CDR
- Very low NIR

The stage for most of human history, because of unpredictable food supply, as well as war and disease.

During most of stage 1, people depended on hunting and gathering for food. A region's population increased when food was easily obtained and declined when it was not. No country remains in stage 1 today.

STAGE 2

- Still high CBR
- Rapidly declining CDR
- Very high NIR

In MDCs 200 years ago, because the Industrial Revolution generated wealth and technology, some of which was used to make communities healthier places to live.

In LDCs 50 years ago, because transfer of penicillin, vaccines, insecticides, and other medicines from MDCs controlled infectious diseases such as malaria and tuberculosis.

STAGE 3

- Rapidly declining CBR
- Moderately declining CDR
- Moderate NIR

In MDCs 100 years ago. People chose to have fewer children, in part a delayed reaction to the decline in mortality in stage 2, and in part because a large family was no longer an economic asset when families moved from farms to cities.

STAGE 4

- Very low CBR
- Low, slightly increasing CDR
- 0 or negative NIR

In MDCs in recent years, because increased access to birth control methods, as well as increased number of women working in the labor force outside the home, induce families to choose to have fewer children.

2.5.1 **SIERRA LEONE**
In stage 2 of the demographic transition

A country that has passed through all four stages of the demographic transition has completed a cycle from little or no natural increase in ***stage 1*** to little or no natural increase in ***stage 4.***

Two crucial differences:

1. CBR and CDR are high in ***stage 1*** and low in ***stage 4.***
2. Total population is much higher in ***stage 4*** than in ***stage 1.***

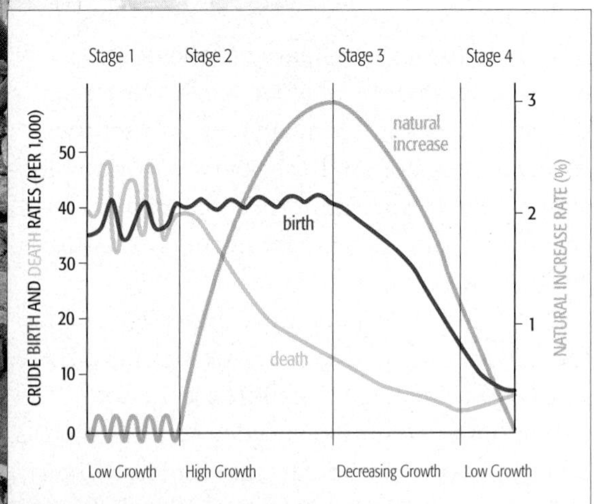

2.5.2 **DEMOGRAPHIC TRANSITION**

CAPE VERDE: STAGE 2 (HIGH GROWTH)

Cape Verde, a collection of 12 small islands in the Atlantic Ocean off the coast of West Africa, moved from ***stage 1*** to ***stage 2*** about 1950. Cape Verde was a colony of Portugal until it became independent in 1975, and the Portuguese administrators left better records of births and deaths than are typical for a colony in ***stage 1.***

Cape Verde's population actually declined during the first half of the twentieth century because of several severe famines, an indication that the country was still in ***stage 1***. Suddenly, in 1950, Cape Verde moved to ***stage 2.*** The reason: an anti-malarial campaign launched that year caused the CDR to sharply decline.

Cape Verde's population pyramid shows a large number of females nearing their prime childbearing years. For Cape Verde to enter ***stage 3***, these females must bear considerably fewer children than did their mothers.

CHILE: STAGE 3 (MODERATE GROWTH)

Chile's CDR declined sharply in the 1930s, moving the country into ***stage 2*** of the demographic transition. As elsewhere in Latin America, Chile's CDR was lowered by the infusion of medical technology from MDCs such as the United States.

Chile has been in ***stage 3*** of the demographic transition since about 1960. It moved to ***stage 3*** of the demographic transition primarily because of a vigorous government family-planning policy, initiated in 1966.

Chile's government reversed its policy and renounced support for family planning during the 1970s. Further reduction in the CBR is also hindered by the fact that most Chileans belong to the Roman Catholic Church, which opposes the use of what it calls artificial birth-control techniques. Therefore, the country is unlikely to move into ***stage 4*** of the demographic transition in the near future.

DENMARK: STAGE 4 (LOW GROWTH)

Denmark, like most Western European countries, has reached ***stage 4*** of the demographic transition. The country entered ***stage 2*** of the demographic transition in the nineteenth century, when the CDR began its permanent decline. The CBR then dropped in the late nineteenth century, and the country moved on to ***stage 3***.

Since the 1970s, Denmark has been in ***stage 4***, with roughly equal CBR and the CDR. Denmark's CDR has actually increased somewhat in recent years because of the increasing number of elderly people. The CDR is unlikely to decline unless a medical breakthrough, such as a cure for cancer, keeps elderly people alive much longer.

Denmark's population pyramid shows the impact of the demographic transition. Instead of a classic pyramid shape, Denmark has a column, demonstrating that the percentages of young and elderly people are nearly the same.

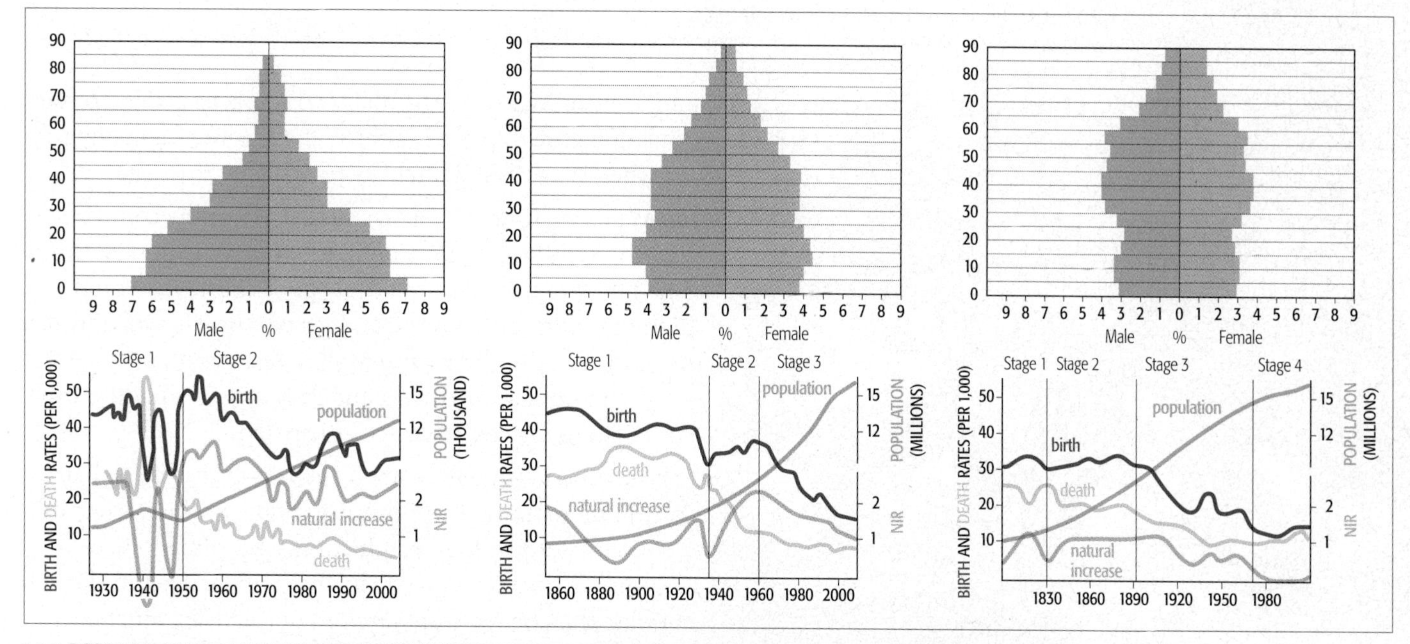

2.5.3 **POPULATION PYRAMID AND DEMOGRAPHIC TRANSITION FOR (left) CAPE VERDE, (center) CHILE, (right) DENMARK**

OVERPOPULATION THREATS

2.6 Malthus's Grim Forecast

- Malthus predicted that population would increase faster than resources.
- Contemporary geographers are divided on the validity of Malthus's thesis.

English economist Thomas Malthus (1766–1834) was one of the first to argue that the world's rate of population increase was far outrunning the development of food supplies. In *An Essay on the Principle of Population,* published in 1798, Malthus claimed that the population was growing much more rapidly than Earth's food supply because population increased ***geometrically***, whereas food supply increased ***arithmetically***. He concluded that population growth would deplete available resources, unless what he called "moral restraint" produced lower CBRs, or unless disease, famine, war, or other disasters produced higher CDRs. Malthus's views remain influential today.

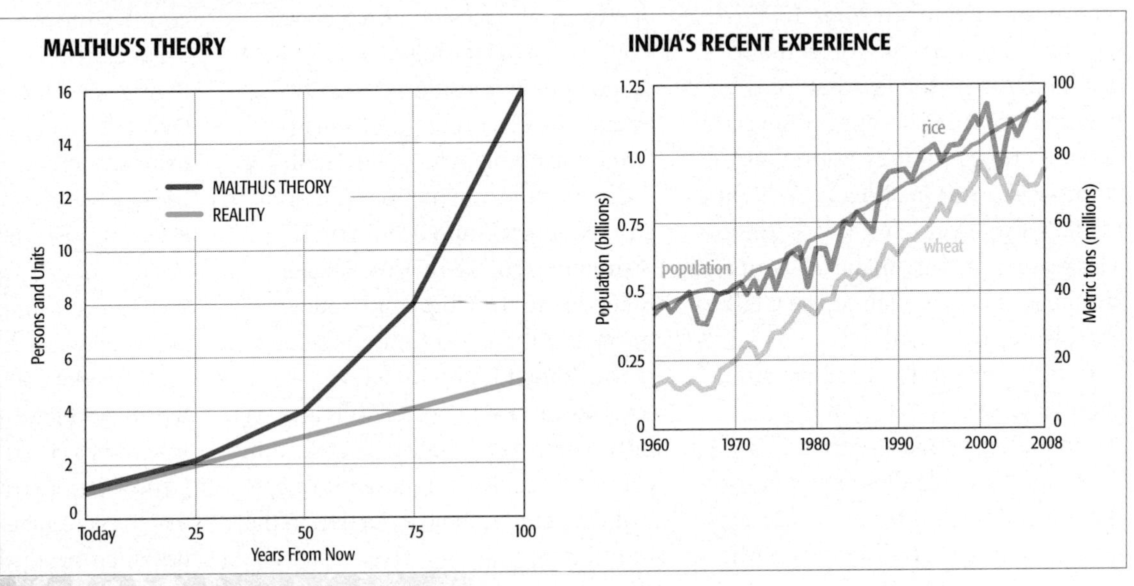

2.6.1 **POPULATION GROWTH AND FOOD SUPPLY**
(left) Malthus's theory was that population grows more rapidly than food supply. (right) In India since 1960, an increase in food production has kept pace with the increase in population.

Supporters of Malthus argue that two characteristics of recent population growth make Malthus's thesis more frightening than when it was first written more than 200 years ago

1. In Malthus's time only a few relatively wealthy countries had entered stage 2 of the demographic transition. Malthus failed to anticipate that relatively poor countries would have the most rapid population growth because of transfer of medical technology from MDCs. As a result, the gap between population growth and resources is wider in some countries than even Malthus anticipated.
2. The second argument made by neo-Malthusians is that world population growth is outstripping a wide variety of resources, not just food production. According to neo-Malthusians, wars and civil violence will increase in the coming years because of scarcities of clean air, suitable farmland, and fuel, as well as food.

MALTHUS'S CRITICS

Many geographers criticize Malthus's theory that population growth depletes resources. To the contrary, a larger population could stimulate economic growth and, therefore, production of more food. Population growth could generate more customers and more ideas for improving technology.

Some theorists maintain that poverty, hunger, and other social welfare problems associated with lack of economic development are a result of unjust social and economic institutions, not population growth. The world possesses sufficient resources to eliminate global hunger and poverty, if only these resources were shared equally.

Some political leaders, especially in Africa, argue that high population growth is good for a country because more people will result in greater power. Population growth is desired in order to increase the supply of young men who could serve in the armed forces. On the other side of the coin, more developed countries are viewed as pushing for lower population growth as a means of preventing further expansion in the percentage of the world's population living in poorer countries.

MALTHUS'S THEORY AND REALITY

On a global scale, conditions during the past half-century have not supported Malthus's theory. Even though the human population has grown at its most rapid rate ever, world food production has consistently grown at a faster rate than the NIR since 1950, according to geographer Vaclav Smil. Malthus was close to the mark on food production, but much too pessimistic on population growth.

Food production increased during the last half of the twentieth century somewhat more rapidly than Malthus predicted. Better growing techniques, higher-yielding seeds, and cultivation of more land all contributed to the expansion in food supply (see Chapter 10). Many people in the world cannot afford to buy food or do not have access to sources of food, but these are problems of distribution of wealth rather than insufficient global production of food, as Malthus theorized.

Malthus's model expected world population to quadruple between 1950 and 2000, from 2.5 billion to 10 billion people, but world population actually grew during this period to only 6 billion. Malthus did not foresee critical cultural, economic, and technological changes that would induce societies sooner or later to move on to stages 3 and 4 of the demographic transition.

OVERPOPULATION THREATS

2.7 Declining Birth Rates

- Some LDCs have lowered birth rates through economic development.
- In other LDCs, distribution of contraceptives have reduced birth rates.

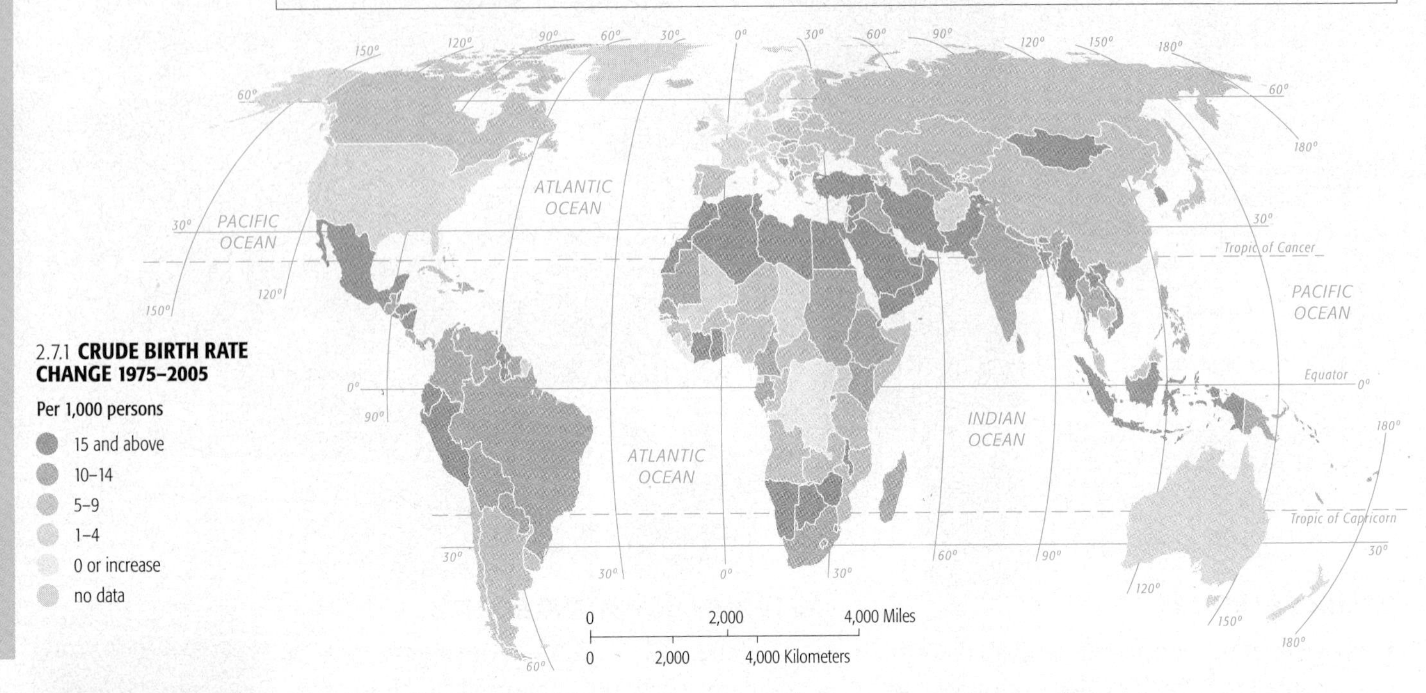

2.7.1 **CRUDE BIRTH RATE CHANGE 1975–2005**

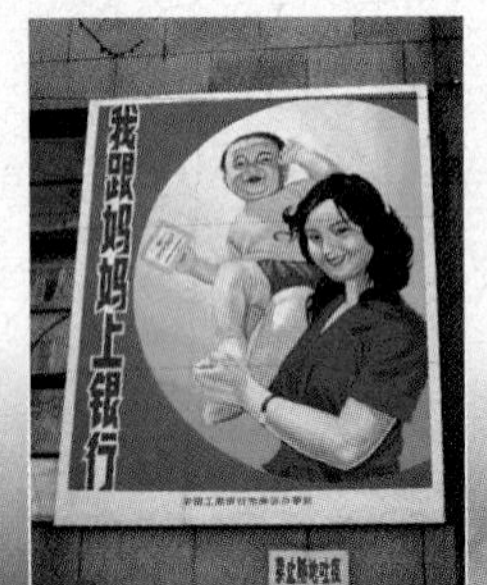

Population has been increasing at a much slower rate during the past quarter-century than it was during the previous half-century. The NIR declined during the 1990s from 1.8 to 1.3 for the world as a whole, from 2.1 to 1.6 in LDCs, and from 0.5 to 0.1 in MDCs. In contrast, during the 1980s, the world NIR rose from 1.7 in 1980 to 1.8 in 1990 because of an increase from 2.0 to 2.1 in LDCs.

In most countries, the decline in the NIR has occurred because of lower birth rates. Two strategies have been successful in reducing birth rates. One alternative emphasizes reliance on economic development, the other on distribution of contraceptives. Because of varied economic and cultural conditions, the most effective method varies among countries.

2.7.2 **BILLBOARDS PROMOTING ONE-CHILD POLICY IN CHINA**

LOWERING BIRTH RATES THROUGH ECONOMIC DEVELOPMENT

One approach to lowering birth rates emphasizes the importance of improving local economic conditions. A wealthier community has more money to spend on education and health-care programs that would promote lower birth rates.

- According to this approach, if more women are able to attend school and to remain in school longer, they are more likely to learn employment skills and gain more economic control over their lives.
- With better education, women would better understand their reproductive rights, make more informed reproductive choices, and select more effective methods of contraception.
- With improved health-care programs, IMRs would decline through such programs as improved prenatal care, counseling about sexually transmitted diseases, and child immunization.
- With the survival of more infants ensured, women would be more likely to choose to make more effective use of contraceptives to limit the number of children.

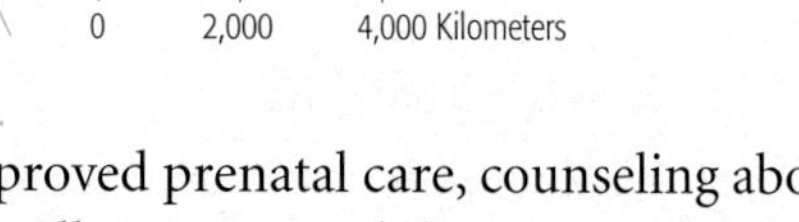

2.7.3 **PERCENTAGE OF WOMEN USING FAMILY PLANNING**

Percent
- 75 and above
- 50–74
- 25–49
- below 25
- no data

LOWERING BIRTH RATES THROUGH CONTRACEPTION

The other approach to lowering birth rates emphasizes the importance of rapidly diffusing modern contraceptive methods. Supporters of this approach agree that economic development may promote lower birth rates in the long run, but they argue that the world cannot wait around for that alternative to take effect. Putting resources into family-planning programs can reduce birth rates much more rapidly.

In LDCs, demand for contraceptive devices is greater than the available supply. Therefore, the most effective way to increase their use is to distribute more of them, cheaply and quickly. According to this approach, contraceptives are the best method for lowering the birth rate.

- Bangladesh is an example of a country that has had little improvement in the wealth and literacy of its people, but 58 percent of the women in the country used contraceptives in 2006 compared to 6 percent a quarter-century earlier. Similar growth in the use of contraceptives has occurred in other LDCs, including Colombia, Morocco, and Thailand.
- The percentage of women using contraceptives is especially low in Africa, so the alternative of distributing contraceptives could have an especially strong impact there. About one-fourth of African women employ contraceptives, compared to three-fourths in Latin America and two-thirds in Asia. The reason for this is partly economics, religion, and education.
- Very high birth rates in Africa and Southwestern Asia also reflect the relatively low status of women. In societies where women receive less formal education and hold fewer legal rights than do men, women regard having a large number of children as a measure of their high status, and men regard it as a sign of their own virility.

Regardless of which alternative is more successful, many oppose birth-control programs for religious and political reasons. Adherents of several religions, including Roman Catholics, fundamentalist Protestants, Muslims, and Hindus, have religious convictions that prevent them from using some or all birth-control devices. Opposition is strong within the United States to terminating pregnancy by abortion, and the U.S. government has at times withheld aid to countries and family-planning organizations that advise abortion, even when such advice is only a small part of the overall aid program.

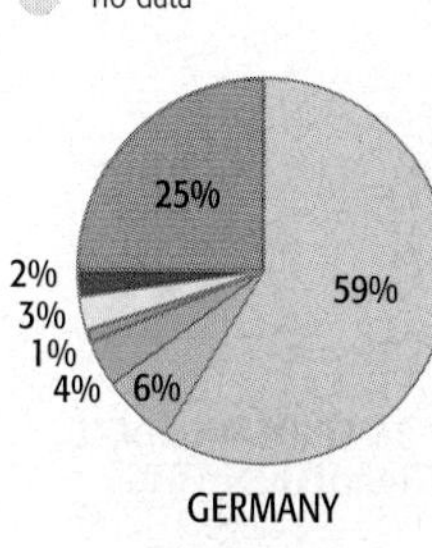

GERMANY

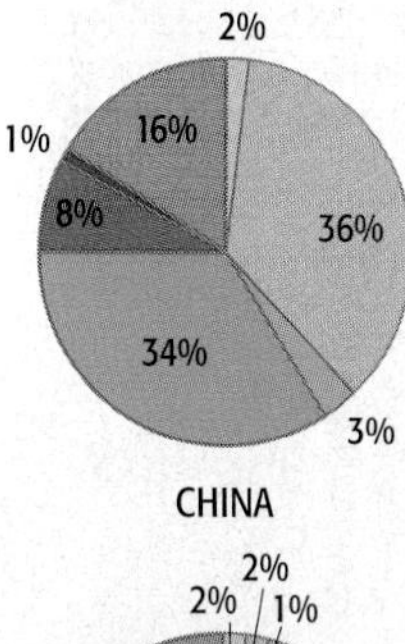

CHINA

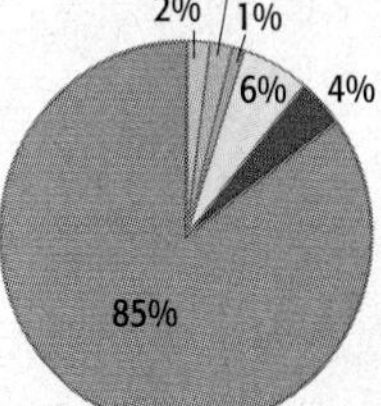

NIGERIA

2.7.4 **FAMILY PLANNING METHOD**

- Pill
- IUD
- Condom
- Female sterilization
- Male sterilization
- Periodic abstinence and withdrawal
- Other
- Not using a method

OVERPOPULATION THREATS

2.8 The Epidemiologic Transition

- Each stage of the demographic transition has distinctive causes of death.
- The leading causes of death shift through the demographic transition.

Medical researchers have identified an **epidemiologic transition** that focuses on distinctive causes of death in each stage of the demographic transition. The term *epidemiologic transition* comes from **epidemiology**, which is the branch of medical science concerned with the incidence, distribution, and control of epidemics, or diseases that affect large numbers of people. Epidemiologists rely heavily on geographic concepts such as scale and connection, because measures to control and prevent an epidemic derive from understanding its distinctive distribution and method of diffusion.

STAGE 1: PESTILENCE AND FAMINE

In stage 1 of the epidemiologic transition, infectious and parasitic diseases were the principal causes of human deaths, along with accidents and attacks by animals and other humans. Malthus called these causes of deaths "natural checks" on the growth of the human population. Epidemiologist Abdel Omran in 1971 called stage 1 the stage of pestilence and famine.

Well documented is the origin and diffusion of history's most violent stage 1 epidemic—the Black Plague, or bubonic plague, which was probably transmitted to humans by fleas from migrating infected rats. The Black Plague originated in Kyrgyzstan in present-day Asia and was brought from there by a Tatar army when it attacked an Italian trading post on the Black Sea in present-day Ukraine. Italians fleeing the trading post then carried the infected rats on ships west to the major coastal cities of Southeastern Europe in 1347. The plague spread from the coast to inland towns and then to rural areas.

The plague reached Western Europe in 1348 and Northern Europe in 1349. About 25 million Europeans—more than half of the continent's population—died between 1347 and 1350. In China, 13 million died from the plague in a single year, 1380.

The plague wiped out entire villages and families, leaving farms with no workers and estates with no heirs. Churches were left without priests and parishioners, schools without teachers and students. Ships drifted aimlessly at sea after entire crews succumbed to the plague.

2.8.1 **EPIDEMIOLOGIC TRANSITION STAGE 2**
Cholera, a stage 2 disease, has been a threat in Iraq, such as this location in the Baghdad suburb of Fdailiyah, where drinking water is being drawn from a water pipe that crosses a canal that carries raw sewage.

STAGE 2: RECEDING PANDEMICS

A **pandemic** is disease that occurs over a wide geographic area and affects a very high proportion of the population. In stage 2, improved sanitation, nutrition, and medicine during the Industrial Revolution reduced the spread of infectious diseases. But death rates did not decline immediately and universally. Poor people crowded into rapidly growing industrial cities had especially high death rates during the Industrial Revolution.

Geographic methods played a key role in understanding the cause of one of the worst stage 2 pandemics, cholera. Dr. John Snow (1813–1858) mapped the distribution of deaths from cholera in 1854 in the poor London neighborhood of Soho. Many in the nineteenth century believed that epidemic victims were being punished for sinful behavior and that most victims were poor because poverty was considered a sin.

Dr. Snow showed that cholera was not distributed uniformly among the poor. Predating GIS by more than a century, Dr. Snow overlaid a map of the distribution of cholera victims with a map of the distribution of water pumps—for poor people the source of water for drinking, cleaning, and cooking. Dr. Snow found that a large percentage of cholera victims were clustered around one pump, on Broad Street. Tests proved that the water at the Broad Street pump was contaminated, and further investigation revealed that contaminated sewage was getting into the water supply near the pump.

Construction of water and sewer systems eradicated cholera by the late nineteenth century. However, cholera reappeared a century later in rapidly growing cities of LDCs as they moved into stage 2 of the demographic transition.

2.8.2 **JOHN SNOW**
John Snow's map of cholera in Soho, London, 1854.

KEY
- Water pump
- Cholera victim

STAGE 3: DEGENERATIVE AND HUMAN-CREATED DISEASES

Stage 3 is characterized by a decrease in deaths from infectious diseases and an increase in chronic disorders associated with aging. The two especially important chronic disorders in stage 3 are cardiovascular diseases, such as heart attacks, and various forms of cancer.

The decline in infectious diseases such as polio and measles has been sharp in stage 3 countries. Effective vaccines were responsible for these declines.

STAGE 4: DELAYED DEGENERATIVE DISEASES

The epidemiologic transition was extended by S. Jay Olshansky and Brian Ault to stage 4, the stage of delayed degenerative diseases. The major degenerative causes of death—cardiovascular diseases and cancers—linger, but the life expectancy of older people is extended through medical advances. Through medicine, cancers spread more slowly or are removed altogether. Operations such as bypasses repair deficiencies in the cardiovascular system. Also improving health are behavior changes such as better diet, reduced use of tobacco and alcohol, and exercise.

Possible STAGE 5: REEMERGENCE OF INFECTIOUS DISEASES

Some medical analysts argue that the world is moving into stage 5 of the epidemiologic transition, the stage of reemergence of infectious and parasitic diseases. Infectious diseases thought to have been eradicated or controlled have returned, and new ones have emerged. A consequence of stage 5 would be higher CDRs. Other epidemiologists dismiss recent trends as a temporary setback in a long process of controlling infectious diseases.

Three reasons help to explain the possible emergence of a stage 5 of the epidemiologic transition:

1. **Evolution**. Infectious disease microbes have continuously evolved and changed in response to environmental pressures by developing resistance to drugs and insecticides; antibiotics and genetic engineering contribute to the emergence of new strains of viruses and bacteria.
2. **Poverty**. Infectious diseases such as tuberculosis (TB) is more prevalent in poor areas because the long, expensive treatment poses a significant economic burden.
3. **Travel**. As they travel, people carry diseases with them and are exposed to the diseases of others.

2.8.3 **EPIDEMIOLOGIC TRANSITION STAGE 5**
Tourists walk past a thermal scanner used to detect passengers with fevers at Manila International airport.

OVERPOPULATION THREATS

2.9 Global Health Threats

- Some infectious diseases have returned and new ones have emerged.
- The most lethal global-scale epidemic has been AIDS.

The most lethal epidemic in recent years has been AIDS (acquired immunodeficiency syndrome). Worldwide, 22 million people had died of AIDS as of 2007, and 33 million were living with HIV (human immunodeficiency virus, the cause of AIDS).

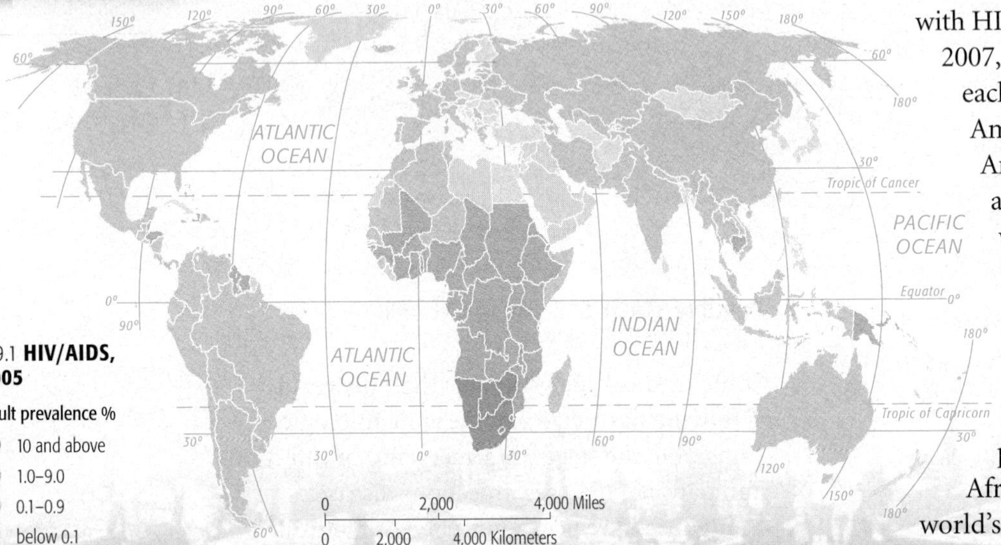

There were 23 million people infected with HIV in sub-Saharan Africa in 2007, 5 million in Asia, 2 million each in Eastern Europe and Latin America, 1 million each in North America and Western Europe, and 1 million elsewhere in the world. Thus, more than 90 percent were in LDCs.

The impact of AIDS has been felt most strongly in sub-Saharan Africa. With one-tenth of the world's population, sub-Saharan Africa had two-thirds of the world's total HIV-positive population

2.9.1 HIV/AIDS, 2005

Adult prevalence %
- 10 and above
- 1.0–9.0
- 0.1–0.9
- below 0.1
- no data

New cases in 1981

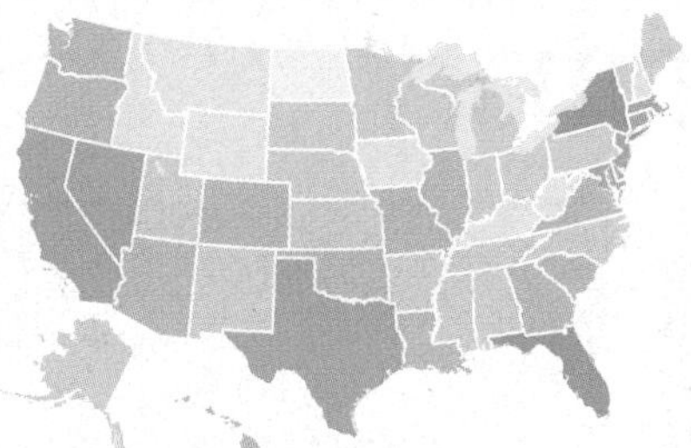

New cases in 1993

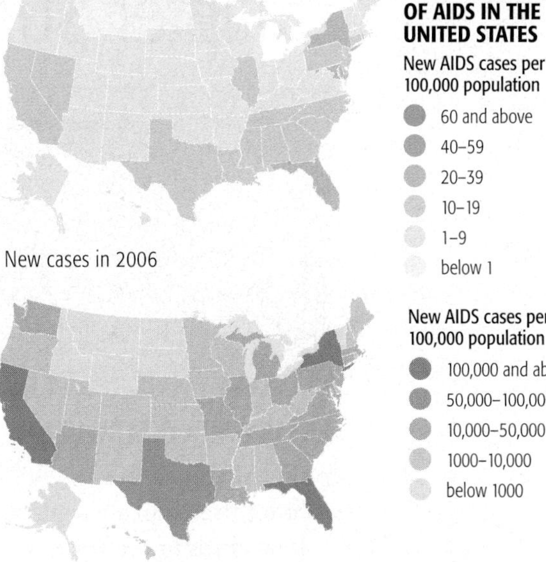

New cases in 2006

Cumulative cases 1981–2006

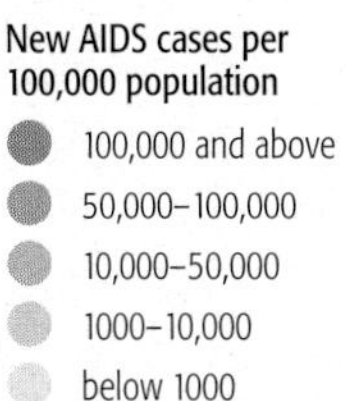

2.9.2 **DIFFUSION OF AIDS IN THE UNITED STATES**

and nine-tenths of the world's infected children. Even harder hit are the countries in the southern-most portion of Africa. With 1 percent of the world's population, the region has more than 30 percent of the world's AIDS cases.

CDRs in southern African countries have risen sharply in the past decade as a result of AIDS, from the mid-teens to the low twenties. The populations of Botswana, Lesotho, and Swaziland are forecast to decline between now and 2050 as a result of AIDS. Life expectancy has declined in these three countries from the mid-50s during the 1980s to the mid-30s in the twenty-first century.

AIDS IN UNITED STATES

New York, California, and Florida were the nodes of origin for the disease within the United States during the early 1980s. New AIDS cases diffused to every state during the 1980s and early 1990s, although New York, California, and Florida remained the focal points. These three states, plus Texas, accounted for half of the nation's new AIDS cases in the peak year of 1993. The rapid decline in new cases thereafter resulted from rapid diffusion of preventive methods and medicines such as AZT.

2.9.3 **THE AIDS MEMORIAL QUILT** assembled as a memorial to those who have died of AIDS.

Chapter Summary

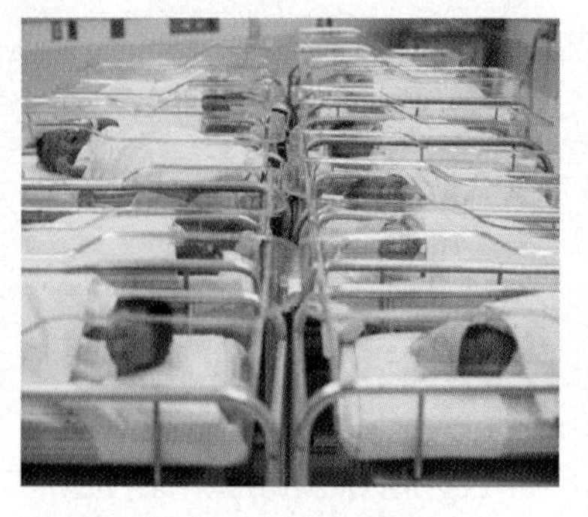

POPULATION DISTRIBUTION

- Global population is concentrated in a few places. Two-thirds of the world's people live in four clusters: East Asia, South Asia, Europe, and Southeast Asia.
- Human beings tend to avoid those parts of Earth's surface that they consider to be too wet, too dry, too cold, or too mountainous.

POPULATION GROWTH

- A population increases because of fertility and decreases because of mortality. The natural increase rate measures the difference between fertility and mortality.
- Virtually all the world's natural increase is concentrated in the LDCs of Africa, Asia, and Latin America, whereas the population is growing slowly or actually declining in MDCs.
- The demographic transition is a process of change in a country's population from a condition of high birth and death rates, with little population growth, to a condition of low birth and death rates, with low population growth.
- MDCs in Europe and North America have reached or neared the end of the demographic transition, whereas LDCs in Africa, Asia, and Latin America are at the stages of rapid population growth.

OVERPOPULATION THREATS

- More than 200 years ago, Thomas Malthus argued that population was increasing more rapidly than the food supply. With the rate of world population growth during the second half of the twentieth century at the highest level in history, some contemporary analysts believe that Malthus's prediction was accurate.
- Accounting for the record high natural increase rate during the twentieth century was a dramatic decline in death rates, especially from infectious diseases.
- Since the 1990s, birth rates have declined sharply, slowing world population growth and reducing fear of overpopulation in most regions. Meanwhile, death rates have increased in MDCs because of chronic disorders associated with aging and in some LDCs because of new infectious diseases, especially AIDS.

SHIBUYA, A DISTRICT OF TOKYO, JAPAN

Geographic Consequences of Change

Worldwide population increased rapidly during the second half of the twentieth century because few countries were in the two stages of the demographic transition that have low population growth—no country remains in stage 1, and few have reached stage 4.

The four-stage demographic transition is characterized by two big breaks with the past. The first break—the sudden drop in the death rate that comes from technological innovation—has been accomplished everywhere. The second break—the sudden drop in the birth rate that comes from changing social customs—has yet to be achieved in many countries.

If most countries in Europe and North America have reached—or at least are approaching—stage 4 of the demographic transition, why aren't other countries elsewhere in the world? LDCs moved into stage 2 of the demographic transition in the twentieth century for different reasons than did MDCs in the nineteenth century. In the past, the CDR declined in MDCs as part of the Industrial Revolution. In contrast, the sudden drop in the CDR in LDCs occurred because of injection of medical technology invented in MDCs.

Having caused the first break with the past through diffusion of medical technology, MDCs now urge LDCs to complete the second break with the past—the reduction in the CBR. However, reducing the CBR is difficult. A decline in the CDR can be induced through introduction of new technology by outsiders, but the CBR will drop only when people decide for themselves to have fewer children.

Geographers fear that as a result, LDCs, especially in Sub-Saharan Africa, are threatened by overpopulation—too many people for the available resources. Geographers caution that the number of people living in a region is not by itself an indication of overpopulation. Some densely populated regions are not overpopulated, whereas some sparsely inhabited areas are. The capacity of the land to support life derives partly from characteristics of the natural environment and partly from human actions to modify the environment through agriculture, industry, and exploitation of raw materials.

SHARE YOUR VOICE Student Essay

By Julieta Gomez de Mello
Florida International University

ONE TOO MANY

I had a history teacher once who joked that South Florida should be proclaimed as an autonomous state, separate from the rest of Florida. While this may sound a bit far-fetched, people who live or have lived down here know how isolated we are from the rest of the country in terms of population and culture.

Simply put, we are one too many.

Thousands upon thousands of people fill Miami's streets under the scorching sun either in cars, riding bicycles or simply walking by foot. Classrooms that are equipped for about 20 students are sometimes replete with 30, even 40 students.

Though most people originate from a Spanish culture, it is meshed with English as well as other languages to create an entirely new dialect that is particular to the South Florida niche. So, how does one distinguish themselves amidst such a large crowd? The answer might not be so simple, but with an open mind, anything is possible.

As a native Spanish speaker myself, I'm always eager to take in a new culture and new experiences. When I was in fourth grade I met a Muslim girl. She was polite, funny and intelligent. We became friends and soon were inseparable. She came to my birthday parties and helped me with homework assignments and was the best friend I had always wanted. Aside from any cultural differences we had, what I can recall the most is how good it felt to have someone who cared and was there for me everyday. I admired her for that.

When I was in middle school and 9/11 came about, I understood the difference between people like my friend and terrorists who, as she herself told me once, manipulated the Islamic religion, molded it to their liking, until it twisted like breadsticks. Living in Miami and associating with people from different countries and cultures has really shaped who I am today and allowed me to understand others. Therefore, the best solution for finding one's identity is to perhaps mesh with the crowd and learn from different cultures, languages and religions to then decide for yourself who you are among the many.

Chapter **Resources**

INDIA VERSUS CHINA

The world's two most populous countries, will heavily influence future prospects for global overpopulation.

These two countries—together encompassing more than one-third of the world's population—have adopted different family-planning programs. As a result of less effective policies, India adds 11 million more people each year than does China. Current projections show that India would surpass China as the world's most populous country during the 2030–40 decade.

INDIA

Like most countries in Africa, Asia, and Latin America, India remained in stage 1 of the demographic transition until the late 1940s. After independence from England in 1947, India's death rate declined sharply whereas the CBR remained high, so the country moved into stage 2 characterized by a very high NIR.

In response to this rapid growth, India became the first country to embark on a national family-planning program, in 1952. The most controversial element was setting up camps in 1971 to perform sterilizations, surgical procedures by which people were made incapable of reproduction. At the height of the program, 8.3 million sterilizations were performed during a 6-month period in 1976, mostly on women.

Widespread opposition to the sterilization program grew in the country, because people feared that they would be forcibly sterilized. Government-sponsored family-planning programs have instead emphasized education. The government provides information at clinics about alternative methods, distributes free birth-control devices, and performs several million legal and safe abortions annually. Nonetheless, India has made only modest progress in reducing births, and remains in stage 2 of the demographic transition.

CHINA

In contrast with India, China has made substantial progress in reducing its rate of growth. The core of the Chinese government's family-planning program has been the One Child Policy, adopted in 1980. Under the One Child Policy, couples needed a permit to have a child.

Couples received financial subsidies, a long maternity leave, better housing, and (in rural areas) more land if they agree to have just one child. The government prohibited marriage for men until they are 22 and women until they are 20. To further discourage births, people received free contraceptives, abortions, and sterilizations. Rules were enforced by a government agency, the State Family Planning Commission.

As China moved toward a market economy in the twenty-first century and Chinese families became wealthier, the harsh rules in the One Child Policy were relaxed, especially in urban areas. Fears that relaxing the One Child Policy would produce a large increase in the birth rate proved unfounded. After a quarter-century of intensive educational programs, the Chinese people have accepted the benefits of family planning. As a result, China is likely to maintain a much lower NIR than India in the twenty-first century.

ON THE INTERNET

The Population Reference Bureau (PRB) provides authoritative demographic information for every country and world region at its web site **www.prb.org**. The PRB also provides electronic access to many of the articles and reports it publishes, including those listed here under Further Readings.

The U.S. Bureau of the Census provides access to the results of the 2000 census at its web site **www.census.gov**. Tables and maps can be generated for a variety of geographic scales, ranging from tracts (equivalent to neighborhoods) to the country as a whole. Access is also provided to data from earlier censuses.

The World Resources Institute (WRI) undertakes policy research and analysis on relationships among population, environment, and development. Among its contributions is an extensive data base on a wide number of demographic, environmental, and development-related indicators for every country of the world covering several decades. WRI's data base, called Earth Trends, is accessed at **http://earthtrends**.wri.org/.

FURTHER READINGS

Bailey, Adrian. *Making Population Geography*. London: Hodder Arnold, 2005.

Kinsella, Kevin, and David R. Phillips. "Global Aging: The Challenge of Success." *Population Bulletin* 60 (1). Washington, DC: Population Reference Bureau, 2005.

Lamptey, Peter, Merywen Wigley, Dara Carr, and Yvette Collymore. "Facing the HIV/AIDS Pandemic." *Population Bulletin* 57 (3). Washington, DC: Population Reference Bureau, 2002.

Malthus, Thomas. *An Essay on the Principles of Population*. 1978 (reprint). London: Royal Economic Society, 1926 (first published 1798).

McFalls, Joseph A., Jr. "Population: A Lively Introduction." *Population Bulletin* 58 (4). Washington, DC: Population Reference Bureau, 2003.

Population Reference Bureau Staff. "Transitions in World Population." *Population Bulletin* 59 (1). Washington, DC: Population Reference Bureau, 2004.

Riley, Nancy E. "China's Population: New Trends and Challenges." *Population Bulletin* 59 (2). Washington, DC: Population Reference Bureau, 2004.

Simon, Julian. *Theory of Population and Economic Growth*. Oxford and New York: Blackwell, 1986.

Smil, Vaclav. "How Many People Can the Earth Feed?" *Population and Development Review* 20 (1994): 255–92.

World Bank. *World Bank Development Report*. New York: Oxford University Press, published annually.

KEY TERMS

Agricultural density
The ratio of the number of farmers to the total amount of land suitable for agriculture.

Arable land
Land suitable for agriculture

Arithmetic density
The total number of people divided by the total land area.

Crude birth rate (CBR)
The total number of live births in a year for every 1,000 people alive in the society.

Crude death rate (CDR)
The total number of deaths in a year for every 1,000 people alive in the society.

Demographic transition
The process of change in a society's population from a condition of high crude birth and death rates and low rate of natural increase to a condition of low crude birth and death rates, low rate of natural increase, and a higher total population.

Dependency ratio
The number of people under the age of 15 and over age 64, compared to the number of people active in the labor force.

Doubling time
The number of years needed to double a population, assuming a constant rate of natural increase.

Epidemiologic transition
Distinctive causes of death in each stage of the demographic transition.

Epidemiology
Branch of medical science concerned with the incidence, distribution, and control of diseases that affect large numbers of people.

Infant mortality rate (IMR)
The total number of deaths in a year among infants under 1 year old for every 1,000 live births in a society.

Life expectancy
The average number of years an individual can be expected to live, given current social, economic, and medical conditions. Life expectancy at birth is the average number of years a newborn infant can expect to live.

Natural increase rate (NIR)
The percentage growth of a population in a year, computed as the crude birth rate minus the crude death rate.

Overpopulation
The number of people in an area exceeds the capacity of the environment to support life at a decent standard of living.

Pandemic
Disease that occurs over a wide geographic area and affects a very high proportion of the population.

Physiological density
The number of people per unit of area of arable land, which is land suitable for agriculture.

Total fertility rate (TFR)
The average number of children a woman will have throughout her childbearing years.

Chapter 3 **MIGRATION**

Diffusion was defined in Chapter 1 as a process by which a characteristic spreads from one area to another and relocation diffusion as the spread of a characteristic through the bodily movement of people from one place to another. The subject of this chapter is a specific type of relocation diffusion called **migration**, which is a permanent move to a new location.

Emigration is migration from a location; **immigration** is migration to a location. The difference between the number of immigrants and the number of emigrants is the **net migration.** Immigrants—both legal and illegal—have been pouring into Western Europe and North America by the millions. They have been emigrating from Latin America, Asia, Africa, and Eastern Europe. Most migrate in search of better job prospects than available at home.

A FAMILY MIGRATES IN 1939 DURING THE GREAT DEPRESSION
They walked from Phoenix, Arizona, to San Diego, California, photographed here en route on US 99 near Brawley, California.

THE KEY ISSUES IN THIS CHAPTER

INTERNATIONAL MIGRATION

3.1 Reasons to Migrate

3.2 Characteristics of Migrants

3.3 Global Migration Patterns

3.4 Guest Workers

MIGRATION TO UNITED STATES

3.5 Changing Origin of U.S. Immigrants

3.6 U.S. Immigration Patterns

3.7 Undocumented U.S. Immigrants

INTERNAL MIGRATION

3.8 Interregional Migration

3.9 Intraregional Migration

INTERNATIONAL MIGRATION

3.1 Reasons to Migrate

- A combination of push and pull factors influences migration decisions.
- Most people migrate for economic reasons.

Geography has no comprehensive theory of migration, although a nineteenth-century outline of 11 migration "laws" written by E. G. Ravenstein is the basis for contemporary geographic migration studies.

Ravenstein's "laws" can be organized into three groups:

- The reasons why migrants move (discussed in this section).
- The characteristics of migrants (see Section 3.2).
- The distance migrants typically move (see Section 3.3).

People migrate because of push factors and pull factors. A **push factor** induces people to move out of their present location, whereas a **pull factor** induces people to move into a new location. We can identify three major kinds of push and pull factors: economic, cultural, and environmental.

ECONOMIC PUSH AND PULL FACTORS

International migration is permanent movement from one country to another, whereas **internal migration** is permanent movement within the same country. Most people migrate from one country to another for ***economic reasons***. People think about emigrating from places that have few job opportunities, and they immigrate to places where the jobs seem to be available. Because of economic restructuring, job prospects often vary from one country to another and within regions of the same country.

An area that has valuable natural resources, such as petroleum or uranium, may attract miners and engineers. A new industry may lure factory workers, technicians, and scientists. Construction workers, restaurant employees, and public-service officials may move to areas where rapid population growth stimulates demand for additional services and facilities.

The United States and Canada have been especially prominent destinations for economic migrants. This same perception of economic plenty now lures people to the United States and Canada from Latin America and Asia.

The relative attractiveness of a region can shift with economic change. Similarly, Scotland and Ireland have attracted migrants in recent years after decades of net out-migration. Following the discovery of petroleum in the North Sea off the coast of northeast Scotland, thousands of people have been lured to jobs in the drilling or refining of petroleum or in supporting businesses.

3.1.1 **ECONOMIC FACTORS**
Mexicans in search of work in Texas.

3.1.2 **CULTURAL FACTORS**
Escaping by boat during the 1970s after the Communist victory in the Vietnam War.

CULTURAL PUSH AND PULL FACTORS

Cultural factors can be especially compelling push factors, forcing people to emigrate from a country. **Forced migration** between countries has historically occurred for two main ***cultural reasons***: slavery and political instability. Millions of people were shipped to other countries as slaves or as prisoners, especially from Africa to the Western Hemisphere, during the eighteenth and early nineteenth centuries.

Large groups of people were no longer forced to migrate as slaves in the twentieth century, but forced international migration increased because of political instability resulting from cultural diversity. Boundaries of newly independent states often have been drawn to segregate two ethnic groups. Because at least some intermingling among ethnicities inevitably occurs, members of an ethnic group caught on the "wrong" side of a boundary may be forced to migrate to the other side. Wars have also forced large-scale migration of ethnic groups in the twentieth and twenty-first centuries, especially in Europe and Africa. Forced migration of ethnicities is discussed in more detail in Chapter 7.

According to the United Nations, **refugees** are people who have been forced to migrate from their homes and cannot return for fear of persecution because of their race, religion, nationality, membership in a social group, or political opinion. Refugees have no home until another country agrees to allow them in or improving conditions make possible a return to their former home. In the interim, they must camp out in tents, board in shelters, or find other temporary homes.

ENVIRONMENTAL PUSH AND PULL FACTORS

People also migrate for ***environmental reasons,*** pulled toward physically attractive regions and pushed from hazardous ones. In an age of improved communications and transportation systems, people can live in environmentally attractive areas that are relatively remote and still not feel too isolated from employment, shopping, and entertainment opportunities.

Attractive environments for migrants include mountains, seasides, and warm climates. Proximity to the Rocky Mountains lures Americans to the state of Colorado, and the Alps pull French people to eastern France. Coastal regions with warm winters, such as southern Spain and southern United States, attract migrants from harsher climates.

Migrants are also pushed from their homes by adverse physical conditions. Water—either too much or too little—poses the most common environmental threat. Many people are forced to move by water-related disasters because they live in a vulnerable area, such as a **floodplain.** More than 1 million people were forced to leave Gulf Coast states in 2005 because of Hurricane Katrina, the largest forced migration in U.S. history. Some soon returned to rebuild their homes, but several hundred thousand relocated permanently to other communities. Katrina also resulted in 1,836 confirmed fatalities and 705 missing.

A lack of water pushes others from their land. Hundreds of thousands have been forced to move from the Sahel region of northern Africa because of drought conditions. The capacity of the Sahel to sustain human life—never very high—has declined recently because of population growth and several years of unusually low rainfall.

3.1.3 **ENVIRONMENTAL FACTORS**
Escape from flooding in Bangladesh.

INTERNATIONAL MIGRATION

3.2 Characteristics of Migrants

- Historically, most migrants were males.
- Families with children comprise an increasing share of migrants.

A century ago, Ravenstein noted distinctive gender and family-status patterns in his migration theories:

- Most long-distance migrants were male.
- Most long-distance migrants were adult individuals rather than families with children.

Since the late twentieth century, these characteristics have changed. Women and children now constitute a majority of migrants.

GENDER OF MIGRANTS

Males historically accounted for most migrants because most people migrate for economic reasons, and men once constituted the overwhelming majority of the labor force. During the nineteenth and much of the twentieth centuries, for example, about 55 percent of immigrants to the United States were male. But the gender pattern reversed in the 1990s, and women now constitute about 55 percent of U.S. immigrants.

Mexicans who come to the United States without proper immigration documents—currently the largest group of U.S. immigrants—show similar gender changes. As recently as the late 1980s, males constituted 85 percent of the Mexican migrants arriving in the United States without proper documents, according to U.S. Bureau of the Census and Department of Homeland Security estimates. But since the 1990s, women have accounted for about half of the undocumented immigrants from Mexico.

The increased female migration to the United States partly reflects the changing role of women in Mexican society: in the past, rural Mexican women were obliged to marry at a young age and to remain in the village to care for children. Now some Mexican women are migrating to the United States to join husbands or brothers already in the United States, but most are seeking jobs. At the same time, women also feel increased pressure to get a job in the United States because of poor economic conditions in Mexico.

3.2.1 **MALE IMMIGRANTS**
The Mara Salvatrucha gang smuggles immigrants from Mexico into the United States.

3.2.2 **FEMALE IMMIGRANTS**
Mexican immigrant picks watermelons at a Michigan farm.

3.2.3 **FAMILIES OF IMMIGRANTS**
Mexican children watch television at a U.S. migrant labor camp while their parents work.

FAMILY STATUS OF MIGRANTS

Ravenstein also stated that most long-distance migrants were young adults seeking work, rather than children or elderly people. For the most part, this pattern continues for the United States. About 40 percent of immigrants are between the ages of 25 and 39, compared to about 23 percent of the entire U.S. population. Immigrants are less likely to be elderly people; only 5 percent of immigrants are over age 65, compared to 12 percent of the entire U.S. population.

An increasing percentage of U.S. immigrants are children—16 percent of immigrants are under age 15, compared to 21 percent for the total U.S. population. With the increase in women migrating to the United States, more children are coming with their mothers.

Recent immigrants to the United States have attended school for fewer years and are less likely to have high school diplomas than are U.S. citizens. The typical undocumented Mexican immigrant has attended school for 4 years, less than the average American, but a year more than the average Mexican.

Similarly, immigrants to Europe from Africa were once predominantly males, but now an increasing number of them are women and children.

3.2.4 **FAMILIES OF IMMIGRANTS**
Hispanic children at the Wesley Community Center after-school program in Phoenix.

INTERNATIONAL MIGRATION

3.3 Global Migration Patterns

- Most international migration is from LDCs to MDCs.
- The United States is the leading destination for international migrants.

Ravenstein made two main points about the distance that migrants travel to their new homes:

- Most migrants relocate a short distance and remain within the same country (see Sections 3.8. and 3.9).
- Long-distance migrants to other countries head for major centers of economic activity.

Consistent with the distance-decay principle presented in Chapter 1, the farther away a place is located, the less likely that people will migrate to it. Thus international migrants are much less numerous than internal migrants.

At a global scale, Asia, Latin America, and Africa have net out-migration, whereas North America, Europe, and Oceania have net in-migration. The three largest flows of migrants are to Europe from Asia and to North America from Asia and from Latin America.

Substantial in-migration also occurs from Europe to North America and from Asia to Oceania. Lower levels of net migration occur from Latin America to Oceania and from Africa to Europe, North America, and Oceania.

The global pattern reflects the importance of migration from less developed countries to more developed countries. Migrants from countries with relatively low incomes and high natural increase rates head for relatively wealthy countries, where job prospects are brighter.

3.3.1 **MIGRANTS IN LATIN AMERICA**
Mexicans on a cargo train en route to the border city of Nuevo Laredo, where they can cross into the United States.

3.3.2 **IMMIGRATION BY COUNTRY**
The United States plays a special role in the study of international migration. The world's third most populous country is inhabited overwhelmingly by direct descendants of immigrants. About 70 million people have migrated to the United States since 1820. The population of the United States includes about 35 million individuals born in other countries, about 12 percent of the population.

Several less populous countries have higher percentages of immigrants than does the United States. One-fourth of the Australian population and one-sixth of the Canadian population are immigrants. The overall percentage of immigrants in Europe is around 5 percent, lower than in the United States, but is much higher in smaller European countries such as Luxembourg and Switzerland.

The highest percentage of immigrants can be found in the Middle East, at about one-half of the region's total population. The population of the United Arab Emirates is made up of approximately 74 percent immigrants, and Kuwait 68 percent. These countries and other petroleum-exporting countries of the Middle East attract immigrants primarily from poorer Middle Eastern countries and from Asia to perform many of the dirty and dangerous functions in the oil fields.

Average annual net migration 2000-2005 (000s)

Gain
- above 100.0
- 20 to 100
- 0 to 20

Loss
- 0 to 20
- 20 to 100
- below 100

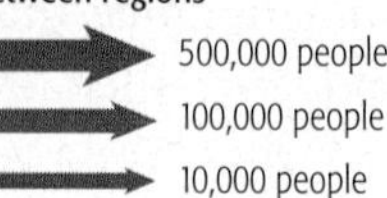

Geographer Wilbur Zelinsky identified a **migration transition**, which consists of changes in a society comparable to those in the demographic transition. A society in stage 1 of the demographic transition has high daily or seasonal **mobility** in search of food. International migration is primarily a phenomenon of countries in stage 2 of the demographic transition, whereas internal migration is more important in stages 3 and 4.

3.3.3 **MIGRANTS FROM ASIA**
Chinese migrant workers board a train in Shanghai to return for a week-long holiday.

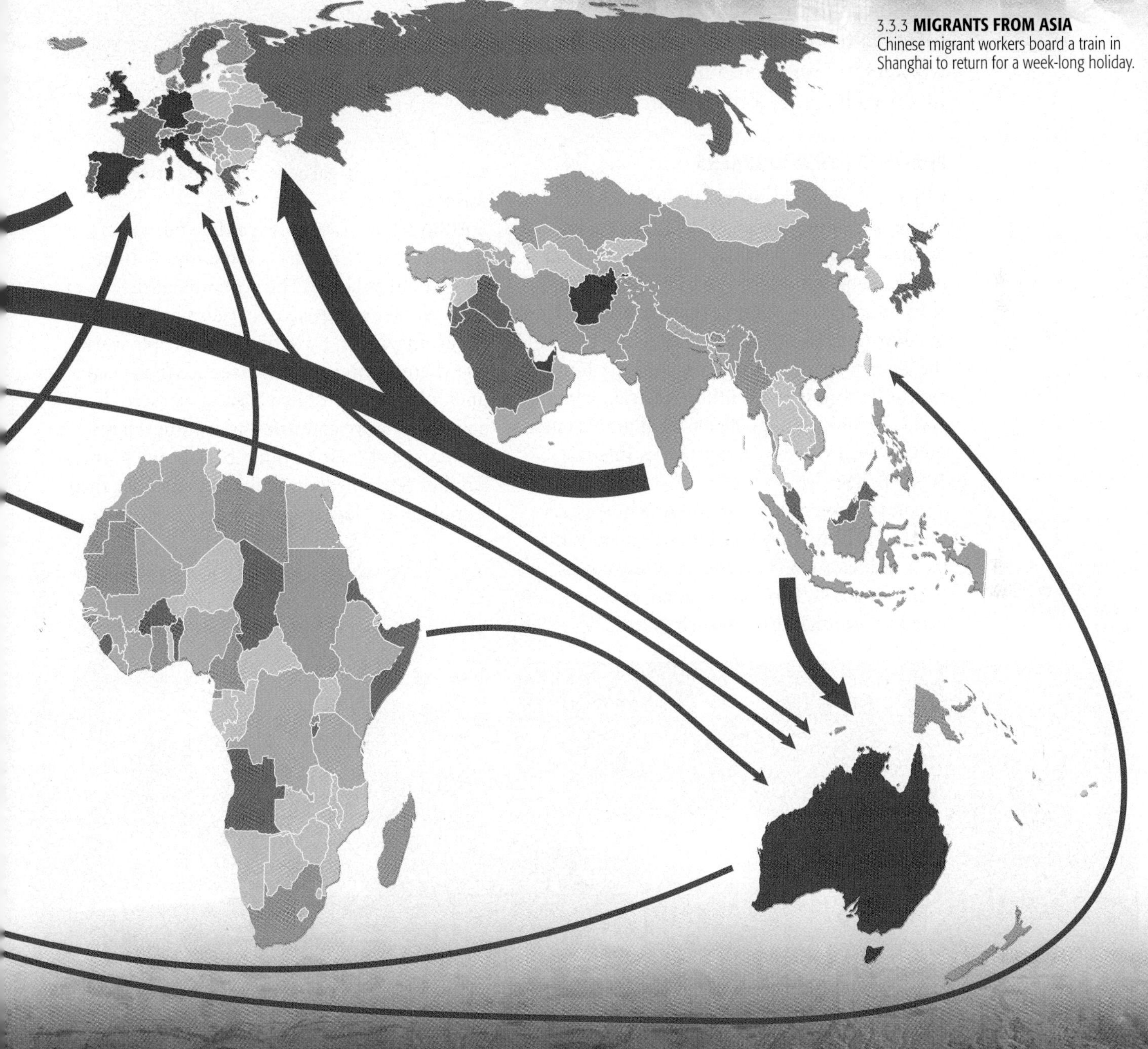

INTERNATIONAL MIGRATION

3.4 Guest Workers

- Guest workers migrate from LDCs to Europe and the Middle East.
- They hold low-paying unskilled jobs that local citizens don't want.

People unable to migrate permanently to a new country for employment opportunities may be allowed to migrate temporarily. Prominent forms of temporary-work migrants include guest workers in Europe and the Middle East and, historically, time-contract workers in Asia.

3.4.2 **GUEST WORKER FROM SURINAME IN THE NETHERLANDS**

EUROPE'S GUEST WORKERS

Citizens of poor countries who obtain jobs in Western Europe and the Middle East are known as **guest workers**. In Europe, guest workers are protected by minimum-wage laws, labor union contracts, and other support programs. Foreign-born workers comprise more than one-half of the labor force in Luxembourg; one-sixth in Switzerland; and one-tenth in Austria, Belgium, and Germany. About 700,000 immigrants enter Europe legally each year, plus an estimated 500,000 illegally.

Guest workers serve a useful role in Western Europe, because they take low-paying jobs that local residents won't accept. In cities such as Berlin, Brussels, Paris, and Zurich, guest workers provide essential services, such as driving buses, collecting garbage, repairing streets, and washing dishes.

Although relatively low paid by European standards, guest workers earn far more than they would at home. The economy of the guest worker's native country also gains from the arrangement. By letting their people work elsewhere, poorer countries reduce their own unemployment problems. Guest workers also help their native countries by sending a large percentage of their earnings back home to their families. The injection of foreign currency then stimulates the local economy.

3.4.1 **GUEST WORKERS FROM MOROCCO IN FRANCE**

The United Kingdom severely restricts the ability of foreigners to obtain work permits. However, British policy is complicated by the legacy of the country's former worldwide empire. When some of the United Kingdom's former colonies were granted independence, residents there could choose between remaining British citizens and becoming citizens of the new country. Millions of former colonials in India, Ireland, Pakistan, and the West Indies retained their British citizenship and eventually moved to the United Kingdom. However, spouses and other family members who are citizens of the new countries do not have the right to come to Britain.

Most guest workers in Europe come from North Africa, the Middle East, Eastern Europe, and Asia. Distinctive migration routes have emerged among the exporting and importing countries. Turkey sends a large number of guest workers to Northern Europe, especially to Germany as a result of government agreements. Three-quarters of a million Turks are employed in Germany, by far the largest movement of guest workers from one country to another within Europe. Many guest workers in France come from former French colonies in North Africa, such as Algeria and Morocco.

3.4.3 **GUEST WORKERS FROM TURKEY IN GERMANY**

3.4.4 **WORKERS FROM CHINA BUILDING A RAILROAD IN THE UNITED STATES**

TIME-CONTRACT WORKERS

Millions of Asians migrated in the nineteenth century as time-contract laborers, recruited for a fixed period to work in mines or on plantations. When their contracts expired, many would settle permanently in the new country. Indians went as time-contract workers to Burma (Myanmar), Malaysia, British Guiana (present-day Guyana in South America), eastern and southern Africa, and the islands of Fiji, Mauritius, and Trinidad. Japanese and Filipinos went to Hawaii, and Japanese also went to Brazil. Chinese worked on the U.S. West Coast and helped build the first railroad to span the United States, completed in 1869.

More than 33 million ethnic Chinese currently live permanently in other countries, for the most part in Asia. Chinese comprise three-fourths of the population in Singapore, one-third in Malaysia, and one-tenth in Thailand. Most migrants were from southeastern China. Migration patterns vary among ethnic groups of Chinese. Chiu Chownese migrate to Cambodia, Laos, and Singapore; Hakka to Indonesia, Malaysia, and Thailand; and Hokkien to Indonesia and the Philippines.

In recent years, people have immigrated illegally in Asia to find work in other countries. Estimates of the number of illegal foreign workers in Taiwan range from 20,000 to 70,000. Most are Filipinos, Thais, and Malaysians who are attracted by employment in textile manufacturing, construction, and other industries. These immigrants accept half the pay demanded by Taiwanese, for the level is much higher than what they are likely to get at home, if they could even find employment.

3.4.5 **MIGRANT WORKERS IN TAIWAN**

MIGRATION TO THE UNITED STATES

3.5 Changing Origin of U.S. Immigrants

- The United States has had three main eras of immigration.
- The principal source of migrants has changed in each era.

The United States has had three main eras of immigration: seventeenth and eighteenth centuries, mid-nineteenth to early twentieth centuries, and late twentieth and early twenty-first centuries. The three eras have drawn migrants from different regions.

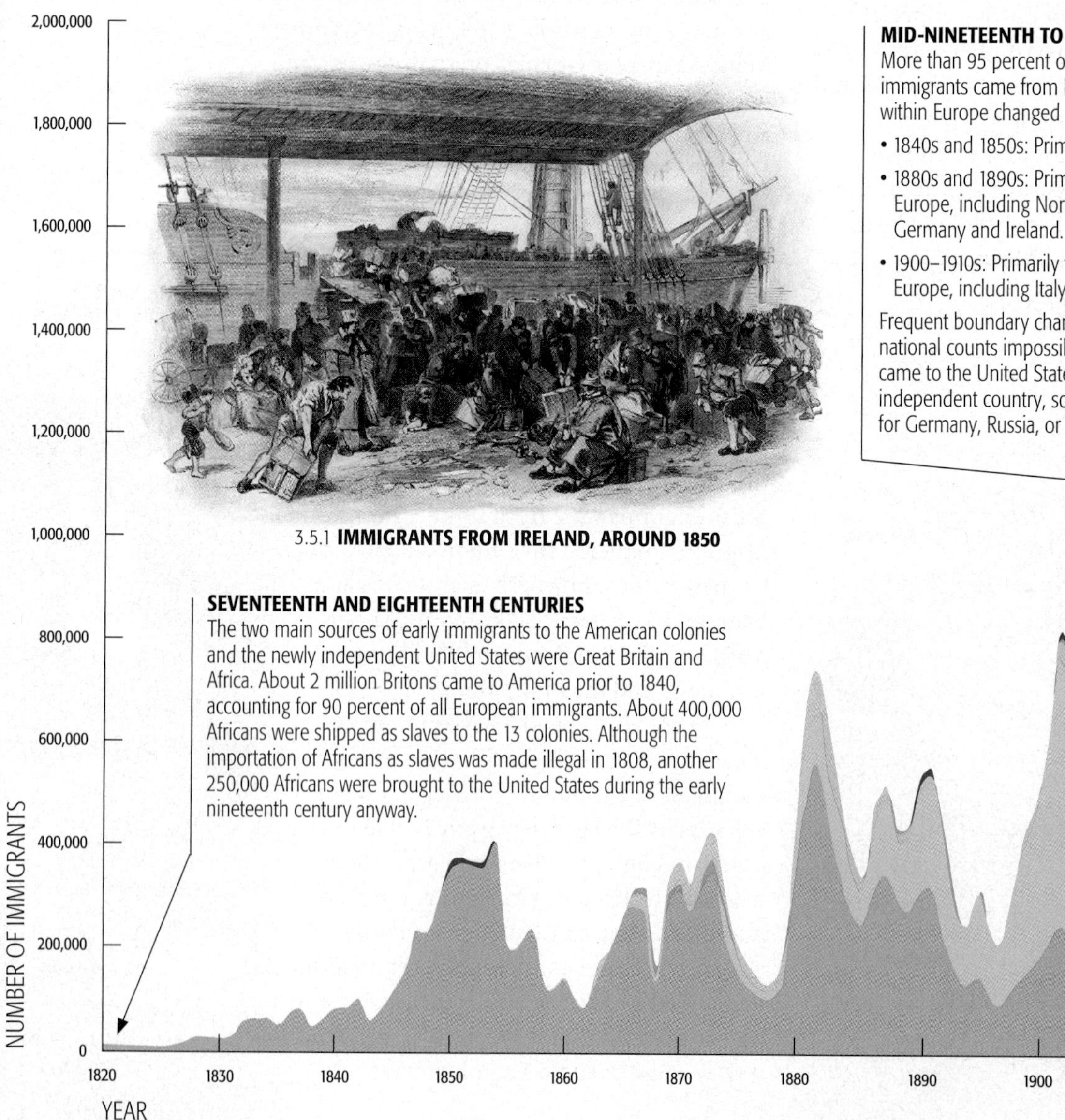

3.5.1 **IMMIGRANTS FROM IRELAND, AROUND 1850**

SEVENTEENTH AND EIGHTEENTH CENTURIES

The two main sources of early immigrants to the American colonies and the newly independent United States were Great Britain and Africa. About 2 million Britons came to America prior to 1840, accounting for 90 percent of all European immigrants. About 400,000 Africans were shipped as slaves to the 13 colonies. Although the importation of Africans as slaves was made illegal in 1808, another 250,000 Africans were brought to the United States during the early nineteenth century anyway.

MID-NINETEENTH TO EARLY TWENTIETH CENTURY

More than 95 percent of nineteenth-century U.S. immigrants came from Europe, but the principal sources within Europe changed during the century.

- 1840s and 1850s: Primarily from Ireland and Germany.
- 1880s and 1890s: Primarily from Northern and Western Europe, including Norway and Sweden, as well as Germany and Ireland.
- 1900–1910s: Primarily from Southern and Eastern Europe, including Italy and Russia.

Frequent boundary changes in Europe make precise national counts impossible. For example, most Poles came to the United States when Poland did not exist as an independent country, so they were included in the totals for Germany, Russia, or Austria.

3.5.3 **IMMIGRANTS FROM EUROPE, AROUND 1900**

LATE TWENTIETH AND EARLY TWENTY-FIRST CENTURIES

The two leading sources of immigrants since the late twentieth century have been Latin America and Asia. About 13 million Latin Americans and 7 million Asians have migrated to the United States in the past half-century, compared to only 2 million and 1 million, respectively, in the two preceding centuries. Officially, Mexico passed Germany in 2006 as the country that has sent to the United States the most immigrants ever. The four leading sources of U.S. immigrants from Asia have been China (including Hong Kong), the Philippines, India, and Vietnam.

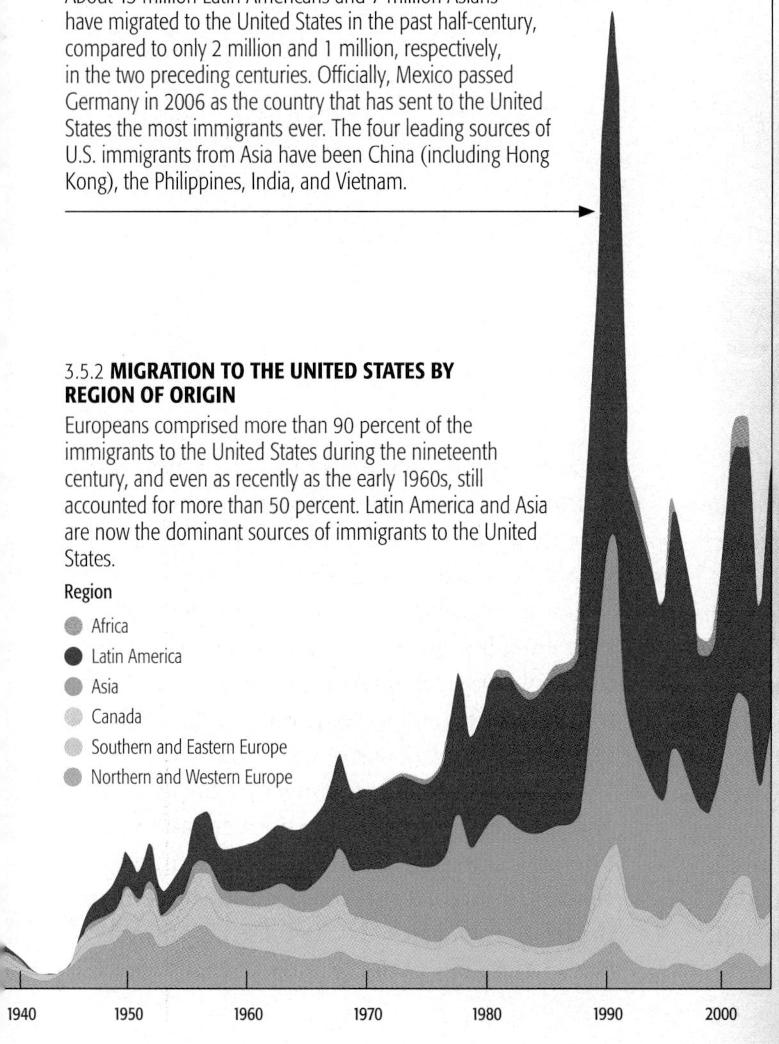

3.5.4 **IMMIGRANTS FROM ASIA, AROUND 2000**

3.5.2 **MIGRATION TO THE UNITED STATES BY REGION OF ORIGIN**

Europeans comprised more than 90 percent of the immigrants to the United States during the nineteenth century, and even as recently as the early 1960s, still accounted for more than 50 percent. Latin America and Asia are now the dominant sources of immigrants to the United States.

MIGRATION TO THE UNITED STATES

3.6 U.S. Immigration Patterns

- Quota laws restrict immigration to the United States.
- Immigrants are not distributed uniformly within the United States.

Although national origins have varied, the push factor for migrating to the United States has remained essentially the same: rapid population growth has discouraged economic prospects at home. Europeans left when their countries entered stage 2 of the demographic transition in the nineteenth century, and Latin Americans and Asians began to leave in large numbers in recent years after their countries entered stage 2. But Europeans arriving in the United States in the nineteenth century found a very different country than did Latin Americans and Asians who have recently arrived.

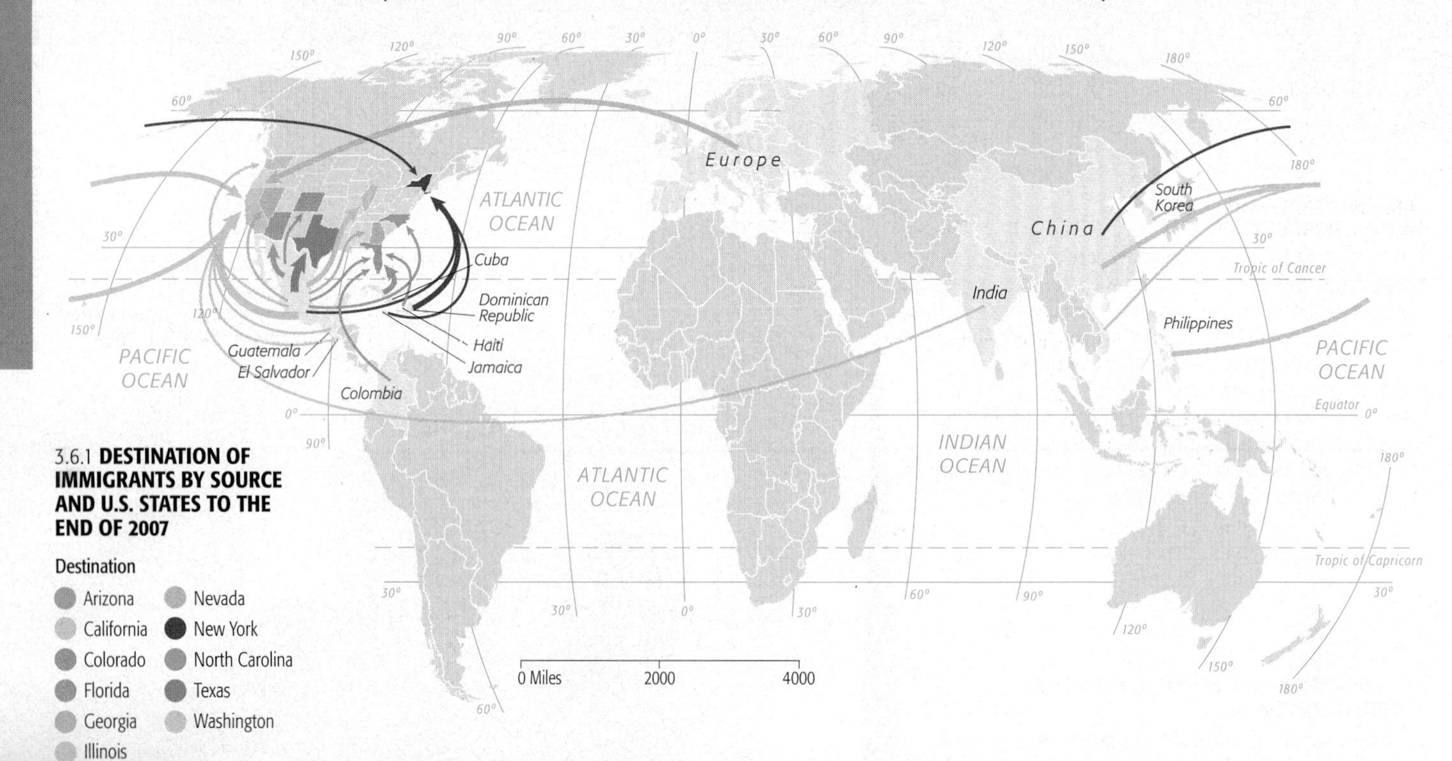

3.6.1 **DESTINATION OF IMMIGRANTS BY SOURCE AND U.S. STATES TO THE END OF 2007**

Destination
- Arizona
- California
- Colorado
- Florida
- Georgia
- Illinois
- Nevada
- New York
- North Carolina
- Texas
- Washington

- 200,000 to 499,999
- 500,000 to 999,999
- More than 1 million

ATTITUDES TOWARD IMMIGRANTS

Unrestricted immigration to the United States ended when Congress passed the Quota Act in 1921 and the National Origins Act in 1924. These laws established **quotas**, or maximum limits on the number of people who could immigrate to the United States from each country during a 1-year period.

Quota laws were originally designed to ensure that most immigrants to the United States continued to be Europeans. A government study in 1911 reflected widespread prejudice when it concluded that immigrants from Southern and Eastern Europe were racially inferior, "inclined toward violent crime," resisted assimilation, and "drove old-stock citizens out of some lines of work." Opposition to immigration further intensified when Americans believed that millions of Asians would flood into the country.

New arrivals have often been regarded with suspicion, but Americans tempered their dislike during the nineteenth century, because immigrants helped to settle the frontier and extend U.S. control across the continent.

European immigrants converted the forests and prairies of the vast North American interior into productive farms. By the early twentieth century, though, most Americans believed that the frontier had closed. When the U.S. frontier closed, the gates to the country partially closed as well.

QUOTAS

According to the original quota, for each country that had native-born persons already living in the United States, 2 percent of their number (based on the 1910 census) could immigrate each year. This limited the number of immigrants from the Eastern Hemisphere to 150,000 per year, virtually all of whom had to be from Europe. The system continued with minor modifications until the 1960s.

Following passage of the Immigration Act of 1965, quotas for individual countries were eliminated in 1968 and replaced with hemisphere quotas. The annual number of U.S. immigrants was restricted to 170,000 from the Eastern Hemisphere and 120,000 from the Western Hemisphere. In 1978, the hemisphere quotas were replaced by a global quota of 290,000, including a maximum of 20,000 per country. The current law has a global quota of 620,000, with no more than 7 percent from one country, but numerous qualifications and exceptions can alter the limit considerably.

Because the number of applicants for admission to the United States far exceeds the quotas, Congress has set preferences. The current law permits up to 480,000 family-sponsored immigrants plus 140,000 employment-related immigrants, again with numerous exceptions. About three-fourths of the immigrants are admitted to reunify families, primarily spouses or unmarried children of people already living in the United States. The typical wait for a spouse to gain entry is currently about 5 years. A handful of brothers and sisters or other relatives of noncitizens are also admitted, although the chance of being selected is as slim as winning the lottery. Skilled workers and exceptionally talented professionals receive most of the remaining one-fourth of the visas. Others are admitted by lottery under a diversity category for people from countries that historically sent few people to the United States.

The quota does not apply to refugees, who are admitted if they are judged genuine refugees. Also admitted without limit are spouses, children, and parents of U.S. citizens. The number of immigrants can vary sharply from year to year, primarily because numbers in these two groups are unpredictable.

3.6.2 **CHINATOWN, SAN FRANCISCO**

DESTINATION OF IMMIGRANTS IN THE UNITED STATES

Recent immigrants are not distributed uniformly throughout the United States. More than one-half are clustered in four states, including more than one-fourth in California and more than one-fourth in New York, Florida, and Texas. Coastal states were once the main entry points for immigrants, because most arrived by ship. Today, nearly all arrive by motor vehicle or airplane, but coastal states continue to attract migrants. California and Texas are the two most popular states for entry of motor vehicles from Mexico, and these states have the country's busiest airports for international arrivals.

Individual states attract immigrants from different countries. Immigrants from Mexico head for California, Texas, or Illinois, whereas immigrants from Caribbean island countries head for New York or Florida. Chinese and Indians immigrate primarily to New York or California, and other Asians immigrate to California (Figure 3.6.1).

Proximity clearly influences some decisions, such as Mexicans preferring California or Texas and Cubans preferring Florida. But proximity is not a factor in Poles heading for Illinois or Iranians for California. Immigrants cluster in communities where people from the same country previously settled. **Chain migration** is the migration of people to a specific location, because relatives or members of the same nationality previously migrated there.

Job prospects affect the states to which immigrants head. The South and West have attracted a large percentage of immigrants, because the regions have had more rapid growth in jobs than the Northeast or Midwest. In recent years, though, many immigrants—especially Mexicans—have migrated to the Midwest to take industrial jobs shunned by Americans, such as in meatpacking and related food processing.

3.6.3 **LITTLE INDIA**
Along 74th Street in the Jackson Heights neighborhood of New York City's Borough of Queens.

MIGRATION TO THE UNITED STATES

3.7 Undocumented U.S. Immigrants

- Some immigrants to the United States arrive without legal documentation.
- Americans do not agree on policies to address undocumented immigrants.

People who cannot legally enter the United States may enter instead without proper documents and thus are called **undocumented immigrants** or unauthorized immigrants. People enter or remain in the United States without authorization primarily because they wish to work but do not have permission to do so from the government.

Foreigners who fail to receive work visas have two choices if they still wish to work in the United States:

- Approximately half of the undocumented residents legally enter the country as students or tourists and then remain after they are supposed to leave.
- The other half simply slip across the border without showing a passport and visa to a border guard.

No one knows how many people immigrate to the United States without proper documents. The Urban Institute placed the total number in the United States in 2005 at around 9.3 million, including 5.3 million from Mexico, 2.2 million from other Latin American countries, 1 million from Asia, one-half million from Europe and Canada, and one-half million from the rest of the world. The Pew Hispanic Center estimated a higher level of 11.1 million in 2005 and 11.9 million in 2008.

About 7.2 million of the 11.1 million unauthorized immigrants in 2005 were employed, according to the Pew Hispanic Center, accounting for 5 percent of the total U.S. civilian labor force. They constituted 24 percent of workers in farming, 17 percent in cleaning, 14 percent in construction, and 12 percent in food preparation.

Most illegal Mexican immigrants have jobs in their home villages but migrate to the United States to earn more money. The largest number work in agriculture, picking fruits and vegetables, although some work in clothing factories. Even those who work long hours for a few dollars a day as farm laborers or factory workers prefer to earn relatively low wages by American standards than to live in poverty at home.

3.7.1 **U.S. BORDER PATROL ROUNDING UP UNDOCUMENTED IMMIGRANTS**

3.7.2 **U.S.-MEXICO BORDER BETWEEN SAN DIEGO (left) AND TIJUANA (right)**

CROSSING THE BORDER

Most undocumented residents have no difficulty finding jobs in the United States. Some employers like to hire immigrants who do not have visas that permit them to work in the United States because they can pay lower wages and do not have to provide health care, retirement plans, and other benefits. Unsatisfactory or troublesome workers can be fired and threatened with deportation.

Guards heavily patrol the official border crossings, most of which are located in urban areas such as El Paso, Texas, and San Diego, California, or along highways. However, the border is 3,600 kilometers (2,000 miles) long. It runs through sparsely inhabited regions and is guarded by only a handful of agents. A fence runs along the border but is broken in many places.

Many undocumented immigrants from Mexico have more difficulty reaching the U.S. border than crossing it. The journey from central and southern Mexico to remote desert border regions can be physically challenging and expensive.

Actually finding the border is difficult in some remote areas. A joint U.S.-Mexican International Boundary and Water Commission is responsible for keeping official maps, on the basis of a series of nineteenth-century treaties. The commission is also responsible for marking the border by maintaining 276 6-foot-tall iron monuments erected in the late nineteenth century, as well as 440 15-inch-tall markers added in the 1970s.

Once in the United States, undocumented immigrants can become "documented" by purchasing forged documents for as little as $25, including a birth certificate, alien registration card, and Social Security number.

Americans are divided concerning whether undocumented migration helps or hurts the country. Most Americans recognize that undocumented immigrants take jobs that no one else wants, and a majority would support some type of work-related program to make them legal. At the same time, Americans would like more effective border patrols so that fewer undocumented immigrants can get into the country.

3.7.3 **PATROLLING THE U.S.-CANADA BORDER NEAR BEECHER FALLS, VERMONT**

3.7.4 **AMERICANS ARE DIVIDED ON IMMIGRATION POLICY**

INTERNAL MIGRATION

3.8 Interregional Migration

- Long-distance migration can open regions for development.
- Long-distance migrants settled the western United States.

Most people find migration within a country less traumatic than international migration because they find familiar language, foods, broadcasts, literature, music, and other social customs after they move. Moves within a country also generally involve much shorter distances than those in international migration. However, internal migration can involve long-distance moves in large countries, such as in the United States and Russia.

INTERREGIONAL MIGRATION WITHIN THE UNITED STATES

Interregional migration is movement from one region of a country to another, whereas **intraregional migration** is movement within one region. The most famous example of interregional migration was the opening the American West. The changing location of the center of the U.S. population graphically demonstrates the interregional migration of the American people across the North American continent over the past 200 years.

American colonists clustered in coastal locations because they depended on shipping links with Europe to receive products and to export raw materials. The Appalachian Mountains blocked western development because of their steep slopes, thick forests, and few gaps that allowed easy passage. The native people who lived in the areas covered by the settlers often resisted the encroachment on their lands, slowing the pace of the expansion.

The center of the U.S. population moved west rapidly during the first half of the nineteenth century. Transportation improvements, especially canals, helped to open the American interior in the early 1800s. Most important was the Erie Canal, which enabled people to travel inexpensively by boat between New York City and the Great Lakes. The Gold Rush lured migrants to California beginning in the late 1840s.

3.8.1 **HOMESTEADERS MIGRATING TO THE U.S. PRAIRIES, AROUND 1890**

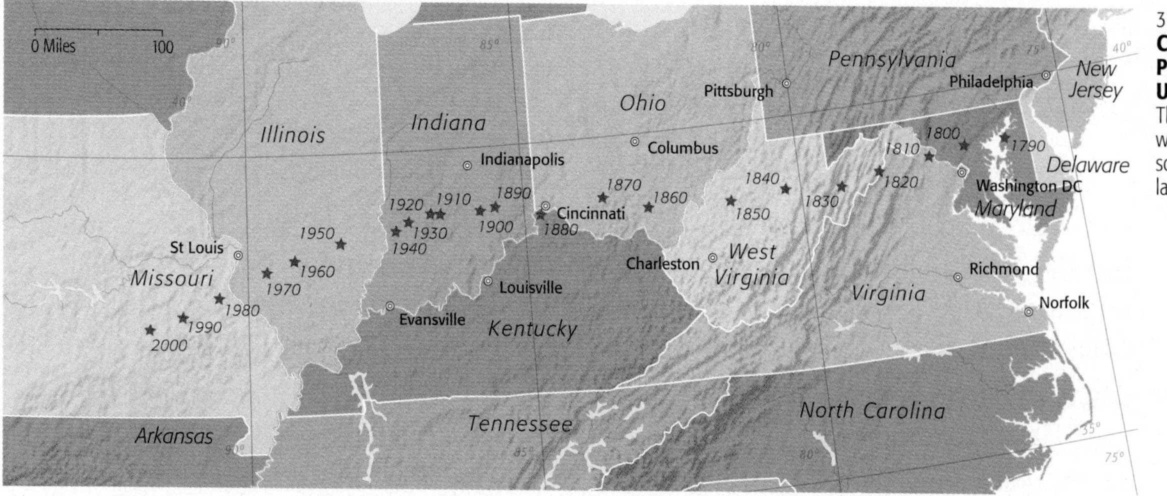

3.8.2 **CHANGING CENTER OF POPULATION IN THE UNITED STATES**
The center has shifted westward every decade, and southward beginning in the late twentieth century.

The westward movement of the U.S. population center slowed after 1880, as more immigrants settled in the Great Plains rather than pass through the region on the way to the West Coast as did earlier pioneers. Pioneers in the early nineteenth century labeled the Great Plains the Great American Desert because they lacked the know-how to farm in a treeless, dry climate with no source of wood to build homes, barns, and fences.

Advances in agricultural technology in the late nineteenth century enabled people to cultivate the Great Plains, which has become one of the world's richest farming areas. Farmers used barbed wire to reduce dependence on wood fencing, the steel plow to cut the thick sod, and windmills and well-drilling equipment to pump more water.

The expansion of the railroads encouraged western settlement beginning in the 1840s. By the 1880s, an extensive rail network permitted settlers on the Great Plains to transport their products to the large concentrations of customers in East Coast cities. The railroad companies also promoted western settlement by selling land to farmers.

Since the late twentieth century, the U.S. population center has drifted southward as well as westward. Americans have migrated to the South primarily for job opportunities and the temperate climate.

INTERREGIONAL MIGRATION IN LARGE COUNTRIES

Long-distance interregional migration has opened new regions for development in large countries other than the United States.

Russia. Migration has been encouraged to remote resource-rich regions in Asia through construction of mines, steel mills, power plants, and other industrial enterprises. When controlled by the former Soviet Union, some of the migration was forced.

Brazil. Migration has been encouraged from the large cities along the Atlantic coast to the sparsely settled tropical interior. In 1960, Brazil's capital was moved from the coastal city of Rio de Janeiro to Brasilia, a newly constructed city in the interior.

Indonesia. Since 1969, the Indonesian government has paid for the migration of more than 5 million people, primarily from the island of Java, where nearly two-thirds of its people live, to less populated islands.

India. To migrate to India's State of Assam, Indians are required to obtain a permit. Outsiders are limited in order to protect the ethnic identity of Assamese.

3.8.3 **INTERREGIONAL MIGRATION IN BRAZIL**
Brasilia ia a planned city designed to encourage migration to Brazil's interior. Most of the housing is in high-rise apartments.

INTERNAL MIGRATION

3.9 Intraregional Migration

- Most intraregional migration traditionally has been from rural to urban areas.
- In MDCs, most intraregional migration is now from cities to suburbs.

Intraregional migration is much more common than interregional or international migration. Most intraregional migration has been from rural to urban areas or from cities to suburbs.

RURAL TO URBAN MIGRATION

As a result of intraregional migration, nearly 50 percent of the world's people live in an urban area today, compared to only 5 percent 200 years ago. Worldwide, more than 20 million people are estimated to migrate each year from rural to urban areas. Migration from rural to urban areas has skyrocketed in recent years in the LDCs of Africa, Asia, and Latin America.

Like interregional migrants, most people who move from rural to urban areas seek economic advancement. They are pushed from rural areas by declining opportunities in agriculture and are pulled to the cities by the prospect of work in factories or in service industries. Many of these migrants cannot find housing in the city and must live in squatter settlements that lack electricity, running water, and paved streets (see Chapter 13).

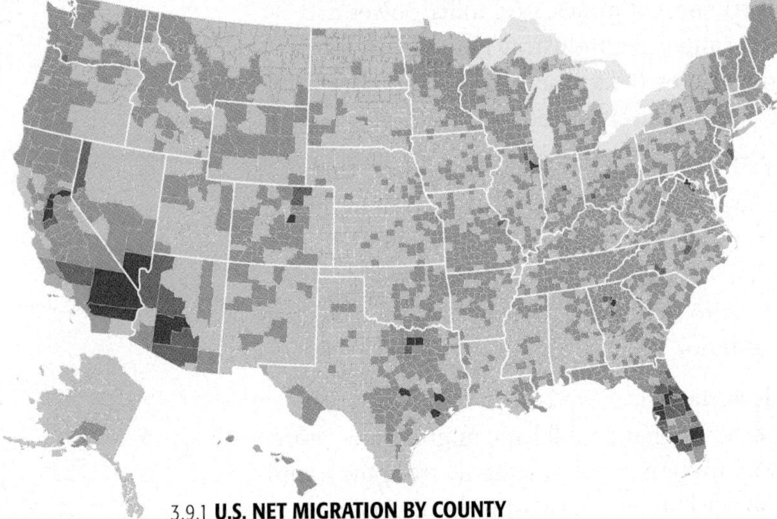

3.9.1 **U.S. NET MIGRATION BY COUNTY**

Annual change 2000-2004

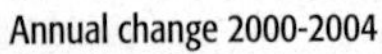

Gain	Loss
10,000 to 56,719	10,000 to 94,896
5,000 to 9,999	5,000 to 9,999
0 to 4,999	1 to 4,999

3.9.2 **RURAL TO URBAN MIGRATION**
Manila, the Philippines.

3.9.3 **URBAN TO SUBURBAN MIGRATION**
Suburbs around Phoenix.

URBAN TO SUBURBAN MIGRATION

In MDCs, most intraregional migration is from central cities out to the suburbs. The population of most central cities has declined in North America and Western Europe, whereas suburbs have grown rapidly.

The major reason for the large-scale migration to the suburbs is not related to employment, as was the case with other forms of migration. For most people, migration to suburbs does not coincide with changing jobs. Instead, people are pulled by a suburban lifestyle.

Suburbs offer the opportunity to live in a detached house rather than an apartment, surrounded by a private yard where children can play safely. A garage or driveway on the property guarantees space to park motor vehicles at no charge. Suburban schools tend to be more modern, better equipped, and safer than those in cities.

As a result of suburbanization, the territory occupied by urban areas has rapidly expanded. To accommodate suburban growth, farms on the periphery of urban areas are converted to housing developments, where new roads, sewers, and other services must be built.

REVERSE MIGRATION

Some MDCs are seeing a reversal of traditional migration patterns. Net migration to suburbs has slowed, whereas migration back into cities and back to rural areas has increased.

Like suburbanization, people move to cities and to rural areas for lifestyle reasons. People are lured to rural areas by the prospect of swapping the frantic pace of urban life for the opportunity to live on a farm. However, most people who move to farms do not earn their living from agriculture. Instead, they work in nearby factories, small-town shops, or other services. Net migration from urban to rural areas is called **counterurbanization**.

3.9.4 **REVERSE MIGRATION**
Inner-city apartment in Manhattan.

Similarly, others have been attracted to inner city neighborhoods that offer cultural activities, nightlife, and ethnic diversity. Especially attracted are recent college graduates and "empty nesters" with grown children.

With modern communications and transportation systems, no location in MDCs is truly isolated, either economically or socially. Computers enable us to work anywhere and still have access to an international network. We can obtain money at any time from a conveniently located electronic transfer machine rather than by going to a bank building. We can select clothing from a mail-order catalog, place the order by telephone, pay by credit card, and have the desired items delivered within a few days.

3.9.5 **URBAN TO SUBURBAN MIGRATION**
Suburbs of Phoenix sprawl across the desert.

Chapter **Summary**

INTERNATIONAL MIGRATION

- Immigration involves a combination of push factors that induce them to leave a place and pull factors that attract them to a new location.
- Three types of push and pull factors are economic, cultural, and environmental.
- Economic factors: People emigrate from places with few job opportunities and immigrate to places where jobs seem abundant.
- Cultural factors: People are forced to migrate because of political instability resulting from cultural diversity.
- Environmental factors: People are pushed from adverse physical conditions and pulled to places with attractive environmental features.

MIGRATION TO THE UNITED STATES

- The United States has been the leading destination for international migrants.
- The principal source of immigrants to the United States has varied over time, but the principal reason for migrating to the United States has remained economic.
- Great Britain and Africa were the principal sources of immigrants prior to the nineteenth century, Europe was the principal source of immigrants to the United States until the twentieth century, and most immigrants are now coming from Latin America and Asia.

INTERNAL MIGRATION

- Interregional migration is migration between regions within a country. Historically, interregional migration was especially important in settling the frontier of large countries such as the United States, Russia, and Brazil.
- Interregional migration persists today because of differences among regions of a country in economic conditions, climate, and other environmental factors.
- Intraregional migration is migration within a region of a country. The most important intraregional migration patterns are from rural to urban areas within LDCs and from cities to suburbs within MDCs.
- Within some MDC regions, migration has increased back to rural areas and to inner cities. Lifestyle choices have motivated these intraregional migration decisions.

IMMIGRANTS ON ELLIS ISLAND WAITING TO BE TRANSFERRED BY FERRY TO MANHATTAN IN 1912

Geographic Consequences of Change

Many European immigrants to North America in the nineteenth century truly expected to find streets paved with gold. While not literally so gilded, the United States and Canada did offer Europeans prospects for economic advancement.

Desire to migrate to the United States has shifted to LDCs in Asia and Latin America. Latin American countries in stage 2 of the demographic transition export a large percentage of their population growth to the United States. The same perception of economic plenty now lures people to the United States and Canada from Latin America and Asia.

Residents of the United States in the nineteenth and early twentieth centuries did not greet European immigrants with open arms. However, immigrants were given the opportunity to enter the country and make new lives. Some of them were successful. Even recently arrived undocumented immigrants stand a good chance of success if given the chance.

The most famous symbol of migration in the world is surely the Statue of Liberty. Its inscription, written by Emma Lazarus, includes the famous words, "Give me your tired, your poor, your huddled masses yearning to breathe free." The statue stands at the mouth of New York Harbor, near Ellis Island, which was for many years the initial landing and processing point for millions of immigrants from Europe. The Statue of Liberty was the first landmark seen by many European immigrants when they sailed into the United States.

In the twenty-first century, most immigrants arrive in the United States by crossing the Mexican border. And the United States no longer asks for immigration of the world's tired, poor, huddled masses yearning to be free.

For many people, the only way to enter the United States or Canada is illegally. In the twenty-first century, the tradition of universal ability to migrate to North America no longer exists. Paradoxically, in an era when human beings have invented easy means of long-distance transport, the right of free migration has been replaced by human barriers.

SHARE YOUR VOICE Student Essay

Patrick Cieslewicz
University of Wisconsin-Stevens Point

PATRICK CIESLEWICZ

Variety has been described as the spice of life. If this proclamation holds true, then I would contend that Wisconsin is one of the most flavorful places in the United States. Wisconsin has a rich heritage of settlers from several backgrounds. I grew up in Green Bay, a port city on Lake Michigan. The magnitude of migration is easily evidenced by the differing ethnicities and cultures of Wisconsin.

Migration played a large role in my life by casting my foundation in this area of our country. My last name comes from Polish ancestors who came to Wisconsin years ago. They were from a small farming community in western Poland. My ancestors were well suited to their new homeland, since the land patterns, climate, and local industries were similar to those left behind in their native land.

The Cieslewicz family settled into a northeastern Wisconsin town named Poland in the late 1800s. They spoke Polish when at home and made traditional foods. The farm that was settled by my early family is still in our possession. My mother's side of the family settled into a northern town called Niagra. Her family worked in mills and factories.

One of my favorite foods that we would have during reunions and holidays was pirogi. It resembles a small pizza and was taken to work by miners who didn't have much space in which they could store their lunch.

Even though my Polish roots were strong, our family had a commitment to America and its ideals of freedom. My grandfather served in the U.S. Army during WWII in the Pacific theater. I remember seeing the pictures of him in the jungles. My father was an Air Force aircraft mechanic during Vietnam, and my brother Nick carried the family military tradition into contemporary times by deploying to Iraq with the U.S. Marine Corps.

It is interesting to consider how, despite the continuous evolution and movement of human life, I am still able to appreciate the richness of my Polish heritage as a contemporary young professional. The will, determination, and hopes of my early family are stitched into my soul. I think that we as people secure our future when we preserve our past.

Chapter **Resources**

UNDOCUMENTED IMMIGRATION

From the United States, the view to the south may seem straightforward, the view from Mexico is more complex.

Millions of Mexicans are trying to cross the border by whatever means, legal or otherwise, in search of employment, family reunification, and a better way of life in the United States.

The view from Mexico is more complex. Along its northern border with the United States, Mexico is the **source** of the undocumented immigrants.

At the same time, along its southern border with Guatemala, Mexico is the **destination** for undocumented immigrants. When talking about Guatemala, Mexicans urge stronger security along the border.

THE U.S.–MEXICAN BORDER

Along the U.S.-Mexican border the contrast in wealth between the two countries is apparent, even in satellite imagery. Small houses packed close together on the Mexican side face large houses with wooded lots and swimming pools on the American side.

THE MEXICAN-GUATEMALAN BORDER

Contrasts also exist along Mexico's southern border. Some cross into Mexico from Guatemala because they can get higher-paying jobs in tropical fruit plantations. The Suchiate River, which marks the border between Hidalgo, Mexico, and Tecum Uman, Guatemala, is sometimes only ankle deep. Immigrants from other Latin American countries, especially El Salvador and Honduras, travel through Guatemala without need of a passport in order to cross into Mexico. Although a passport is needed to cross the border from Guatemala into Mexico, hundreds of thousands do so illegally.

THE ULTIMATE DESTINATION

The ultimate destination for most undocumented immigrants into Mexico is the U.S. border. Ironically, it is easier for undocumented immigrants in the United States if they are not Mexican. Undocumented Mexicans apprehended in the United States are usually bused back across the border into Mexico and released, but those from other countries who are apprehended in the United States are usually arrested and released with orders to appear at a court hearing. Once released, they are free to travel within the United States, blending in with other immigrants, and few show up at scheduled court hearings several months later.

Meanwhile, the millions of Mexicans living legally and illegally in the United States have constituted a powerful political and economic force back in Mexico. The Inter-American Development Bank estimated that immigrants in the United States sent $17 billion back to Mexico in 2005, and $28 million to other Latin American countries. Most of these remittances were used by relatives for food, clothing, and shelter, but government officials have tried to channel some of the money into development projects. The Mexican government has also faced pressure to make it easier for the millions of its citizens living in the United States to vote in elections back home.

KEY TERMS

Chain migration
Migration of people to a specific location because relatives or members of the same nationality previously migrated there.

Counterurbanization
Net migration from urban to rural areas in more developed countries.

Emigration
Migration from a location.

Floodplain
The area subject to flooding during a given number of years according to historical trends.

Forced migration
Permanent movement compelled usually by cultural factors.

Guest workers
Workers who migrate to the more developed countries of Northern and Western Europe, usually from Southern and Eastern Europe or from North Africa, in search of higher-paying jobs.

Immigration
Migration to a new location.

Internal migration
Permanent movement within a particular country.

International migration
Permanent movement from one country to another.

Interregional migration
Permanent movement from one region of a country to another.

Intraregional migration
Permanent movement within one region of a country.

Migration
Form of relocation diffusion involving a permanent move to a new location.

Migration transition
Change in the migration pattern in a society that results from industrialization, population growth, and other social and economic changes that also produce the demographic transition.

Mobility
All types of movement from one location to another.

Net migration
The difference between the level of immigration and the level of emigration.

Pull factor
Factor that induces people to move to a new location.

Push factor
Factor that induces people to leave old residences.

Quotas
In reference to migration, laws that place maximum limits on the number of people who can immigrate to a country each year.

Refugees
People who are forced to migrate from their home country and cannot return for fear of persecution because of their race, religion, nationality, membership in a social group, or political opinion.

Undocumented immigrants
People who enter a country without proper documents.

FURTHER READINGS

Bankston, Carl L., III, Danielle Antoinette Hidalgo, and R. Kent Rasmussen, eds. *Immigration in U.S. History*. Pasadena, CA: Salem Press, 2006.

Bartram, David. *International Labor Migration: Foreign Workers and Public Policy*. New York: Palgrave Macmillan, 2005.

Bohon, Stephanie A. "Occupational Attainment of Latino Immigrants in the United States." *Geographical Review* 95 (2005): 249–66.

Eltis, David, ed. *Coerced and Free Migration: Global Perspectives*. Stanford, CA: Stanford University Press, 2002.

Grigg, D. B. "E. G. Ravenstein and the 'Laws of Migration.'" *Journal of Historical Geography* 3 (1977): 41–54.

Martin, Philip, Susan Martin, and Patrick Weil. *Managing Migration: The Promise of Cooperation*. Lanham, MD: Lexington Books, 2006.

Messina, Anthony M., ed. *West European Immigration and Immigrant Policy in the New Century*. Westport, CT: Praeger, 2002.

Nevins, Joseph. *Operation Gatekeeper: The Rise of the "Illegal Alien" and the Making of the U.S.–Mexico Boundary*. New York: Routledge, 2002.

Özden, Caglar, and Maurice Schiff, eds. *International Migration, Remittances, and the Brain Drain*. New York: Palgrave Macmillan, 2006.

Perry, Mark J. "Domestic Net Migration in the United States: 2000 to 2004," *Current Population Reports*, April 2006.

Willis, Katie, and Brenda Yeoh, eds. *Gender and Migration*. Northampton, MA: Edward Elgar, 2000.

Zelinsky, Wilbur. "The Hypothesis of the Mobility Transition." *Geographical Review* 61 (1971): 219–49.

Zolberg, Aristide R. *A Nation by Design: Immigration Policy in the Fashioning of America*. Cambridge, MA: Harvard University Press, 2006.

ON THE INTERNET

Yearbook of Immigration Statistics is published annually by the Office of Immigration Statistics in the U.S. Department of Homeland Security (**http://www.dhs.gov/ximgtn/statistics/publications/yearbook.shtm).**

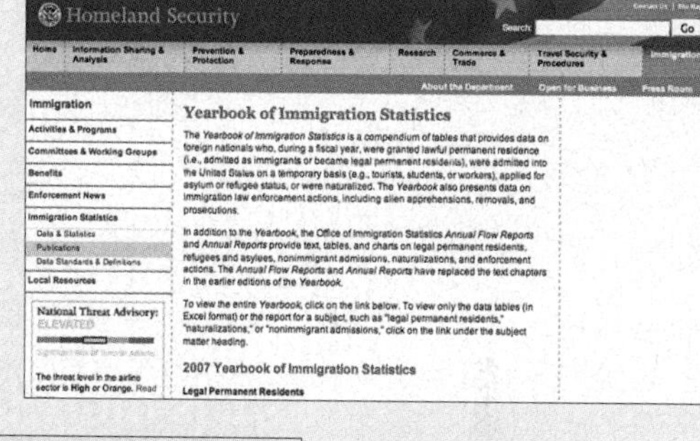

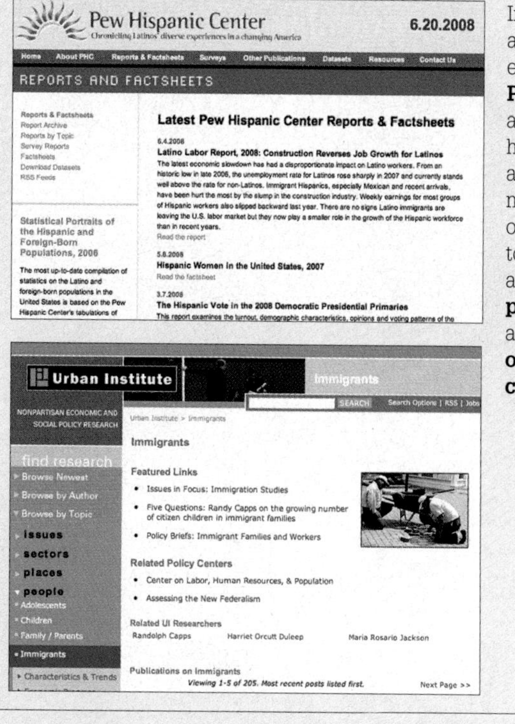

Immigration statistics are available by country of origin extending back to 1820. **Pew Hispanic Center** and the **Urban Institute** have both examined in a number of reports the number and characteristics of undocumented immigrants to the United States. These are available at **http://pewhispanic.org/reports/** and **http://www.urban.org/immigrants/index.cfm**, respectively.

Chapter 4 FOLK AND POPULAR CULTURE

Culture is the body of customary beliefs, social forms, and material traits that together constitute a group of people's distinct tradition. Culture derives from the survival activities of everyone's daily life—food, clothing, and shelter. It also involves leisure activities, such as the arts and recreation. Each cultural group provides for these needs and activities in its own way. Culture can be distinguished from a **habit** (a repetitive act that a particular individual performs) or a **custom** (a repetitive act of a group).

Geographers study a particular social custom's origin, its diffusion, and its integration with other social characteristics. Social customs falls into two basic categories:

- **Folk culture** is traditionally practiced primarily by small, homogeneous groups living in isolated rural areas.
- **Popular culture** is found in large, heterogeneous societies that share certain habits.

Landscapes dominated by a collection of folk customs change relatively little over time, whereas modern communications facilitate frequent changes in popular customs. As a result, folk culture is more likely to vary from place to place at a given time, whereas popular culture is more likely to vary from time to time at a given place.

STARBUCKS IN THE FORBIDDEN CITY, BEIJING, CHINA

THE KEY ISSUES IN THIS CHAPTER

ORIGIN AND DIFFUSION OF CULTURE

4.1 Origin and Diffusion of Folk Customs

4.2 Origin and Diffusion of Popular Culture

GEOGRAPHIC PATTERNS OF CULTURE

4.3 Folk and Popular Sports

4.4 Origin and Diffusion of House Types

4.5 Folk and Popular Food Customs

4.6 Popular Clothing Preferences

LANDSCAPES OF POPULAR CULTURE

4.7 Diffusion of Popular Media

4.8 Landscapes of Popular Culture

4.9 Environmental Impact of Popular Culture

ORIGIN AND DIFFUSION OF CULTURE

4.1 Origin and Diffusion of Folk Customs

- Folk customs typically have anonymous origins.
- Folk customs typically diffuse slowly through relocation migration.

A social custom originates at a hearth, a center of innovation. Folk customs often have anonymous hearths, originating from anonymous sources, at unknown dates, through unidentified originators. They may also have multiple hearths, originating independently in isolated locations.

Relocation diffusion was defined in Chapter 1 as the spread of a characteristic through bodily movement of people from one place to another. This is the principal way that folk customs diffuse—slowly and on a small scale, primarily through migration.

FOLK MUSIC

Music exemplifies the origins of folk customs. According to a Chinese legend, music was invented in 2697 B.C., when the Emperor Huang Ti sent Ling Lun to cut bamboo poles that would produce a flutelike sound matching the call of the phoenix bird.

4.1.1 **FOLK MUSIC** Performed by the Gullah people of South Carolina's Low Country.

Folk songs are usually composed anonymously and transmitted orally. A song may be modified from one generation to the next as conditions change, but the content is most often derived from events in daily life that are familiar to the majority of the people. Folk songs tell a story or convey information about daily activities such as farming, life-cycle events (birth, death, and marriage), or mysterious events such as storms and earthquakes.

4.1.2 **ALAN LOMAX (1915–2002).** Collected information about U.S. folk music.

In Vietnam, where most people are farmers, information about agricultural technology is conveyed through folk songs. For example, the following folk song provides advice about the difference between seeds planted in summer and seeds planted in winter:

Ma chiêm ba tháng không già
Ma mùa tháng rưỡi ắt là không non[1]

This song can be translated as follows:

While seedlings for the summer crop are not old when they are three months of age
Seedlings for the winter crop are certainly not young when they are one-and-a-half months old

The song hardly sounds lyrical to a Western ear. But when English-language folk songs appear in print, similar themes emerge, even if the specific information conveyed about the environment differs.

THE AMISH

Amish customs illustrate how folk customs are distributed through relocation diffusion. The Amish have distinctive clothing, farming, religious practices, and other customs. They leave a unique pattern on landscapes where they settle. Shunning mechanical and electrical power, the Amish still travel by horse and buggy and continue to use hand tools for farming.

Although the Amish population in the United States numbers only about 227,000, a mere 0.07 percent of the total population, Amish folk culture remains visible on the landscape in at least 28 states. The distribution of Amish folk culture across a major portion of the U.S. landscape is explained by examining the diffusion of their culture through migration.

In the 1600s a Swiss Mennonite bishop named Jakob Ammann gathered a group of followers who became known as the Amish. The Amish originated in Bern, Switzerland; the Alsace region in northeastern France; and the Palatinate region of southwestern Germany. They migrated to other portions of northwestern Europe in the 1700s, primarily for religious freedom. In Europe the Amish did not develop distinctive language, clothing, or farming practices and gradually merged with various Mennonite church groups.

Several hundred Amish families migrated to North America in two waves. The first group, primarily from Bern and the Palatinate, settled in Pennsylvania in the early 1700s, enticed by William Penn's offer of low-priced land. Because of lower land prices, the second group, from Alsace, settled in Ohio, Illinois, and Iowa in the United States and Ontario, Canada, in the early 1800s. From these core areas, groups of Amish migrated to other locations where inexpensive land was available.

Living in rural and frontier settlements relatively isolated from other groups, Amish communities retained their traditional customs, even as other European immigrants to the United States adopted new ones. We can observe Amish customs on the landscape in such diverse areas as southeastern Pennsylvania, northeastern Ohio, and east-central Iowa. These communities are relatively isolated from each other but share cultural traditions distinct from those of other Americans.

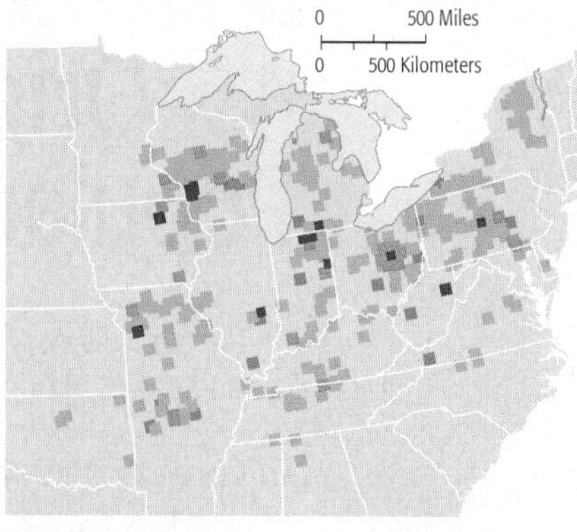

Percent of total population
- 5% and above
- 1%–5%
- below 0.05%
- -0.05%

4.1.3 **THE AMISH**
Amish settlements are distributed throughout the northeastern United States. The counties with the three largest concentrations are Holmes (Ohio), Lancaster (Pennsylvania), and LaGrange (Indiana).

Amish folk culture continues to diffuse slowly through interregional migration within the United States. In recent years a number of Amish families have sold their farms in Lancaster County, Pennsylvania—the oldest and at one time largest Amish community in the United States—and migrated to Christian and Todd counties in southwestern Kentucky.

According to Amish tradition, every son is given a farm when he is an adult, but land suitable for farming is expensive and hard to find in Lancaster County because of its proximity to growing metropolitan areas. With the average price of farmland in southwestern Kentucky less than one-fifth that in Lancaster County, an Amish family can sell its farm in Pennsylvania and acquire enough land in Kentucky to provide adequate farmland for all their sons. Amish families are also migrating from Lancaster County to escape the influx of tourists who come from the nearby metropolitan areas to gawk at the distinctive folk culture.

ORIGIN AND DIFFUSION OF CULTURE

4.2 Origin and Diffusion of Popular Culture

- Popular culture typically originates at specific times and places of origin.
- Popular culture typically has a known originator.

In contrast to folk customs, popular culture is most often a product of the economically more developed countries (MDCs), especially North America, Western Europe, and Japan. Popular music and fast food are good examples. They arise from a combination of advances in industrial technology and increased leisure time.

Industrial technology permits the uniform reproduction of objects in large quantities (CDs, T-shirts, pizzas). Many of these objects help people enjoy leisure time, which has increased as a result of the widespread change in the labor force from predominantly agricultural work to predominantly service and manufacturing jobs.

In contrast to folk music, popular music is written by specific individuals for the purpose of being sold to a large number of people. It displays a high degree of technical skill and is frequently capable of being performed only in a studio with electronic equipment.

Popular music as we know it today originated around 1900. At that time, the main form of popular musical entertainment in the United States and Western Europe was the variety show, called the music hall in the United Kingdom and vaudeville in the United States.

To provide songs for music halls and vaudeville, a music industry was developed in New York, in a district that became known as Tin Pan Alley. The name derived from the sound of pianos being furiously pounded by people called song pluggers, who were demonstrating tunes to publishers. Tin Pan Alley was home to songwriters, music publishers, orchestrators, and arrangers. Companies in Tin Pan Alley originally tried to sell as many printed songsheets as possible. After World War II, Tin Pan Alley disappeared as recorded music became more important than printed songsheets.

PUNK BAND PERFORMS IN BEIJING, CHINA

JAPANESE MUSICIAN PERFORMS ROCK MUSIC IN OKINAWA, JAPAN

Hip hop is a more recent form of popular music that also originated in New York. Whereas the music industry of Tin Pan Alley originated in Manhattan office buildings, hip hop originated in the late 1970s in the South Bronx, a neighborhood predominantly populated by low-income African American and Puerto Rican people (a changeover from its predominant population of middle-class white people of European origin). Rappers in other low-income New York City neighborhoods of Queens, Brooklyn, and Harlem adopted the style with local twists—"thug" rap in Queens and clever lines in Brooklyn. Hip hop remained predominantly a New York phenomenon until the late 1980s, when it spread to Oakland and Atlanta and then to other large cities in the South, Midwest, and West.

Hip hop demonstrates well the interplay between globalization and local diversity that is a prominent theme of this book:

- On the one hand, hip hop is a return to a very local form of musical expression rather than a form that is studio manufactured. Lyrics make local references and represent a distinctive hometown scene. A long-standing clash over the precise neighborhood in New York City where hip-hop music originated was known as the Bridge Wars, because rapper KRS-One was based in the South Bronx and rival Marley Marl's Juice Crew was based in Queenbridge, located on the other side of the Triborough Bridge from the Bronx.
- At the same time, hip hop has diffused rapidly around the world through instruments of globalization: the music is broadcast online and sold through Web marketing. Artists are expressing a sense of a specific place across the boundless space of the Internet.

4.2.1 **POPULAR MUSIC**
This "map," prepared by Marc Smith and Andrew Fiore, shows the hierarchy of popularity of artists and types of music as reflected in the rec.music newsgroup (accessed at http://groups.google.com/group/rec.music.info). The larger the box in the map, the more popular the artist.

TEENAGERS AT A POP MUSIC CONCERT, HONG KONG, CHINA

GEOGRAPHIC PATTERNS OF CULTURE

4.3 Folk and Popular Sports

- Modern spectator sports are good examples of popular culture.
- Some sports retain their folk custom roots.

Many sports originated as isolated folk customs and were diffused like other folk culture through the migration of individuals. The contemporary diffusion of organized sports, however, displays the characteristics of popular culture.

SOCCER'S FOLK ORIGINS

Soccer originated as a folk custom in England during the eleventh century and was transformed into a part of global popular culture in the nineteenth century.

The origin of soccer (called football outside North America) is obscure. According to football historians, the earliest contest took place after Denmark invaded England between 1018 and 1042. Workers excavating a building site encountered a Danish soldier's head, which they began to kick. "Kick the Dane's head" was imitated by boys, one of whom got the idea of using an inflated cow bladder.

In 1863, several British football clubs formed the Football Association to standardize the rules and to organize professional leagues. *Association* was shortened to assoc, which ultimately became twisted around into the word soccer. The clubs had been formed, often by churches, to provide factory workers with organized recreation during leisure hours. Organization of the sport into a formal structure in Great Britain marks the transition of football from folk custom to popular culture.

SOCCER AS POPULAR CULTURE

Association football diffused from England first to continental Europe in the late 1800s and then to other countries in the twentieth century:

- Football was first played in continental Europe in the late 1870s by Dutch students who had been in Britain.
- Football got to Spain via English engineers working in Bilbao in 1893 and was quickly adopted by local miners.
- British citizens further diffused the game throughout the worldwide British Empire.
- Soccer diffused to Russia when the English manager of a textile factory near Moscow organized a team at the factory in 1887 and advertised in London for workers who could play football. After the Russian Revolution in 1917, both the factory and its football team were absorbed into the Soviet Electric Trade Union. The team, renamed the Moscow Dynamo, became the country's most famous.
- The global popularity of soccer is seen in the World Cup, in which national soccer teams compete every 4 years. The World Cup was held in Germany in 2006 and South Africa in 2010. Thanks to television, each final match breaks the record for the most spectators of any event in world history.

SURVIVING FOLK SPORTS

Cultural groups still have their own preferred sports, which are often unfamiliar to people elsewhere:

- Cricket is popular primarily in Britain and former British colonies.
- Ice hockey prevails, logically, in colder climates, especially in Canada, Northern Europe, and Russia.
- China's most popular sports include martial arts, known as wushu, such as archery, fencing, wrestling, and boxing.
- Baseball, once confined to North America, became popular in Japan after it was introduced by American soldiers who occupied the country after World War II.
- The first college football game played in the United States, between Princeton and Rutgers in 1869, was really soccer, and officials of several colleges met 4 years later to adopt football rules consistent with those of British soccer. But Harvard's representatives successfully argued for adoption of rugby rules instead. Rugby was so thoroughly modified by U.S. colleges that an entirely new game—American football—emerged.
- Seventeenth-century European explorers observed the Iroquois nation playing lacrosse, known in their language as *guhchigwaha*, which means "bump hips." European colonists in Canada picked up the game from the Iroquois and diffused it to a handful of U.S. communities, especially in Maryland, upstate New York, and Long Island. In recent years, lacrosse has fostered cultural identity among the Iroquois Confederation of Six Nations (Cayugas, Mohawks, Oneidas, Onondagas, Senecas, and Tuscaroras), because they have been invited by the International Lacrosse Federation to participate in the Lacrosse World Championships, along with teams from Australia, Canada, England, and the United States.

FOLK SPORTS
Mohawk Indian tribe members playing lacrosse (above), Gaelic football in Belfast, Northern Ireland (left), and Australian Rules football in Melbourne, Australia (below)

Despite the diversity in the distribution of sports across Earth's surface and the anonymous origin of some games, organized spectator sports today are part of popular culture. The common element in professional sports is the willingness of people throughout the world to pay for the privilege of viewing, in person or on TV, events played by professional athletes.

GEOGRAPHIC PATTERNS OF CULTURE

4.4 Origin and Diffusion of House Types

- Housing was once a good example of a folk custom.
- Distinctive regional differences remain in U.S. folk housing.

Older houses in the United States display local folk-culture traditions. When families migrated westward in the 1700s and 1800s, they cut trees to clear fields for planting and used the wood to build houses, barns, and fences. The style of pioneer homes reflected whatever style was prevailing at the place on the East Coast from which the pioneers migrated. In contrast, houses built in the United States during the past half-century display popular culture influences.

FOLK HOUSE FORMS

Cultural geographer Fred Kniffen identified three major hearths or nodes of folk house forms in the United States: New England, Middle Atlantic, and Lower Chesapeake. Migrants carried house types from New England northward to upper New England and westward across the southern Great Lakes region; from the Middle Atlantic westward across the Ohio Valley and southwestward along the Appalachian trails; and from the lower Chesapeake southward along the Atlantic Coast.

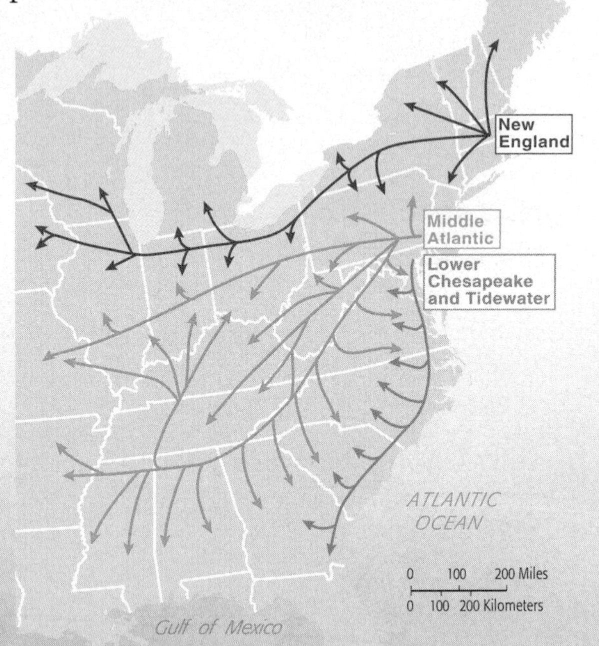

4.4.1 **SOURCE AREAS OF U.S. HOUSE TYPES**

THE SPREAD OF HOUSE TYPES

Four major house types were popular in New England at various times during the eighteenth and early nineteenth centuries: two-chimney, saltbox, Cape Cod, and front gable & wing. When settlers from New England migrated westward, they took their house type with them. The New England house type can be found throughout the Great Lakes region as far west as Wisconsin, because this area was settled primarily by migrants from New England. As the house preferred by New Englanders changed over time, the predominant form found on the landscape varied based on the date of initial settlement.

The major house type in the Middle Atlantic region was known as the I-house, typically two full stories in height, with gables to the sides. The I-house resembled the letter *I*—it was only one room deep and at least two rooms wide. The I-house became the most extensive style of construction in much of the eastern half of the United States, especially in the Ohio Valley and Appalachia. Settlers built I-houses in much of the Midwest because most of them had migrated from the Middle Atlantic region.

The Lower Chesapeake or Tidewater style of house typically comprised one story, with a steep roof and chimneys at either end. These houses spread from the Chesapeake Bay–Tidewater, Virginia area along the southeast coast. As was the case with the Middle Atlantic I-house, the form of housing that evolved along the southeast coast typically was only one room deep. In wet areas, houses in the coastal southeast were often raised on piers or on a brick foundation.

Today, such distinctions are relatively difficult to observe in the United States. The style of housing does not display the same degree of regional distinctiveness because rapid communication and transportation systems provide people throughout the country with knowledge of alternative styles. Furthermore, most people do not build the houses in which they live. Instead, houses are usually mass produced by construction companies.

TWO-CHIMNEY

4.4.2 **DIFFUSION OF NEW ENGLAND HOUSE TYPES**

GEOGRAPHIC PATTERNS OF CULTURE

4.5 Folk and Popular Food Customs

- Food customs are influenced in part by environmental conditions.
- People embrace or avoid specific foods for cultural reasons.

Humans eat mostly plants and animals—living things that spring from the soil and water of a region. For example, rice demands a milder, moist climate, whereas wheat thrives in colder, drier regions.

People adapt their food preferences to conditions in the environment.

- In Asia, soybeans are widely grown, but raw they are toxic and indigestible. Lengthy cooking renders them edible, but in Asia fuel is scarce. Asians have adapted to this environmental dilemma by deriving foods from soybeans that do not require extensive cooking. These include bean sprouts (germinated seeds), soy sauce (fermented soybeans), and bean curd (steamed soybeans).
- In Southern Europe, traditional preferences for quick-frying foods resulted in part from fuel shortages. In Northern Europe, an abundant wood supply encouraged the slow stewing and roasting of foods over fires, which also provided home heat in the colder climate.

TABOOS

People refuse to eat particular plants or animals that are thought to embody negative forces in the environment. Such a restriction on behavior imposed by social custom is a **taboo.** Some folk cultures may establish food taboos because of concern for the natural environment. These taboos may help to protect endangered animals or to conserve scarce natural resources.

Taboos against consumption of certain foods are embedded in religions.

- As prescribed in the Bible, the ancient Hebrews were prohibited from eating some animals. These taboos arose partially from concern for the environment by the Hebrews, who lived as pastoral nomads in the dry lands of the Middle East. Biblical taboos were developed through oral tradition and by rabbis into the kosher laws observed today by some Jews.
- Muslims, like Jews, embrace the taboo against pork. The pig is unsuited to pastoral nomadism, and its meat spoils relatively quickly in the region's hot dry climate.
- In India, Hindu sanctions against consuming cows is explained in part by the need to maintain a large supply of oxen (castrated male cows), the traditional choice for pulling plows as well as carts. A large supply of oxen must be maintained in India because every field has to be plowed at approximately the same time—when the monsoon rains arrive.

4.5.1 **ANNUAL WINE PRODUCTION**

WINE

The spatial distribution of wine production demonstrates that the environment plays a role in the distribution of popular as well as folk food customs. The distinctive character of a wine derives from a unique combination of soil, climate, and other physical characteristics at the place where the grapes are grown.

Vineyards are best cultivated in temperate climates of moderately cold, rainy winters and fairly long, hot summers. Hot, sunny weather is necessary in the summer for the fruit to mature properly, whereas winter is the preferred season for rain, because plant diseases that cause the fruit to rot are more active in hot, humid weather. Vineyards are planted on hillsides, if possible, to maximize exposure to sunlight and to facilitate drainage. A site near a lake or river is also desirable because water can temper extremes of temperature.

Grapes can be grown in a variety of soils, but the best wine tends to be produced from grapes grown in soil that is coarse and well drained—a soil not necessarily fertile for other crops. For example, the soil is generally sandy and gravelly in the Bordeaux wine region, chalky in Champagne country, and of a slate composition in the Moselle Valley. The distinctive character of each region's wine is especially influenced by the unique combination of trace elements, such as boron, manganese, and zinc, in the rock or soil. In large quantities these elements could destroy the plants, but in small quantities they lend a unique taste to the grapes.

VINEYARDS
Vineyards in Australia (upper left), France (upper right), and Sonoma County California (bottom)

GEOGRAPHIC PATTERNS OF CULTURE

4.6 Popular Clothing Preferences

- People adopt distinctive clothing customs in part because of environmental conditions.
- With globalization, some clothing preferences have been adopted by people around the world.

People living in folk cultures wear clothing appropriate for distinctive agricultural practices. In arctic climates, fur-lined boots protect against the cold, and snowshoes permit walking on soft, deep snow without sinking in. People living in warm and humid climates may not need any footwear if heavy rainfall and time spent in water discourage such use.

On the other hand, individual clothing preferences reveal how popular culture can be distributed across the landscape with little regard for distinctive physical features:

- In the MDCs of North America and Western Europe, clothing preferences generally reflect occupations rather than particular environments. A lawyer in California is more likely to dress like a lawyer in New York than like a steelworker in California.
- Wealth also influences clothing preferences. For social purposes, people with sufficient income may update their wardrobe frequently with the latest fashions.

GLOBALIZATION OF CLOTHING STYLES

Improved communications have permitted the rapid diffusion of clothing styles from one region of Earth to another. Original designs for women's dresses, created in Paris, Milan, London, or New York, are reproduced in large quantities at factories in Asia and sold for relatively low prices in North American and European chain stores. Speed is essential in manufacturing copies of designer dresses because fashion tastes change quickly.

Until recently, a year could elapse from the time an original dress was displayed to the time that inexpensive reproductions were available in the stores. Now the time lag is less than 6 weeks because of the diffusion of fax machines, computers, and satellites. Sketches, patterns, and specifications are sent instantly from European fashion centers to American corporate headquarters and then on to Asian factories. Buyers from the major retail chains can view the fashions on large, high-definition televisions linked by satellite networks.

The globalization of clothing styles has involved increasing awareness by North Americans and Europeans of the variety of traditional clothing around the world. Increased travel and the diffusion of television have exposed people in MDCs to other forms of dress, just as people in other parts of the world have come into contact with Western dress. The poncho from South America, the dashiki of the Yoruba people of Nigeria, and the Aleut parka have been adopted by people elsewhere in the world. The continued use of traditional clothing in some parts of the globe may persist not because of distinctive environmental conditions or traditional cultural values, but to preserve past memories or to attract tourists.

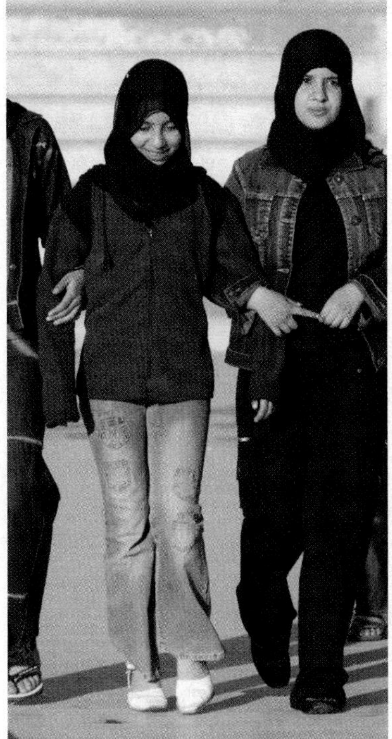

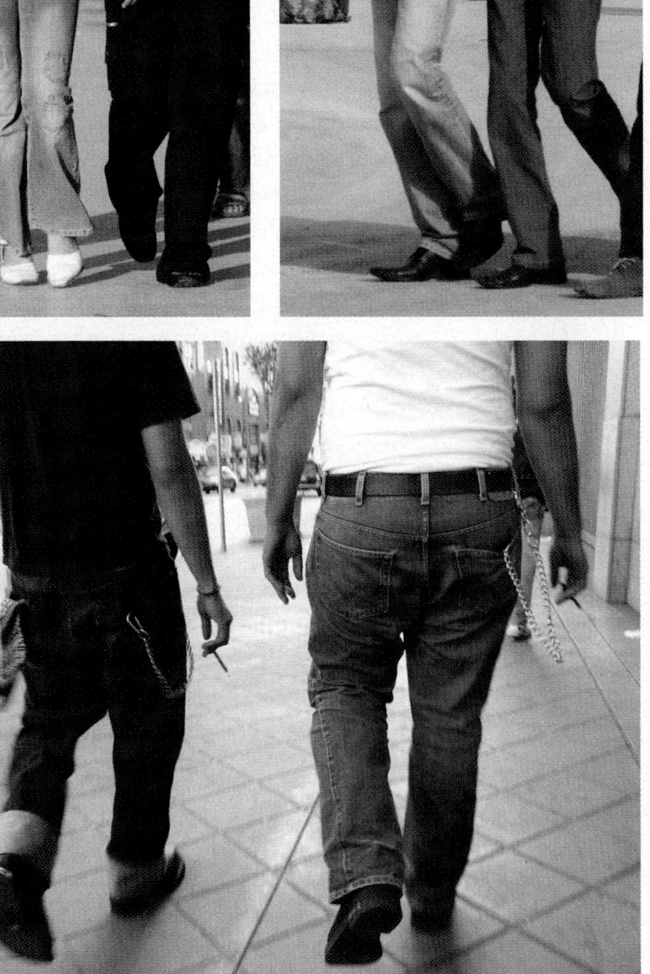

JEANS

An important symbol of the diffusion of Western popular culture is jeans, which is a prized possession for young people throughout the world. In the late 1960s, jeans acquired an image of youthful independence in the United States as young people adopted a style of clothing previously associated with low-status manual laborers and farmers.

Locally made denim trousers are available throughout Europe and Asia for under $10, but "genuine" jeans made by Levi Strauss, priced at $50 to $100, are preferred as a status symbol. Millions of second-hand Levis are sold each year in Asia, especially in Japan and Thailand, priced between $100 and $1,000.

Even in the face of the globalization of popular culture such as wearing jeans, some local variation persists:

- Asians especially prefer Levi's 501 jeans with a button fly rather than a zipper.
- Within the United States, the button fly is more common on the West Coast, whereas easterners prefer the zipper fly because it doesn't let in cold air.

Jeans became an obsession and a status symbol among youth in the former Soviet Union when the Communist government prevented their import. Gangs would attack people to steal their American-made jeans, and authentic jeans would sell for $400 on the black market. The scarcity of high-quality jeans was just one of many consumer problems that were important motives in the dismantling of Communist governments in Eastern Europe around 1990.

With the end of communism, jeans can now be imported freely into Russia. Levi Strauss opened a store in the center of Moscow that sells jeans for about $50, about a week's wage for a typical Russian. In an integrated global economy, prominent symbols of popular culture have diffused around the world. Access to these products is now limited primarily by lack of money rather than government regulation.

Ironically, as access to Levi's increased around the world, American consumers turned away from the brand. Sales plummeted from $7 billion in 1996 to $4 billion in 2004, the year Levi's closed its last U.S. factory. To reclaim lost consumers in the United States, Levi's has tried to market jeans equipped with an iPod remote control and docking station fitted in the pocket.

LANDSCAPES OF POPULAR CULTURE

4.7 Diffusion of Popular Media

- TV is an especially important reflector and transmitter of culture around the world.
- The Internet is diffusing around the world more rapidly than did TV a half-century ago.

Watching TV is an especially significant popular custom for two reasons.

- TV is the most popular leisure activity in MDCs throughout the world.
- TV is the most important mechanism by which knowledge of popular culture, such as professional sports, is rapidly diffused across Earth.

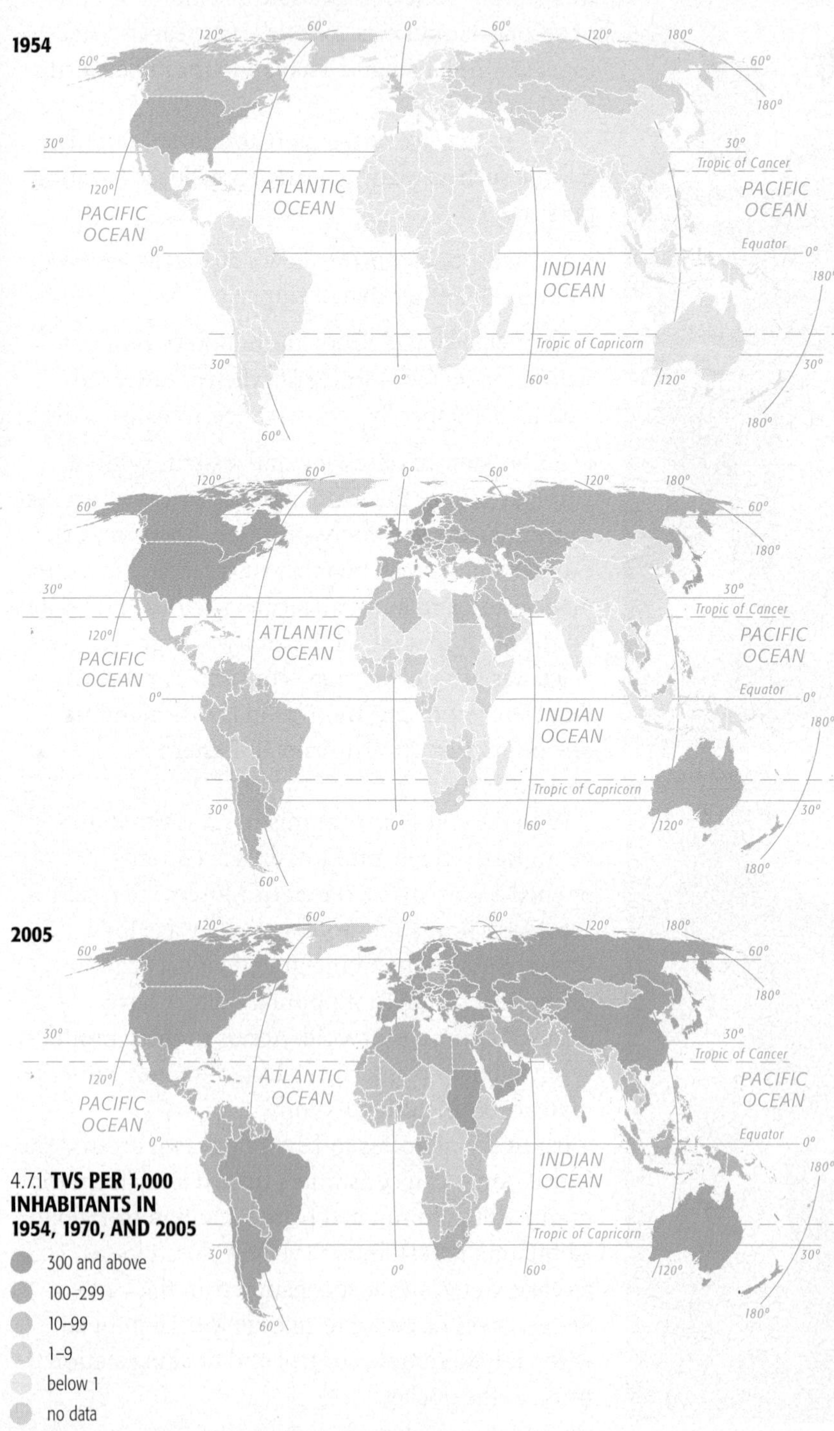

4.7.1 **TVS PER 1,000 INHABITANTS IN 1954, 1970, AND 2005**

DIFFUSION OF TV

The U.S. public first saw television in the 1930s, but its diffusion was blocked for a number of years when broadcasting was curtailed or suspended entirely during World War II. With the end of World War II, the number of television sets increased rapidly in the United States, from 10,000 in 1945 to 1 million in 1949, 10 million in 1951, and 50 million in 1959.

In 1954, the first year that the United Nations published data on the subject, the United States had 86 percent of the world's 37 million TV sets, the United Kingdom 9 percent, the Soviet Union and Canada 2 percent each, and a handful of other countries (primarily Cuba, Mexico, France, and Brazil) the remainder. The United States had approximately 200 TV sets per 1,000 inhabitants in 1954, and the rest of the world had approximately 2 per 1,000.

In 1970, the United States still had far more TV sets per capita than any other country except Canada. However, rapid growth of ownership in Europe meant that the share of the world's sets in the United States had declined to one-fourth. Still, in 1970, half of the countries in the world, including most of those in Africa and Asia, had little if any TV broadcasting.

International differences in TV ownership had diminished, although had not disappeared altogether. The United States still had a much higher level of television ownership than the world as a whole, but so did Canada, Europe, Australia, New Zealand, and Japan. Meanwhile, ownership rates climbed sharply between 1970 and 2000 in LDCs, such as in China from less than 1 per 1,000 to 304 per 1,000 and in Indonesia from less than 1 per 1,000 to 154 per 1,000.

DIFFUSION OF THE INTERNET

The diffusion of Internet service follows the pattern established by television a generation earlier, but at a more rapid pace. There were 40 million Internet users worldwide in 1995, including 25 million in the United States, and Internet service had not yet reached most countries.

Between 1995 and 2000, Internet usage increased rapidly in the United States, from 9 percent to 44 percent of the population. But the increase was much greater in the rest of the world, from 40 million Internet users in 1995 to 400 million in 2000. As Internet usage diffused rapidly, the U.S. share also declined rapidly, from 62 percent of the world total in 1995 to 31 percent in 2000.

In the first years of the twenty-first century, Internet usage continued to diffuse rapidly. World usage more than doubled in 4 years, from 400 million in 2000 to 880 million in 2004. U.S. usage continued to increase, but at a more modest rate, to 63 percent of the population. Eight countries now had more Internet users per capita than the United States. Given the history of television, the Internet is likely to diffuse rapidly to other countries in the years ahead.

Note that all six maps in Section 4.7 use the same classifications. For example, the highest class in all maps is 300 or more per 1,000. What is different is the time interval. The diffusion of television from the United States to the rest of the world took a half-century, whereas the diffusion of the Internet has taken only a decade.

4.7.2 **INTERNET USERS PER 1,000 INHABITANTS IN 1990, 2000, AND 2005**

- 300 and above
- 100–299
- 10–99
- 1–9
- below 1
- no data

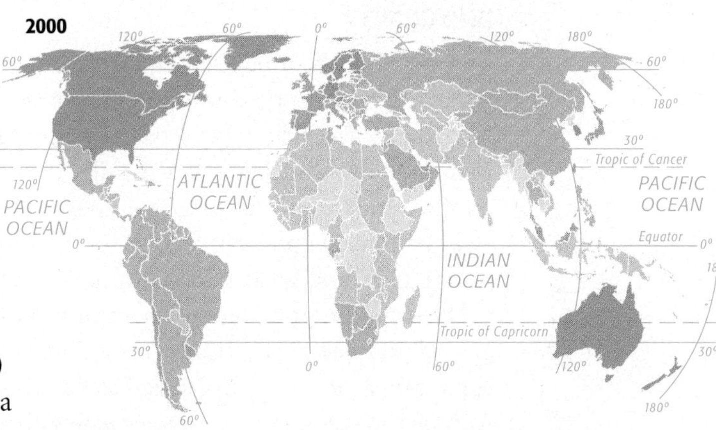

4.8 Landscapes of Popular Culture

- To promote familiarity with products, popular culture can produce uniform landscapes.
- Folk customs may be threatened by the diffusion of popular culture.

Popular culture is less likely than folk culture to be influenced by physical features. The spatial organization of popular culture reflects primarily the distribution of social and economic features. In a global economy and culture, popular culture appears increasingly uniform.

UNIFORMITY

The distribution of popular culture around the world tends to produce more uniform landscapes. The spatial expression of a popular custom in one location will be similar to another. In fact, promoters of popular culture want a uniform appearance to generate "product recognition" and greater consumption.

The diffusion of fast-food restaurants is a good example of such uniformity. Such restaurants are usually organized as franchises. A franchise is a company's agreement with businesspeople in a local area to market that company's product. The franchise agreement lets the local outlet use the company's name, symbols, trademarks, methods, and architectural styles. To both local residents and travelers, the buildings are immediately recognizable as part of a national or multinational company. A uniform sign is prominently displayed.

Uniformity in the appearance of the landscape is promoted by a wide variety of other popular structures in North America, such as gas stations, supermarkets, and motels. These structures are designed so that both local residents and visitors immediately recognize the purpose of the building, even if the name of the company is not recognized.

Physical expression of uniformity in popular culture has diffused from North America to other parts of the world. American motels and fast-food chains have opened in other countries. These establishments appeal to North American travelers, yet most customers are local residents who wish to sample American customs they have seen on television.

Diffusion of popular culture across Earth is not confined to products that originate in North America. With faster communications and transportation, customs from any place on Earth can rapidly diffuse elsewhere. Japanese vehicles and electronics, for example, have diffused in recent years to the rest of the world, including North America. Until the 1970s, vehicles produced in North America, Europe, and Japan differed substantially in appearance and size, but in recent years styling has become more uniform, largely because of consumer preference around the world for Japanese vehicles. Carmakers such as General Motors, Ford, Toyota, and Honda now manufacture similar models in North and South America, Europe, and Asia, instead of separately designed models for each continent.

MEDIA COVERAGE OF REFUGEE CRISIS IN RWANDA
The crisis was covered with electronic communications in a location where most inhabitants didn't even have electricity.

THREAT TO FOLK CUSTOMS

Many fear the loss of folk culture, especially because rising incomes can fuel demand for the possessions typical of popular culture. When people turn from folk to popular culture, they may also turn away from the society's traditional values. And the diffusion of popular culture from MDCs can lead to dominance of Western perspectives.

Leaders of some LDCs consider the dominance of popular customs by MDCs as a threat to their independence. The threat is posed primarily by the media, especially news-gathering organizations and TV.

The process of gathering news worldwide is expensive, and most newspapers and broadcasters are unable to afford their own correspondents. Instead, they buy the right to use the dispatches of one or more of the main news organizations. These news-gathering organizations are owned by companies based in MDCs.

Three MDCs—the United States, the United Kingdom, and Japan—dominate the TV industry in LDCs. The Japanese operate primarily in South Asia and East Asia, selling their electronic equipment. British companies have invested directly in management and programming for television in Africa. U.S. corporations own or provide technical advice to many Latin American stations. These three countries are also the major exporters of programs. Even in Europe, the United States has been the source of imports for two-thirds of entertainment programs.

Leaders of many LDCs view the spread of TV as a new method of economic and cultural imperialism on the part of the United States and other MDCs. American TV, like other media, presents characteristically American beliefs and social forms, such as upward social mobility, relative freedom for women, glorification of youth, and stylized violence. These attractive themes may conflict with and drive out traditional social customs.

To avoid offending traditional values, many satellite broadcasters in Asia do not carry MTV or may allow governments to censor unacceptable videos. Cartoons featuring Porky Pig may be banned in Muslim countries, where people avoid pork products. Instead, entertainment programs emphasize family values and avoid controversial cultural, economic, and political issues.

LANDSCAPES OF POPULAR CULTURE

4.9 Environmental Impact of Popular Culture

- Popular culture can increase demand for use of natural resources.
- Popular culture can cause pollution of air, land, and water.

The diffusion of some popular customs can adversely impact environmental quality in two ways—depletion of scarce natural resources and pollution of the landscape.

INCREASED DEMAND FOR NATURAL RESOURCES

Diffusion of some popular customs increases demand for raw materials, such as minerals and other substances found beneath Earth's surface. The depletion of resources used to produce energy, especially petroleum, is discussed in Chapter 14.

Popular culture may demand a large supply of certain animals, resulting in depletion or even extinction of some species. For example, some animals are killed for their skins, which can be shaped into fashionable clothing and sold to people living thousands of kilometers from the animals' habitat. The skins of the mink, lynx, jaguar, and leopard have been heavily consumed for various articles of clothing, to the point that the survival of these species is endangered. This unbalances ecological systems of which the animals are members. Folk culture may also encourage the use of animal skins, but the demand is usually smaller than for popular culture.

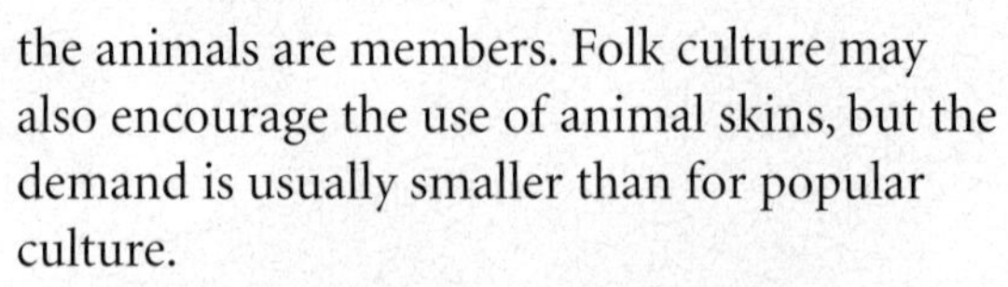

Increased demand for some products can strain the capacity of the environment. An important example is increased meat consumption. This has not caused extinction of cattle and poultry; we simply raise more. But animal consumption is an inefficient way for people to acquire calories—90 percent less efficient than if people simply ate grain directly.

To produce 1 kilogram (2.2 pounds) of beef sold in the supermarket, nearly 10 kilograms (22 pounds) of grain are consumed by the animal. For every kilogram of chicken, nearly 3 kilograms (6.6 pounds) of grain are consumed by the fowl. This grain could be fed to people directly, bypassing the inefficient meat step. With a large percentage of the world's population undernourished, some question this inefficient use of grain to feed animals for eventual human consumption.

POLLUTION

Popular culture also can pollute the environment. The environment can accept and assimilate some level of waste from human activities. But popular culture generates a high volume of waste—solids, liquids, and gases—that must be absorbed into the environment. Although waste is discharged in all three forms, the most visible is solid waste—cans, bottles, old cars, paper, and plastics. These products are often discarded rather than recycled. With more people adopting popular customs worldwide, this problem grows.

Folk culture, like popular culture, can also cause environmental damage, especially when natural processes are ignored. A widespread belief exists that indigenous peoples of the Western Hemisphere practiced more "natural," ecologically sensitive agriculture before the arrival of Columbus and other Europeans. Geographers increasingly question this. In reality, pre-Columbian folk customs included burning grasslands for planting and hunting, cutting extensive forests, and overhunting some species. Very high rates of soil erosion have been documented in Central America from the practice of folk culture.

The MDCs that produce endless supplies for popular culture have created the technological capacity both to create large-scale environmental damage and to control it. However, a commitment of time and money must be made to control the damage. Adverse environmental impact of popular culture is further examined in Chapter 14.

GOLF

Golf courses, because of their large size (80 hectares, or 200 acres), provide a prominent example of imposing popular culture on the environment. A surge in U.S. golf popularity has spawned construction of roughly 200 courses during the past two decades. Geographer John Rooney attributes this to increased income and leisure time, especially among recently retired older people and younger people with flexible working hours.

According to Rooney, the provision of golf courses is not uniform across the United States. Although perceived as a warm-weather sport, the number of golf courses per person is actually greatest in north-central states, from Kansas to North Dakota, as well as the northeastern states abutting the Great Lakes, from Wisconsin to upstate New York. People in these regions have a long tradition of playing golf, and social clubs with golf courses are important institutions in the fabric of the regions' popular customs.

In contrast, access to golf courses is more limited in the South, in California, and in the heavily urbanized Middle Atlantic region between New York City and Washington, D.C. Rapid population growth in the South and West and lack of land on which to build in the Middle Atlantic region have reduced the number of courses per capita. However, selected southern and western areas, such as coastal South Carolina, southern Florida, and central Arizona, have high concentrations of golf courses as a result of the arrival of large numbers of golf-playing northerners, either as vacationers or as permanent residents.

Golf courses are designed partially in response to local physical conditions. Grass species are selected to thrive in the local climate and still be suitable for the needs of greens, fairways, and roughs. Existing trees and native vegetation are retained if possible (few fairways in Michigan are lined by palms). Yet, like other popular customs, golf courses remake the environment—creating or flattening hills, cutting grass or letting it grow tall, carting in or digging up sand for traps, and draining or expanding bodies of water to create hazards.

Chapter **Summary**

ORIGIN AND DIFFUSION OF CULTURE

- Folk culture is more likely to have an anonymous origin and to diffuse slowly through migration.
- Popular culture is more likely to be invented and diffused rapidly with the use of modern communications.

GEOGRAPHIC PATTERNS OF CULTURE

- Unique regions of folk customs, such as food preferences and house types, arise because of lack of interaction among groups, even those living nearby.
- Popular culture, such as sports and clothing styles, diffuses rapidly; differences in popular culture are more likely to be observed in one place at different points in time than among different places at one point in time.

LANDSCAPES OF POPULAR CULTURE

- Rapid diffusion of popular culture has been facilitated by modern communications, especially TV in the twentieth century and the Internet in the twenty-first century.
- The diffusion of popular culture can produce uniform landscapes and loss of distinctive local folk customs.
- The diffusion of some popular customs can adversely impact environmental quality through depletion of scarce natural resources and pollution of the landscape

INTERNET ACCESS IN JAISALMER FORT, INDIA

Geographic Consequences of Change

Material culture can be divided into two types—folk and popular. Folk culture most often exists among small, homogeneous groups living in relative isolation at a low level of economic development. Popular culture is characteristic of societies with good communications and transportation, which enable rapid diffusion of uniform concepts.

Geographers study an array of thousands of social customs with distinctive spatial distributions. Groups display preferences in providing material needs such as food, clothing, and shelter, and in leisure activities such as performing arts and recreation. Examining where various social customs are practiced helps us to understand the extent of cultural diversity in the world.

Folk customs are especially interesting to geographers because they provide a unique identity to each group of people who occupy a specific area of Earth's surface. Folk customs have relatively clustered distributions, and preserving them can be seen as enhancing diversity in the world.

Popular culture is important, too, because it derives from the high levels of material wealth characteristic of societies that are economically developed. As societies seek to improve their economic level, they may abandon traditional folk culture and embrace popular culture associated with MDCs.

Underlying the patterns of material culture are differences in the ways people relate to their environment. Material culture contributes to the modification of the environment, and in turn, nature influences the cultural values of an individual or a group. Geographers, then, classify culture as popular and folk based on differences in the ways the environment is modified, and meaning is derived from environmental conditions. Popular culture makes relatively extensive modifications of the environment, given society's greater technological means and inclination to do so.

Popular culture is becoming more dominant, threatening the survival of unique folk cultures. The disappearance of local folk customs reduces diversity and the intellectual stimulation that arises from differences in backgrounds.

The dominance of popular culture can also threaten the quality of the environment. Folk culture derived from local natural elements may be more sensitive to the protection and enhancement of the environment. Popular culture is less likely to reflect the diversity of local physical conditions and is more likely to modify the environment in accordance with global values.

SHARE YOUR VOICE Student Essay

Anya Yermakova
Northwestern University

¡Olé! or the Art of Being Human

I open the door to a small bar in the gypsy neighborhood of Sevilla with my notebook in hand, ready to annotate the "communicative and improvisational aspects of flamenco performance." I freeze. My notebook comes down, my pen is slowly put away in the back pocket of my jeans, and I find myself glued to the face of the singing gypsy woman, colored in absolute misery, while my own expression morphs into stupor.

Her semi-tonal melismatic voice and convulsed facial expression in combination with the sharp, impulsive rasguéo of the guitar is uncomfortable, almost socially unacceptable, making the performance agonizing and even sexual. Meanwhile, I am apparently still standing in the doorway, in an involuntary state of trance, realizing that I was foolish to think I can simply observe this energy– I, too, am enthralled in it.

A week later, the street beside the river Guadalquivir in Triana brings me to a heated debate with a flamenco fusion street guitarist about the role of sharp-11 in flamenco, when a passerby rudely interrupts: "That's no bulerías, man; that's no flamenco that you're doing." He responds, 'Well, then you tell me what flamenco is, and you'll see that it's exactly what I'm doing."

Challenge is immediately insinuated by the initiation of palmas (hand claps) and is quickly accepted through the explosion of a ringing flat-9 chord that retrieves a unanimous "Olé!" As the intensity grows, the palmas accelerate the rhythm, and, oh my, my hands are evidently partaking! And just when you feel that the force has augmented enough to break through the air into chaos, it all comes together in a violent ending, reminding me what it feels like to breathe.

Performing art is much more than entertainment – it is pure emotion that spills from the performer and carries to an unrelated individual in the audience, establishing a powerful, intangible connection. I yearn for the sound to smother me in energy every time I play or hear music, and I am devoted to communicating that beautiful feeling of being human.

Chapter **Resources**

TRADITION AND CHANGE FOR WOMEN

The global diffusion of popular culture threatens the subservience of women to men that is embedded in many folk customs.

Women were traditionally relegated to performing household chores, such as cooking and cleaning, and to bearing and raising large numbers of children. Those women who worked outside the home were likely to be obtaining food for the family, either through agricultural work or by trading handicrafts.

Advancement of women was limited by low levels of education and high rates of victimization from violence, often inflicted by male family members. The concepts of legal equality and availability of economic and social opportunities outside the home have become widely accepted in MDCs, even where women continue to suffer from discriminatory practices.

Yet, contact with popular culture can also bring negative impacts for women in LDCs.

ON THE INTERNET

The *Atlas of Cyberspace* (**www.cybergeography.org**) includes items that depict information on familiar maps, such as the growth in Internet usage by country or U.S. state. Other "maps" in the atlas are actually diagrams or graphic representations such as frequency of visits to web sites or chat rooms, as well as other versions of the popular music "map" on page 79.

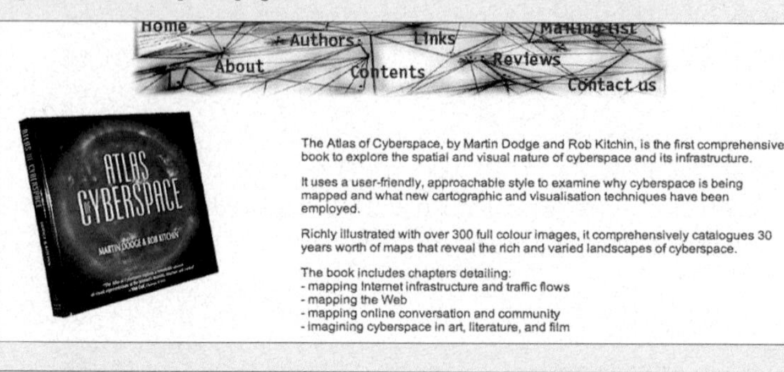

KEY TERMS

Culture
The body of customary beliefs, social forms, and material traits that together constitute a group of people's distinct tradition.

Custom
The frequent repetition of an act, to the extent that it becomes characteristic of the group of people performing the act.

Folk culture
Culture traditionally practiced by a small, homogeneous, rural group living in relative isolation from other groups.

Habit
A repetitive act performed by a particular individual.

Popular culture
Culture found in a large, heterogeneous society that shares certain habits despite differences in other personal characteristics.

Taboo
A restriction on behavior imposed by social custom.

PROSTITUTION

Men from MDCs, especially Germany, Japan, Netherlands, and Norway, purchase "sex tours" to LDCs, particularly South Korea, the Philippines, and Thailand, that include airfare, hotels, and the use of a predetermined number of women. International prostitution is encouraged in these countries as a major source of foreign currency.

DOWRIES

In India the groom traditionally provided a small dowry or gift to the bride's family as a sign of respect. Now, the custom has reversed, and the family of a bride is often expected to provide a substantial dowry to the husband's family. A dowry can take the form of either cash or expensive consumer goods, such as motor vehicles, electronics, and household appliances. If the bride's family is unable to pay a promised dowry or installments, the groom's family may cast the bride out on the street, and her family may refuse to take her back. Husbands and in-laws angry over the small size of dowry payments have killed 5,000 to 7,000 women during the past two decades, according to government statistics, and the practice still exists. Dowries have been illegal in India since 1961 because of the adverse impact on women.

In a highly publicized dowry case, just before the start of a wedding ceremony in 2003, a groom's family demanded a dowry of $25,000 in cash, in addition to two televisions, two home theater sets, two refrigerators, two air conditioners, and one car that had already been paid. The bride halted the ceremony and called the police on her cell phone. The family was arrested for violation of the 1961 anti-dowry law. The story appeared in *The Times of India* with the headline "It Takes Guts to Send Your Groom Packing."

FURTHER READINGS

Bale, John. *Sports Geography*, 2d ed. London and New York: Routledge, 2003.

Bennett, Merril K. *The World's Foods*. New York: Arno Press, 1954.

Blacking, John, and Joann W. Kealiinohomoku. *The Performing Arts: Music and Dance*. The Hague: Mouton, 1979.

Crowley, William K. "Old Order Amish Settlement: Diffusion and Growth." *Annals of the Association of American Geographers* 68 (1978): 249–65.

DeBlij, Harm J. *A Geography of Viticulture*. Miami: University of Miami Geographical Society, 1981.

Denevan, William E. "The Pristine Myth: The Landscape of the Americas in 1492." *Annals of the Association of American Geographers* 82 (1992): 367–85.

Hall, C. Michael, and Stephen J. Page. *The Geography of Tourism and Recreation*, 2d ed. London and New York: Routledge, 2002.

Kniffen, Fred B. "Folk-Housing: Key to Diffusion." *Annals of the Association of American Geographers* 55 (1965): 549–77.

Lomax, Alan. *The Folk Songs of North America*. Garden City, NY: Doubleday, 1960.

McAlester, Virginia, and Lee McAlester. *A Field Guide to American Houses*. New York: Alfred A. Knopf, 1984.

Rooney, John F., Jr., Wilbur Zelinsky, and Dean R. Louder, eds. *This Remarkable Continent: An Atlas of United States and Canadian Society and Culture*. College Station: Texas A & M University Press for the Society for the North American Cultural Survey, 1982.

CREDIT

page 76: [1]From John Blacking and Joann W. Kealiinohomoku, eds., *The Performing Arts: Music and Dance* (The Hague: Mouton, 1979), 144. Reprinted by permission of the publisher.

Chapter 5 **LANGUAGE**

Language is a system of communication through speech, a collection of sounds that a group of people understands to have the same meaning. Many languages also have a literary tradition or a system of written communication. However, hundreds of spoken languages lack a literary tradition.

The lack of written records makes it difficult to document the distribution of many languages. *Ethnologue: Languages of the World,* one of the most authoritative sources of languages (at www.ethnologue.com) estimates that the world has 6,912 living languages.

Earth's heterogeneous collection of languages is one of the most obvious examples of cultural diversity. Language is like luggage: people carry it with them when they move from place to place. They incorporate new words into their own language when they reach new places, and they contribute words brought with them to the existing language at the new location.

IN MANY LANGUAGES
Newspapers published in (clockwise from upper left) Italian, Arabic, Chinese, Spanish, Polish, Gujarati, Greek, and Russian.

THE KEY ISSUES IN THIS CHAPTER

LANGUAGES OF THE WORLD

5.1 Classifying Languages

5.2 Distribution of Languages

5.3 Indo-European Languages

DISTRIBUTION OF ENGLISH

5.4 Origin and Diffusion of English

5.5 English Dialects

5.6 Global Dominance of English

GEOGRAPHIC DIVERSITY OF LANGUAGES

5.7 Multilingual States

5.8 Preserving Endangered Languages

5.9 French and Spanish in North America

LANGUAGES OF THE WORLD

5.1 Classifying Languages

- The world's 6,000-plus languages can be classified into families, branches, and groups.
- Only around 100 of these languages are used by more than 5 million people.

Approximately 100 languages are spoken by at least 5 million people, another 70 by between 2 million and 5 million people. The remaining 6,000 or more languages are spoken by fewer than 2 million people.

The world's languages can be divided into **language families**, which are a collection of languages related through a common ancestral language that existed long before recorded history. Within a language family, a **language branch** is a collection of languages related through a common ancestral language that existed several thousand years ago. Differences are not as extensive or as old as in language families, and archaeological evidence can confirm that the branches derived from the same family. A **language group** is a collection of languages within a branch that share a common origin in the relatively recent past and display relatively few differences in grammar and vocabulary.

Figure 5.1.1 attempts to depict relationships among language families, branches, and groups. Language families form the trunks of the trees, whereas individual languages are displayed as leaves. The larger the trunks and leaves, the greater the number of speakers of those families and languages. Some trunks divide into several branches, which logically represent language branches.

Numbers on the tree are in millions of native speakers. Native speakers are people for whom the language is their first language. The totals exclude those who speak the language as a second language. Figure 5.1.1 displays each language family as a separate tree at ground level, because differences among families predate recorded history. Linguists speculate that language families were joined together as a handful of superfamilies tens of thousands of years ago. Superfamilies are shown as roots below the surface, because their existence is highly controversial and speculative.

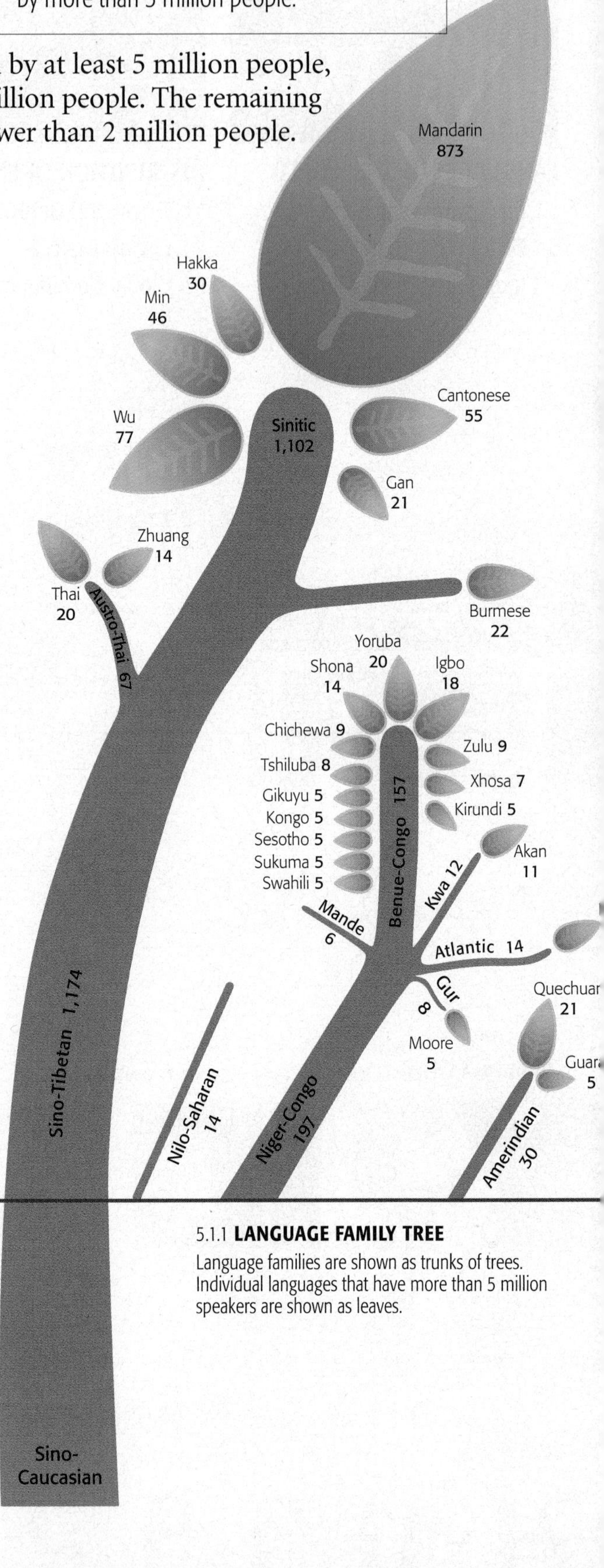

5.1.1 **LANGUAGE FAMILY TREE**

Language families are shown as trunks of trees. Individual languages that have more than 5 million speakers are shown as leaves.

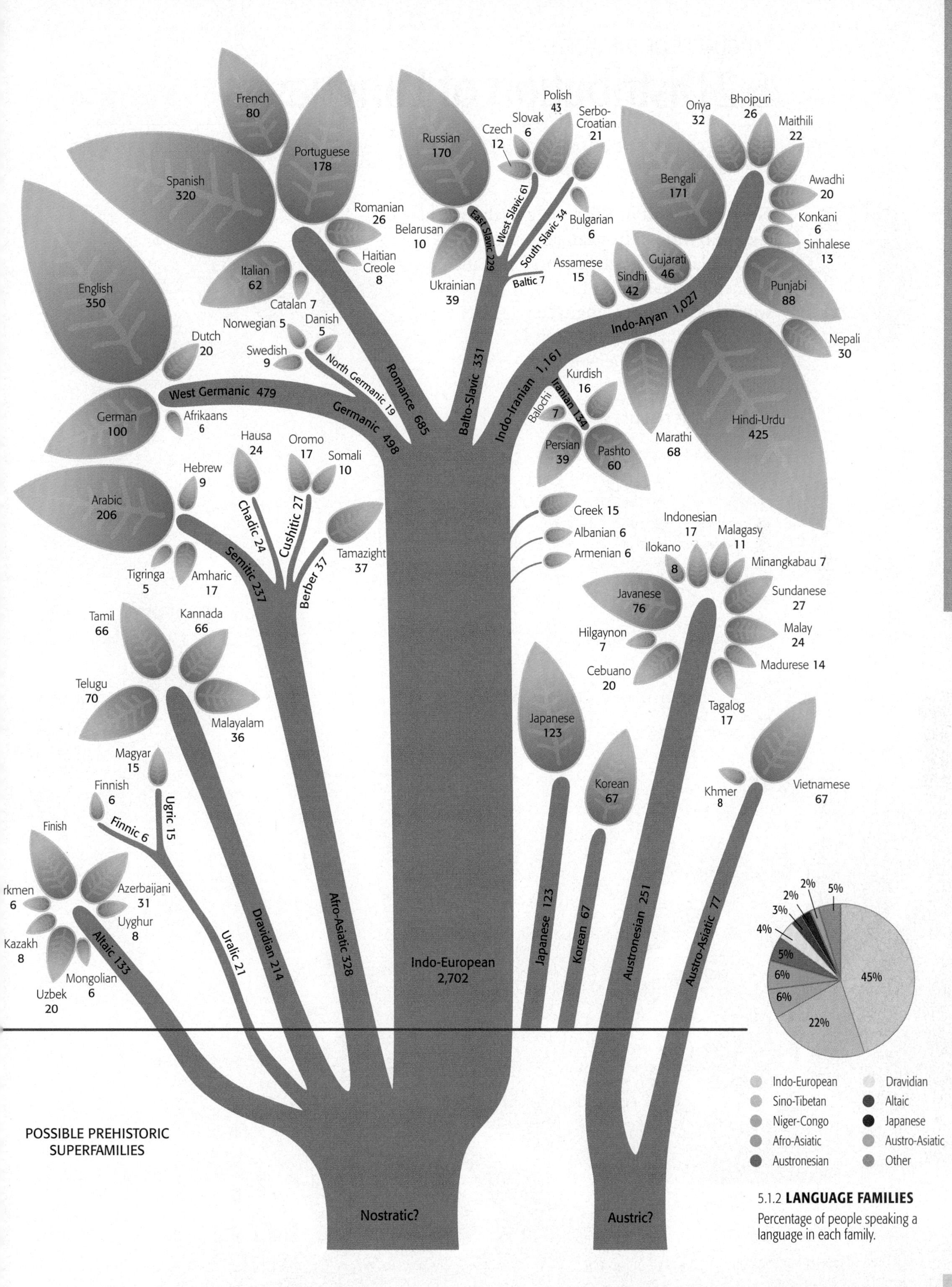

5.1.2 LANGUAGE FAMILIES

Percentage of people speaking a language in each family.

LANGUAGES OF THE WORLD

5.2 Distribution of Languages

- Two language families are used by nearly three-fourths of the world.
- Seven language families are used by most of the remainder.

Nearly one-half of the people in the world speak an Indo-European language. The second-largest family is Sino-Tibetan, spoken by nearly one-fourth of the world. Seven other language families are used by between 2 and 6 percent of the world.

5.2.1 **LANGUAGE FAMILIES**
Most languages can be classified into one of a handful of language families.

Spanish
Languages with more than 100 million speakers

French
Languages with 50–100 million speakers

INDO-EUROPEAN FAMILY
The world's most widely spoken family, discussed in more detail on the next page. Indo-European is the language family of English.

ALTAIC FAMILY
Spoken across an 8,000-kilometer (5,000-mile) band of Asia between Turkey and China. Turkish, by far the most widely used Altaic language, was once written with Arabic letters. In 1928 the Turkish government, led by Kemal Ataturk, ordered that the language be written with the Roman alphabet as a symbol of modernization of the culture and economy.

SINO-TIBETAN FAMILY
Encompasses the languages of China. There is no single Chinese language. Rather, the most important is Mandarin (or, as the Chinese call it, *pu tong hua*, or common speech). Chinese languages are based on 420 one-syllable words, which in turn can be combined to form multisyllabic words. Chinese languages are written with a collection of thousands of characters, mostly **ideograms**, which represent ideas or concepts rather than specific pronunciations.

JAPANESE
An example of an isolated language, unrelated to other language families. Japanese is written in part with Chinese ideograms, but it also uses two systems of phonetic symbols like Western languages.

AUSTRO-ASIATIC FAMILY
Based in Southeast Asia. Vietnamese, the most spoken Austro-Asiatic language, is written with the Roman alphabet. The Vietnamese alphabet was devised in the seventh century by Roman Catholic missionaries from Europe, who brought with them their form of writing.

DRAVIDIAN FAMILY
Languages spoken in southern India and northern Sri Lanka. Between 35 and 70 million speak four languages in this family. Origins of Dravidian are unknown, but scholars generally believe that the language family was once spoken across much of South Asia.

AUSTRONESIAN FAMILY
Languages spoken mostly in Indonesia. The people of Madagascar speak Malagasy, which also belongs to the Austronesian family. This is evidence of migration to Madagascar from Indonesia, 3,000 kilometers (1,900 miles) across the Indian Ocean, apparently in small boats, roughly 2,000 years ago.

AFRO-ASIATIC FAMILY
Includes Arabic and Hebrew. Arabic is the major Afro-Asiatic language, an official language in two dozen countries of the Middle East, and the language of Islam's holiest book, the Quran. Hebrew, the language of much of the Jewish Bible and Christian Old Testament, is a rare case of an extinct language that was revived in the twentieth century as a modern language used in Israel (see page 114).

NIGER-CONGO FAMILY
More than 1,000 distinct languages have been documented in Africa, but no one knows the precise number, and scholars disagree on classifying those known into families. Most lack a written tradition and only 8 are spoken by more than 10 million people. More than 95 percent of the people in sub-Saharan Africa speak languages that are generally classified as belonging to the Niger-Congo family.

LANGUAGES OF THE WORLD

5.3 Indo-European Languages

- Four branches of Indo-European have relatively large numbers of speakers.
- These branches have distinctive spatial distributions.

Indo-European is divided into eight branches. The four branches described below are spoken by large numbers of people. The four less extensively used Indo-European branches are Albanian, Armenian, Greek, and Celtic.

5.3.1 **GERMANIC BRANCH OF INDO-EUROPEAN**

North Germanic: Danish, Faeroese, Icelandic, Norwegian, Swedish

West Germanic: English, Frisian, German, Netherlandish (Dutch), Mixed with non-Germanic

5.3.2 **BRANCHES OF INDO-EUROPEAN FAMILY**

Albanian, Armenian, Balto-Slavic, Celtic, Germanic, Greek, Indo-Iranian, Romance, Non-Indo-European Languages

GERMANIC BRANCH

English is part of the Germanic branch because of the language spoken by Germanic tribes that invaded England 1,500 years ago (see next page). Scandinavian languages are also classified as Germanic.

ROMANCE BRANCH

Romance languages evolved from Latin spoken by the Romans 2,000 years ago. As the conquering Roman armies occupied the provinces of their vast empire, they brought the Latin language with them.

Following the collapse of the Roman Empire in the fifth century, communication among the former provinces declined, creating regional variations in spoken Latin. After several hundred years of isolation from each other, residents of various parts of the former empire spoke distinct languages, including Spanish, Portuguese, French, Italian, and Romanian. Spanish and Portuguese have achieved worldwide importance because of the colonial activities of their European speakers in Latin America.

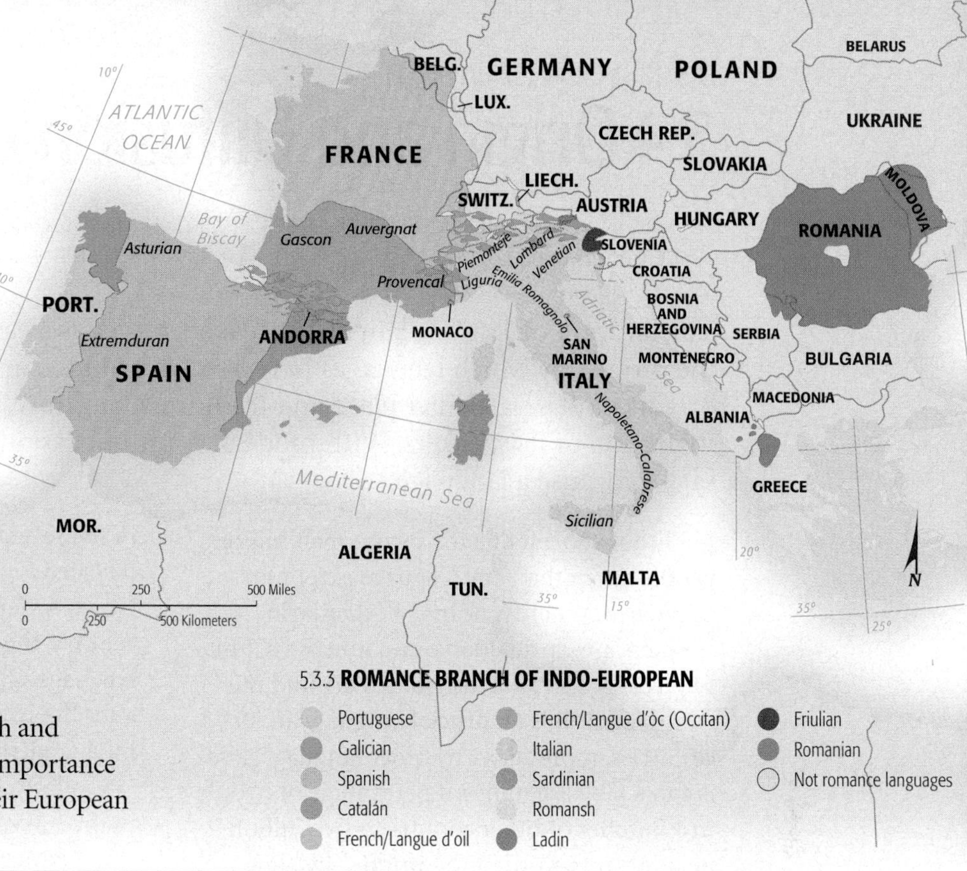

5.3.3 **ROMANCE BRANCH OF INDO-EUROPEAN**

BALTO-SLAVIC BRANCH

The Balto-Slavic branch is the predominant one in Eastern Europe. Slavic was once a single language, but differences developed in the seventh century A.D. when several groups of Slavs migrated from Asia to different areas of Eastern Europe and thereafter lived in isolation from one another.

The most widely used Slavic language is Russian. The importance of Russian increased in the twentieth century with the Soviet Union's rise to power. Soviet officials forced native speakers of other languages to learn Russian as a way of fostering cultural unity among the country's diverse peoples. With the demise of the Soviet Union, the newly independent republics adopted official languages other than Russian.

INDO-IRANIAN BRANCH

The branch of Indo-European with the most speakers, including the most widely used languages in South Asia, is Indo-Iranian. Hindi is spoken by one-third of Indians, but in many different ways that are sometimes identified as distinct languages, though there is only one official way to write it. Pakistan's principal language, Urdu, is spoken much like Hindi, but uses the Arabic alphabet. The two languages are sometimes considered a single one known as Hindustani.

One of the main elements of cultural diversity among the 1 billion plus residents of India is language. After India became an independent state in 1947, Hindi was proposed as the official language, but Dravidian speakers from southern India strongly objected. Therefore, India's constitution as amended recognizes 18 official languages, including 13 Indo-European, 4 Dravidian, and 1 Sino-Tibetan.

5.3.4 **LANGUAGES AND LANGUAGE FAMILIES OF SOUTH ASIA**

Indo-European
- Dialect of Hindi
- Other Indo-European
- **HINDI** Official language of India
- **Kachi** Other language

Families other than Indo-European
- Sino-Tibetan
- Austro-Asiatic
- Dravidian
- Burushaski

DISTRIBUTION OF ENGLISH

5.4 Origin and Diffusion of English

- English is in the Germanic branch because German speakers invaded England 1,500 years ago.
- English differs from German because of the invasion of French-speaking Normans in 1066.

The location of English-language speakers serves as a case study for understanding the process by which languages have been distributed around the world. A language originates at a particular place and traditionally diffuses to other locations through the migration of its speakers. With modern communications, migration is no longer necessary to diffuse a language.

English is spoken fluently by one-half billion people, more than any language except for Mandarin. Whereas nearly all Mandarin speakers are clustered in one country—China—English speakers are distributed around the world. English is an **official language** in 50 countries, more than any other language, and is spoken by a significant percentage of people in a number of other countries. Two billion people—one-third of the world—live in a country where English is an official language, even if they cannot speak it.

ENGLISH COLONIES

The contemporary distribution of English speakers around the world exists because the people of England migrated with their language when they established colonies during the past four centuries. English first diffused west from England to North American colonies in the seventeenth century. After England defeated France in a battle to dominate the North American colonies during the eighteenth century, the position of English as the principal language of North America was assured, even after the United States and Canada became independent countries.

Similarly, the British took control of Ireland in the seventeenth century, South Asia in the mid-eighteenth century, the South Pacific in the late eighteenth and early nineteenth centuries, and southern Africa in the late nineteenth century. More recently, the United States has been responsible for diffusing English to several places, most notably the Philippines, which Spain ceded to the United States in 1899, a year after losing the Spanish-American War. In each case, English became an official language, even if only the colonial rulers and a handful of elite local residents could speak it.

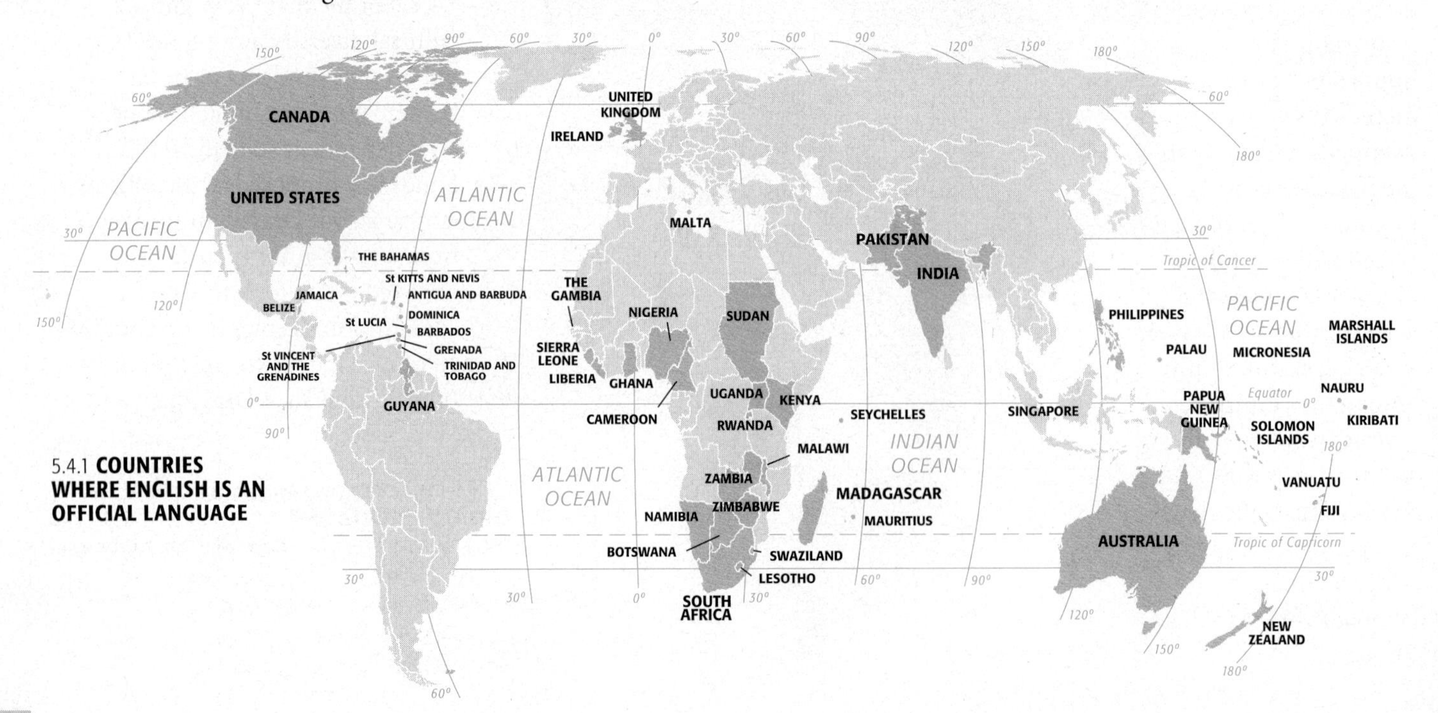

5.4.1 **COUNTRIES WHERE ENGLISH IS AN OFFICIAL LANGUAGE**

ORIGIN OF ENGLISH IN ENGLAND

The global distribution of English may be a function primarily of migration from England since the seventeenth century. Similarly, English is the language of England because of migration to England at earlier times.

- **Celtic tribes around 2000 B.C.** The Celts spoke languages classified as Celtic. We know nothing of earlier languages.
- **Angles, Saxons, and Jutes around A.D. 450.** These tribes from northern Germany and southern Denmark pushed the Celtic tribes to remote northern and western parts of Britain, including Cornwall and the highlands of Scotland and Wales. The name England comes from Angles' land, and English people are often called Anglo-Saxons.
- **Vikings in the ninth century.** Vikings from present-day Norway landed on the northeast coast of England. Although unable to conquer Britain, Vikings remaining in the country contributed words from their language.
- **The Normans in 1066.** The Normans, from present-day Normandy in France, conquered England in 1066 and established French as the official language for the next 300 years.

During the 300-year period that French was the official language of England, the royal family, nobles, judges, and clergy spoke French, whereas most of the people continued to speak English. The Germanic language used by the common people and the French used by the leaders mingled to form a new language, which we know as English.

Modern English owes its simpler, straightforward words, such as *sky, horse, man,* and *woman,* to its Germanic roots, and fancy, more elegant words, such as *celestial, equestrian, masculine,* and *feminine* to its French invaders.

The British Parliament enacted the Statute of Pleading in 1362 to change the official language of court business from French to English, though Parliament itself continued to conduct business in French until 1489.

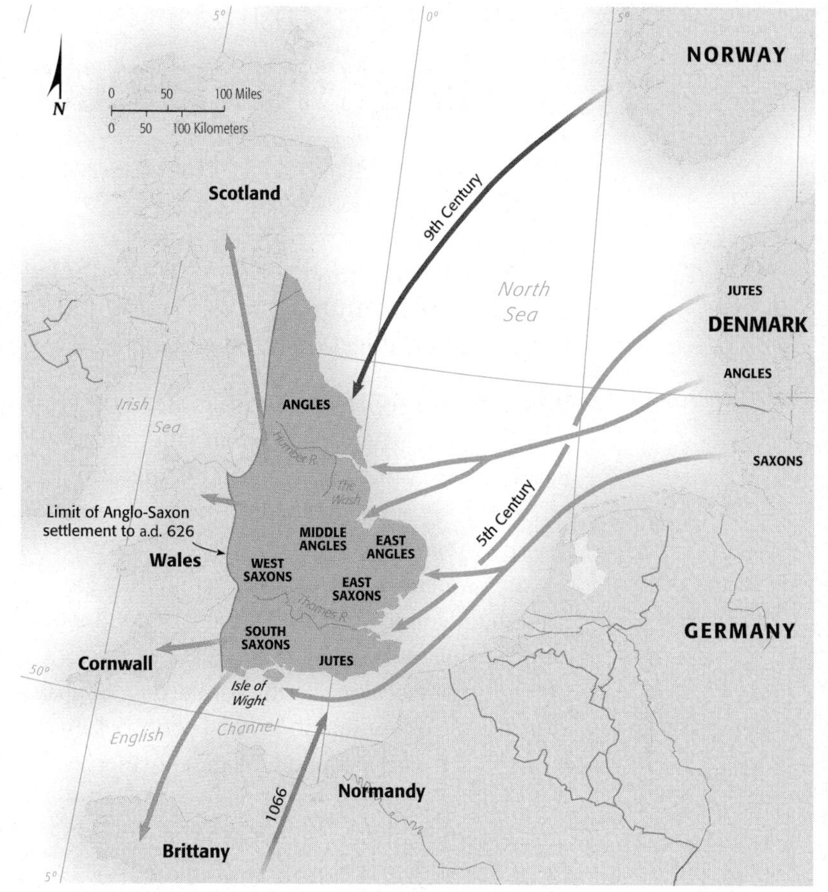

5.4.2 **INVASIONS OF ENGLAND**

Invasion and Migration Routes
Celts
Normans
Germans
Vikings

5.4.3 **NORMAN CONQUEST OF ENGLAND IN 1066**
The Norman Conquest made French the official language of England for around 300 years.

DISTRIBUTION OF ENGLISH

5.5 English Dialects

- American and British dialects differ in vocabulary, spelling, and pronunciation.
- Regional dialects exist within the United States and England

A **dialect** is a regional variation of a language distinguished by distinctive vocabulary, spelling, and pronunciation. Geographers are especially interested in differences in dialects because they reflect distinctive features of the environments in which groups live.

When speakers of a language migrate to other locations, various dialects of that language may develop. This was the case with the migration of English speakers to North America several hundred years ago.

North Americans are well aware that they speak English differently from the British, not to mention people living in India, Pakistan, Australia, and other English-speaking countries. Further, English varies by regions within individual countries. In both the United States and England, northerners sound different from southerners.

Why is the English language in the United States so different from that in England? As is so often the case with languages, the answer is isolation. Separated by the Atlantic Ocean, English in the United States and England evolved independently during the eighteenth and nineteenth centuries, with little influence on one another. Few residents of one country could visit the other, and the means to transmit the human voice over long distances would not become available until the twentieth century.

DIALECTS WITHIN ENGLAND

Dialects in England can be grouped into three main ones—Northern, Midland, and Southern. As already discussed, English originated with three invading Germanic tribes who settled in different parts of England—the Angles in the north, the Jutes in the southeast, and the Saxons in the south and west. The language each spoke was the basis of distinct regional dialects of Old English and Middle English.

People in the south of England pronounce words like *grass* and *path* with an */ah/* sound, whereas people in the Midlands and North use a short */a/* as do most people in the United States. People in the Midlands and North pronounce *butter* and *Sunday* with the */oo/* sound of words like *boot*. Northerners pronounce *ground* and *pound* like "grund" and "pund", with the */uh/* sound similar to the word *punt* in U.S. football.

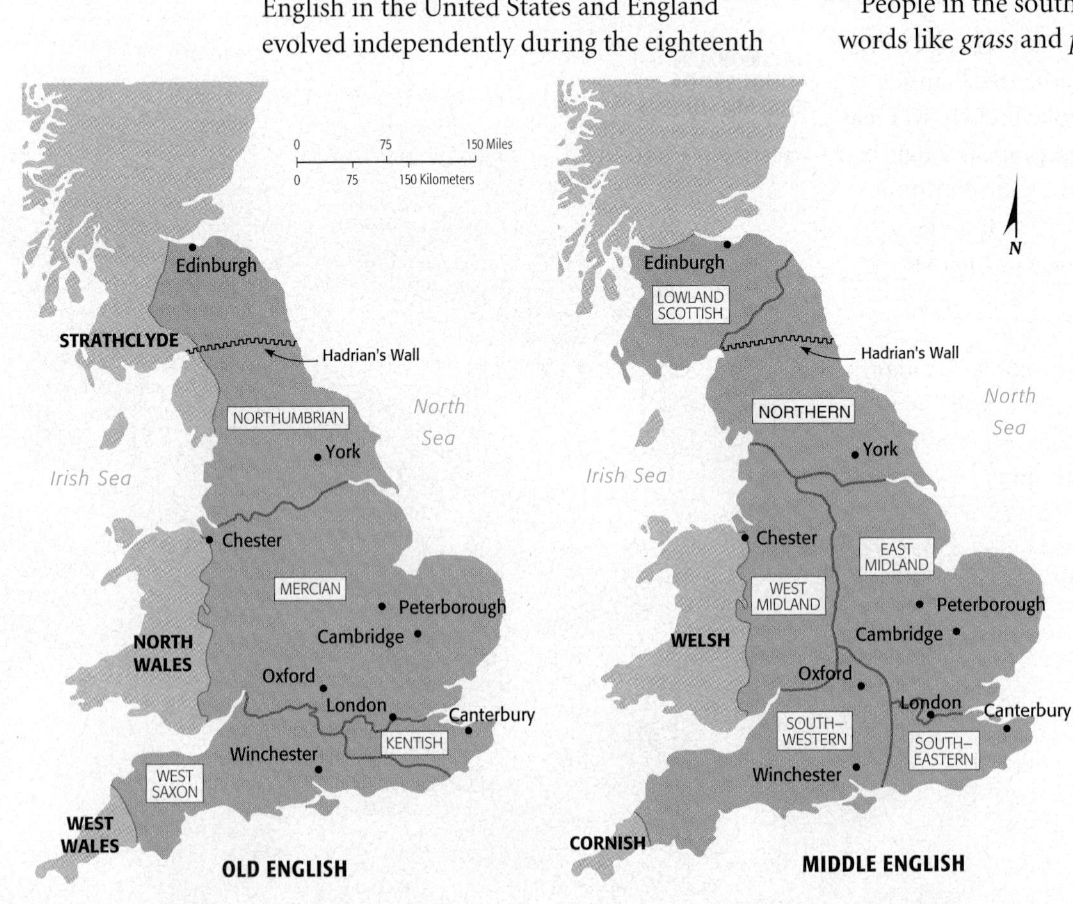

5.5.1 **(Left) OLD ENGLISH DIALECTS, BEFORE THE NORMAN INVASION OF 1066 (Right) MIDDLE ENGLISH DIALECTS (1150–1500)**

English
Celtic
Dialect boundary
MERCIAN Dialect

DIFFERENCES BETWEEN BRITISH AND AMERICAN ENGLISH

U.S. English differs from that of England in three significant ways.

- The vocabulary is different largely because settlers in America encountered many new objects and experiences. The new continent contained physical features, such as large forests and mountains, which had to be given new names. New animals were encountered, including the *moose, raccoon,* and *chipmunk*, all of which were given names borrowed from Native Americans.
- Spelling diverged because of a strong national feeling in the United States for an independent identity. Noah Webster, the creator of the first comprehensive American dictionary and grammar books, was determined to develop uniquely American rules of spelling.
- From the time of their arrival in North America, colonists began to pronounce words differently from the British. Surprisingly, pronunciation has changed more in England than in the United States.

The American pronunciation of *half* (where the *a* is pronounced like the *a* in man in American and like the *a* in father in England) and lord (rhymes with *board* in American and sounds like *laud* in England) is closer to the way they used to be pronounced in England than is the current British pronunciation.

5.5.2 **DIFFERENCE IN BRITISH AND AMERICAN ENGLISH NAMES FOR CAR AND MOTORING WORDS**

(British words are listed first in capital letters)

- **PETROL** Gas
- **LORRY** Truck
- **SLEEPING POLICEMAN** Speed Bump
- **CAR PARK** Parking Lot
- **CAR JOURNEY** Road Trip
- **ZEBRA CROSSING** Cross Walk
- **MOTORWAY** Freeway
- **SALOON** Sedan
- **PETROL STATION** Gas Station
- **BONNET** Hood
- **WINDSCREEN** Windshield
- **BOOT** Trunk
- **REVERSING LIGHTS** Back-up Lights
- **EXHAUST PIPE** Tail Pipe
- **DUAL CARRIAGEWAY** Divided Highway
- **NUMBER PLATE** License Plate
- **FLYOVER** Overpass
- **MULTI-STOREY CAR PARK** Parking Garage
- **CAT'S EYE** Raised Pavement Marker
- **CARAVAN/CAMPERVAN** RV
- **PAVEMENT** Sidewalk
- **ESTATE CAR** Station Wagon
- **MANUAL CAR** Stickshift Car
- **GEAR STICK** Stick
- **INDICATORS** Turn Signal
- **TRAFFIC LIGHTS** Stoplight
- **AMBER LIGHT (TRAFFIC LIGHTS)** Yellow Light

DIALECTS WITHIN THE UNITED STATES

Major dialect differences exist in the United States, primarily on the East Coast. Three major dialect regions are known as Northern, Midlands, and Southern. The English dialects now spoken in the U.S. Southeast and New England are easily recognizable. Current distinctions result from the establishment of independent and isolated colonies in the seventeenth century.

Pronunciations that distinguish the Southern dialect include making such words as *half* and *mine* into two syllables ("ha-af" and "mi-yen"), pronouncing *poor* as "po-ur," and pronouncing *Tuesday* and *due* with a /y/ sound ("Tyuesday" and "dyue"). The New England accent is well known for dropping the /r/ sound, so that *heart* is pronounced "hot" and *ear* ends with /ah/. This characteristic dropping of the /r/ sound is shared with speakers from the south of England and reflects the place of origin of most New England colonists.

Some words are commonly used within one of the three major dialect areas, but rarely in the other two. In most instances, these words relate to rural life, food, and objects from daily activities. Language differences tend to be greater in rural areas than in cities, because farmers are relatively isolated from interaction with people from other dialect regions.

For example, a container commonly used on farms is known as a *pail* in the north and a *bucket* in the Midlands and South. A small stream is known as a *brook* in the North, a *run* in the Midlands, and a *branch* in the South.

The diffusion of particular English dialects into the middle and western parts of the United States is a result of the westward movement of colonists from the three dialect regions of the East. As more of the West was opened to settlement during the nineteenth century, people migrated from all parts of the East Coast. The California gold rush attracted people from throughout the East, many of whom subsequently moved to other parts of the West. The mobility of Americans has been a major reason for the relatively uniform language that exists throughout much of the West.

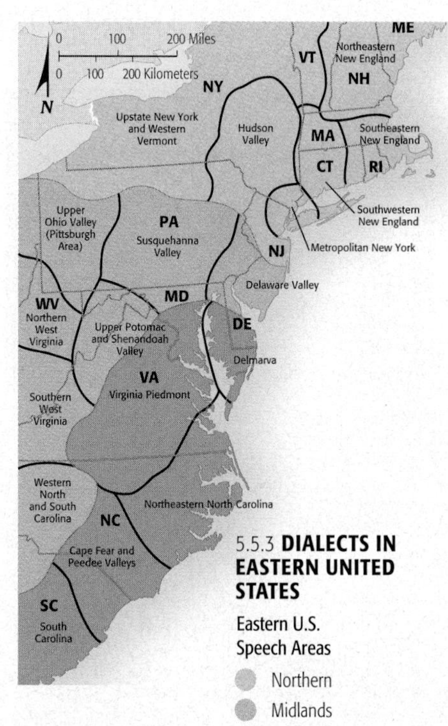

5.5.3 **DIALECTS IN EASTERN UNITED STATES**

Eastern U.S. Speech Areas

- Northern
- Midlands
- Southern

DISTRIBUTION OF ENGLISH

5.6 Global Dominance of English

- English is the world's leading lingua franca.
- English is increasingly being combined with other languages.

Increasingly in the modern world, the language of international communication is English. A Polish airline pilot who flies over France speaks to the traffic controller on the ground in English.

The rapid growth in importance of English is reflected in the percentage of students learning English as a second language. More than 90 percent of students in the European Union learn English in middle or high school. Around the world, some 500 million people speak English as a second language, and an unknown number have some working knowledge of it.

LINGUA FRANCA

A language of international communication, such as English, is known as a **lingua franca**. Other contemporary lingua franca languages include Swahili in East Africa, Hindustani in South Asia, Indonesian in Southeast Asia, and Russian in the former Soviet Union.

In the past, a lingua franca achieved widespread distribution through relocation diffusion: in other words, migration and conquest. Two thousand years ago, use of Latin spread through Europe along with the Roman Empire, and in recent centuries use of English spread around the world through the British Empire.

The recent dominance of English is a result of expansion diffusion, the spread of a trait through the snowballing effect of an idea rather than through the relocation of people. Diffusion of English-language popular culture, as well as global communications such as TV and the Internet, have made English increasingly familiar to speakers of other languages. English has also diffused through integration of vocabulary with other languages. The combination of English with French is called **franglais**, with Spanish **Spanglish**, and with German **Denglish**.

ENGLISH ON THE INTERNET

English has been the leading language of the Internet since its inception. During the 1990s, three-fourths of the people online and three-fourths of Web sites used English. Into the twenty-first century, though, other languages have been catching up to English. Mandarin is set to pass English as the leading language of online users.

English speakers have been accustomed to searching the Internet with Google, or perhaps Yahoo. Searchers in France similarly can use the French-language Google.fr, which may yield similar results as English-language Google. But increasingly, speakers of other languages may prefer other search engines, such as French-language Viola.fr, German-language Web.de, or Japanese-language Dragon.co.jp.

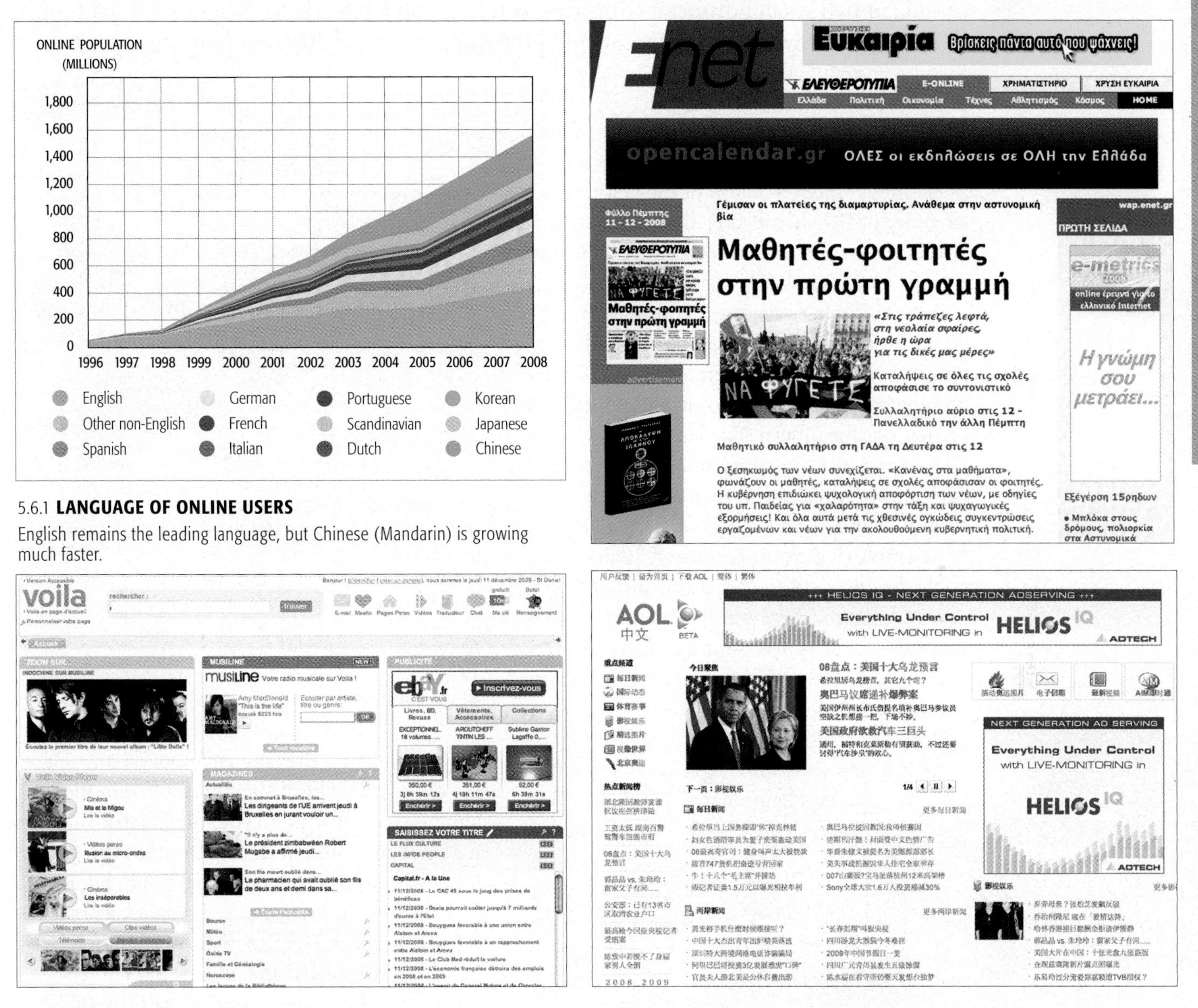

5.6.2 **SEARCH ENGINES IN GREEK (top), FRENCH (lower left), AND JAPANESE (lower right)**

5.6.1 **LANGUAGE OF ONLINE USERS**

English remains the leading language, but Chinese (Mandarin) is growing much faster.

ISOLATED LANGUAGES

An **isolated language** is a language unrelated to any other and therefore not attached to any language family. The best example in Europe is Basque, apparently the only language currently spoken in Europe that survives from the period before the arrival of Indo-European speakers. No attempt to link Basque to the common origin of the other European languages has been successful. Basque was probably once spoken over a wider area but was abandoned where its speakers came in contact with Indo-Europeans. Basque is spoken by 600,000 people in the Pyrenees Mountains of northern Spain and southwestern France. Basque's lack of connection to other languages reflects the isolation of the Basque people in their mountainous homeland.

GEOGRAPHIC DIVERSITY OF LANGUAGES

5.7 Multilingual States

- Belgium and Switzerland are examples of multilingual states within Europe.
- Nigeria is an example of an African country with language diversity.

Difficulties can arise at the boundary between two languages. Note that the boundary between the Romance and Germanic branches of Indo-European runs through the middle of two small European countries, Belgium and Switzerland. Belgium has had more difficulty than Switzerland in reconciling the interests of the different language speakers.

BELGIUM

Southern Belgians (known as Walloons) speak French, whereas northern Belgians (known as Flemings) speak a dialect of the Germanic language of Dutch, called Flemish. The language boundary sharply divides the country into two regions.

Antagonism between the Flemings and Walloons is aggravated by economic and political differences. Historically, the Walloons dominated Belgium's economy and politics, and French was the official state language.

In response to pressure from Flemish speakers, Belgium was divided into two independent regions, Flanders and Wallonia. Each elects an assembly that controls cultural affairs, public health, road construction, and urban development in its region. The national government turns over approximately 15 percent of its tax revenues to pay for the regional governments.

Motorists in Belgium clearly see the language boundary on expressways. Heading north, the highway signs suddenly change from French to Flemish at the boundary between Wallonia and Flanders. Brussels, the capital city, is an exception. Although located in Flanders, Brussels is officially bilingual and signs are in both French and Flemish.

Belgium had difficulty fixing a precise boundary between Flemish and French speakers, because people living near the boundary may actually use the language spoken on the other side. This difficulty resulted in the jailing of one town's mayor and collapse of the national government during the 1980s. The town in question is named Voeren in Flemish and Fourons in French. Jose Happart, its mayor, refused to speak Flemish, which is required by national law because the town is in Flanders. Happart had been elected on a platform of returning the town to French Wallonia, from which it had been transferred in 1963, when the national government tried to clear up the language boundary. After refusing to be tested on his knowledge of Dutch, Happart (who in fact knew Dutch) was jailed and removed from office. In protest, French-speaking members quit the coalition governing the country, forcing the Belgian prime minister to resign.

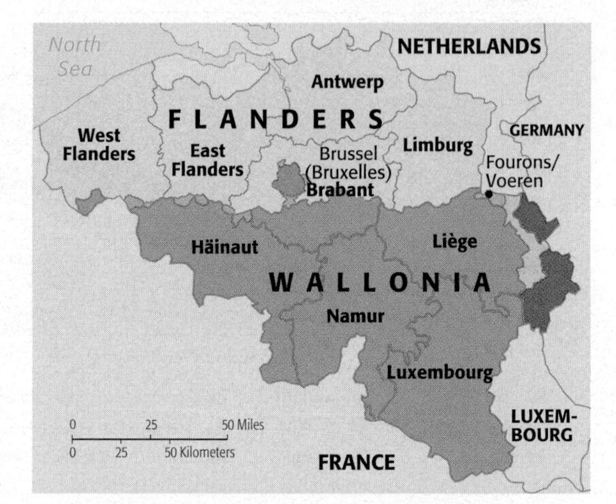

5.7.1 **LANGUAGES IN WALLONIA, BELGIUM**
Road signs such as the one on the lower left, show the names of cities in French first and Flemish second. The order is reversed in Flanders.

Ethnicities
- Flemings (speaking Dutch dialects)
- Walloons (speaking French)
- Germans
- Flemings and Walloons (Legally bilingual)

Protected Minorities
- Walloons in Flanders
- Flemings in Wallonia
- Germans in Wallonia

SWITZERLAND

In contrast, Switzerland peacefully exists with multiple languages. The key is a decentralized government, in which local authorities hold most of the power and decisions are frequently made by voter referenda. Switzerland has four official languages—German (used by 65 percent of the population), French (18 percent), Italian (10 percent), and Romansh (1 percent). Swiss voters made Romansh an official language in a 1938 referendum, despite the small percentage of people who use the language.

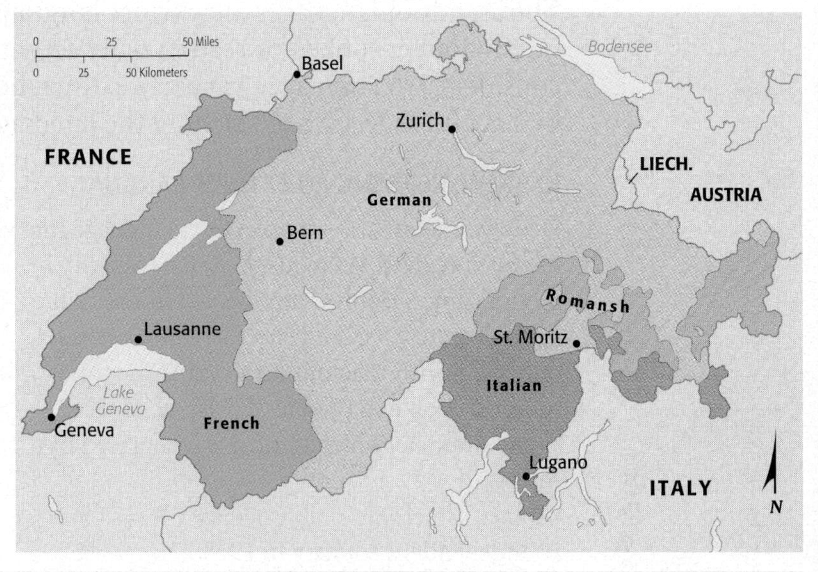

5.7.3 **LANGUAGES OF SWITZERLAND**
Switzerland has four official languages, shown on the sign above: German (top left), French (top right), Italian (lower left), and Romansh (lower right). The sign prevents hikers, vehicles, and horses from entering the forest because of timber cutting.

NIGERIA

Africa's most populous country, Nigeria, displays problems that can arise from the presence of many speakers of many languages. Nigeria has 493 distinct languages, according to *Ethnologue*, only 3 of which have widespread use. Hausa, Yoruba, and Igbo are spoken by approximately 15 percent each, and the remaining 55 percent of the population use one of the other 490 languages.

Groups living in different regions of Nigeria have often battled. The southern Igbos attempted to secede from Nigeria during the 1960s, and northerners have repeatedly claimed that the Yorubas discriminate against them. To reduce these regional tensions, the government has moved the capital from Lagos in the Yoruba-dominated southwest to Abuja in the center of Nigeria.

Nigeria reflects the problems that can arise when great cultural diversity—and therefore language diversity—is packed into a relatively small region. Nigeria also illustrates the importance of language in identifying distinct cultural groups at a local scale. Speakers of one language are unlikely to understand any of the others in the same family, let alone languages from other families.

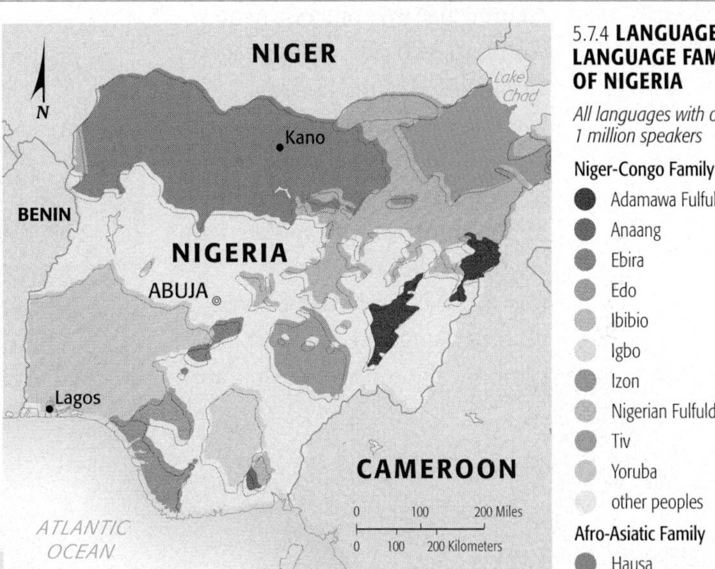

5.7.4 **LANGUAGES AND LANGUAGE FAMILIES OF NIGERIA**

All languages with over 1 million speakers

Niger-Congo Family
- Adamawa Fulfulde
- Anaang
- Ebira
- Edo
- Ibibio
- Igbo
- Izon
- Nigerian Fulfulde
- Tiv
- Yoruba
- other peoples

Afro-Asiatic Family
- Hausa
- other peoples

Nilo-Saharan Family
- Kanuri
- other peoples

GEOGRAPHIC DIVERSITY OF LANGUAGES

5.8 Preserving Endangered Languages

- Many languages have become extinct.
- Other languages are being preserved by governments.

Thousands of languages are **extinct languages**, once in use—even in the recent past—but no longer spoken or read in daily activities by anyone in the world. *Ethnologue* considers 516 languages as nearly extinct, because only a few older speakers are still living, and they are not teaching the languages to their children.

HEBREW: REVIVING AN EXTINCT LANGUAGE

Hebrew is a rare case of an extinct language that has been revived. A language of daily activity in biblical times, Hebrew diminished in use in the fourth century B.C. and was thereafter retained only for Jewish religious services. At the time of Jesus, people in present-day Israel generally spoke Aramaic, which in turn was replaced by Arabic.

When Israel was established as an independent country in 1948, Hebrew became one of the new country's two official languages, along with Arabic. Hebrew was chosen because the Jewish population of Israel consisted of refugees and migrants from many countries who spoke many languages. Because Hebrew was still used in Jewish prayers, no other language could so symbolically unify the disparate cultural groups in the new country.

The task of reviving Hebrew as a living language was formidable. Words had to be created for thousands of objects and inventions unknown in biblical times, such as telephones, cars, and electricity. Eliezer Ben-Yehuda, who lived in Palestine before the creation of the state of Israel, is credited with initiating the revival and inventing 4,000 new Hebrew words.

5.8.1 **LANGUAGES IN ISRAEL**
Street signs are in Hebrew, Arabic, and English. The McDonalds sign is in Hebrew only.

CELTIC: PRESERVING ENDANGERED LANGUAGES

Some endangered languages are being preserved. The European Union has established the European Bureau for Lesser Used Languages (EBLUL), based in Dublin, Ireland, to provide financial support for the preservation of 60 indigenous, regional, and minority languages spoken by some 50 million Europeans.

Among the European languages being preserved are several belonging to the Celtic branch of Indo-European. Celtic is of particular interest to English speakers because it was the major language in the British Isles before the Germanic Angles, Jutes, and Saxons invaded. Today, Celtic languages survive in remoter parts of Scotland, Wales, and Ireland, and on the Brittany peninsula of France.

- **Ireland** Irish Gaelic is one of the Republic of Ireland's official languages, along with English. One-half million speak Irish on a daily basis, and another million people regard themselves as competent in it. More than half of the residents of Ireland's remote western counties speak Irish. The Official Languages Act of 2003 requires that government publications be in Irish as well as English. Bilingual signs are posted in many shops. Irish language TV and radio stations, as well as newspapers, are generally available. Irish is a compulsory subject in school.
- **Scotland** Gaelic was carried from Ireland to Scotland about 1,500 years ago, but Scottish Gaelic is much less used today than Irish. Only 58,652 or 1 percent of the people speak Scottish Gaelic, according to the 2001 national census. Most of the speakers are in remote highlands and islands of northern Scotland. An extensive body of literature exists in Gaelic languages, including the Robert Burns poem Auld Lang Syne ("old long since"), the basis for the popular New Year's Eve song.
- **Wales** Roughly one-half million people can read, write, and converse in Welsh, around one-sixth of the population of Wales. As is the case with Ireland and Scotland, the percentage of residents who can speak Welsh is highest in the most remote portions of Wales. Several government acts have elevated Welsh to equal status with English in the public sector. Teaching Welsh in schools has been compulsory since 2000. Road signs are bilingual. Welsh-language coins circulate, and a television and radio station broadcast in Welsh.

5.8.2. CELTIC ROAD SIGNS

English appears above Welsh on the top photo marking the England-Wales border and underneath Scottish Gaelic on the bottom photo on Scotland's Isle of Skye.

- **Cornwall** Cornish became extinct in 1777, with the death of the language's last known native speaker, Dolly Pentreath, who lived in Mousehole (pronounced "muzzle"). Before Pentreath died, an English historian recorded as much of her speech as possible so that future generations could study the Cornish language. One of her last utterances was later translated as "I will not speak English … you ugly, black toad!" Two thousand people are now said to be fluent in Cornish. Magazines and radio programs use the language, and it is being taught in some schools. A standard written form of Cornish was established in 2008.
- **Brittany** Like Cornwall, Brittany is an isolated peninsula that juts out into the Atlantic Ocean. Breton differs from the other Celtic languages in that it has more French words. One-half million people speak Breton regularly, and another one-half million are said to have knowledge of Breton but do not use it regularly.

The survival of any language depends on the political and military strength of its speakers. The Celtic languages declined because the Celts lost most of the territory they once controlled to speakers of other languages. The long-term decline of languages such as Celtic provides an excellent example of the precarious struggle for survival that many languages experience. Faced with the diffusion of alternatives used by people with greater political and economic strength, speakers of Celtic and other languages must make strong efforts to preserve their linguistic identity.

GEOGRAPHIC DIVERSITY OF LANGUAGES

5.9 French and Spanish in North America

• French is one of Canada's two official languages. • Spanish is increasingly used in the United States.

The French-speaking people of Canada and the Spanish-speaking people of the United States both live on a continent dominated by English speakers.

FRENCH IN CANADA

French is one of Canada's two official languages, along with English. French speakers comprise one-fourth of the country's population. Most are clustered in Québec, where they comprise more than three-fourths of the province's speakers.

Until recently, Québec was one of Canada's poorest and least developed provinces. Its economic and political activities were dominated by an English-speaking minority, and the province suffered from cultural isolation and lack of French-speaking leaders.

The Québec government has made the use of French mandatory in many daily activities. Québec's Commission de Toponyme is renaming towns, rivers, and mountains that have names with English-language origins. The word Stop has been replaced by Arrêt on the red octagonal road signs, even though Stop is used throughout the world, even in France and other French-speaking countries. French must be the predominant language on all commercial signs, and the legislature passed a law banning non-French outdoor signs altogether (ruled unconstitutional by the Canadian Supreme Court).

Many Québécois favored total separation of the province from Canada as the only way to preserve their cultural heritage. Voters in Québec have thus far rejected separation from Canada, but by a slim majority. Alarmed at these pro-French policies, many English speakers and major corporations moved from Montréal, Québec's largest city, to English-speaking Toronto, Ontario.

Confrontation during the 1970s and 1980s has been replaced in Québec by increased cooperation between French and English speakers. Montréal's neighborhoods, once highly segregated between French-speaking residents on the east and English-speaking residents on the west, have become more linguistically mixed.

Although French dominates over English, Québec faces a fresh challenge of integrating a large number of immigrants from Europe, Asia, and Latin America who don't speak French. Many immigrants would prefer to use English rather than French as their lingua franca but are prohibited from doing so by the Québec government.

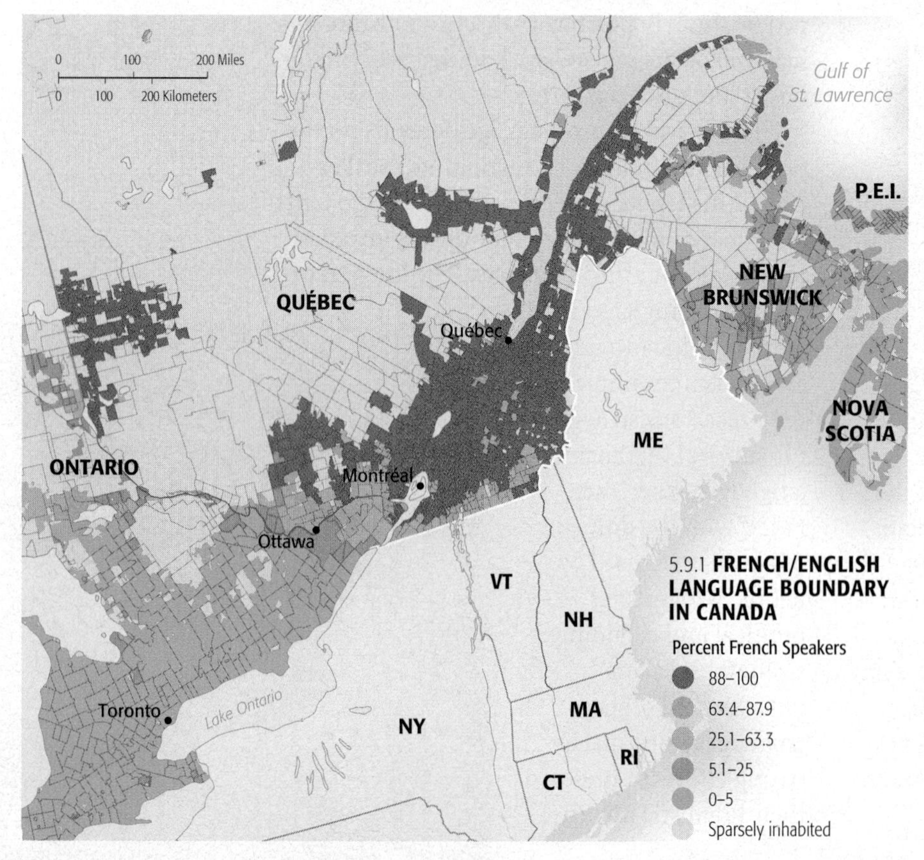

5.9.1 **FRENCH/ENGLISH LANGUAGE BOUNDARY IN CANADA**

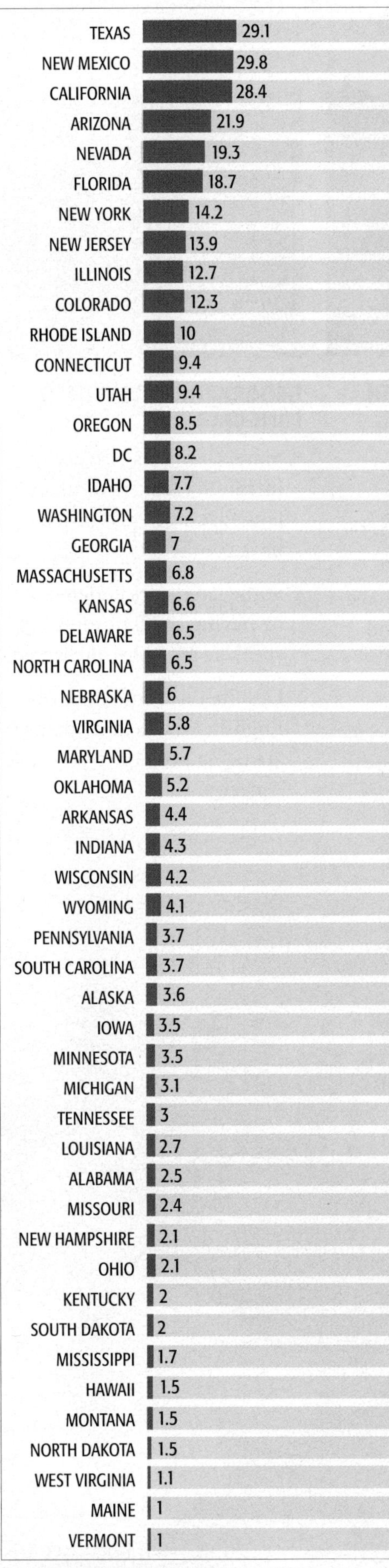

5.9.2 **PERCENT SPANISH-SPEAKING RESIDENTS BY U.S. STATE**

SPANISH IN THE UNITED STATES

Spanish has become an increasingly important language in United States because of large-scale immigration from Latin America, as discussed in Chapter 3. In some communities, government documents and advertisements are printed in Spanish. Several hundred Spanish-language newspapers and radio and TV stations operate in the United States, especially in southern Florida, the Southwest, and large northern cities.

Linguistic unity is an apparent feature of the United States, a nation of immigrants who learn English to become Americans. However, the diversity of languages in the United States is greater than it first appears. A language other than English was spoken at home by 47 million Americans in 2000, 17 percent of the population over age 5. Spanish was spoken at home by 28 million people in the United States. More than 2 million spoke Chinese; at least 1 million each spoke French, German, Italian, Tagalog, and Vietnamese, and at least 500,000 each spoke Arabic, Korean, Polish, Portuguese, and Russian. In reaction against the increasing use of Spanish in the United States, 30 states and a number of localities have laws making English the official language.

Americans have debated whether schools should offer bilingual education. Some people want Spanish-speaking children to be educated in Spanish, because they think that children will learn more effectively if taught in their native language and that this will also preserve their own cultural heritage. Others argue that learning in Spanish creates a handicap for people in the United States when they look for jobs, virtually all of which require knowledge of English.

Promoting the use of English symbolizes that language is the chief cultural bond in the United States in an otherwise heterogeneous society. With the growing dominance of the English language in the global economy and culture, knowledge of English is important for people around the world, not just inside the United States.

Chapter **Summary**

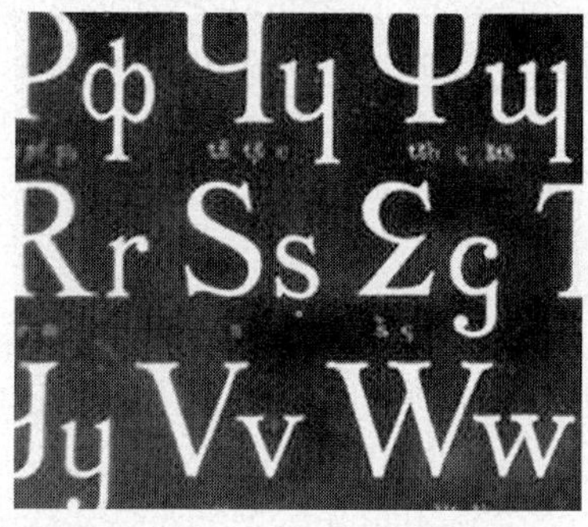

LANGUAGES OF THE WORLD

- Languages can be classified into families, branches, and groups.
- Nearly half of the world speak a language in the Indo-European family and nearly one-fourth speak a Sino-Tibetan language.
- Indo-European is divided into eight branches, including Germanic, the branch to which English belongs.
- Seven other language families encompass the languages of one-fourth of the world.

DISTRIBUTION OF ENGLISH

- The origin of the English language can be traced to invasions of England by Germanic tribes 1,500 years ago.
- From England, the language diffused around the world when English speakers established colonies.
- American and British speak different dialects of English because of relative isolation of the two groups.
- English has become the most important language for international communication.

GEOGRAPHIC DIVERSITY OF LANGUAGES

- Some countries peacefully integrate speakers of many languages, whereas others have conflicts.
- Some languages endangered by having small numbers of speakers are being preserved.
- Despite the dominance of English, French and Spanish are being preserved in North America.

TEACHING ENGLISH IN BEIJING, CHINA

Geographic Consequences of Change

In view of the global dominance of English, many U.S. citizens do not recognize the importance of learning other languages. (Does your own college or university have a foreign-language requirement for graduation?) However, one of the best ways to learn about the beliefs, traits, and values of people living in other regions is to learn their language. The lack of effort by Americans to learn other languages is a source of resentment among people elsewhere in the world, especially when Americans visit or work in other countries.

The inability to speak other languages is also a handicap for Americans who try to conduct international business. Successful entry into new overseas markets requires knowledge of local culture, and officials who can speak the local language are better able to obtain important information. Japanese businesses that wish to expand in the United States send English-speaking officials, but American businesses that wish to sell products to the Japanese are rarely able to send a Japanese-speaking employee.

The dominance of English as an international language has facilitated the diffusion of popular culture and science and the growth of international trade. The emergence of the Internet as an important means of communication has further strengthened the dominance of English.

People in smaller countries need to learn English to participate more fully in a global economy and culture. All children learn English in the schools of countries such as the Netherlands and Sweden to facilitate international communication. This may seem culturally unfair, but obviously it is more likely that several million Dutch people will learn English, than that several hundred million English speakers around the world will learn Dutch.

SHARE YOUR VOICE Student Essay

Maria Khodorkovsky
Georgia State University

Bilingual in America

It isn't uncommon for an American child to grow up bilingual. In a country of immigrants, if you haven't moved directly from a distant land, you have by a young age at least attended bar and bat mitzvahs, dinners in broken English, and been privy to the array of strange and wonderful customs of schoolmates and neighbors.

I grew up in such a nucleus of cultures, in metro-Atlanta's Brookhaven neighborhood. There, a cross-section would reveal an enormous Hispanic population spanning from the tip of Chile to El Paso, Texas, a Korean microcosm, Russian Jew refugees, and a tight-knit Anglo-African community.

Many years later, when I moved from the dynamic Brookhaven neighborhood to Atlanta and began working as an interpreter, I had only a vague and formulaic awareness that my vocation was of use to others and rewarding to myself. But as I was immersed, little by little, into the world I knew as a child, I became conscious that in the simple act of translation I could transcend the long thread of history and experience that separated me from the people for whom I translated. Knowledge of a particular community's mother tongue became for me an unconditional and immutable letter of invitation – to their dinner tables, to their wedding feasts, to their joys and woes.

It has been over four years now that I have been engaged in legal, medical, and academic interpretation for the Hispanic and Russian-Turkish immigrant populations of Atlanta. The people for whom I translate are doubly isolated, both in language and in proximity to their homes and loved ones. It is perhaps this isolation that leads them to bring a professional interpreter into their families as they would an adopted daughter, to send care packages of ethnic goodies, to line up their sons and implore me to pick one (I kid you not!) and to make of me the mast on which to hoist their dreams of Americanization.

Language is not just a variety of ways to say the same thing, as I had originally assumed. Language is a bind as warm and familiar as a nation's dress, cuisine, or customs, and in situations where its lack may mean penury, language is the most precious of commodities.

Chapter Resources

LANGUAGE ORIGINS: WAR OR PEACE?

If Germanic, Romance, Balto-Slavic, and Indo-Iranian languages are all part of the same Indo-European language family, then they must be descended from a single common ancestral language.

Unfortunately, the existence of a single ancestor cannot be proved with certainty, because it would have existed thousands of years before the invention of writing or recorded history.

So where did Indo-European originate? Linguists and anthropologists generally accept that a single Indo-European language must have existed, but they disagree on when and where the language originated and the process and routes by which it diffused.

THE "WAR" THESIS

The first Indo-European speakers may have been the Kurgan people, who lived near the border of present-day Russia and Kazakhstan. The Kurgans were nomads, among the first to domesticate horses and cattle around 5,000 years ago. In search of grasslands for their animals, Kurgan warriors conquered much of Europe and South Asia, using their domesticated horses as weapons.

According to an influential hypothesis by Marija Gimbutas, Indo-European language spread with the Kurgan conquests. Branches of Indo-European originated from the migration of some Kurgans westward through Europe, some eastward to Siberia, and others southeastward to Iran and South Asia.

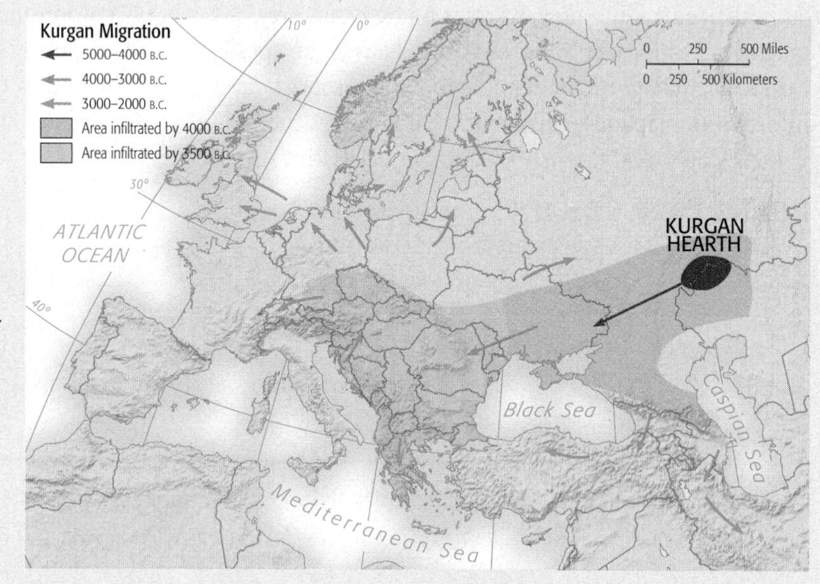

KEY TERMS

Denglish
Combination of German and English.

Dialect
A regional variety of a language distinguished by vocabulary, spelling, and pronunciation.

Extinct language
A language that was once used by people in daily activities but is no longer used.

Franglais
A term used by the French for English words that have entered the French language; a combination of français and anglais, the French words for "French" and "English," respectively.

Ideograms
The system of writing used in China and other East Asian countries in which each symbol represents an idea or a concept rather than a specific sound, as is the case with letters in English.

Isolated language
A language that is unrelated to any other languages and therefore not attached to any language family.

Language
A system of communication through the use of speech, a collection of sounds understood by a group of people to have the same meaning.

Language branch
A collection of languages related through a common ancestor that existed several thousand years ago. Differences are not as extensive or as old as with language families, and archaeological evidence can confirm that the branches derived from the same family.

Language family
A collection of languages related to each other through a common ancestor long before recorded history.

Language group
A collection of languages within a branch that share a common origin in the relatively recent past and display relatively few differences in grammar and vocabulary.

Lingua franca
A language mutually understood and commonly used in trade by people who have different native languages.

Literary tradition
A language that is written as well as spoken.

Official language
The language adopted for use by the government for the conduct of business and publication of documents.

Spanglish
Combination of Spanish and English, spoken by Hispanic Americans.

THE "PEACE" THESIS

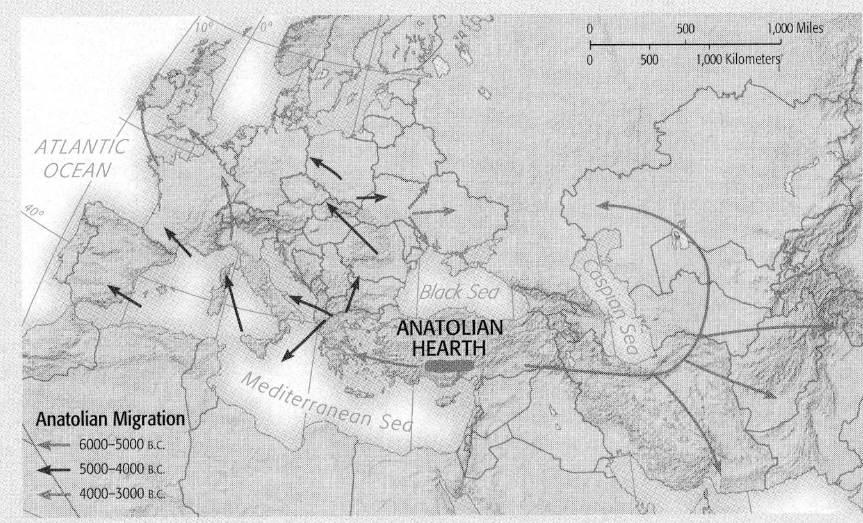

Archaeologist Colin Renfrew argues that the first Indo-European speakers lived 2,000 years before the Kurgans in eastern Anatolia, part of present-day Turkey. Biologist Russell D. Gray supports the Renfrew position, and dates the first speakers even earlier, at around 6700 B.C. Renfrew believes that Indo-European diffused into Europe and South Asia along with agricultural practices rather than by military conquest. The language triumphed because its speakers became more numerous and prosperous by growing their own food instead of relying on hunting.

According to Renfrew the first Indo-European speakers migrated from Anatolia westward to Greece (the origin of the Greek language branch) and from Greece westward toward Italy, Sicily, Corsica, the Mediterranean coast of France, Spain, and Portugal (the origin of the Romance language branch). From the Mediterranean coast, the speakers migrated northward toward central and northern France and on to the British Isles (perhaps the origin of the Celtic language branch).

Indo-European also diffused northward from Greece toward the Danube River (Romania) and westward to central Europe, according to Renfrew. From there the language diffused northward toward the Baltic Sea (the origin of the Germanic language branch) and eastward toward the Dnestr River near Ukraine (the origin of the Slavic language branch). From the Dnestr River, speakers migrated eastward to the Dnepr River (the homeland of the Kurgans).

The Indo-Iranian branch of the Indo-European language family originated either directly through migration from Anatolia along the south shores of the Black and Caspian seas by way of Iran and Pakistan, or indirectly by way of Russia north of the Black and Caspian seas.

Regardless of how Indo-European diffused, communication was poor among different peoples, whether warriors or farmers. After many generations of complete isolation, individual groups evolved increasingly distinct languages.

FURTHER READINGS

Aitchison, John, and Harold Carter. *Language, Economy, and Society: The Changing Fortunes of the Welsh Language in the Twentieth Century*. Cardiff: University of Wales Press, 2000.

Baugh, Albert C., and Thomas Cable. *A History of the English Language*. 5th ed. London and New York: Routledge, 2002.

Cardona, George, Henry M. Hoeningswald, and Alfred Senn, eds. *Indo-European and Indo-Europeans*. Philadelphia: University of Pennsylvania Press, 1970.

Delgado de Carvalho, C. M. "The Geography of Languages." In *Readings in Cultural Geography*, eds. Philip L. Wagner and Marvin W. Mikesell. Chicago: University of Chicago Press, 1962.

Gordon, Raymond G., Jr., ed. *Ethnologue: Languages of the World*, 15th ed. Dallas: SIL International, 2005.

Kirk, John M., Stewart Sanderson, and J. D. A. Widdowson, eds. *Studies in Linguistic Geography: The Dialects of English in Britain and Ireland*. London: Croom Helm, 1985.

Krantz, Grover S. *Geographical Development of European Languages*. New York: Peter Lang, 1988.

Kurath, Hans. *Word Geography of the Eastern United States*. Ann Arbor: University of Michigan Press, 1949.

Renfrew, Colin. *Archaeology and Language*. Cambridge: Cambridge University Press, 1988.

Trudgill, Peter. "Linguistic Geography and Geographical Linguistics." *Progress in Geography* 7 (1975): 227–52.

Williams, Colin H., ed. *Language in Geographic Context*. Clevedon, UK: Multilingual Matters, 1988.

ON THE INTERNET

The book *Ethnologue: Languages of the World*, listed in the Further Readings, is available online at **www.ethnologue.com**. Every country has a map showing the distribution of languages within the country. Detailed information is also provided for every language of the world.

Chapter 6 **RELIGION**

Religion interests geographers because it is essential for understanding how humans occupy Earth. People care deeply about their religion and draw from religion their core values and beliefs, essential elements of culture. Unfortunately, intense identification with one religion can sometimes lead adherents into conflict with followers of other religions.

Geographers, though, are not theologians, so they stay focused on those elements of religions that are geographically significant. Some religions appeal to people throughout the world, whereas other religions appeal primarily to people in geographically limited areas. Like language, migrants take their religion with them to new locations, but although migrants typically learn the language of the new location, they retain their religion.

Religious values are also important in understanding the meaningful ways that people organize the landscape. As a major facet of culture, religion leaves a strong imprint on the physical environment.

STONEHENGE, ENGLAND
ERECTED AROUND 2500 B.C.
No conclusive evidence exists for how and why Stonehenge was constructed.

THE KEY ISSUES IN THIS CHAPTER

RELIGIONS OF THE WORLD

6.1 Origins of Religions

- Geographers distinguish between ethnic and universalizing religions.
- The two types of religions have different origins and distributions.

Geographers distinguish between two types of religions:

- An **ethnic religion** appeals primarily to one group of people living in one place. Hinduism is the largest ethnic religion.
- A **universalizing religion** attempts to be global, to appeal to all people, wherever they may live in the world. The three with the largest number of adherents are Christianity, Islam, and Buddhism.

HINDUISM

Universalizing religions have precise places of origin, based on events in the life of a man. Ethnic religions have unknown or unclear origins, not tied to single historical individuals.

The largest ethnic religion, Hinduism, did not originate with a specific founder. The word *Hinduism* originated in the sixth century B.C. to refer to people living in what is now India. Whereas the origins of Christianity, Islam, and Buddhism are recorded in the relatively recent past, Hinduism existed prior to recorded history. The earliest surviving Hindu documents were written around 1500 B.C., although archaeological explorations have unearthed objects relating to the religion from 2500 B.C. Aryan tribes from Central Asia invaded India about 1400 B.C. and brought with them Indo-European languages, as discussed in Chapter 5. In addition to their language, the Aryans brought their religion.

6.1.1 **Mt. KAILAS**
Hindus believe that this mountain is the home of Siva. Thousands of Hindus make an annual pilgrimage to the base of the mountain.

6.1.2 **CHURCH OF THE HOLY SEPULCHRE, JERUSALEM**
Most Christians believe that the church was constructed on the site of Christ's crucifixion, burial, and Resurrection.

CHRISTIANITY

Christianity was founded upon the teachings of Jesus, who was born in Bethlehem between 8 and 4 B.C. and died on a cross in Jerusalem about A.D. 30. Raised as a Jew, Jesus gathered a small band of disciples and preached the coming of the Kingdom of God. He was referred to as *Christ*, from the Greek word for the Hebrew word *messiah*, which means "anointed."

In the third year of his mission, he was betrayed to the authorities by one of his companions, Judas Iscariot. After sharing the Last Supper (the Jewish Passover seder) with his disciples in Jerusalem, Jesus was arrested and put to death as an agitator. On the third day after his death, his tomb was found empty. Christians believe that Jesus died to atone for human sins, that he was raised from the dead by God, and that his Resurrection from the dead provides people with hope for salvation.

BUDDHISM

The founder of Buddhism, Siddhartha Gautama, was born about 563 B.C. in Lumbini, in present-day Nepal. The son of a lord, Gautama led a privileged life, with a beautiful wife, palaces, and servants.

According to Buddhist legend, Gautama's life changed after a series of four trips. He encountered a decrepit old man on the first trip, a disease-ridden man on the second trip, and a corpse on the third trip. After witnessing these scenes of pain and suffering, Gautama began to feel he could no longer enjoy his life of comfort and security. Then, on a fourth trip, Gautama saw a monk, who taught him about withdrawal from the world. Gautama lived in a forest for six years, thinking and experimenting with forms of meditation, and emerged as the Buddha, the "awakened or enlightened one."

6.1.3 **MAHABODHI TEMPLE, BODH GAYA, INDIA**
The temple was built where Buddha found perfect wisdom.

6.1.4 **PROPHET'S MOSQUE, MADINAH**
The Tomb of Muhammad is in this mosque, in Islam's second holiest city.

ISLAM

The word *Islam* in Arabic means "submission to the will of God," and it has a similar root to the Arabic word for *peace*. An adherent of the religion of Islam is known as a Muslim, which in Arabic means "one who surrenders to God."

The Prophet of Islam, Muhammad, was born in Makkah about the year 570. At age 40, while engaged in a meditative retreat, Muhammad received his first revelation from God through the Angel Gabriel. The Quran, the holiest book in Islam, is a record of God's words, as revealed to the Prophet Muhammad through Gabriel.

Muhammad was a descendent of Ishmael, who was the son of Abraham and Hagar. Jews and Christians trace their history through Abraham's wife Sarah and their son Isaac. Sarah prevailed upon Abraham to banish Hagar and Ishmael, who wandered through the Arabian desert, eventually reaching Makkah , more commonly known to Westerners as Mecca.

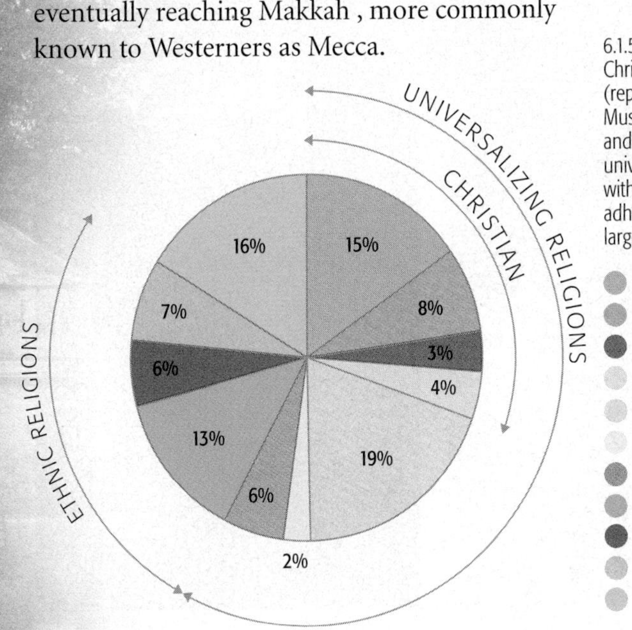

6.1.5 **WORLD RELIGIONS**
Christianity, Islam (represented by Sunni Muslim and Shiite Muslim), and Buddhism are the three universalizing religions with the largest number of adherents. Hinduism is the largest ethnic religion.

- Roman Catholic
- Protestant
- Orthodox / Eastern Christian
- Other Christian
- Sunni Muslim
- Shiite Muslim
- Buddhist
- Hindu
- Chinese traditional
- Other traditional
- Nonreligious

RELIGIONS OF THE WORLD

6.2 Geographic Branches of Religions

- The three largest universalizing religions have different branches.
- Branches have distinctive regional distributions.

Each of the three largest universalizing religions is subdivided into branches, denominations, and sects.

- A **branch** is a large and fundamental division within a religion.
- A **denomination** is a division of a branch that unites a number of local congregations in a single legal and administrative body.
- A **sect** is a relatively small group that has broken away from an established denomination.

6.2.1 **CHRISTIAN PLACES OF WORSHIP**
(Top) Eastern Orthodox church in Gifhorn, Germany. (Lower right) Roman Catholic Cathedral in Pisa, Italy. (Lower left) Protestant church in Edgartown, Massachusetts.

BRANCHES OF CHRISTIANITY

Christianity has three major branches:

- **Roman Catholicism.**
  God conveys His grace directly to humanity through seven sacraments, including Baptism, Confirmation, Penance, Anointing the Sick, Matrimony, Holy Orders, and the Eucharist (the partaking of bread and wine that repeats the actions of Jesus at the Last Supper). The Eucharist literally and miraculously become the body and blood of Jesus while keeping only the appearances of bread and wine, an act known as transubstantiation.
- **Eastern Orthodoxy.**
  A collection of 14 self-governing churches derive from the faith and practices in the Eastern part of the Roman Empire. The split between the Roman and Eastern churches dates to the fifth century and became final in A.D. 1054. The Russian Orthodox Church has more than 40 percent of all Eastern Orthodox Christians, the Romanian Church 20 percent, the Bulgarian, Greek, and Serbian Orthodox churches approximately 10 percent each, and nine others have the remaining 10 percent.
- **Protestantism.**
  The Protestant Reformation movement is regarded as beginning when Martin Luther posted 95 theses on the door of the church at Wittenberg, Germany, on October 31, 1517. According to Luther, individuals had primary responsibility for achieving personal salvation through direct communication with God. Grace is achieved through faith rather than through sacraments performed by the Church.

6.2.2 **MUSLIM PLACES OF WORSHIP**
(Above) Sunni Mosque in Manama, Bahrain. (Below) Shiite Mosque in Samarra, Iraq, which was destroyed in 2006.

BRANCHES OF BUDDHISM

The two largest branches of Buddhism are Theravada and Mahayana.

- *Theravada*, which means "the way of the elders," emphasizes Buddha's life of wisdom, self-help, and solitary introspection.
- *Mahayana* ("the bigger ferry" or "raft"), which split from Theravada Buddhism about 2,000 years ago, emphasizes Buddha's life of teaching, compassion, and helping others.

6.2.3 **BUDDHIST MONKS**
(Above) Theravada Buddhist at Tooth Temple in Kandy, Sri Lanka. (Left) Mahayana Buddhist at Great Buddha statue in Kamakura, Japan.

BRANCHES OF ISLAM

Islam is divided into two important branches: *Sunni* (from the Arabic word for *orthodox*) and *Shiite* (from the Arabic word for *sectarian*, sometimes written Shia in English). Differences between the two main branches go back to the earliest days of Islam and reflect disagreement over the line of succession in Islamic leadership after the Prophet Muhammad, who had no surviving son nor a follower of comparable leadership ability.

DEITIES IN HINDUISM

Hinduism does not have a central authority or a single holy book, so each individual selects suitable rituals. The average Hindu has allegiance to a particular god or concept within a broad range of possibilities:

- The manifestation of God with the largest number of adherents—an estimated 68 percent—is Vaishnavism, which worships the god Vishnu, a loving god incarnated as Krishna.
- An estimated 27 percent adhere to Sivaism, dedicated to Siva, a protective and destructive god.
- Shaktism is a form of worship dedicated to the female consorts of Vishnu and Siva.

In India some geographic concentration of support for these deities exists: Siva and Shakti in the north, Shakti and Vishnu in the east, Vishnu in the west, Siva and some Vishnu in the south. However, holy places for Siva and Vishnu are dispersed throughout India.

6.2.4 **HINDUISM**
Bathing in Ganges

RELIGIONS OF THE WORLD

6.3 Global Distribution

- Christianity predominates in the Western Hemisphere and Europe.
- Islam and Buddhism predominate in Asia.

ANGLO AMERICA
2% Other religions
25% Catholic
10% Nonreligious
2% Sunni Muslim
2% Jewish
29% Protestant
30% Other Christian

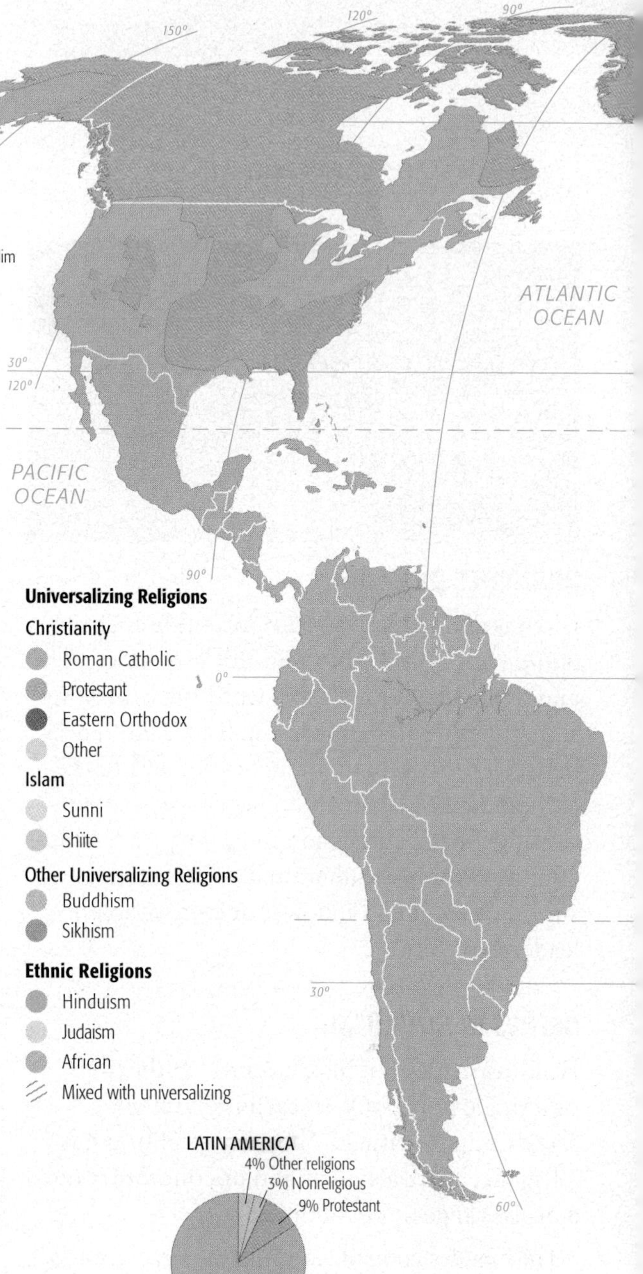

ATLANTIC OCEAN
PACIFIC OCEAN

Universalizing Religions
Christianity
Roman Catholic
Protestant
Eastern Orthodox
Other
Islam
Sunni
Shiite
Other Universalizing Religions
Buddhism
Sikhism
Ethnic Religions
Hinduism
Judaism
African
Mixed with universalizing

LATIN AMERICA
4% Other religions
3% Nonreligious
9% Protestant
84% Catholic

UNIVERSALIZING RELIGIONS

- **Christianity.**
  With more than 2 billion adherents, Christianity is the predominant religion in North America, South America, Europe, and Australia. Within Europe, Roman Catholicism is the dominant Christian branch in the southwest and east, Protestantism in the northwest, and Eastern Orthodoxy in the east and southeast. In the Western Hemisphere, Roman Catholicism predominates in Latin America and Protestantism in North America.
- **Islam.**
  The religion of 1.3 billion people, Islam is the predominant religion of the Middle East from North Africa to Central Asia. One-half of the world's Muslims live outside the Middle East in Indonesia, Pakistan, Bangladesh, and India. The Sunni branch comprises 83 percent of Muslims and is the largest branch in most Muslim countries. The Shiite branch is clustered in Iran, Pakistan, and Iraq.
- **Buddhism.**
  With nearly 400 million adherents, Buddhism is predominant mainly in China and Southeast Asia. Mahayanists account for about 56 percent of Buddhists, primarily in China, Japan, and Korea. Theravadists comprise about 38 percent of Buddhists, especially in Cambodia, Laos, Myanmar, Sri Lanka, and Thailand. The remaining 6 percent are Tantrayanists, found primarily in Tibet and Mongolia.
- **Other universalizing religions.**
  Sikhism and Bahá'í are the next two largest universalizing religions. All but 3 million of the world's 25 million Sikhs are clustered in the Punjab region of India. The 8 million Bahá'ís are dispersed among many countries, primarily in Africa and Asia.

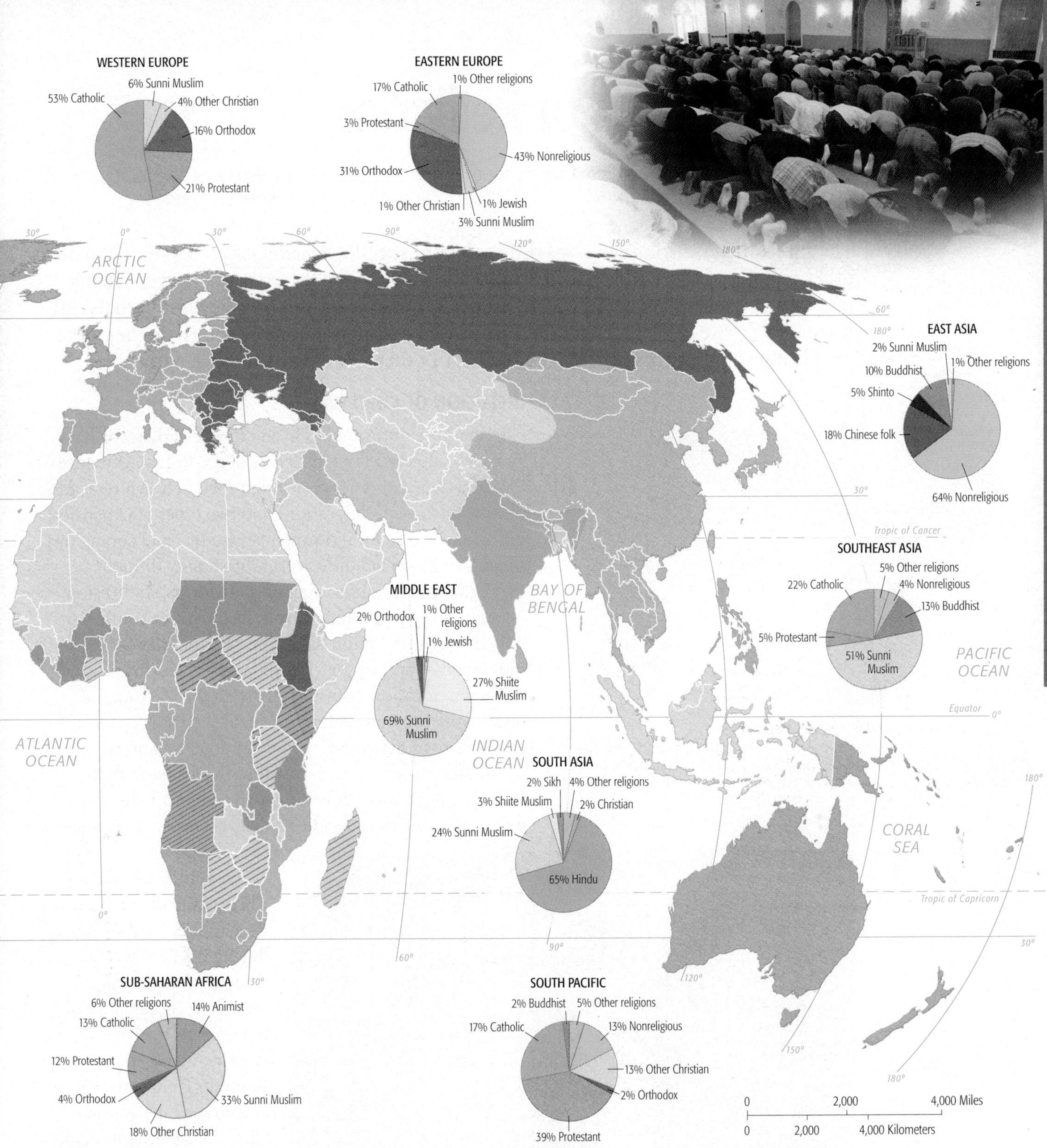

ETHNIC RELIGIONS:

- **Hinduism.**
  Hindus account for more than 80 percent of the population of India and Nepal. All but 3 percent of the world's Hindus are concentrated in India, and most of the remainder in India's neighbor, Nepal.
- **Other ethnic religions.**
  Several hundred million people practice ethnic religions in East Asia, especially Confucianism and Daoism in China and Shintoism in Japan. Approximately 100 million Africans, 12 percent of the continent's population, follow traditional ethnic religions, sometimes called **animism**. Judaism has about 6 million adherents in the United States, 5 million in Israel, 2 million in Europe, and 1 million each in Asia and Latin America.

RELIGIONS OF THE WORLD

6.4 Diffusion of Religions

- Universalizing religions have diffused beyond their places of origin.
- Missionaries and military conquest have been important methods of diffusing universalizing religions.

The three universalizing religions diffused from specific hearths, or places of origin, to other regions of the world. The hearths where each of the three largest universalizing religions originated are based on the events in the lives of the three key individuals. All three hearths are in Asia. Followers transmitted the messages preached in the hearths to people elsewhere, diffusing them across Earth's surface along distinctive paths.

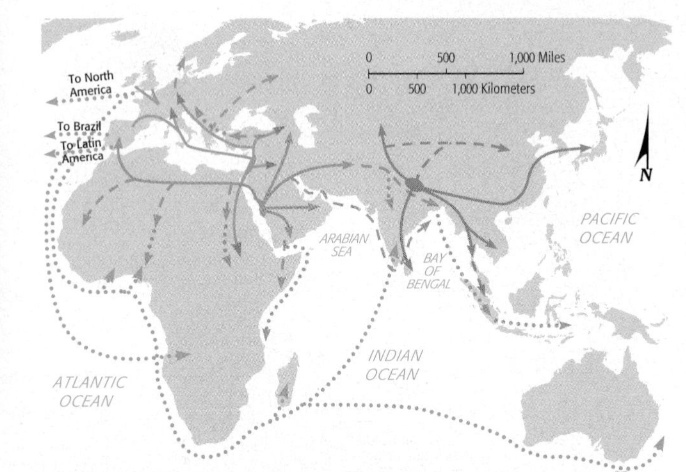

6.4.1 **DIFFUSION OF UNIVERSALIZING RELIGIONS**
Buddhism's hearth is in present-day Nepal and northern India. Christianity's is in present-day Israel, and Islam's is in present-day Saudi Arabia. Buddhism diffused primarily east toward Europe, and Islam west toward northern Africa and east toward southwestern Asia.

	Buddhism	Christianity	Islam
Hearth	●	●	●
Diffusion by 8th century	→	→	→
Diffusion by 12th century	- -→	- -→	- -→
Diffusion after 12th century	···→	···→	···→

DIFFUSION OF CHRISTIANITY

Christianity's diffusion has been rather clearly recorded since Jesus first set forth its tenets in the Roman province of Palestine. In Chapter 1 we distinguished between relocation diffusion (through migration) and two types of expansion diffusion (hierarchical through key leaders and contagious through widespread diffusion). Christianity diffused through a combination of all of these forms of diffusion.

Christianity first diffused from its hearth in Southwest Asia through relocation diffusion. Missionaries—individuals who help to transmit a universalizing religion through relocation diffusion—carried the teachings of Jesus along the Roman Empire's protected sea routes and excellent road network to people in other locations.

Christianity spread widely within the Roman Empire through contagious diffusion—daily contact between believers in the towns and nonbelievers in the surrounding countryside.

The dominance of Christianity throughout the Roman Empire was assured during the fourth century through hierarchical diffusion—acceptance of the religion by the empire's key elite figure, the emperor. Emperor Constantine encouraged the spread of Christianity by embracing it in 313, and Emperor Theodosius proclaimed it the empire's official religion in the year 380. Migration and missionary activity by Europeans since the year 1500 has extended Christianity to other regions of the world.

6.4.2 **DIFFUSION OF CHRISTIANITY**
Constantine's *Vision of the Cross*, fresco in the Vatican designed by Raphael and completed by his students in 1524.

6.4.3 **DIFFUSION OF BUDDHISM**
Dhamek Stupa, Deer Park, Sarnath, India. The stupa (Sanskrit for "heap" or "mound") was built around A.D. 500 to mark the spot where Buddha gave his first sermon.

DIFFUSION OF BUDDHISM

Most responsible for the spread of Buddhism was Asoka, emperor of the Magadhan Empire from about 273 to 232 B.C. About 257 B.C., at the height of the Magadhan Empire's power, Asoka became a Buddhist and thereafter attempted to put into practice Buddha's social principles.

Emperor Asoka's son, Mahinda, led a mission to the island of Ceylon (now Sri Lanka), where the king and his subjects were converted to Buddhism. As a result, Sri Lanka is the country that claims the longest continuous tradition of practicing Buddhism. Missionaries were also sent in the third century B.C. to Kashmir, the Himalayas, Burma (Myanmar), and elsewhere in India.

In the first century A.D., merchants along the trading routes from northeastern India introduced Buddhism to China. Chinese rulers allowed their people to become Buddhist monks during the fourth century A.D., and in the following centuries Buddhism turned into an important Chinese religion. Buddhism further diffused from China to Korea in the fourth century and from Korea to Japan two centuries later. During the same era, Buddhism lost its original base of support in India.

DIFFUSION OF ISLAM

Muhammad's successors organized followers into armies that extended the region of Muslim control over an extensive area of Africa, Asia, and Europe. Within a century of Muhammad's death, Muslim armies conquered Palestine, the Persian Empire, and much of India, resulting in the conversion of many non-Arabs to Islam, often through intermarriage.

To the west, Muslims captured North Africa, crossed the Strait of Gibraltar, and conquered part of Western Europe, retaining it, particularly much of present-day Spain, until 1492. During the same century in which the Christians regained all of Western Europe, Muslims took control of much of southeastern Europe and Turkey.

As was the case with Christianity, Islam, as a universalizing religion, diffused well beyond its hearth in Southwest Asia through relocation diffusion of missionaries to portions of sub-Saharan Africa and Southeast Asia. Although it is spatially isolated from the Islamic core region in Southwest Asia, Indonesia, the world's fourth most populous country, is predominantly Muslim because Arab traders brought the religion there in the thirteenth century.

6.4.4 **DIFFUSION OF ISLAM**
Mezquita de Cordoba, Spain. This was the second largest mosque in the world, until 1236 when the Muslims were expelled from this part of Spain, and the structure was reconsecrated as a cathedral.

THE RELIGIOUS LANDSCAPE

6.5 Diversity of Universalizing Religions

- Other universalizing religions and Christian churches exist in addition to the largest ones.
- The United States has strong regional differences in the distribution of religions.

Christianity, Islam, and Buddhism are the three universalizing religions with the largest number of adherents, and Roman Catholicism, Protestantism, and Eastern Orthodoxy are the three largest branches within Christianity. Other universalizing religions and other Christian churches have flourished in addition to the largest ones.

OTHER CHRISTIAN CHURCHES

Several Christian churches developed independent of the three main branches of Christianity.

- Two small Christian churches survive in northeast Africa—the Coptic Church of Egypt and the Ethiopian Church.
- The Armenian Church originated in Antioch, Syria, and was important in diffusing Christianity to South and East Asia between the seventh and thirteenth centuries.
- The Maronites are an example of a small Christian sect that plays a disproportionately prominent role in political unrest. They are clustered in Lebanon, which has suffered through a long civil war fought among religious groups.
- The Church of Jesus Christ of Latter-day Saints (Mormons) regards its church as a branch of Christianity separate from other branches. About 3 percent of Americans are members of the Latter-day Saints, a large percentage clustered in Utah and surrounding states.

6.5.1 **MORMON TEMPLE, SALT LAKE CITY**

OTHER UNIVERSALIZING RELIGIONS

Sikhism and Bahá'í are the two universalizing religions other than Christianity, Islam, and Buddhism with the largest numbers of adherents.

- **Sikhism.** God was revealed to Sikhism's first guru (religious teacher or enlightener) Nanak (1469–1538) as The One Supreme Being, or Creator, who rules the universe by divine will. Sikhism's most important ceremony, introduced by the tenth guru, Gobind Singh (1666–1708), is the Amrit (or Baptism), in which Sikhs declare they will uphold the principles of the faith. Gobind Singh also introduced the practice of men wearing turbans on their heads and never cutting their beards or hair. Wearing a distinctive clothing helped to give Sikhs a disciplined outlook and a sense of unity of purpose.
- **Bahá'í.** Siyyid 'Ali Muhammad, known as the Báb (Persian for "gateway"), founded the Bábi movement, a precursor of the "Baha'i" faith in Shíráz, Iran, in 1844. Bahá'ís believe that one of the Báb's disciples, Husayn 'Ali Nuri, known as Bahá'u'lláh (Arabic for "Glory of God"), was the prophet and messenger of God. Bahá'u'lláh's function was to overcome the disunity of religions and establish a universal faith through abolition of racial, class, and religious prejudices.

6.5.2 **COPTIC CHURCH, EGYPT**

DISTRIBUTION OF U.S. RELIGIONS

The United States displays regional variations in adherence to religions. Roman Catholics are most numerous in the Northeast and Southwest, a function of migration from predominantly Roman Catholic countries in Europe in the nineteenth century and Latin America in recent years. Among Protestant denominations, Baptists are most numerous in the Southeast, Lutherans in the Upper Midwest, and Methodists in the Northeast.

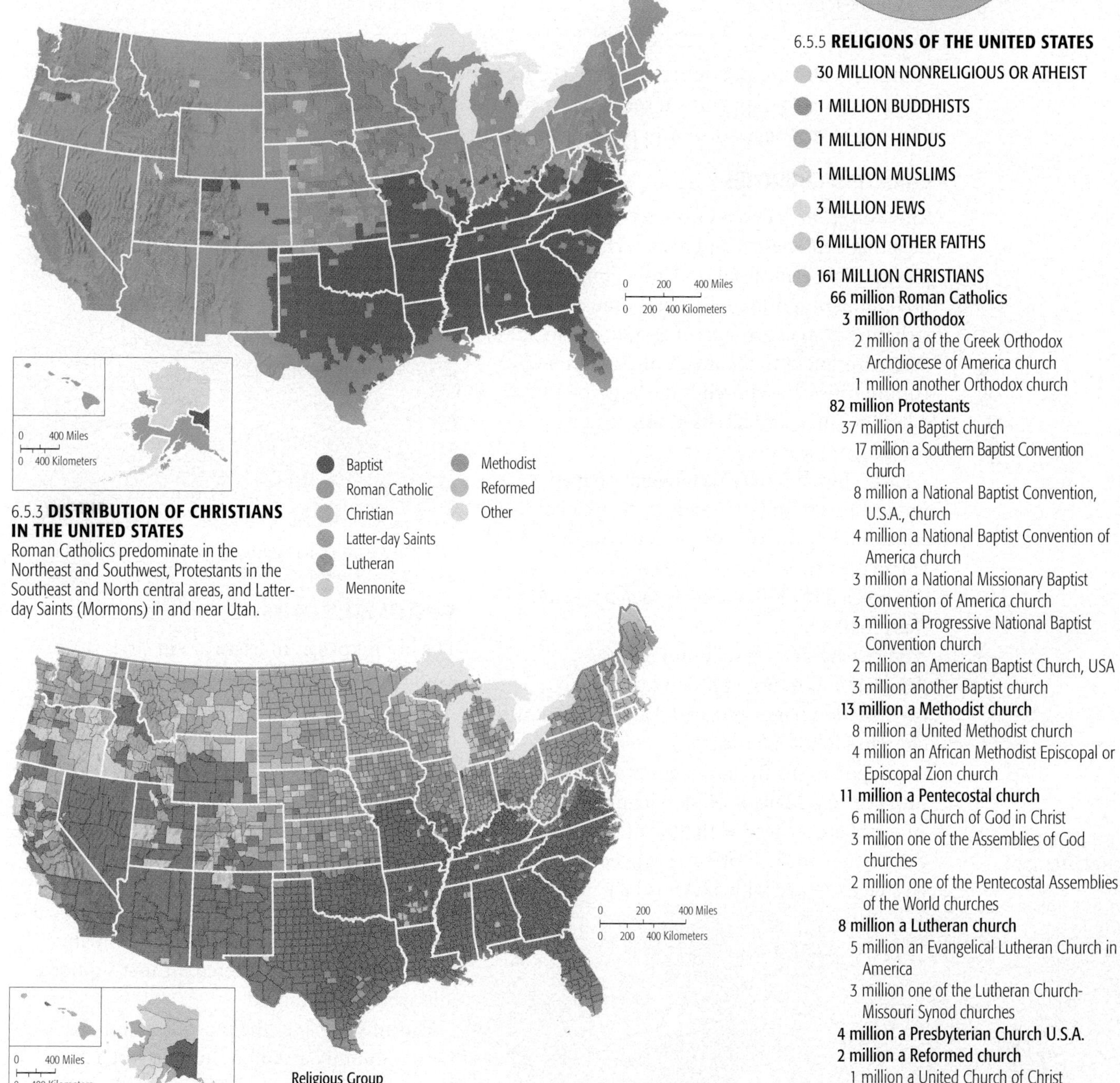

6.5.3 **DISTRIBUTION OF CHRISTIANS IN THE UNITED STATES**
Roman Catholics predominate in the Northeast and Southwest, Protestants in the Southeast and North central areas, and Latter-day Saints (Mormons) in and near Utah.

6.5.4 **DISTRIBUTION OF PROTESTANTS IN THE UNITED STATES**
Baptists predominate in the South, Methodists in the Mid-Atlantic, Lutherans in the Upper Midwest, and members of the Church of Christ in New England.

6.5.5 **RELIGIONS OF THE UNITED STATES**

- 30 MILLION NONRELIGIOUS OR ATHEIST
- 1 MILLION BUDDHISTS
- 1 MILLION HINDUS
- 1 MILLION MUSLIMS
- 3 MILLION JEWS
- 6 MILLION OTHER FAITHS
- 161 MILLION CHRISTIANS
  - 66 million Roman Catholics
  - 3 million Orthodox
    - 2 million a of the Greek Orthodox Archdiocese of America church
    - 1 million another Orthodox church
  - 82 million Protestants
    - 37 million a Baptist church
      - 17 million a Southern Baptist Convention church
      - 8 million a National Baptist Convention, U.S.A., church
      - 4 million a National Baptist Convention of America church
      - 3 million a National Missionary Baptist Convention of America church
      - 3 million a Progressive National Baptist Convention church
      - 2 million an American Baptist Church, USA
      - 3 million another Baptist church
    - 13 million a Methodist church
      - 8 million a United Methodist church
      - 4 million an African Methodist Episcopal or Episcopal Zion church
    - 11 million a Pentecostal church
      - 6 million a Church of God in Christ
      - 3 million one of the Assemblies of God churches
      - 2 million one of the Pentecostal Assemblies of the World churches
    - 8 million a Lutheran church
      - 5 million an Evangelical Lutheran Church in America
      - 3 million one of the Lutheran Church-Missouri Synod churches
    - 4 million a Presbyterian Church U.S.A.
    - 2 million a Reformed church
      - 1 million a United Church of Christ
      - 1 million another Reformed Church
    - 2 million an Episcopal church
    - 3 million one of the Churches of Christ
    - 1 million a Christian Church (Disciples of Christ)
    - 1 million a Seventh-day Adventist
  - 10 million other Christians
    - 6 million a Church of Jesus Christ of Latter-day Saints
    - 1 million a Jehovah's Witness church
    - 3 million other Christians

THE RELIGIOUS LANDSCAPE

6.6 Holy Places in Universalizing Religions

- Universalizing religions have holy places associated with the founder's life.
- Structures play distinctive roles in each of the universalizing religions.

Religions elevate particular places to a holy position. A universalizing religion endows with holiness cities and sacred structures associated with the founder's life. Its holy places are not related to any particular feature of the physical environment.

CHRISTIAN CHURCHES

The word *church* derives from a Greek term meaning *lord, master,* and *power*. The church plays a more critical role in Christianity than do buildings in other religions, because the structure is an expression of religious principles, an environment in the image of God. In many communities, the church is the largest and tallest building and has been placed at a prominent location.

Early churches were rectangular-shaped, modeled after Roman buildings for public assembly, known as *basilicas*. A raised altar, where the priest conducted the service, symbolized the hill of Calvary, where Jesus was crucified.

Since Christianity split into many denominations, no single style of church construction has dominated. Eastern Orthodox churches follow an ornate architectural style that developed in the Byzantine Empire during the fifth century. Many Protestant churches in North America are austere, with little ornamentation, a reflection of the Protestant conception of a church as an assembly hall for the congregation.

6.6.1 **HOLY PLACES IN CHRISTIANITY**
Basilica of St Boniface, Munich, Germany

6.6.2 **HOLY PLACES IN ISLAM.**
al-Haram al-Sharif Mosque, Makkah, Saudi Arabia

MUSLIM HOLY CITIES

The holiest places in Islam are in cities associated with the life of the Prophet Muhammad. The holiest city for Muslims is Makkah (Mecca), the birthplace of Muhammad. The holiest object in the Islamic landscape, al-Ka'ba, a cubelike structure encased in silk, stands at the center of the Great Mosque, al-Haram al-Sharif. Every healthy Muslim who has adequate financial resources is expected to undertake a pilgrimage, called a hajj, to Makkah. The second most holy geographic location is Madinah, where Muhammad gathered his first supporters and where he is buried.

Muslims consider the mosque as a space for community assembly, but it is not a consecreated place like a Roman Catholic or Eastern Orthodox Christian church. The mosque is organized around a central courtyard. The pulpit is placed at the end of the courtyard facing Makkah, the direction toward which all Muslims pray. A minaret or tower is where a man known as a muzzan calls people to prayer.

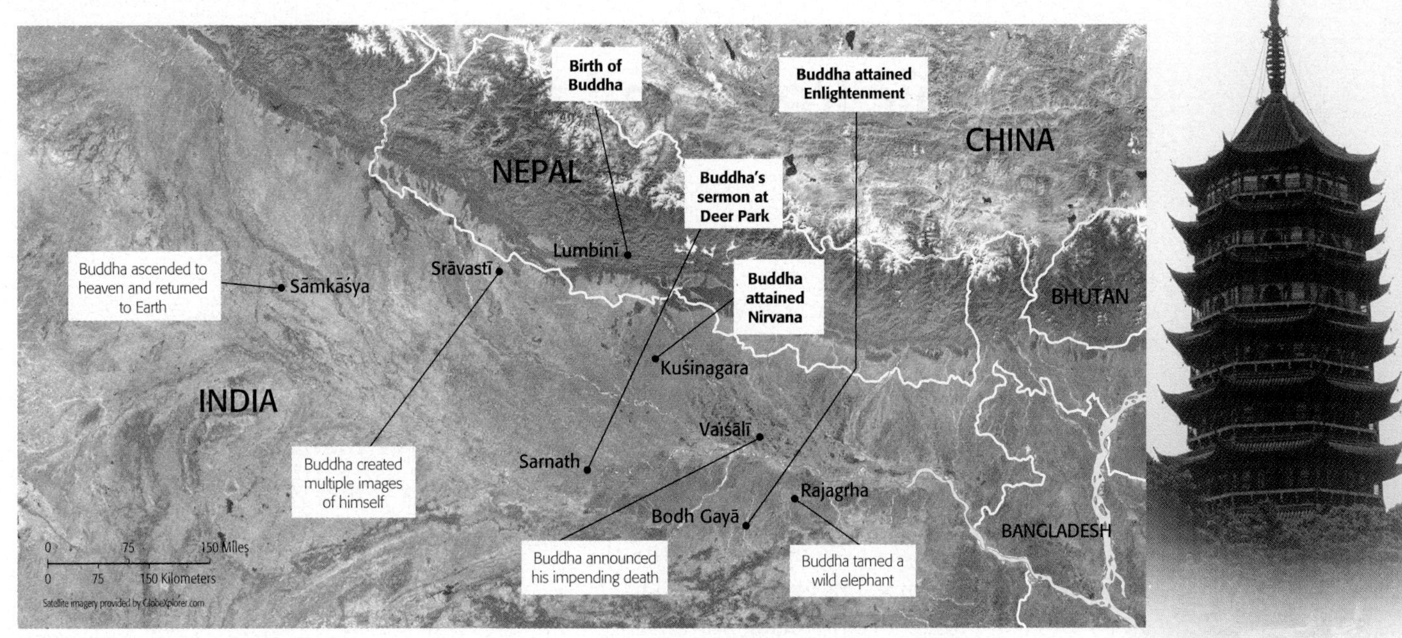

6.6.3 **HOLY PLACES IN BUDDHISM**
Eight places are holy to Buddhists because they were the locations of important events in Buddha's life. The four most important of the eight places are concentrated in a small area of northeastern India and southern Nepal.

The pagoda is a prominent and visually attractive element of the Buddhist landscape. Pagodas contain relics that Buddhists believe to be a portion of Buddha's body or clothing. Pagodas are not designed for congregational worship. Individual prayer or meditation is more likely to be undertaken at an adjacent temple, a remote monastery, or in a home.

HOLY PLACES IN BAHÁ'Í

Bahá'ís have built Houses of Worship in every continent to dramatize that Bahá'í is a universalizing religion with adherents all over the world. Sites include Wilmette, Illinois, in 1953; Sydney, Australia, and Kampala, Uganda, both in 1961; Lagenhain, near Frankfurt, Germany, in 1964; Panama City, Panama, in 1972; Tiapapata, near Apia, Samoa, in 1984; and New Delhi, India, in 1986. Additional ones are planned in Tehran, Iran; Santiago, Chile; and Haifa, Israel. The first Bahá'í House of Worship, built in 1908 in Ashgabat, Russia, now the capital of Turkmenistan, was turned into a museum by the Soviet Union and demolished in 1962 after a severe earthquake.

HOLY PLACES IN SIKHISM

Sikhism's most holy structure, the Darbar Sahib (Golden Temple) was built at Amritsar during the seventh century. Sikhs seeking autonomy from India used the Golden Temple as a base to attack the Indian army. In retaliation, the Indian army in 1984 killed a thousand Sikhs who were defending the Temple. India's Prime Minister Indira Gandhi in turn was assassinated later that year by two of her guards, who were Sikhs.

6.6.4 **HOLY PLACES IN SIKHISM**
Darbar Sahib (Golden Temple) at Amritsar, India

THE RELIGIOUS LANDSCAPE

6.7 Ethnic Religions and the Landscape

- In ethnic religions, the calendar, pilgrimages, and beliefs in the origin of the universe are grounded in the physical environment.
- Ethnic religions are tied to the physical environment of a particular place.

Ethnic religions differ from universalizing religions in their understanding of relationships between human beings and nature. A variety of events in the physical environment are more likely to be incorporated into the principles of an ethnic religion.

THE CALENDAR IN JUDAISM

Ethnic religions celebrate the seasons—the calendar's annual cycle of variation in climatic conditions. Knowledge of the calendar is critical to successful agriculture. Prayers are offered to hope for favorable environmental conditions or to give thanks for past success.

Judaism is classified as an ethnic, rather than a universalizing, religion in part because its major holidays are based on events in the agricultural calendar of the religion's homeland in present-day Israel. The name *Judaism* derives from *Judah*, one of the patriarch Jacob's 12 sons; *Israel* is another biblical name for Jacob.

Israel—the only country where Jews are in the majority—uses a lunar rather than a solar calendar. The lunar month is only about 29 days long, so a lunar year of about 350 days quickly becomes out of step with the agricultural seasons. The Jewish calendar solves the problem by adding an extra month seven out of every 19 years, so that its principal holidays are celebrated in the same season every year.

Fundamental to Judaism is belief in one all-powerful God. It was the first recorded religion to espouse **monotheism**, belief that there is only one God. Judaism offered a sharp contrast to the **polytheism** practiced by neighboring people, who worshipped a collection of gods.

6.7.1. **JEWISH HOLIDAY OF SUKKOTH**
On the holiday of Sukkoth, Jews carry branches of date palm, myrtle, and willow to symbolize gratitude for the many agricultural bounties offered by God.

COSMOGONY IN CHINESE ETHNIC RELIGIONS

Cosmogony is a set of religious beliefs concerning the origin of the universe. The cosmogony underlying Chinese ethnic religions, such as Confucianism and Daoism, is that the universe is made up of two forces, yin and yang, which exist in everything. The force of yin (earth, darkness, female, cold, depth, passivity, and death) interact with the force of yang (heaven, light, male, heat, height, activity, and life) to achieve balance and harmony. An imbalance results in disorder and chaos.

Confucianism, based on the sayings of the philosopher and teacher Confucius (551–479 B.C.) emphasizes the importance of the ancient Chinese tradition of *li*, which can be translated roughly as "propriety" or "correct behavior," such as following traditions, fulfilling obligations, and treating others with sympathy and respect.

Daoism, organized by a government administrator Lao-Zi (604 to ca 531 B.C.), emphasizes the mystical and magical aspects of life. Daoists seek *dao* (or *tao*), which means the "way" or "path." *Dao* cannot be comprehended by reason and knowledge, because everything is not ultimately subject to rational analysis, so myths and legends develop to explain events.

6.7.2 **CONFUCIUS**

SPIRITS IN INANIMATE OBJECTS

To animists, God's powers are mystical, and only a few people on Earth can harness these powers for medical or other purposes. God can be placated, however, through prayer and sacrifice. Rather than attempting to transform the environment, animists accept environmental hazards as normal and unavoidable.

Animists believe that such inanimate objects as plants and stones, or such natural events as thunderstorms and earthquakes, are "animated," or have discrete spirits and conscious life. African animist religions are said to be based on monotheistic concepts, although below the supreme god there is a hierarchy of divinities. These divinities may be assistants to the supreme god or personifications of natural phenomena, such as trees or rivers.

As recently as 1980, some 200 million Africans—half the population of the region at the time—were classified as animists. Some atlases and textbooks persist in classifying Africa as predominantly animist, even though the actual percentage is small and declining. The rapid decline in animists in Africa has been caused by diffusion of the two largest universalizing religions, Christianity and Islam.

6.7..3 **AFRICAN ANIMIST RELIGIONS** The character and form of the Odo-Kuta have evolved from the animistic origins of this once hunter-gatherer people. The circular design represents a model of the world and the individual's place in it. The masks' power to enforce this model of order is considered absolute, since mask wisdom comes from beyond the human realm and renders their authority beyond the questioning of humans.

PILGRIMAGES IN HINDUISM

Hindus consider a **pilgrimage** (known as a tirtha) to a holy place to be an act of purification. As an ethnic religion of India, holy places in Hinduism derive from the physical geography of India. According to a survey conducted by the geographer Surinder Bhardwaj, the natural features most likely to rank among the holiest shrines in India are riverbanks or coastlines.

Hindu holy places are organized into a hierarchy. Particularly sacred places attract Hindus from all over India, despite the relatively remote locations of some, whereas less important shrines attract primarily local pilgrims. Because Hinduism has no central authority, the relative importance of shrines is established by tradition, not by doctrine.

6.7.4 **HINDU PILGRIMAGE** The most common form of disposal of bodies in India is cremation. In middle-class families, bodies are more likely to be cremated in an electric oven at a crematorium. A poor person may be cremated in an open fire, such as this one within sight of the Taj Mahal. High-ranking officials and strong believers in traditional religious practices may also be cremated on an outdoor fire.

RELIGIOUS CONFLICTS

6.8 Territorial Conflicts: Ireland

- Northern Ireland and the city of Belfast are divided among Roman Catholics and Protestants.
- Conflicts between the two religious groups have flared periodically in North Ireland.

The attempt by intense adherents of one religion to organize Earth can conflict with the spatial expression of other religious or nonreligious ideas. Contributing to more intense religious conflict has been a resurgence of religious **fundamentalism**, which is a literal interpretation and a strict and intense adherence to basic principles of a religion. A group convinced that its religious view is *the* correct one may spatially intrude upon the territory controlled by others.

The most troublesome religious boundary in Western Europe lies on the island of Eire (Ireland). The Republic of Ireland, which occupies five-sixths of the island, is 92 percent Roman Catholic, but the island's northern one-sixth, which is part of the United Kingdom rather than Ireland, is about 58 percent Protestant and 42 percent Roman Catholic

The entire island was an English colony for many centuries and was made part of the United Kingdom in 1801. Agitation for independence from Britain increased in Ireland during the nineteenth century, especially after poor economic conditions and famine in the 1840s led to mass emigration, as described in Chapter 3. Following a succession of bloody confrontations, Ireland became a self-governing dominion within the British Empire in 1921. Complete independence was declared in 1937, and a republic was created in 1949.

When most of Ireland became independent, a majority in 26 northern districts voted to remain in the United Kingdom. Protestants, who comprised the majority in Northern Ireland, preferred to be part of the predominantly Protestant United Kingdom rather than join the predominantly Roman Catholic Republic of Ireland.

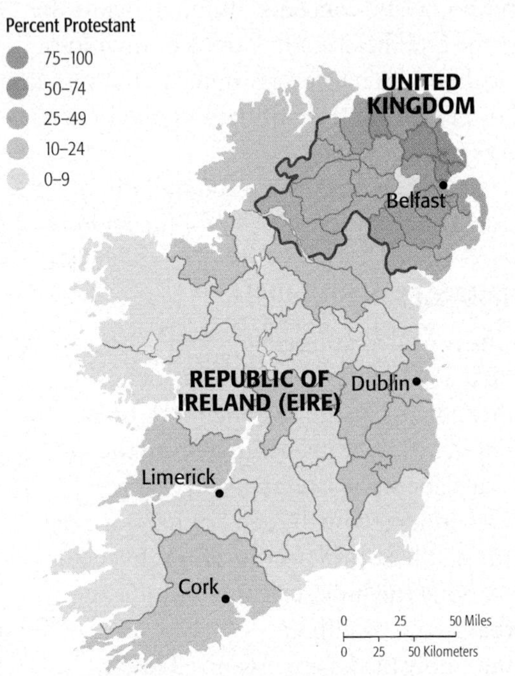

6.8.1 **DISTRIBUTION OF PROTESTANTS IN IRELAND, 1911**
Long a colony of England, Ireland became a self-governing dominion within the British Empire in 1921. In 1937, it became a completely independent country, but 26 districts in the north of Ireland chose to remain part of the United Kingdom. The Republic of Ireland today is more than 95 percent Roman Catholic, whereas Northern Ireland has a Protestant majority. The boundary between Roman Catholics and Protestants does not coincide precisely with the international border, so Northern Ireland includes some communities that are predominantly Roman Catholic. This is the root of a religious conflict that continues today.

Roman Catholics in Northern Ireland have been victimized by discriminatory practices, such as exclusion from higher-paying jobs and better schools. Demonstrations by Roman Catholics protesting discrimination began in 1968. Since then, more than 3,000 have been killed in Northern Ireland—both Protestants and Roman Catholics—in a never-ending cycle of demonstrations and protests.

A small number of Roman Catholics in both Northern Ireland and the Republic of Ireland joined the Irish Republican Army (IRA), a militant organization dedicated to achieving Irish national unity by whatever means available, including violence. Similarly, a scattering of Protestants created extremist organizations to fight the IRA, including the Ulster Defense Force (UDF).

Although the overwhelming majority of Northern Ireland's Roman Catholics and Protestants are willing to live peacefully with the other religious group, extremists disrupt daily life for everyone. As long as most Protestants are firmly committed to remaining in the United Kingdom and most Roman Catholics are equally committed to union with the Republic of Ireland, peaceful settlement appears difficult. Peace agreements implemented in 1999 provided for the sharing of power, but the British government has suspended the arrangement several times because of violations.

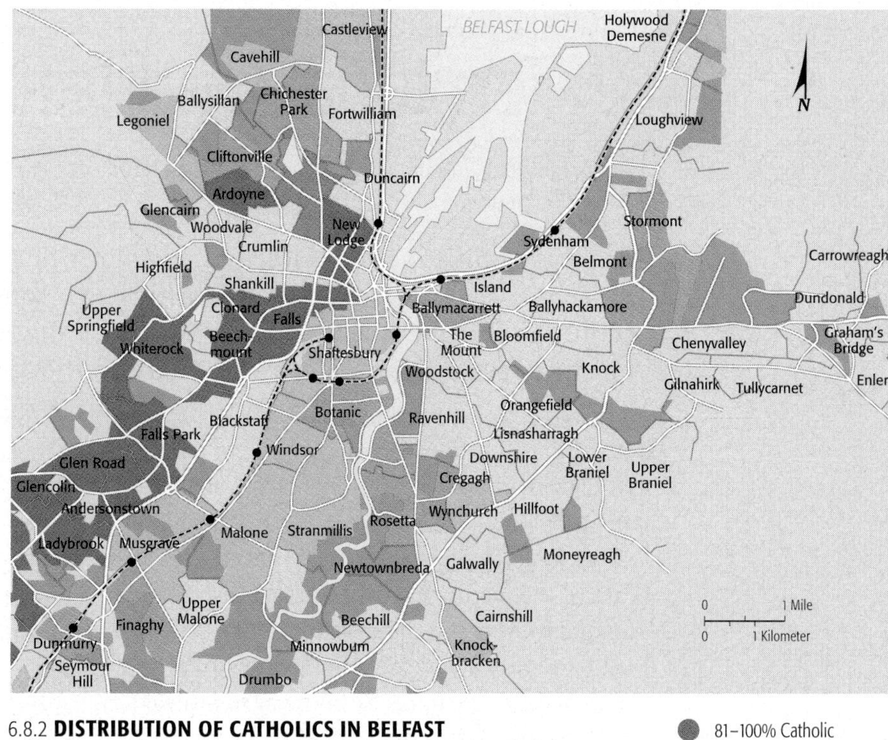

6.8.2 **DISTRIBUTION OF CATHOLICS IN BELFAST**
Belfast, the capital of Northern Ireland, contains predominantly Catholic neighborhoods, primarily west of the river Lagan, and predominantly Protestant neighborhoods to the east of the river. Troubles have been concentrated in areas along the boundaries of the two religions.

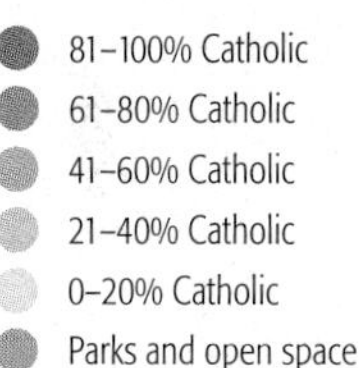

6.8.3 **NORTHERN IRELAND TROUBLES**
(opposite page) Twenty-six unarmed Roman Catholic civil rights protesters in Derry were shot by British soldiers on January 30, 1972, now known in Northern Ireland as Bloody Sunday. (Below) Protestants known as Apprentice Boys march through Ballymena, Northern Ireland.

RELIGIOUS CONFLICTS

6.9 Territorial Conflicts: The Middle East

- Jews, Muslims, and Christians have fought to control Israel/Palestine.
- Places holy to all three religions are clustered in Jerusalem.

Religious conflict in the Middle East is among the world's longest standing and most intractable. Jews, Christians, and Muslims have fought on and off for 2,000 years to control the same small strip of land, which the Romans called Palestine after the Philistines, seafaring invaders who occupied the area in the twelfth century B.C.

All three groups trace their origins to Abraham in the Bible, but the religions diverge in ways that have made it difficult for them to share the same territory.

As an ethnic religion, Judaism makes a special claim to the territory it calls the Promised Land. The major events in the development of Judaism took place there, and the religion's customs and rituals acquire meaning from the agricultural life of the ancient Hebrew tribe.

Christians consider Palestine the Holy Land and Jerusalem the Holy City because the major events in Jesus's life, death, and Resurrection were concentrated there.

Muslims regard Jerusalem as their third holy city, after Makkah and Madinah, because it is the place from which Muhammad is thought to have ascended to heaven.

Palestine was incorporated into a succession of empires, culminating with the British after World War I. In1947, the United Nations partitioned Palestine into two independent states, one Jewish-controlled state named Israel and one Arab Muslim-controlled state named Jordan. When the British withdrew in 1948, Jordan and other neighboring Arab Muslim states attacked Israel unsuccessfully. Israel won three more wars with its neighbors, in 1956, 1967, and 1973.

Especially important was the 1967 Six-Day War, when Israel captured territory, including Jerusalem, from its neighbors. Israel returned the Sinai Peninsula to Egypt in exchange for a peace treaty in 1979. The West Bank (formerly part of Jordan) and Gaza (formerly part of Egypt) have been joined to create what is now known as Palestine, with its own Arab Muslim government, but under Israeli military control.

The major obstacle to a lasting peace in the region is the status of Jerusalem. Jerusalem is especially holy to Jews as the location of the Temple, their center of

6.9.1 **BOUNDARY CHANGES IN PALESTINE/ISRAEL**

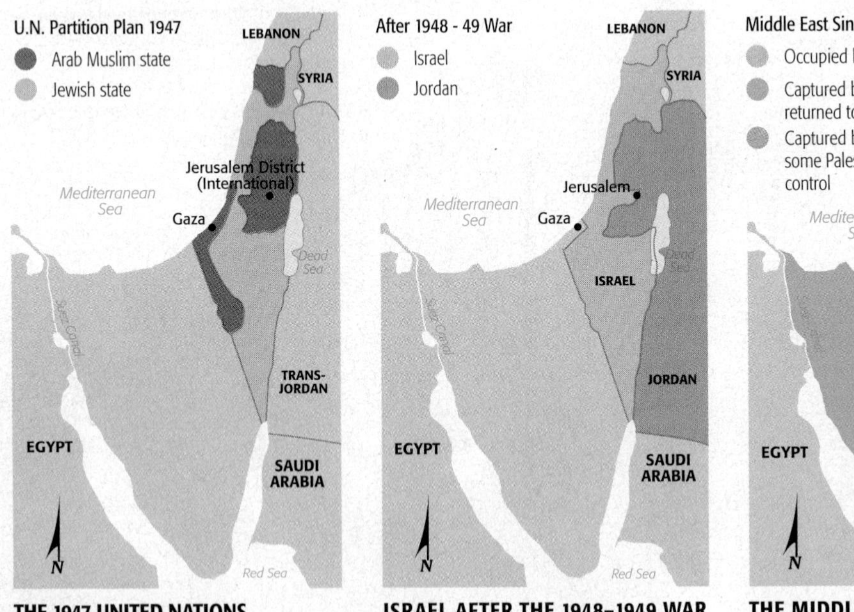

THE 1947 UNITED NATIONS PARTITION PLAN
Two countries were created, with the boundaries drawn to separate the predominantly Jewish areas from the predominantly Arab Muslim areas. Jerusalem was intended to be an international city, run by the United Nations.

ISRAEL AFTER THE 1948–1949 WAR
The day after Israel declared its independence, several neighboring states began a war, which ended in an armistice. Israel's boundaries were extended beyond the UN partition to include the western suburbs of Jerusalem. Jordan gained control of the West Bank and East Jerusalem, including the Old City, where holy places are clustered.

THE MIDDLE EAST SINCE THE 1967 WAR
Israel captured the Golan Heights from Syria, the West Bank and East Jerusalem from Jordan, and the Sinai Peninsula and Gaza Strip from Egypt. Israel returned Sinai to Egypt in 1979 and turned over Gaza and a portion of the West Bank to the Palestinians in 1994. Israel still controls the Golan Heights, most of the West Bank, and East Jerusalem.

worship in ancient times. The Second Temple was destroyed by the Romans in A.D. 70, but its Western Wall survives as a site for daily prayers by observant Jews.

The most important Muslim structure in Jerusalem is the Dome of the Rock, built in A.D. 691. Muslims believe that the large rock beneath the building's dome is the place from which Muhammad ascended to heaven, as well as the altar on which Abraham prepared to sacrifice his son Isaac. Next to the Dome of the Rock is al-Aqsa Mosque, finished in A.D. 705.

The challenge facing Jews and Muslims is that al-Aqsa was built on the site of the ruins of the Jewish Second Temple. Through a complex arrangement of ramps, Muslims have free access to the Mosque without passing in front of the Wall. But with holy Muslim structures sitting literally on top of holy Jewish structures, the two cannot be logically divided by a line on a map.

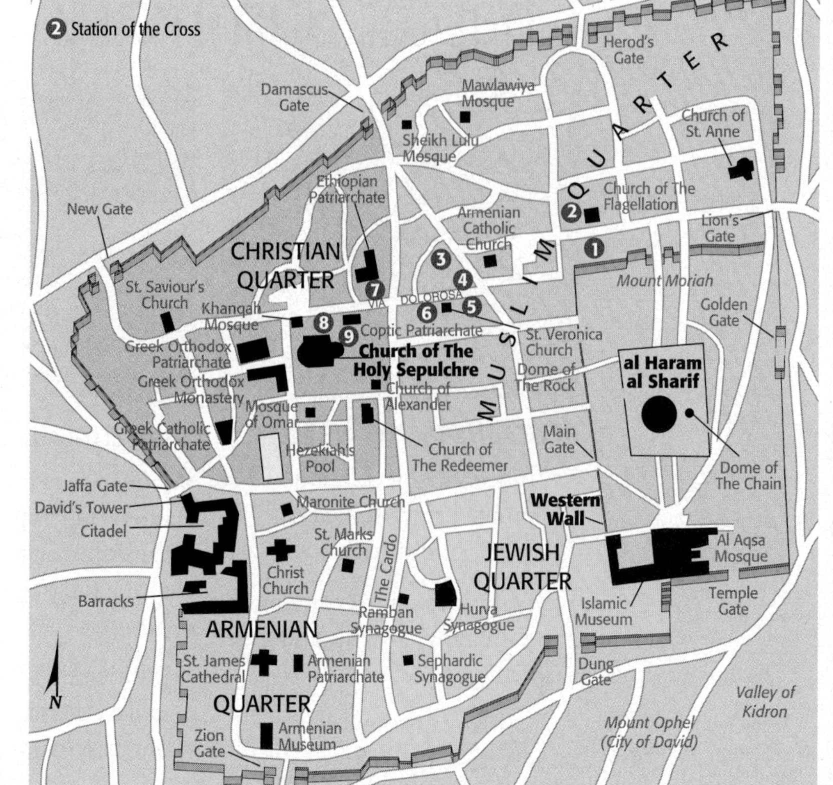

6.9.2 **JERUSALEM**. Less than 1/4 square mile, the Old City of Jerusalem contains religious structures important to Jews (the Western Wall), Muslims (Dome of the Rock and al-Aqsa Mosque), and Christians (Church of the Holy Sepulchre and Stations of the Cross). In the photo, Dome of the Rock is in the left rear, al-Aqsa Mosque right rear, and the Western Wall right foreground.

Chapter **Summary**

RELIGIONS OF THE WORLD

- A universalizing religion has a known origin and clear patterns of diffusion.
- An ethnic religion typically has unknown origins and little diffusion.
- Christianity is the universalizing religion with the most adherents in Europe and the Western Hemisphere.
- Islam and Buddhism are the universalizing religions with the most adherents in Asia.
- Hinduism is the largest ethnic religion.

THE RELIGIOUS LANDSCAPE

- Holy places and holidays in universalizing religions are typically related to events in the life of its founder or prophet.
- Ethnic religions are closely tied to the physical geography of a particular place.

RELIGIOUS CONFLICTS

- Ireland has experienced conflicts between Roman Catholics and Protestants.
- The Middle East has long been a battleground for Jews, Muslims, and Christians.
- Peaceful coexistence among religions is especially challenging in Jerusalem because important religious structures share the same space.

POTALA PALACE, LHASA, TIBET
home of the Dalai Lama before he was forced to flee after the Chinese takeover in the 1950s.

Geographic Consequences of Change

The Dalai Lama, the spiritual leader of Tibetan Buddhists, is as important to that religion as the Pope is to Roman Catholics. Traditionally, the Dalai Lama—which translates as *oceanic teacher*—was not only the spiritual leader of Tibetan Buddhism, but also the head of the government of Tibet.

When the Dalai Lama dies, Tibetan Buddhists believe that his spirit enters the body of a child. In 1937 a group of priests located and recognized a two-year-old child named Tenzin Gyatso as the 14th Dalai Lama, the incarnation of the deceased 13th Dalai Lama, Bodhisattva Avalokiteshvara. The child was brought to Lhasa in 1939 when he was four and enthroned a year later. Priests trained the young Dalai Lama to assume leadership and sent him to college when he was 16.

China, which had ruled Tibet from 1720 until its independence in 1911, invaded the rugged, isolated country in 1950, turned it into a province named Xizang in 1951, and installed a Communist government in Tibet in 1953. After taking control of Tibet, the Chinese Communists sought to reduce the domination of Buddhist monks in the country's daily life by destroying monasteries and temples. Farmers were required to join agricultural communes unsuitable for their nomadic style of raising livestock, especially yaks. After crushing a rebellion in 1959, China executed or imprisoned tens of thousands and forced another 100,000, including the Dalai Lama, to emigrate.

In recent years, the Chinese have built new roads and power plants to help raise the low standard of living in Tibet. The Chinese argue that they have brought modern conveniences to Tibet, including paved roads, hospitals, schools, and agricultural practices. Some monasteries have been rebuilt, but no new monks are being trained. Tibet has been given a small degree of autonomy to operate local government.

The Dalai Lama has become an articulate spokesperson for religious freedom, and in 1989 he was awarded the world's most prestigious award for peace, the Nobel Peace Prize. Despite the efforts of the Dalai Lama and other Buddhists, though, when the current generation of priests dies, many Buddhist traditions in Tibet may be lost forever.

The conflict between traditional Buddhism and the Chinese government is one of many examples of the importance of religion. In the modern world of global economics and culture, local religious belief continues to play a strong role in people's lives.

SHARE YOUR VOICE Student Essay

by Jeniffer Moliner
Miami Dade College, Kendall Campus

Religion (Or the Lack Thereof)

It got harder and harder to believe what they'd say at that church. But for the sake of my grandmother, I kept quiet for a while—staring off into space, counting the ribs on poor Jesus up on the cross, or the sunspots on the priest's balding old head. But ultimately my guilt and exasperation coerced my confession.

I went straight to the source of my once tenacious Catholicism: Grandma. It would have been simpler, I believe, if she had exploded in self-righteous condemnation, or attempted an exorcism, or done anything other than look at me with this pitying expression, like I was some poor lost soul. That day I had unknowingly caused a rift between us that would only grow wider and deeper as time went on.

It began with small back-handed remarks that progressed to snide comments and finally blatant ignorance. "It's the music you listen to." Art always gets attacked first the minute an individual challenges tradition. "You aren't queer, are you?" I honestly believe that she wished it were as simple as reciting a line or two of Leviticus. "Well you can't live your life without anything spiritual to turn to."

Not true. In fact, I had been living such a life for years. I just felt that there was nothing around for me to feel particularly spiritual about. I couldn't see the face of God in a city dump, I couldn't feel His presence in an empty parking lot, or be assured of His message by reading a bumper sticker on the back of some SUV. God, whose existence is so ubiquitous for some, was suddenly noticeably absent from my own life.

But, she has never understood that. In a lot of ways I envied my grandmother, who had never thought to question her faith; it was as if her belief was instinctual. I wish I could say that eventually we learned to accept each other's beliefs, but when it comes to religion, there are always eggshells around which we must carefully tread.

Chapter Resources

CONFLICTING GEOGRAPHIES OF THE HOLY LAND

The map of the Holy Land looks very different to Muslims and Jews

PALESTINIAN PERSPECTIVES

MUSLIM MILITANTS ATTACK ON JERUSALEM, FEBRUARY 2001

Palestinians emerged as Israel's principal opponent after the 1973 war. Egypt and Jordan renounced their claims to Gaza and the West Bank, respectively, and recognized the Palestinians as the legitimate rulers of these territories. Five groups of people consider themselves Palestinians:

- People living in the territories captured by Israel in 1967.
- Muslim citizens of Israel.
- People who fled from Israel after Israel was created in 1948.
- People who fled from the occupied territories after the 1967 war.
- Citizens of other countries who identify themselves as Palestinians.

Palestinians see repeated efforts by Jewish settlers to increase the territory under their control. Islam had been the principal religion of the Holy Land since the seventh century. Few Jews had lived in the area since A.D. 70, when the Romans expelled or killed most of them. When the British gained control of Palestine after World War I, it permitted Jews to buy land and move there, allegedly displacing Muslim residents.

Most Muslim-majority countries refused to recognize the 1947 UN partition of Palestine and the resulting creation of a Jewish-majority state of Israel a year later. Muslim hostility increased after the 1967 war, when Israel occupied portions of neighboring countries and then permitted some of its citizens to build settlements in the occupied territories.

Some Palestinians are willing to recognize Israel with its Jewish majority in exchange for return of all territory taken in the 1967 war. Others still do not recognize the right of Israel to exist and want to continue fighting for control of the entire territory between the Jordan River and the Mediterranean Sea.

PORTION OF BARRIER BUILT BY ISRAEL

ISRAELI PERSPECTIVES

Israel sees itself as a very small country—20,000 square kilometers (8,000 square miles)—with a Jewish majority surrounded by a region of hostile Muslim Arabs populating more than 25 million square kilometers (10 million square miles).

Repeated attacks by its neighbors and by Palestinians have led Israel to construct a barrier near its borders. The West Bank barrier is 670 kilometers (420 miles) in length. About 20 percent of it follows the Green Line, which was the boundary between Israel and Jordan between 1949 and 1967. The remaining 80 percent is between 20 meters (65 feet) and 20 kilometers (12 miles) inside the West Bank.

Most of the barrier is a 60-meter (200-foot) area with several obstacles, including barbed wire, a trench, sand, and an electronic fence. The barrier is controversial because it places on Israel's side around 10 percent of the West Bank, home to 49,400 Palestinians according to a UN estimate. The Israeli Supreme Court has twice declared portions of the route illegal because some Palestinians could not reach their fields, water sources, and places of work.

Israeli statistics show that it has drastically reduced suicide bombings and other attacks on Israeli civilians. As a result, the barrier has strong support among the Israeli public.

FURTHER READINGS

Falah, Ghazi-Walid, and Caroline Nagel, eds. *Geographies of Muslim Women: Gender, Religion, and Space.* New York: Guilford Press, 2005.

Fickeler, Paul. "Fundamental Questions in the Geography of Religions." In *Readings in Cultural Geography*, eds. Philip L. Wagner and Marvin W. Mikesell. Chicago: University of Chicago Press, 1962.

Ivakhiv, Adrian. "Toward a Geography of 'Religion': Mapping the Distribution of an Unstable Signifier." *Annals of the Association of American Geographers* 96 (2006): 169–75.

Kay, Jeanne. "Human Dominion over Nature in the Hebrew Bible." *Annals of the Association of American Geographers* 79 (1989): 214–32.

Levine, Gregory J. "On the Geography of Religion." *Transactions of the Institute of British Geographers,* New Series 11, no. 4 (1987): 428–40.

Pacione, Michael. "The Relevance of Religion for a Relevant Human Geography." *Scottish Geographical Journal* 115 (1999): 117–31.

Park, Chris C. *Sacred Worlds: An Introduction to Geography and Religion.* London and New York: Routledge, 1994.

Sopher, David E. "Geography and Religions." *Progress in Human Geography* 5 (1981): 510–24.

Stump, Roger W. *Boundaries of Faith: Geographical Perspectives on Religious Fundamentalism.* Lanham, MD: Rowman and Littlefield, 2000.

KEY TERMS

Animism
Belief that objects, such as plants and stones, or natural events, such as thunderstorms and earthquakes, have a discrete spirit and conscious life.

Branch
A large and fundamental division within a religion.

Cosmogony
A set of religious beliefs concerning the origin of the universe.

Denomination
A division of a branch that unites a number of local congregations in a single legal and administrative body.

Ethnic religion
A religion with a relatively concentrated spatial distribution whose principles are likely to be based on the physical characteristics of the particular location in which its adherents are concentrated.

Fundamentalism
Literal interpretation and strict adherence to basic principles of a religion (or a religious branch, denomination, or sect).

Missionary
An individual who helps to diffuse a universalizing religion.

Monotheism
The doctrine or belief of the existence of only one god.

Pilgrimage
A journey to a place considered sacred for religious purposes.

Polytheism
Belief in or worship of more than one god.

Sect
A relatively small group that has broken away from an established denomination.

Universalizing religion
A religion that attempts to appeal to all people, not just those living in a particular location.

ON THE INTERNET

Statistics on the number of adherents to religions, branches, and denominations can be controversial. An authoritative nondenominational source of statistical information is maintained on the Web at **www.adherents.com**. Statistics are provided by religion and by location. The site also notes when different sources sharply disagree about the numbers.

Glenmary Research Center is the principal source of information about adherents within the United States. The center, which is affiliated with the Roman Catholic Church, provides maps of the largest branch or denomination by county at **www.glenmary.org.**

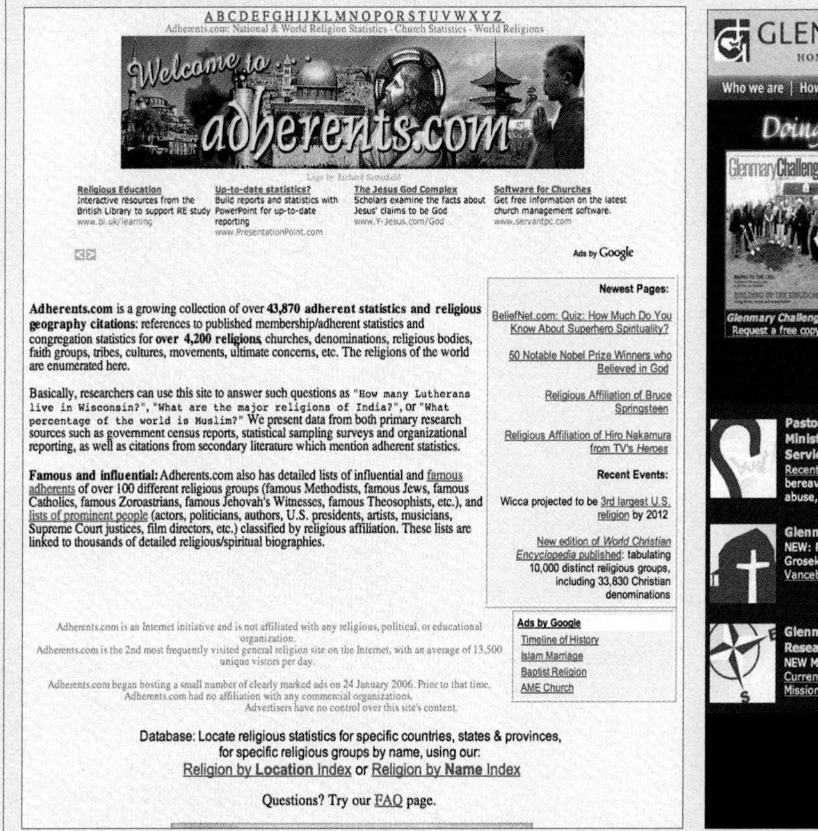

Chapter 7 **ETHNICITY**

People are members of groups with which they share important attributes. If you are a citizen of the United States of America, you are identified as an American, which is a type of nationality. Many Americans further identify themselves as belonging to an **ethnicity**, a group with which they share a cultural background. One-fourth of Americans identify their ethnicity as African American or Hispanic. Other Americans identify with ethnicities tracing back to Europe or Asia.

The significance of ethnic diversity is controversial in the United States.

- To what extent does discrimination persist against minority ethnicities, especially African Americans and Hispanics?
- Should preferences be given to minority ethnicities to correct past patterns of discrimination?
- To what extent should the distinct cultural identity of ethnicities be encouraged or protected?

Ethnicity is an especially important cultural element of local diversity because our ethnic identity is immutable. We can deny or suppress our ethnicity, but we cannot choose to change it in the same way we can choose to speak a different language or practice a different religion. If our parents come from two ethnic groups or our grandparents from four, our ethnic identity may be extremely diluted, but it never completely disappears.

CHINATOWN CAFE, BOSTON

THE KEY ISSUES IN THIS CHAPTER

Collection of the New York Historical Society #46093

U.S. ETHNICITIES

7.1 U.S. Ethnic Distribution
7.2 African American Migration
7.3 Race Differs from Ethnicity

ETHNICITIES AND NATIONALITIES

7.4 Transforming Ethnicities into Nationalities
7.5 Nation-States and Multinational States
7.6 Combining and Dividing Ethnicities

ETHNIC CONFLICTS

7.7 Ethic Diversity in Southwest Asia
7.8 Ethnic Cleansing
7.9 Ethnic Cleansing in the Balkans

U.S. ETHNICITIES

7.1 U.S. Ethnic Distribution

- The two most numerous U.S. ethnicities are Hispanics and African Americans.
- Ethnic groups cluster in different regions of the United States.

Ethnicity is identity with a group of people who share the cultural traditions of a particular homeland or hearth. Ethnicity comes from the Greek word *ethnikos*, which means "national." Ethnicity is distinct from **race**, which is identity with a group of people who share a biological ancestor. Race comes from a middle-French word for "generation".

The two most numerous ethnicities in the United States are Hispanics (or Latinos), at 15 percent of the total population, and African Americans at 13 percent. In addition, about 5 percent are Asian American and 1 percent American Indian.

7.1.1 **DISTRIBUTION OF HISPANICS**

Percent Hispanic American
- 60 and above
- 30–59
- 10–29
- below 10

0 400 Miles
0 400 Kilometers

7.1.2 **DISTRIBUTION OF AFRICAN AMERICANS**

Percent African American
- 60 and above
- 30–59
- 10–29
- below 10

0 400 Miles
0 400 Kilometers

7.1.3 **DISTRIBUTION OF ASIAN AMERICANS**

Percent Asian American (Including Pacific Islanders)
- 60 and above
- 30–59
- 10–29
- below 10

0 400 Miles
0 400 Kilometers

7.1.4 **DISTRIBUTION OF AMERICAN INDIANS**

Percent American Indian, Eskimo and Aleut
- 60 and above
- 30–59
- 10–29
- below 10

0 400 Miles
0 400 Kilometers

REGIONAL DISTRIBUTION OF ETHNICITIES

Ethnicities have distinctive regional distributions within the United States.

- **Hispanics**
  Hispanics comprise the fastest growing ethnicity in the United States. Mexico is the source of about 64 percent of Hispanics, Puerto Rico 9 percent, and Cuba 4 percent. California is home to 30 percent of all Hispanics, Texas 20 percent, and Florida and New York 15 percent each. New Mexico's population is 44 percent Hispanic.
- **African Americans**
  African Americans are highly clustered in the South, including more than one-third of the population of Mississippi and more than one-fourth of Alabama, Georgia, Louisiana, Maryland, and South Carolina. At the other end of the scale, nine states in New England and the Plains have fewer than 1 percent African Americans.
- **Asian Americans**
  Asian Americans are more than 40 percent of the population of Hawaii. One-half of all Asian Americans live in California, where they comprise 12 percent of the population. Chinese account for 23 percent of Asian Americans, persons from India 19 percent, Filipinos 18 percent, Korean and Vietnamese 10 percent each, Japanese 7 percent, and others 13 percent.
- **American Indians**
  American Indians also includes Native Alaskans. The ethnicity comprises 13 percent of the population of Alaska, 10 percent of New Mexico, and 9 percent of South Dakota.

ETHNICITIES IN CITIES

Ethnicities are highly clustered within cities. In the nineteenth century, immigrants from the various European countries settled their own neighborhoods in U.S. cities. A visible remnant of European ethnic neighborhoods is the clustering of restaurants in such areas as Little Italy and Greektown. The children and grandchildren of European immigrants moved out of most of the original inner-city neighborhoods during the twentieth century. For descendants of European immigrants, ethnic identity is more likely to be retained through religion, food, and other cultural traditions rather than through location of residence. Ethnic clustering in cities is now most likely by the three largest ethnicities identified on the previous page..

- In Chicago, African Americans occupy extensive areas on the South and West sides and Hispanics on the Northwest and Southwest sides.
- In Los Angeles, African Americans cluster to the south, Hispanics to the east, and Asians to the north and outer east.

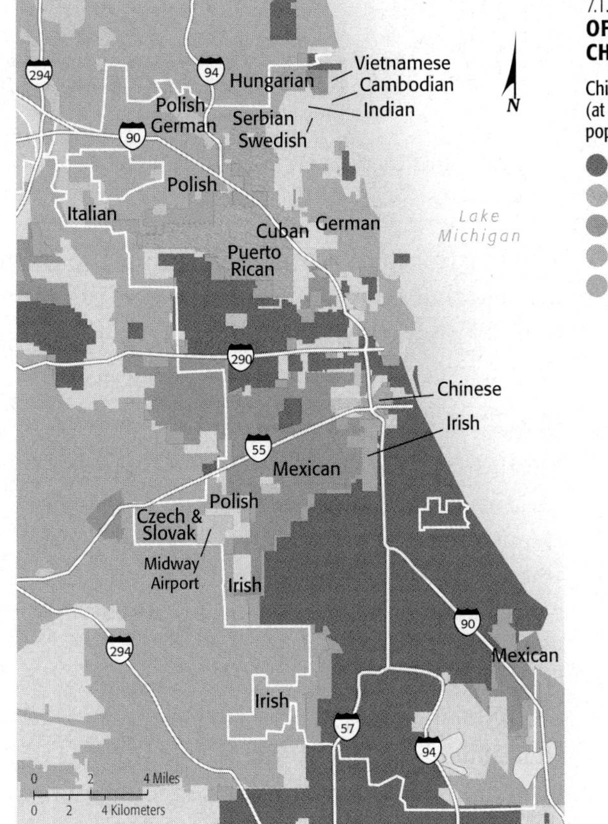

7.1.5 **DISTRIBUTION OF ETHNICITIES IN CHICAGO**

Chicago Ethnicities (at least 60% of the population)

- African American
- Asian American
- Hispanic/Latino(a)
- White
- No majority group

7.1.6 **DISTRIBUTION OF ETHNICITIES IN LOS ANGELES**

Los Angeles Ethnicities (at least 60% of the poulation)

- African American
- Asian American
- Hispanic/Latino(a)
- White
- No majority group

HISPANIC OR LATINO(a)?

- *Hispanic or Hispanic American* is a term that the U.S. government chose in 1973 to describe the group because it was an inoffensive label that could be applied to all people from Spanish-speaking countries. Some Americans of Latin American descent have instead adopted the terms *Latino* (males) and *Latina* (females). A 1995 U.S. Census Bureau survey found that 58 percent of Americans of Latin American descent preferred the term *Hispanic* and 12 percent *Latino/Latina.*
- Mexicans are sometimes called *Chicanos* (males) or *Chicanas* (females). Originally, the term was considered insulting, but in the 1960s Mexican American youths in Los Angeles began to call themselves Chicanos and Chicanas with pride.

U.S. ETHNICITIES

7.2 African American Migration

- African Americans display distinctive immigration patterns.
- The legacy of slavery is a major factor in the distribution of African Americans.

The clustering of ethnicities within the United States is partly a function of the same process that helps geographers to explain the distribution of other cultural factors, namely migration. The migration patterns of African Americans have been especially distinctive.

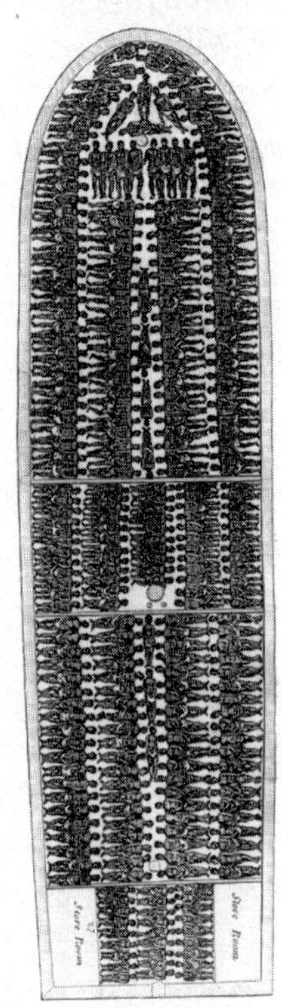

7.2.1 **SLAVE SHIP**
The diagram shows the extremely high density by which Africans were transported to the Americas to be sold as slaves. The image shows human figures packed into the hold of the ship lying next to each other with no room to move. Collection of the New York Historical Society, #46093.

FORCED MIGRATION FROM AFRICA

Most African Americans are descended from Africans forced to migrate to the Western Hemisphere as slaves. Slavery is a system whereby one person owns another person as a piece of property and can force that slave to work for the owner's benefit.

The slave trade was a response to a shortage of labor in the sparsely inhabited Western Hemisphere. Europeans who owned large plantations in the Americas turned to African slaves as a cheap and abundant source of labor. At the height of the slave trade between 1710 and 1810, at least 10 million Africans were uprooted from their homes and sent on European ships to the Western Hemisphere for sale in the slave market. Fewer than 5 percent of the slaves ended up in the United States.

A number of European countries adopted the **triangular slave trade,** an efficient triangular trading pattern:

- Ships left Europe for Africa with cloth and other trade goods used to buy the slaves.
- Slaves and gold went from Africa to the Western Hemisphere.
- To complete the triangle, the same ships carried sugar and molasses from the Caribbean back to Europe.
- Some ships added another step, carrying molasses from the Caribbean to the North American colonies and rum from there to Europe.

The forced migration began when people along the east and west coasts of Africa, taking advantage of their superior weapons, captured members of groups living farther inland and sold the captives to Europeans. Europeans in turn shipped the captured Africans to the Americas, selling them as slaves either on consignment or through auctions.

The large-scale forced migration of Africans obviously caused them unimaginable hardship, separating families and destroying villages. Traders generally seized the stronger and younger villagers, who could be sold as slaves for the highest price. The Africans were packed onto ships at extremely high density, kept in chains, and provided with minimal food and sanitary facilities. Approximately one-fourth died crossing the Atlantic.

During the eighteenth century, the British shipped about 400,000 Africans to the 13 colonies that later formed the United States. Most were destined for large plantations in the South, primarily those growing cotton and tobacco. In 1808 the United States banned bringing in additional Africans as slaves, but an estimated 250,000 were illegally imported during the next half-century.

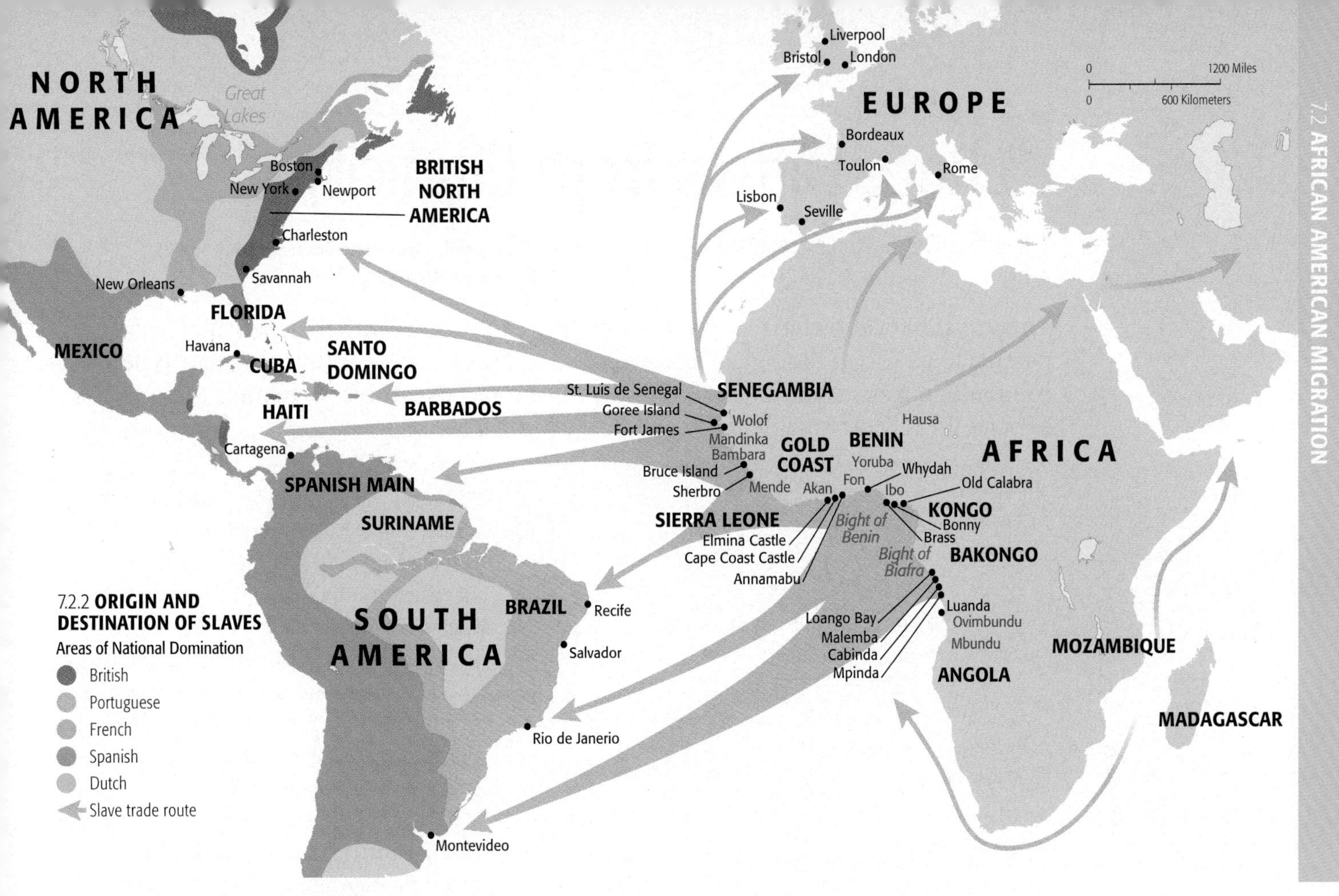

7.2.2 **ORIGIN AND DESTINATION OF SLAVES**

EMIGRATION FROM THE SOUTH

Freed as slaves during the Civil War (1861–1865), most African Americans remained in the rural South during the late nineteenth century working as sharecroppers. A **sharecropper** worked fields rented from a landowner and paid the rent by turning over to the landowner a share of the crops. As sharecroppers, African Americans were burdened with heavy debts and were forced to grow cash crops for the landlord rather than food for themselves to eat.

The introduction of farm machinery and decline in land devoted to cotton reduced demand for labor in the South. At the same time sharecroppers were being pulled by the prospect of jobs in the booming industrial cities of the North and West, especially during the time of the two World Wars. As a result, large numbers of southern African Americans moved to the north and west. The first big wave came in the 1910s and 1920s and the second in the 1940s and 1950s.

African Americans migrated out of the South along several clearly defined channels:

- From the Carolinas and other South Atlantic states north to Baltimore, Philadelphia, New York, and other northeastern cities, along U.S. Route 1 (parallel to present-day I-95)
- From Alabama and eastern Tennessee north to either Detroit, along U.S. Route 25 (present-day I-75), or Cleveland, along U.S. Route 21 (present-day I-77)
- From Mississippi and western Tennessee north to St. Louis and Chicago, along U.S. routes 61 and 66 (present-day I-55)
- From Texas west to California, along U.S. routes 80 and 90 (present-day I-10 and I-20)

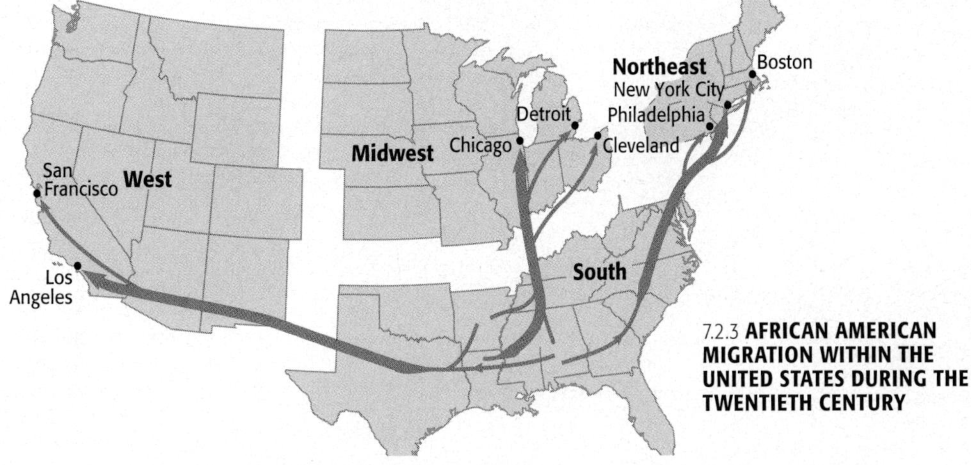

7.2.3 **AFRICAN AMERICAN MIGRATION WITHIN THE UNITED STATES DURING THE TWENTIETH CENTURY**

U.S. ETHNICITIES

7.3 Race Differs from Ethnicity

- Ethnicity and race are often confused.
- In the past, races have been spatially segregated by legal discrimination.

The traits that characterize race are those that can be transmitted genetically from parents to children. Biological features of all humans, such as skin color, hair type and color, blood traits, and shape of body, head, and facial features, were once thought to be scientifically classifiable into a handful of world races.

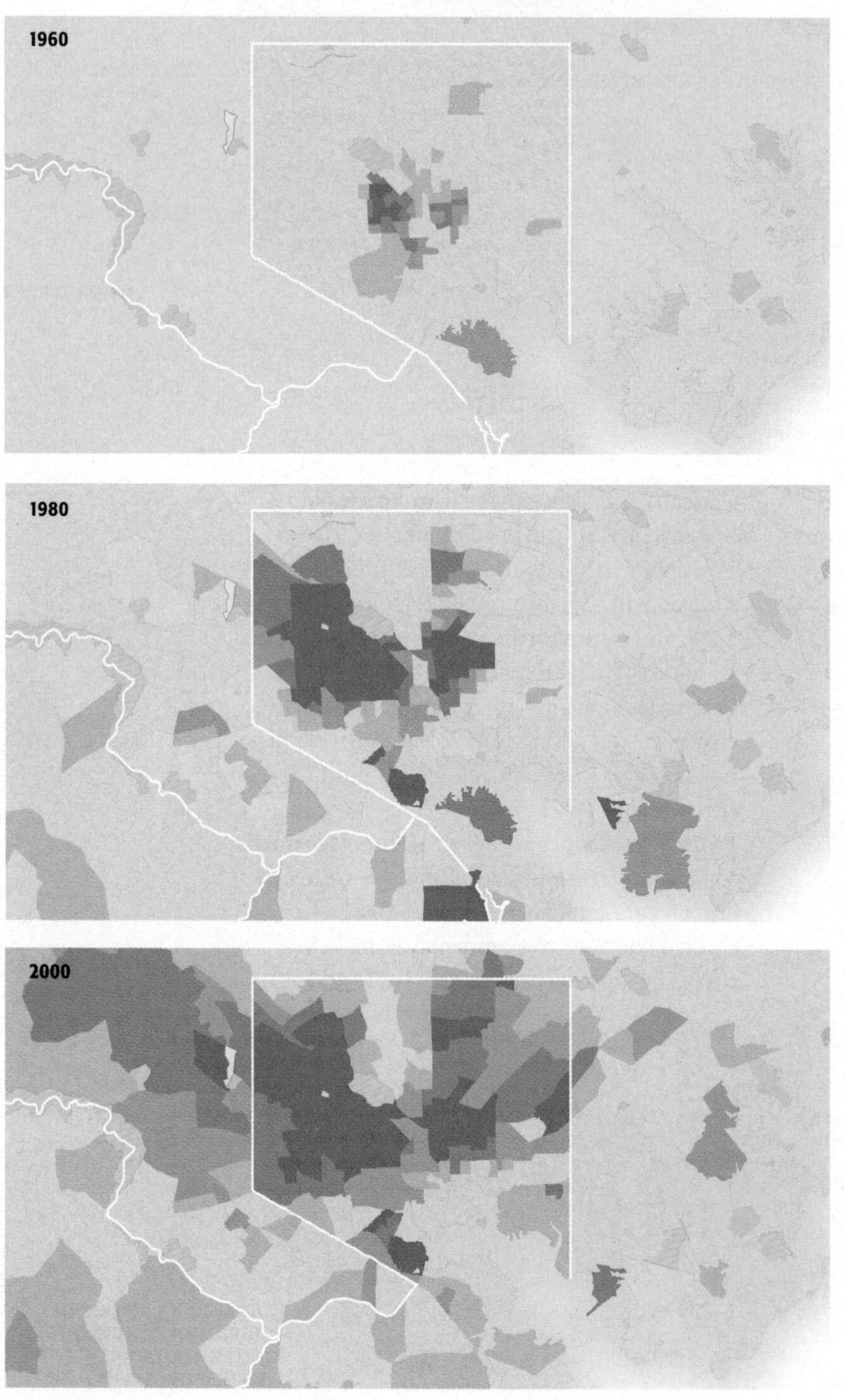

At best, however, biological features are so highly variable among members of a race that any prejudged classification is meaningless. At worst, biological classification by race is the basis for **racism**, which is the belief that race is the primary determinant of human traits and capacities and that racial differences produce an inherent superiority of a particular race. A **racist** is a person who subscribes to the beliefs of racism.

Race and ethnicity are often confused. In the United States, consider three prominent groups:

- Asian is recognized as a distinct race by the U.S. Bureau of the Census, so Asian American as a race and Asian American as an ethnicity encompass basically the same group of people.
- African American and black are different groups, although the 2000 census combined the two. Most black Americans are descended from African immigrants and therefore also belong to an African American ethnicity. Some American blacks, however, trace their cultural heritage to Latin America, Asia, or Pacific islands.
- Hispanic or Latino is not considered a race, so members of the Hispanic or Latino ethnicity select any race they wish on the census.

7.3.1 **EXPANSION OF AFRICAN AMERICAN POPULATION IN BALTIMORE, 1960 (top), 1980 (middle), 2000 (bottom)**

Percent Baltimore African American
- 90 and above
- 60–89
- 30–59
- 10–29
- below 10

"SEPARATE BUT EQUAL" DOCTRINE

A distinctive feature of race relations in the United States has been the strong discouragement of spatial interaction—in the past through legal means, today through cultural preferences or discrimination.

The U.S. Supreme Court in 1896 upheld a Louisiana law that required black and white passengers to ride in separate railway cars. In *Plessy v. Ferguson*, the Supreme Court stated that Louisiana's law was constitutional because it provided separate, *but equal*, treatment of blacks and whites, and equality did not mean that whites had to mix socially with blacks.

Once the Supreme Court permitted "separate but equal" treatment of the races, southern states enacted a comprehensive set of laws to segregate blacks from whites as much as possible. These were called "Jim Crow" laws, named for a nineteenth-century song-and-dance act that depicted blacks offensively. Blacks had to sit in the back of buses, and shops, restaurants, and hotels could choose to serve only whites. Separate schools were established for blacks and whites. Restrictive covenants kept blacks—as well as Catholics and Jews—from moving into some neighborhoods.

Segregation laws were eliminated during the 1950s and 1960s. The landmark Supreme Court decision *Brown v. Board of Education of Topeka, Kansas* in 1954 found that having separate schools for blacks and whites was unconstitutional, because no matter how equivalent the facilities, racial separation branded minority children as inferior and therefore was inherently unequal. A year later the Supreme Court further ruled that schools had to be desegregated "with all deliberate speed."

Rather than integrate, whites fled. The expansion of the black ghettos in American cities was made possible by "white flight," the emigration of whites from an area in anticipation of blacks immigrating into the area.

RACE IN THE U.S. CENSUS

Every 10 years the U.S. Bureau of the Census asks people to classify themselves according to the race with which they most closely identify. Americans were asked in 2000 to identify themselves by checking the box next to one of the following 14 races:

- White
- Black, African American, or Negro
- American Indian or Alaska Native
- Asian Indian
- Chinese
- Filipino
- Japanese
- Korean
- Vietnamese
- Other Asian
- Native Hawaiian
- Guamanian or Chamorro
- Samoan
- Other Pacific Islander
- Other race

If American Indian, Other Pacific Islander, Other Asian, or Other race were selected, the respondent was asked to write in the specific name.

In 2000 about 75 percent of Americans checked that they were white, 12 percent black, 4 percent Asian (Asian Indian, Chinese, Filipino, Japanese, Korean, or Vietnamese), 1 percent American Indian or Alaska Native, 0.1 percent Native Hawaiian or other Pacific Islander (including Guamanian and Samoan), and 6 percent some other race. The 2000 census permitted people to check more than one box, and 2 percent of the respondents did that.

7.3.2 **SEGREGATION IN THE UNITED STATES**
Until the 1960s in the U.S. South, whites and blacks had to use separate drinking fountains, as well as separate restrooms, bus seats, hotel rooms, and other public facilities.

ETHNICITIES AND NATIONALITIES

7.4 Transforming Ethnicities into Nationalities

- Nationalities identify with a particular country.
- Ethnicities have been transformed into nationalities because of desire for self-rule

Ethnicity and race are distinct from nationality, another term commonly used to describe a group of people with shared traits. **Nationality** is identity with a group of people who share legal attachment and personal allegiance to a particular country. It comes from the Latin word *nasci*, which means "to have been born."

Nationality and ethnicity are similar concepts in that membership in both is defined through shared cultural values. In principle, the cultural values shared with others of the same ethnicity derive from religion, language, and material culture, whereas those shared with others of the same nationality derive from voting, obtaining a passport, and performing civic duties. A nationality, once established, must hold the loyalty of its citizens to survive. Politicians and governments try to instill loyalty through **nationalism**, which is loyalty and devotion to a nationality.

Ethnic groups have been transformed into nationalities because desire for self-rule is a very important shared attitude for many of them. To preserve and enhance distinctive cultural characteristics, ethnicities seek to govern themselves without interference. The concept that ethnicities have the right to govern themselves is known as **self-determination**.

During the past 200 years, political leaders have generally supported the right of self-determination for many ethnicities and have attempted to organize Earth's surface into a collection of nation-states. A **nation-state** is a state whose territory corresponds to that occupied by a particular ethnicity that has been transformed into a nationality.

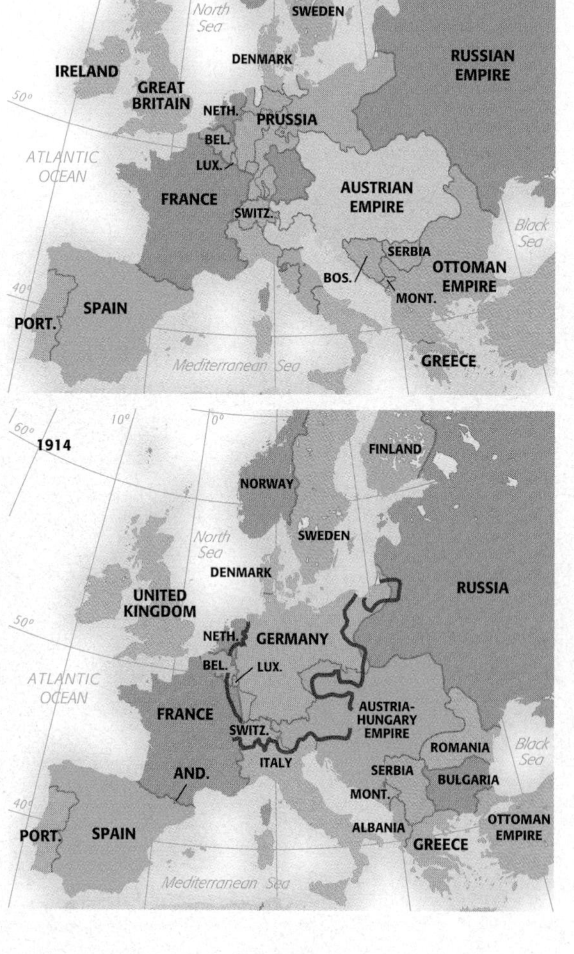

7.4.1 **EUROPE 1848**
Western Europe consisted of nation-states, but Central Europe was a collection of small principalities, and Eastern Europe was divided among several empires.

7.4.2 **EUROPE 1914**
By the outbreak of World War I, Germany and Italy had unified into nation-states.

— German-speaking territory in 1914

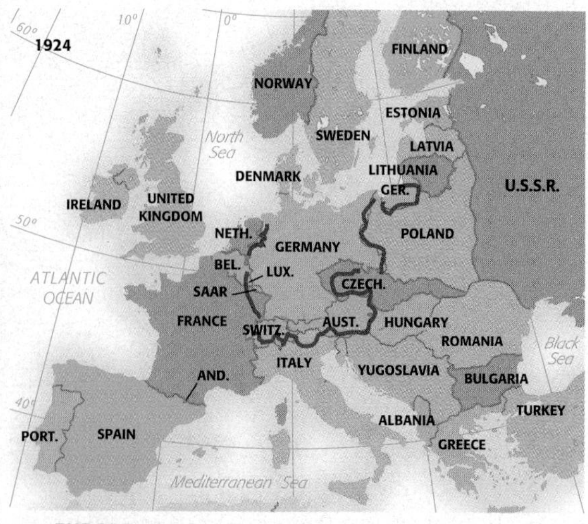

7.4.3 **EUROPE 1924**
After World War I, the Eastern Europe empires were broken up into a collection of nation-states and multinational states.

NATION-STATES IN EUROPE

Most European nationalities lived under the rule of an empire in the early nineteenth century. A century later, at the outbreak of World War I, ethnicities had been transformed into nationalities in much of Western Europe. The states of France, Germany, Italy, Portugal, Spain, and the United Kingdom coincided fairly closely to the territory inhabited by the speakers of French, German, Italian, Portuguese, Spanish, and English, respectively.

After World War I, nation-states were also carved out of Eastern Europe, also in principle according to language boundaries. After losing World War II, the boundaries of Germany bore little relationship to the German-speaking territory of a century earlier; the western portion of the German-speaking region was divided into two states, and the eastern portion was transferred to Poland.

With the collapse of communism in Eastern Europe in the late twentieth century, Czechoslovakia, the Soviet Union, and Yugoslavia broke into numerous nation-states. Yet despite continuing attempts to create nation-states in Europe, the territories of states rarely corresponded precisely to the territories occupied by ethnicities—a recipe for continued conflict.

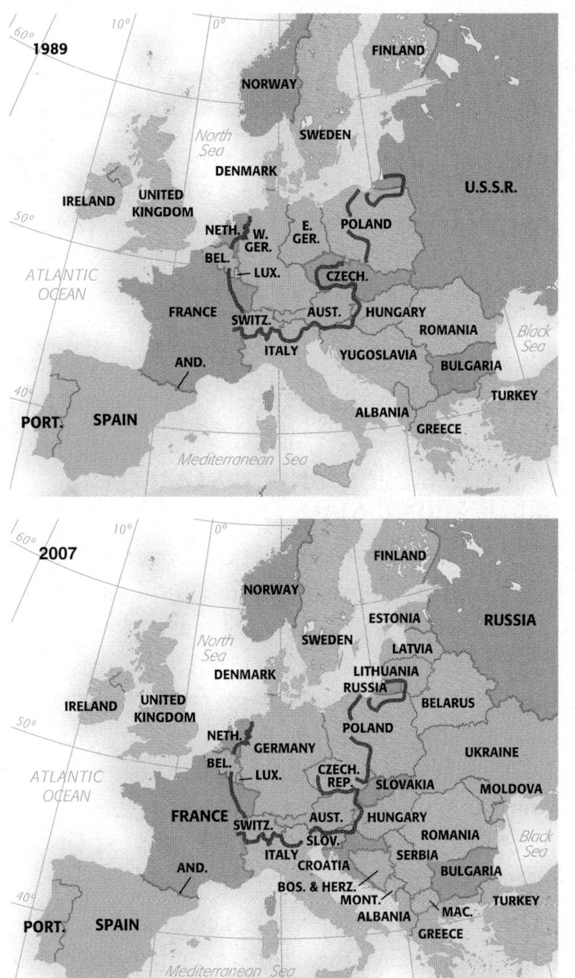

7.4.5 **EUROPE 1989**
After World War II, Germany was divided into two states, neither of which matched the traditional area of German-speaking people.

— German-speaking territory in 1914

7.4.6 **EUROPE 2008**
Multinational states in Eastern Europe were broken up into smaller nation-states.

7.4.4 **NAZI YOUTH RALLY, 1930's**
The Nazi Party claimed that all German-speaking parts of Europe constituted one nationality and should be unified into one state.

ETHNICITIES AND NATIONALITIES

7.5 Nation-States and Multinational States

- The Soviet Union was once the largest multinational state, but after its breakup, Russia became the largest.
- A multinational state has more than one ethnicity recognized as distinct nationalities.

A **multinational state** contains two or more ethnic groups with traditions of self-determination that agree to coexist peacefully by recognizing each other as distinct nationalities.

THE SOVIET UNION: ONCE THE LARGEST MULTINATIONAL STATE

The Soviet Union was an especially prominent example of a multinational state until its collapse in 1991. The Soviet Union's 15 republics were based on its 15 largest ethnicities. Less numerous ethnicities were not given the same level of recognition.

The 15 largest ethnicities of the former Soviet Union are now independent countries that represent varying degrees of nation-states. The 15 other than Russia are:

- **Three Baltic states: Estonia, Latvia, and Lithuania.** These three small neighbors have differences in language and religion and distinct historical traditions. They had been independent until annexed to the Soviet Union in 1940.
- **Three European states: Belarus, Moldova, Ukraine.** Belarusians and Ukrainians became distinctive ethnicities because they were isolated from the main body of Eastern Slavs—the Russians—between the thirteenth and eighteenth centuries. Moldovans are ethnically indistinguishable from Romanians.
- **Five Central Asian states: Kazakhstan, Kyrgyzstan, Tajikistan, Turkmenistan, and Uzbekistan.** The "stans" are predominantly Muslim and speak Altaic languages (except for Tajiks who speak a language similar to Persian).
- **Three Caucasus states: Armenia, Azerbaijan, and Georgia.** Armenians are Eastern Orthodox Christians who speak a separate branch of Indo-European. Azeris (or Azerbaijanis) are Muslims who speak an Altaic language. The two nation-states have clashed over drawing boundaries between them. Georgians are Eastern Orthodox Christians. Two ethnicities within Georgia, the Ossetians and Abkhazianas, are fighting for autonomy and possible reunification with Russia.

7.5.1 **COUNTRIES IN THE UNION OF SOVIET SOCIALIST REPUBLICS**

LITHUANIANS DEMONSTRATE FOR INDEPENDENCE FROM THE SOVIET UNION

7.5.2 **ETHNICITIES IN RUSSIA**

Ethnic Groups

Slavic Peoples
- Russians
- Ukrainians
- Belorussians

Turkic Peoples
- Tatars, Bashkirs, Kazakhs, Kirgiz
- Uzbeks
- Turkmen, Azerbaidzhani
- Other Turkic peoples

Caucasian Peoples
- Georgians, Chechens, Ingush, peoples of Dagestan

Paleo-Siberian Peoples
- Chukchi, Koryaks, Nivkhi
- Eskimos
- Kets
- Uninhabited or sparsely settled

Other Indo-European Peoples
- Lithuanians, Latvians, Armenians, Moldavians, Tadzhiks, Ossetians
- X Germans
- ▲ Jews

Other Uralic and Altaic Peoples
- Estonians, Karelians, Mari, Komi, Mordvins, Udmurts, Mansi, Khanty, Nentsy, Buryats, Kalmyks, Evenki, Eveny, Nganasany

RUSSIA: NOW THE LARGEST MULTINATIONAL STATE

Karl Marx wrote that nationalism was a means for the dominant social classes to maintain power over workers, and he believed that workers would identify with other working-class people instead of with an ethnicity. Ethnicity was thought to have been left behind as an insignificant relic, such as wearing quaint costumes to amuse tourists. But with the breakup of communism and the collapse of the Soviet Union, ethnic identity once again has become more important than nationality.

Russia identifies 21 national republics that are supposed to be the homes of the largest ethnicities, but the government recognizes in some way at least 170 ethnicities. Overall, 20 percent of the country's population is non-Russian. Tatars, Ukrainians, Bashkirs, Chuvashs, Chechens, and Armenians are the five largest ethnicities, with at 1 million each.

Particularly troublesome for the Russians are the Chechens, a group of Sunni Muslims who speak a Caucasian language and practice distinctive social customs. Chechnya was brought under Russian control in the nineteenth century only after a 50-year fight. When the Soviet Union broke up, the Chechens declared independence and refused to join the newly created country of Russia. After ignoring the declaration of independence for 3 years, Russian leaders sent in the army to regain control of Chechnya.

Russia fought hard to prevent Chechnya from gaining independence because it feared that other ethnicities would follow suit. Chechnya was also important to Russia because the region contained deposits of petroleum. Russia viewed political stability in the area as essential for promoting economic development and investment by foreign petroleum companies.

CHECHNYAN WOMAN PREPARED TO FIGHT FOR INDEPENDENCE IN 1995

ETHNICITIES AND NATIONALITIES

7.6 Combining and Dividing Ethnicities

- Countries were created in South Asia to separate two ethnicities.
- Conflicts have arisen because country boundaries have not matched those of ethnicities.

Ethnicities do not always find ways to live together peacefully. Newly independent countries are often created to separate two ethnicities; however, two ethnicities can rarely be segregated completely. Conflicts arise when an ethnicity is split among more than one country.

PARTITION OF SOUTH ASIA

South Asia provides a vivid example of the challenges in creating nation-states for neighboring ethnicities. When the British ended their colonial rule of the Indian subcontinent in 1947, they divided the colony into two irregularly shaped countries—India and Pakistan. Pakistan was split by India into eastern and western portions. East Pakistan became the independent state of Bangladesh in 1971.

The basis for separating West and East Pakistan from India was ethnicity. The people of Pakistan were predominantly Muslim, those in India, Hindu. Antagonism between the two religious groups was so great that the British decided to place the Hindus and Muslims in separate states. The assassination in 1948 of Mahatma Gandhi, the leading Hindu advocate of nonviolence and reconciliation with Muslims, ended the possibility of creating a single state in which Muslims and Hindus lived together peacefully.

The partition of South Asia into two states resulted in massive migration, because the two boundaries did not correspond precisely to the territory inhabited by the two ethnicities. More than 14 million people caught on the wrong side of a boundary felt compelled to migrate during the late 1940s. As they attempted to reach the other side of the new border, Hindus in Pakistan and Muslims in India were killed by people from the rival religion. Extremists attacked small groups of refugees traveling by road and halted trains to massacre the passengers.

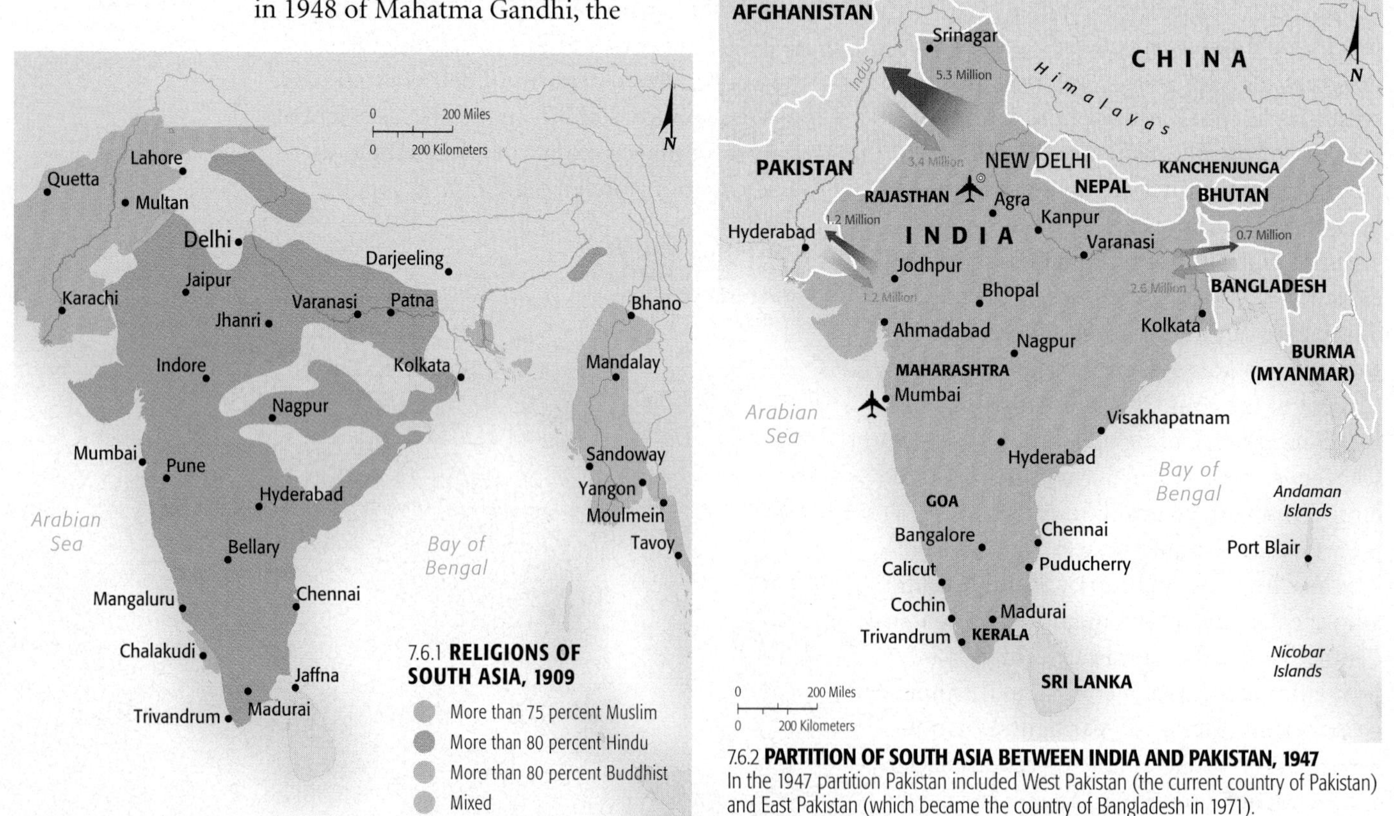

7.6.1 **RELIGIONS OF SOUTH ASIA, 1909**

7.6.2 **PARTITION OF SOUTH ASIA BETWEEN INDIA AND PAKISTAN, 1947**
In the 1947 partition Pakistan included West Pakistan (the current country of Pakistan) and East Pakistan (which became the country of Bangladesh in 1971).

KASHMIR AND PUNJAB

Pakistan and India never agreed on the location of the boundary separating the two countries in the northern region of Kashmir. Since 1972 the two countries have maintained a "line of control" through the region, with Pakistan administering the northwestern portion and India the southeastern portion. Muslims, who comprise a majority in both portions, have fought a guerrilla war to secure reunification of Kashmir, either as part of Pakistan or as an independent country. India blames Pakistan for the unrest and vows to retain its portion of Kashmir. Pakistan argues that Kashmiris on both sides of the border should choose their own future in a vote, confident that the majority Muslim population would break away from India.

India's religious unrest is further complicated by the presence of 25 million Sikhs, who have long resented that they were not given their own independent country when India was partitioned (see Chapter 6). Although they constitute only 2 percent of India's total population, Sikhs comprise a majority in the Indian state of Punjab, situated south of Kashmir along the border with Pakistan. Sikh extremists have fought for more control over the Punjab or even complete independence from India.

7.6.3 **KASHMIR AND PUNJAB: DISPUTED BETWEEN INDIA AND PAKISTAN**

- Jammu and Kashmir
- Azad Kashmir
- Northern Areas
- Siachen Glacier
- Aksai Chin
- Shaksam Valley

SRI LANKA

Sri Lanka (formerly Ceylon), an island country of 20 million inhabitants off the Indian coast, has been torn by fighting between the Sinhalese and the Tamils. Since fighting began in 1983, 60,000 have died in the conflict between the two ethnicities.

Sinhalese, who comprise 74 percent of Sri Lanka's population, migrated from northern India in the fifth century B.C., occupying the southern portion of the island. Three hundred years later, the Sinhalese were converted to Buddhism, and Sri Lanka became one of that religion's world centers. Sinhalese is an Indo-European language, in the Indo-Iranian branch. Tamils—18 percent of Sri Lanka's population—migrated across the narrow 80-kilometer-wide (50-miles-wide) Palk Strait from India beginning in the third century B.C. and occupied the northern part of the island. Tamils are Hindus, and the Tamil language, in the Dravidian family, is also spoken by 60 million people in India.

Since independence in 1948, Sinhalese have dominated the government, military, and most of the commerce. Tamils feel that they suffer from discrimination at the hands of the Sinhalese-dominated government and have received support for a rebellion that began in 1983 from Tamils living in other countries.

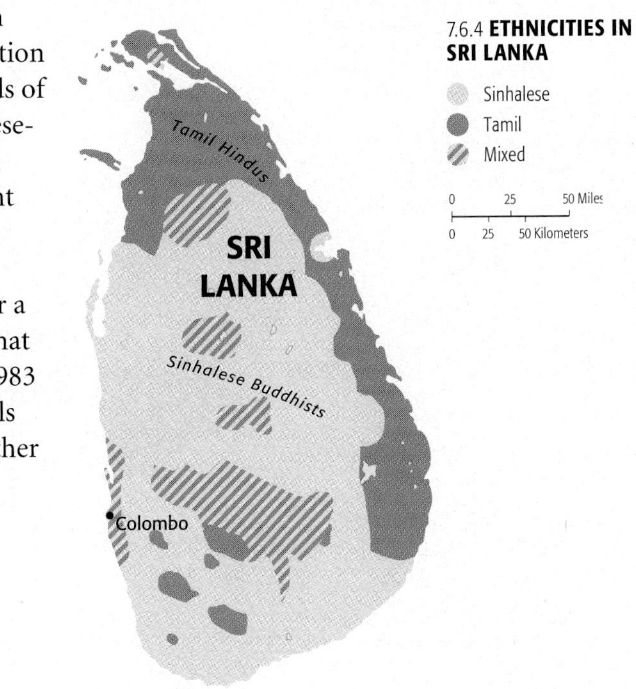

7.6.4 **ETHNICITIES IN SRI LANKA**

ETHNIC CONFLICTS

7.7 Ethnic Diversity in Southwest Asia

- Iraq contains numerous ethnicities that have fought with each other.
- Ethnic conflict in Iraq has occurred at local and regional scales.

U.S.-led wars in Southwest Asia have encountered tensions between globalization trends and realities of local diversity. The global-scale case for the United States sending troops was to fight terrorists in the region, including the leaders of the September 11, 2001, attack on the World Trade Center and Pentagon (see Chapter 8). But once in those countries, the United States was quickly entangled in long-standing local-scale ethnic rivalries that had little to do with a global war on terrorism.

AFGHANISTAN

Afghanistan's largest ethnicities are Pashtun, Tajik, and Hazara, along with significant numbers of Baluchi, Turkmen, and Uzbek. Civil war among the ethnic groups began when the king was overthrown by a military coup in 1973 and was replaced 5 years later in a bloody coup by a government sympathetic to the Soviet Union. The Soviet Union sent 115,000 troops to Afghanistan beginning in 1979 after fundamentalist Muslims, known as *mujahedeen*, or "holy warriors," started a rebellion against the pro-Soviet government. Unable to subdue the *mujahedeen*, the Soviet Union withdrew its troops in 1989, and the Soviet-installed government in Afghanistan collapsed in 1992.

After several years of infighting among the ethnicities that had defeated the Soviet Union, a faction of the Pashtun called the Taliban (which means "religious students") gained control over most of the country in 1995. The Taliban imposed very harsh, strict laws on Afghanistan, according to Islamic values as the Taliban interpreted them.

The United States invaded Afghanistan in 2001 and overthrew the Taliban-led government, because it was harboring al-Qaeda. Removal of the Taliban unleashed a new struggle for control of Afghanistan among the country's many ethnic groups. The Taliban were able to regroup and resume an insurgency against the U.S.-backed Afghanistan government.

7.7.1 **ETHNICITIES IN AFGHANISTAN, IRAN, AND IRAQ**

IRAN

Hostility between the United States and Iran dates from 1979, when a revolution forced abdication of Iran's pro-U.S. Shah Mohammad Reza Pahlavi. Supporters of exiled fundamentalist Shiite Muslim leader Ayatollah Ruholiah Khomeini proclaimed Iran an Islamic republic and rewrote the constitution to place final authority with the Ayatollah. Militant supporters of the Ayatollah seized the U.S. embassy on November 4, 1979, and held 62 Americans hostage until January 20, 1981.

Iran and Iraq fought a war between 1980 and 1988 over control of the Shatt al-Arab waterway, formed by the confluence of the Tigris and Euphrates rivers flowing into the Persian Gulf. An estimated 1.5 million died in the war, until it ended when the two countries accepted a UN peace plan.

As the United States launched its war on terrorism, Iran was a less immediate target than Afghanistan and Iraq. However, the United States accused Iran of harboring al-Qaeda members and of trying to gain influence in Iraq, where, like Iran, the majority of the people were Shiites. More troubling to the international community was evidence that Iran was developing a nuclear weapons program. Prolonged negotiations have sought to dismantle Iran's nuclear capabilities without resorting to yet another war in the region.

IRAQ

The United States led an attack against Iraq in 2003 in order to depose the country's longtime president, Saddam Hussein. U.S. officials justified removing Hussein, because he ran a brutal dictatorship, created weapons of mass destruction, and allegedly had close links with terrorists (see page 192 for further discussion).

Having invaded Iraq and removed Hussein from power, the United States expected an enthusiastic welcome from the Iraqi people. Instead, the United States became embroiled in a complex and violent struggle among religious sects and tribes.

The principal ethnic group in the north, the Kurds, welcomed the United States because they gained more security and autonomy than they had under Hussein. But Sunni Muslims, the majority in the center, opposed the U.S.-led attack because they feared loss of power and privilege given to them by Hussein, who was a Sunni.

Shiite Muslims also opposed the U.S. presence. Although they had been treated poorly by Hussein and controlled Iraq's post-Hussein government, Shiites shared a long-standing hostility toward the United States with their neighbors in Shiite-controlled Iran.

At the local scale, Iraq is divided into around 150 tribes. After Hussein was toppled, tribes stepped into the political vacuum, establishing control over their local territories.

And it gets even more complex. A tribe ('ashira) is divided into several clans (fukhdhs), which in turn encompass several houses (beit), which in turn include several extended families (kham). Tribes are grouped into more than a dozen federations (qabila). Most Iraqis have stronger loyalty to a tribe or clan than to a national government.

7.7.2 **MAJOR TRIBES IN IRAQ**

Tribes of Iraq

- Al-Azza
- Al-Bu Mohammed
- Al-Dhufair
- Al-Duriyeen
- Al-Ghalal
- Al-Hassan
- Al-Jubur
- Al-Khaza'il
- Al-Qarraghul
- Al-Soudan
- Al-Sumaida'
- Al-Suwamra
- Al-Tikarta
- Al-Ubayd
- Al-Umtayr
- Al-Zubayd
- Anniza
- Bani Hachim
- Bani Lam
- Bani Rikab
- Mi'dan Tribes
- Rubai'a
- Shammar
- Shammar Touga
- Tai'
- Zoba'

Groups

- Al-Dulaim Confederation
- Kurdish Tribes
- Al-Muntafiq Confederation
- Mixed tribal groups
- Sparsely populated

ETHNIC CONFLICTS

7.8 Ethnic Cleansing

- Ethnic cleansing is the forcible removal of an ethnic group by a more powerful one.
- Ethnic cleansing has been practiced in several African countries.

Ethnic cleansing is a process in which a more powerful ethnic group forcibly removes a less powerful one in order to create an ethnically homogeneous region.

Ethnic cleansing is undertaken to remove an entire ethnicity from an area so that the surviving ethnic group can be the sole inhabitants. Rather than a clash between armies of male soldiers, ethnic cleansing involves the removal of every member of the less powerful ethnicity—women as well as men, children as well as adults, the frail elderly as well as the strong youth.

Sub-Saharan Africa has been a region especially plagued by conflicts among ethnic groups competing to become dominant within the various countries. Conflict is widespread in Africa largely because the present-day boundaries of states were drawn by European colonial powers about a hundred years ago without regard for the traditional distribution of ethnicities. When the European colonies in Africa became independent states, especially during the 1950s and 1960s, the boundaries of the new states typically matched the colonial administrative units imposed by the Europeans. As a result, many African states contained large numbers of ethnicities.

7.8.1 **ETHNIC CONFLICTS IN AFRICA**

The map is based on a 1980 U.S. Department of State assessment of ethnic groups in Africa.

SAHRAWI ARAB DEMOCRATIC REPUBLIC

Also known as Western Sahara, it was controlled by Spain until 1976, Western Sahara was claimed by neighboring Morocco, while the Polisario Front declared it to be an independent country called the Sahrawi Republic. Both positions have support from other countries. Thousands of refugees from the territory live in camps in nearby Algeria.

DARFUR

An ethnic war erupted in Sudan's western-most Darfur region in 2003. Resenting discrimination and neglect by the national government, Darfur's black Africans rebelled. Marauding Arab nomads, known as janjaweed, with the support of the Sudanese government, crushed Darfur's black population, made up mainly of settled farmers. An estimated 450,000 have been killed and another 2.5 million have been living in dire conditions in refugee camps in the harsh desert environment of Darfur.

SUDAN

A civil war has raged since the 1980s between two ethnicities, the black Christian and animist rebels in the southern provinces and the Arab–Muslim-dominated government forces in the north. The rebels have been resisting government attempts to convert the country from a multi-ethnic society to one nationality tied to Muslim traditions. More than 2 million Sudanese—5 percent of the population—died in the civil war, and another 1 million were forced to migrate from the south to the north or to Ethiopia. An accord in 2005 called for autonomy for the southern Christians and sharing of power in the national government. Many in the south, though, pushed for full independence from Sudan.

ERITREA

Eritrea, located along the Red Sea, became an Italian colony in 1890. Ethiopia, an independent country for more than 2,000 years, was captured by Italy during the 1930s. After World War II, Ethiopia regained its independence, and the United Nations awarded Eritrea to Ethiopia. Ethiopia was expected to give Eritrea considerable autonomy, but when that did not occur, the Eritreans rebelled, beginning a 30-year fight for independence (1961–1991). Eritrean rebels defeated the Ethiopian army in 1991, and 2 years later Eritrea became an independent state. But war between Ethiopia and Eritrea flared up again in 1998 because of disputes over the location of the border.

SOMALIA

Somalia's 9 million inhabitants are divided among six major ethnic groups known as clans, each of which is divided into many sub-clans in control of various portions of the country. When 300,000 died from famine and warfare in 1992, the United States sent several thousand troops to protect food delivery. Islamist militias took control of much of Somalia in 2006, but that has failed to stop fighting among various sub-clans.

ETHIOPIA

Ethiopia is a complex **multi-ethnic state,** even with the loss of Eritrea. From the late nineteenth century until the 1990s, Ethiopia was controlled by the Amharas, who are Christians. After the government defeat in the early 1990s, power passed to a combination of ethnic groups. The Oromo, who are Muslim fundamentalists from the south, are the largest ethnicity in Ethiopia, at 40 percent of the population. Tigres live in the far north, the birthplace of the Ethiopian Orthodox Church. The Amhara had banned the languages and cultures of these groups since conquering Ethiopia in the late nineteenth century.

DEMOCRATIC REPUBLIC OF CONGO

Longtime president Joseph Mobutu had impoverished the country, which he renamed Zaire, while amassing a personal fortune from the sale of its minerals. Rebel groups tried for years to overthrow Mobutu until Tutsis from Rwanda helped them succeed. Tutsis soon split with the new president Laurent Kabila and led a rebellion that gained control of the eastern half of the Congo. Kabila was assassinated in 2001 and succeeded by his son, who negotiated an accord with rebels the following year.

RWANDA

Minority Tutsi cattle herders long dominated the majority Hutu farmers of this small country. When Rwanda gained its independence in 1962, Hutus killed or ethnically cleansed most of the Tutsis out of fear that the Tutsis would seize control. Children of the ethnically cleansed Tutsis, most of whom lived in neighboring Uganda, poured back into Rwanda, defeated the Hutu army, and killed a half-million Hutus, while suffering a half-million casualties of their own.

SOUTH AFRICA

The country's whites—14 percent of the population—created a legal system called **apartheid** that segregated races into different geographic areas. Blacks, who comprised 75 percent of the population, were restricted to certain neighborhoods and jobs. The laws were dismantled in the 1990s, and South Africa has a black president. But the legacy of apartheid lingers in economic inequality between blacks and whites.

ETHNIC CONFLICTS

7.9 Ethnic Cleansing in the Balkans

- Ethnic cleansing has occurred in the Balkans region of Southeastern Europe.
- Multi-ethnic states have been created and torn apart in the Balkans.

Ethnic cleansing has been especially prominent in the Balkan Peninsula of southeastern Europe. The region has long been a hotbed of unrest, a complex assemblage of ethnicities.

Northern portions of the Balkans had been incorporated into the Austria-Hungary Empire, whereas southern portions were ruled by the Ottomans. After World War I the allies created a new country, Yugoslavia, to unite several Balkan ethnicities that spoke similar South Slavic languages. Under the long leadership of Josip Broz Tito, who governed Yugoslavia from 1953 until his death in 1980, Yugoslavia had five recognized nationalities, four official languages, three prominent religions, and two alphabets.

Rivalries among ethnicities resurfaced in Yugoslavia during the 1980s after Tito's death, leading to its breakup into seven small countries.

BOSNIA & HERZEGOVINA

At the time of Yugoslavia's breakup, Bosnia & Herzegovina consisted of 48 percent Bosnian Muslim, 37 percent Serb, and 14 percent Croat. The Muslim population was classified as an ethnicity rather than a nationality.

Rather than live in an independent multi-ethnic country with a Muslim plurality, Bosnia & Herzegovina's Serbs and Croats fought to unite the portions of the republic that they inhabited with neighboring Serbia and Croatia, respectively. To strengthen their cases for breaking away from Bosnia & Herzegovina, Serbs and Croats engaged in ethnic cleansing of Bosnian Muslims.

Ethnic cleansing by Bosnian Serbs against Bosnian Muslims was especially severe, because much of the territory inhabited by Bosnian Serbs was separated from Serbia by areas with Bosnian Muslim majorities. By ethnically cleansing Bosnian Muslims from intervening areas, Bosnian Serbs created one continuous area of Bosnian Serb domination rather than several discontinuous ones.

Accords reached in Dayton, Ohio, in 1996 by leaders of the three ethnicities divided Bosnia & Herzegovina into three regions, one each dominated, respectively, by the Bosnian Croats, Muslims, and Serbs. In recognition of the reality of their ethnic cleansing, Bosnian Serbs and Croats received more land than their share of the population warranted.

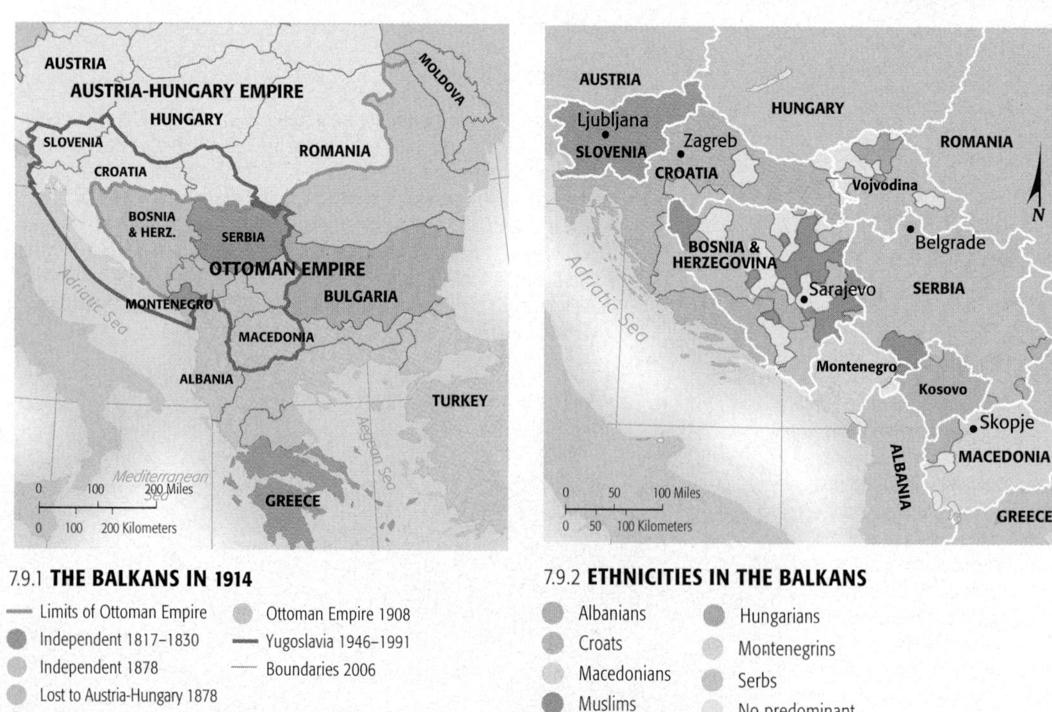

7.9.1 **THE BALKANS IN 1914**

- Limits of Ottoman Empire
- Independent 1817–1830
- Independent 1878
- Lost to Austria-Hungary 1878
- Independent 1908
- Ottoman Empire 1908
- Yugoslavia 1946–1991
- Boundaries 2006

7.9.2 **ETHNICITIES IN THE BALKANS**

- Albanians
- Croats
- Macedonians
- Muslims
- Slovenes
- Bulgarians
- Hungarians
- Montenegrins
- Serbs
- No predominant majority

7.9.3 **LANGUAGES IN THE BALKANS**

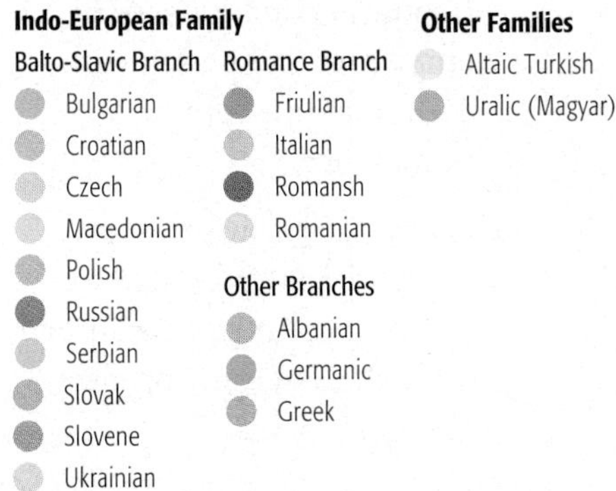

KOSOVO

The population of Kosovo is more than 90 percent ethnic Albanian. At the same time, Serbs consider Kosovo an essential place in the formation of the Serb ethnicity because they fought an important—though losing—battle against the Ottoman Empire there in 1389.

As part of Yugoslavia, Kosovo had been an autonomous province. With the breakup of Yugoslavia, Serbia took direct control of Kosovo and launched a campaign of ethnic cleansing of the Albanian majority. At its peak in 1999, Serb ethnic cleansing had forced 750,000 of Kosovo's 2 million ethnic Albanian residents from their homes, mostly to camps in Albania.

Outraged by the ethnic cleansing, the United States and Western European countries, operating through the North Atlantic Treaty Organization (NATO), launched an air attack against Serbia. The bombing campaign ended when Serbia agreed to withdraw all of its soldiers and police from Kosovo.

Kosovo declared independence from Serbia in 2008. The United States and most European countries have recognized the independence, but countries allied with Serbia, including China and Russia, oppose it.

BALKANIZATION

A century ago, the term **Balkanized** was widely used to describe a small geographic area that could not successfully be organized into one or more stable states because it was inhabited by many ethnicities with complex, long-standing antagonisms toward each other. World leaders at the time regarded **Balkanization**—the process by which a state breaks down through conflicts among its ethnicities—as a threat to peace throughout the world, not just in a small area. They were right: Balkanization led directly to World War I because the various nationalities in the Balkans dragged into the war the larger powers with whom they had alliances.

After two world wars and the rise and fall of communism during the twentieth century, the Balkans have once again become Balkanized in the twenty-first century. Peace has come to the Balkans in a tragic way, through the "success" of ethnic cleansing. Millions of people were rounded up and killed or forced to migrate because they constituted ethnic minorities. Ethnic homogeneity may be the price of peace in areas that once were multi-ethnic.

7.9.4 **ETHNIC CLEANSING IN BOSNIA & HERZEGOVINA**
The Stari Most (old bridge), built by the Turks in 1566 across the Neretva River, was an important symbol and tourist attraction in the city of Mostar (top). The bridge was blown up by Serbs in 1993 (middle), in an attempt to demoralize Bosnian Muslims as part of ethnic cleansing. With the end of the war in Bosnia & Herzegovina, the bridge was rebuilt in 2004 (bottom).

Chapter **Summary**

COLLECTION OF THE NEW YORK HISTORICAL SOCIETY #46093

U.S. ETHNICITIES

- Major ethnicities in the United States include African Americans, Hispanic Americans, Asian Americans, and American Indians.
- Ethnic groups are clustered in regions of the country and within urban neighborhoods.
- African Americans have a distinctive history of forced migration for slavery.
- In the United States, race and ethnicity are often used interchangeably.

ETHNICITIES AND NATIONALITIES

- Nationalities are ethnic groups that possess among their cultural traditions the attachment and loyalty to a particular country.
- A nationality combines an ethnic group's language, religion, and artistic expressions with a country's particular independence movement, history, and patriotism.
- During the past two centuries, many nation-states have been created that attempt to match the boundaries of a nationality.

ETHNIC CONFLICTS

- Multinational states combine more than one nationality in a single state.
- Conflicts can arise when a country contains several ethnicities competing with each other for control or dominance.
- Conflicts also arise when ethnicities are allocated to different countries because of difficulties in drawing clear-cut boundaries.
- Ethnic cleansing by more powerful ethnic groups against weaker ones has been especially widespread in Sub-Saharan Africa and the Balkans.

THE DETROIT INSTITUTE OF ARTS (DIA)

THE DETROIT MUSEUM OF AFRICAN AMERICAN HISTORY

Geographic Consequences of Change

Two major museums standing one block apart in Detroit illustrate the challenges of encouraging respect for different ethnic identities in the United States. One of the museums, the Detroit Institute of Arts (DIA), contains a major collection of paintings by medieval European artists, many of which were donated a century ago by rich Detroit industrialists. The DIA's most famous work is an enormous mural completed in 1932 by the Mexican muralist Diego Rivera, glorifying workers in Detroit's auto factories. The 80-year-old building, the country's fifth-largest art museum, looks like a Greek temple.

The nearby Museum of African American History houses the nation's largest exhibit devoted to the history and culture of African Americans. Founded in 1965, the museum has moved twice to larger buildings, including the current one opened in 1997. The building is designed to reflect the cultural heritage of Africa, including an entry with large bronze doors topped by 14-carat gold-plate decorative masks. The exhibits are primarily photographs, videos, and text.

The financially strapped city of Detroit has had difficulty adequately funding both museums, so it has had to make choices. Which museum should take priority—a crumbling temple of European masterpieces or an emotionally powerful testimony to the rich cultural traditions of an important ethnic minority? Does it matter that Detroit's African American population was 5 percent when the DIA was built and 75 percent when the Museum of African American history was built?

SHARE YOUR VOICE Student Essay

Elizabeth Packer
UCLA sophomore

Discovering Los Angeles

While Los Angeles may bring to mind images of Hollywood stars and sun-drenched beaches, it certainly has not earned a reputation for being easy to navigate, with its notorious traffic and endless miles of freeway. Though there is truth to this characterization—certain areas are very unfriendly to pedestrians and the traffic really can boggle the mind—I have found LA to be accessible, the tangled geography of the city yielding many pleasures, if you just put in a little effort.

I only recently moved to LA, trading my home on the East Coast for UCLA's sunny campus two years ago. Since then, I've experienced the city through the eyes of a new transplant, filtered through my particular academic lens—I'm a geography major, attuned to the relationships we create with the space we inhabit and how we are influenced by our surroundings.

Los Angeles is unlike any other city. Physically, its geography begins on the shores of the Pacific in the west, creeps north from Santa Monica into the Hollywood Hills, encircles the skyscrapers of downtown, moves south into Compton and yields to the Mexican neighborhoods of the East. It is a sprawling urban space that is difficult to define, and I've found it's best to think of LA in terms of its many distinct, yet interconnected, neighborhoods. This mix is LA's most refreshing asset in my mind.

In terms of diversity, whether measured in restaurants, concerts, or people, LA really is sure to have something up your alley. On top of that, the ever present possibility of seeing an A-Lister right around the corner only adds to LA's appeal, if you follow the latest celebrity gossip (as I must confess, I do).

As a college student, I love having all of LA's offerings at my fingertips when campus life gets old. Though the absence of a car can be frustrating at times, I have become quite good at navigating the many bus lines that cover the city, enabling adventures as diverse as a trek into LA's wild side—a hike through Griffith Park that yielded unbelievable views of the Hollywood sign—to enjoying the city's infamous nightlife, with visits to the bars and clubs of the Sunset Strip.

Chapter Resources

DOCUMENTING ETHNIC CLEANSING

Early reports of ethnic cleansing by Serbs in former Yugoslavia were so shocking that many people dismissed them as journalistic exaggeration or partisan propaganda.

It took one of geography's most important analytic tools, aerial-photography interpretation, to provide irrefutable evidence of the process, as well as the magnitude, of ethnic cleansing. The process of ethnic cleansing involved four steps. A series of three photographs taken by NATO air reconnaissance over the village of Glodane, in western Kosovo, illustrated the four steps.

Assessment photograph of Glodane Village, Kosovo, used by Joint Staff Director of Intelligence Rear Adm. Thomas R. Wilson, U.S. Navy, during a press briefing on NATO Operation Allied Force in the Pentagon on April 6, 1999.

- **STEP 1:** Move a large amount of military equipment and personnel into a village with no strategic value. The figure shows the village's houses and farm buildings clustered on the left side, with fields on the outskirts of the village, including the center and right portions of the photograph. The circles in the figure show the location of Serb armored vehicles along the main street of the village.
- **STEP 2:** Round up all the people in the village. In Bosnia, Serbs often segregated men from women, children, and old people. The men were either placed in detention camps or "disappeared"—undoubtedly killed—whereas the others were forced to leave the village. In Kosovo, men were herded together with the others rather than killed. In the photograph of Glodane, the farm field immediately to the east of the main north–south road is filled with the villagers. At the scale that the photograph is reproduced in this book, the people appear as a dark mass. The white rectangles to the north of the people are civilian cars and trucks.
- **STEP 3:** Force the people to leave the village. This step appeared dramatically in the second photograph of the sequence, depicting the same location a short time later. The second photograph showed one major change: the people and vehicles massed in the field in the first photograph were gone—no people and no vehicles. The villagers were forced into a convoy—some in the vehicles, others on foot—heading for the Albanian border 16 kilometers (10 miles) to the west.
- **STEP 4**: Destroy the vacated village. The third photograph of the sequence showed that the buildings in the village had been set on fire.

Aerial photographs such as these not only proved that ethnic cleansing was occurring but also provided critical evidence to prosecute Serb leaders for war crimes.

KEY TERMS

Apartheid
Laws (no longer in effect) in South Africa that physically separated different races into different geographic areas.

Balkanization
Process by which a state breaks down through conflicts among its ethnicities.

Balkanized
A small geographic area that could not successfully be organized into one or more stable states because it was inhabited by many ethnicities with complex, long-standing antagonisms toward each other.

Ethnic cleansing
Process in which a more powerful ethnic group forcibly removes a less powerful one in order to create an ethnically homogeneous region.

Ethnicity
Identity with a group of people that share distinct physical and mental traits as a product of common heredity and cultural traditions.

Multi-ethnic state
State that contains more than one ethnicity.

Multinational state
State that contains two or more ethnic groups with traditions of self-determination that agree to coexist peacefully by recognizing each other as distinct nationalities.

Nationalism
Loyalty and devotion to a particular nationality.

Nationality
Identity with a group of people that share legal attachment and personal allegiance to a particular place as a result of being born there.

Nation-state
A state whose territory corresponds to that occupied by a particular ethnicity that has been transformed into a nationality.

Race
Identity with a group of people descended from a common ancestor.

Racism
Belief that race is the primary determinant of human traits and capacities and that racial differences produce an inherent superiority of a particular race.

Racist
A person who subscribes to the beliefs of racism.

Self-determination
Concept that ethnicities have the right to govern themselves.

Sharecropper
A person who works fields rented from a landowner and pays the rent and repays loans by turning over to the landowner a share of the crops.

Triangular slave trade
A practice, primarily during the eighteenth century, in which European ships transported slaves from Africa to Caribbean islands, molasses from the Caribbean to Europe, and trade goods from Europe to Africa.

FURTHER READINGS

Alberts, Heike C. "Changes in Ethnic Solidarity in Cuban Miami." *Geographical Review* 95 (2005): 231–48.

Allen, James P., and Eugene Turner. "Ethnic Residential Concentrations in United States Metropolitan Areas." *Geographical Review* 95 (2005): 267–85.

Arreola, Daniel D., ed. *Hispanic Spaces, Latino Places: Community and Cultural Diversity in Contemporary America.* Austin: University of Texas Press, 2004.

Christopher, A. J. "'To Define the Undefinable': Population Classification and the Census in South Africa." *Area* 34 (2002): 401–8.

Marriott, Alan. "Nationalism and Nationality in India and Pakistan." *Geography* 85 (2000): 173–77.

Murphy, Alexander B. "Territorial Policies in Multiethnic States." *Geographical Review* 79 (1989): 410–21.

Peake, Linda, and Audrey Kobayashi. "Policies and Practices for an Antiracist Geography at the Millennium." *Professional Geographer* 54 (2002): 50–61.

Skop, Emily, and Wei Li. "Asians in America's Suburbs: Patterns and Consequences of Settlement." *Geographical Review* 95 (2005): 167–88.

ON THE INTERNET

Images of ethnic cleansing in Kosovo and elsewhere are collected by GlobalSecurity.org and available at **www.globalSecurity.org.** GlobalSecurity.org is an organization that is a reliable source of background information on conflicts by providing facts and figures. The U.S. Department of Defense DefenseLINK web site also posts photos of conflicts at **www.defenselink.mil/photos/.**

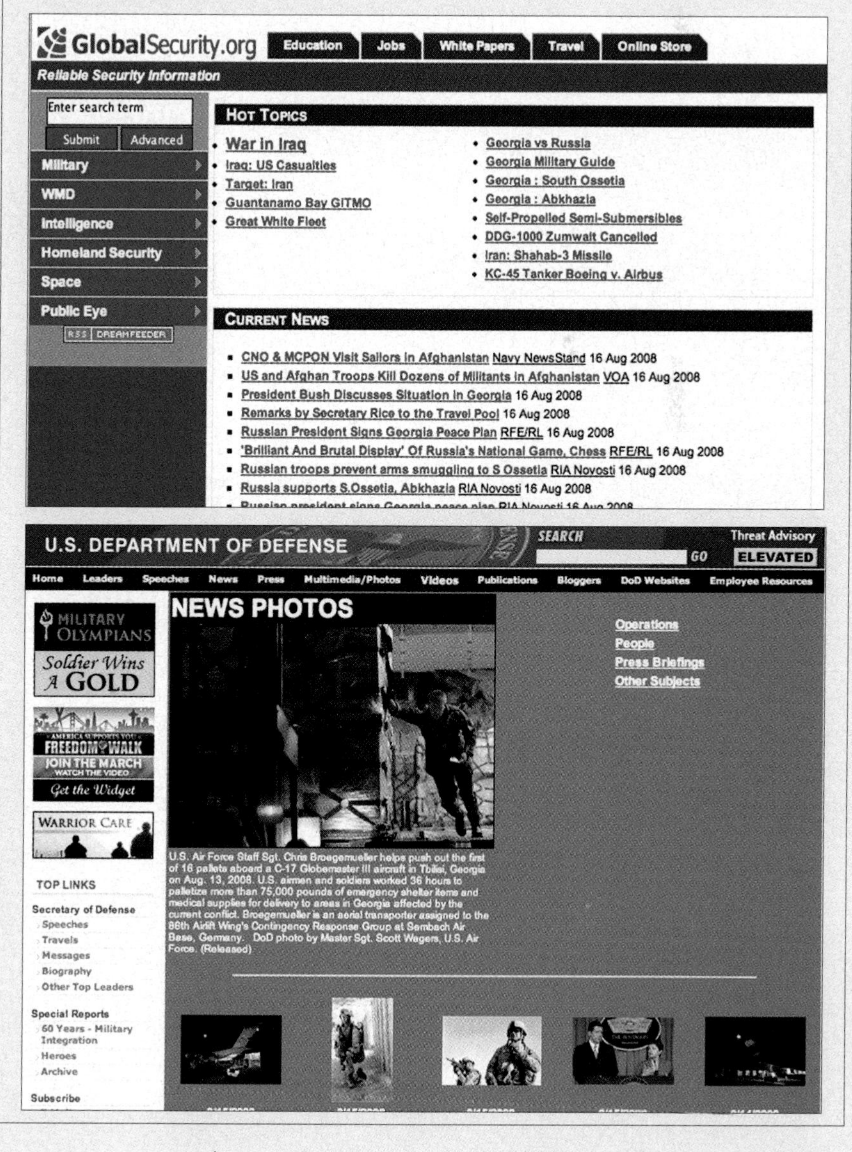

Chapter 8 **POLITICAL GEOGRAPHY**

How many countries can you name? Old-style geography sometimes required memorization of countries and their capitals. Human geographers now emphasize a thematic approach. Despite this change in emphasis, you still need to know the locations of countries. Without such knowledge, you lack a basic frame of reference—knowing where things are. It would be like translating an article in a foreign language by looking up each word in a dictionary.

In recent years, we have repeatedly experienced military conflicts and revolutionary changes in once-obscure places. No one can predict where the next war will erupt, but political geography helps to explain the cultural and physical factors that underlie political unrest in the world. Political geographers study how people have organized Earth's land surface into countries and alliances, reasons underlying the observed arrangements, and the conflicts that result from the organization.

When looking at satellite images of Earth, we easily distinguish landmasses and water bodies, mountains and rivers, deserts and fertile agricultural land, urban areas and forests. What we cannot see are boundaries between countries. Boundary lines are not painted on Earth, but they might as well be, for these national divisions are very real. To many, national boundaries are more meaningful than natural features. One of Earth's most fundamental cultural characteristics—one that we take for granted—is the division of our planet's surface into countries. In the post–Cold War era, though, the familiar division of the world is crumbling. Geographers try to explain why this is happening.

THE KEY ISSUES IN THIS CHAPTER

ATTACK ON NEW YORK WORLD TRADE CENTER, SEPTEMBER 11, 2001

STATES OF THE WORLD

8.1 Defining States

- GIScience applies contemporary technology to create maps.
- The ability to make complex maps is now on everyone's computer desktop.

A **state** is an area organized into a political unit and ruled by an established government that has control over its internal and foreign affairs. A state occupies a defined territory on Earth's surface and contains a permanent population. A state has **sovereignty**, which means independence from control of its internal affairs by other states. Because the entire area of a state is managed by its national government, laws, army, and leaders, it is a good example of a formal or uniform region. The term *country* is a synonym for *state*.

A map of the world shows that virtually all habitable land belongs to states, but for most of history, this was not so. As recently as the 1940s, the world contained only about 50 states, compared to 192 members of the United Nations as of 2009. The largest state not in the United Nations is Taiwan.

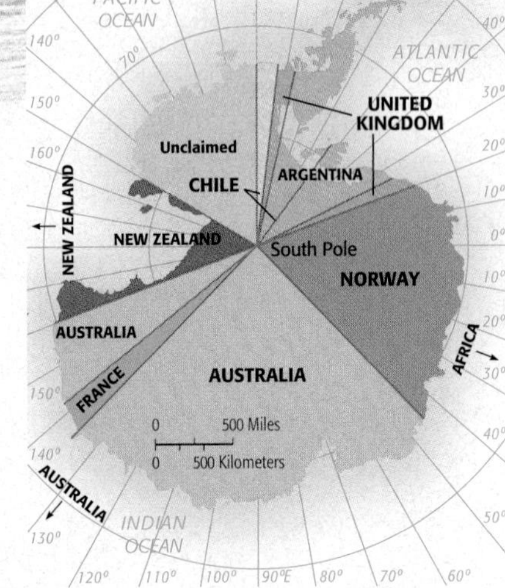

ANTARCTICA: NOT A SOVEREIGN STATE

Antarctica is the only large landmass in the world that is not part of a sovereign state. It comprises 14 million square kilometers (5.4 million square miles), which makes it 50 percent larger than Canada. Portions are claimed by Argentina, Australia, Chile, France, New Zealand, Norway, and the United Kingdom; claims by Argentina, Chile, and the United Kingdom are conflicting. In 1959 these seven countries, plus Belgium, Japan, South Africa, the Soviet Union, and the United States, signed a treaty suspending any territorial claims for 30 years and establishing guidelines for scientific research. In 1991, 24 countries agreed to extend the treaty for another 50 years, established new pollution control standards, and banned mining and oil exploration for 50 years. States may establish research stations on Antarctica for scientific investigations, but no military activities are permitted.

TAIWAN: A SOVEREIGN STATE?

Is the island of Taiwan a sovereign state? According to China's government officials, Taiwan is not a separate sovereign state but is a part of China. Until recently, the government of Taiwan agreed.

This confusing situation arose from a civil war between the Nationalists and the Communists in China during the late 1940s. After losing, Nationalist leaders in 1949 fled to the island of Taiwan, 200 kilometers (120 miles) off the Chinese coast. The Nationalists proclaimed that they were still the legitimate rulers of the entire country of China. Until some future occasion when they could defeat the Communists and recapture all of China, the Nationalists argued, at least they could continue to govern one island of the country.

Most other governments in the world consider Mainland China (officially the People's Republic of China) and the island of Taiwan (officially the Republic of China) as two separate and sovereign states. In recent years, the president and political party in power have also announced their desire to make Taiwan a sovereign independent state. But the government of China views this position as a dangerous departure from the long-standing arrangement between the two entities.

The question of who constituted the legitimate government of China plagued U.S. officials during the 1950s and 1960s. The United States had supported the Nationalists during the civil war, so many Americans opposed acknowledging that China was firmly under the control of the Communists. Consequently, the United States continued to regard the Nationalists as the official government of China until 1971, when U.S. policy finally changed, and the United Nations voted to transfer China's seat from the Nationalists to the Communists. Taiwan is now the most populous state not in the United Nations.

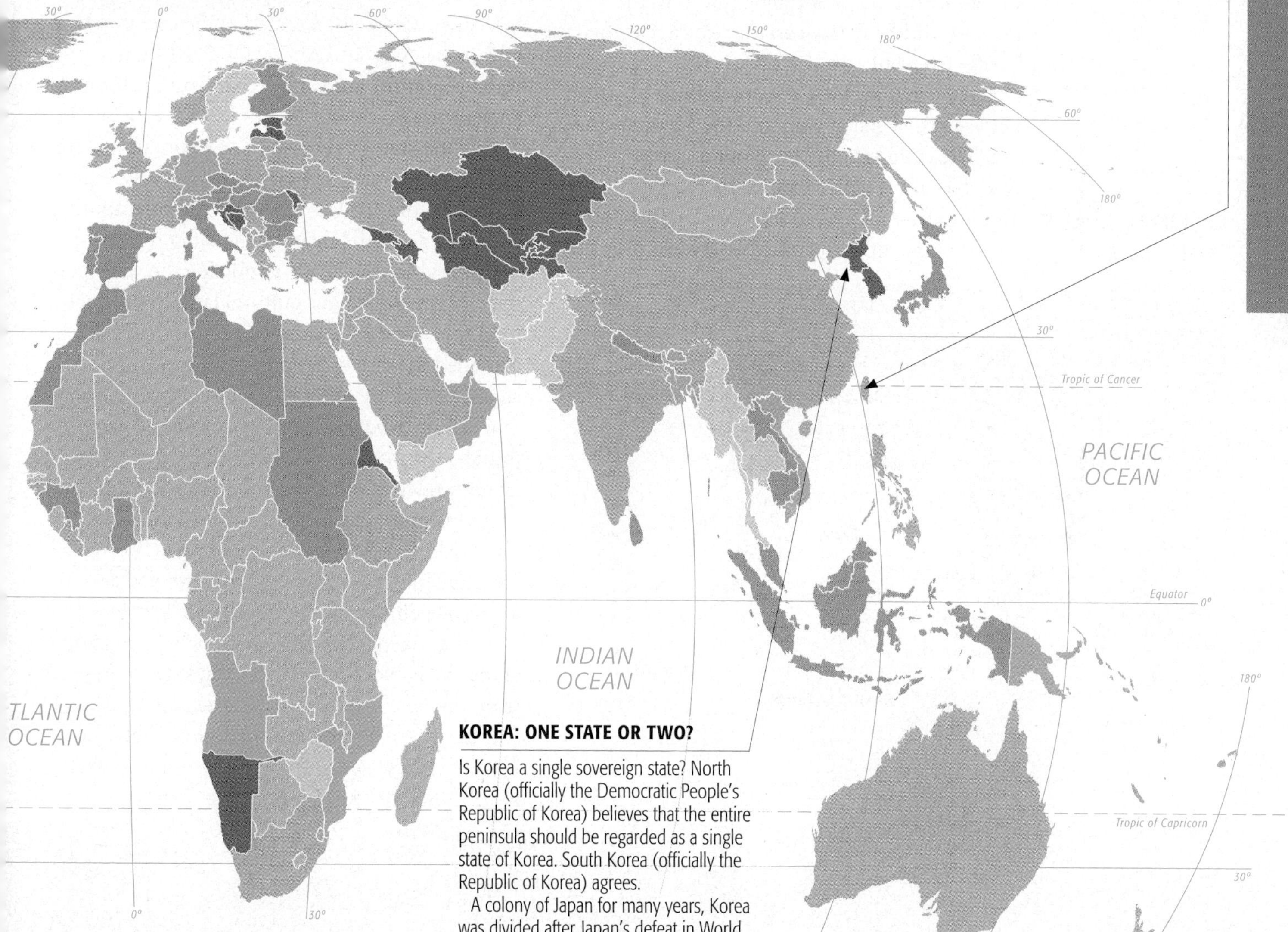

PACIFIC OCEAN

INDIAN OCEAN

TLANTIC OCEAN

Tropic of Cancer

Equator

Tropic of Capricorn

KOREA: ONE STATE OR TWO?

Is Korea a single sovereign state? North Korea (officially the Democratic People's Republic of Korea) believes that the entire peninsula should be regarded as a single state of Korea. South Korea (officially the Republic of Korea) agrees.

A colony of Japan for many years, Korea was divided after Japan's defeat in World War II into northern and southern sections along 38° north latitude. North Korea invaded South Korea in 1950, touching off a 3-year war that ended with a cease-fire.

Both Korean governments are committed to reuniting the country into one sovereign state. In 1992, North Korea and South Korea were admitted to the United Nations as separate countries. A peace agreement was signed in 2007, but economic and political differences between the two make reunification improbable.

8.1.1 **MEMBERS OF THE UNITED NATIONS**

192 members

- original members: 51
- 1940s: added 8
- 1950s: added 24
- 1960s: added 42
- 1970s: added 25
- 1980s: added 7
- 1990s: added 31
- 2000s: added 4
- nonmember

STATES OF THE WORLD

8.2 Development of States

- City-states originated in ancient times in the Fertile Crescent.
- States developed in Europe through consolidation of kingdoms.

The concept of dividing the world into a collection of independent states is recent. Prior to the 1800s, Earth's surface was organized in other ways, such as city-states, empires, and tribes. Much of Earth's surface consisted of unorganized territory.

ANCIENT STATES

The modern movement to divide the world into states originated in Europe. However, the development of states can be traced to the ancient Middle East, in an area known as the Fertile Crescent.

The ancient Fertile Crescent formed an arc between the Persian Gulf and the Mediterranean Sea. The eastern end, Mesopotamia, was centered in the valley formed by the Tigris and Euphrates rivers, in present-day Iraq. The Fertile Crescent then curved westward over the desert, turning southward to encompass the Mediterranean coast through present-day Syria, Lebanon, and Israel. The Nile River valley of Egypt is sometimes regarded as an extension of the Fertile Crescent. Situated at the crossroads of Europe, Asia, and Africa, the Fertile Crescent was a center for land and sea communications in ancient times.

8.2.1 **FERTILE CRESCENT**

— Fertile Crescent

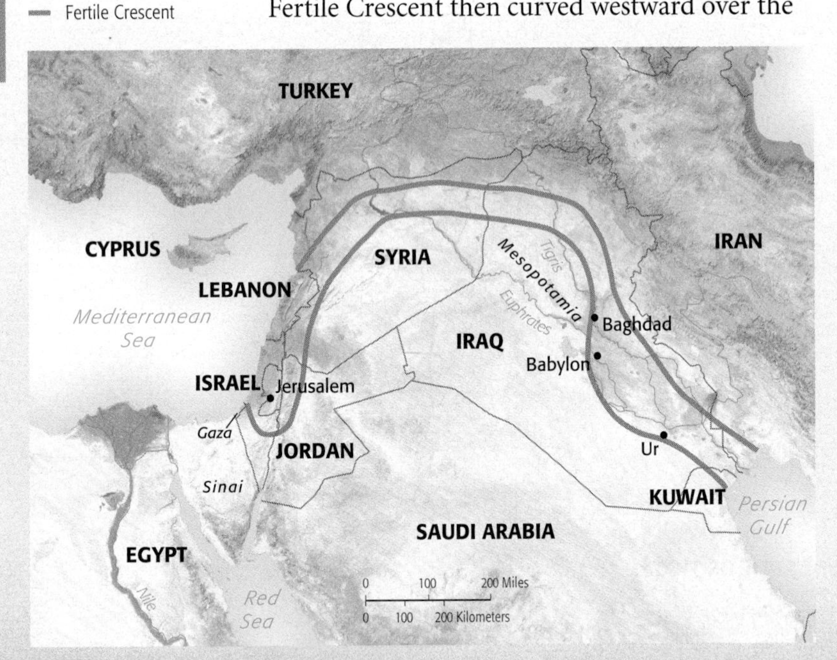

The first states to evolve in Mesopotamia were known as city-states. A **city-state** is a sovereign state that comprises a town and the surrounding countryside. Walls clearly delineated the boundaries of the city, and outside the walls the city controlled agricultural land to produce food for urban residents. The countryside also provided the city with an outer line of defense against attack by other city-states. Periodically, one city or tribe in Mesopotamia would gain military dominance over the others and form an empire. Mesopotamia was organized into a succession of empires by the Sumerians, Assyrians, Babylonians, and Persians.

Meanwhile, the state of Egypt emerged as a separate empire to the west of the Fertile Crescent. Egypt controlled a long, narrow region along the banks of the Nile River, extending from the Nile Delta at the Mediterranean Sea southward for several hundred kilometers. Egypt's empire lasted from approximately 3000 B.C. until the fourth century B.C.

8.2.2 **PARTHENON, ATHENS**

8.2.3. **FORUM, ROME**

EARLY EUROPEAN STATES

Political unity in the ancient world reached its height with the establishment of the Roman Empire, which controlled most of Europe, North Africa, and Southwest Asia, from modern-day Spain to Iran and from Egypt to England. At its maximum extent, the empire comprised 38 provinces, each using the same set of laws that were created in Rome. Massive walls helped the Roman army defend many of the empire's frontiers. The Roman Empire collapsed in the fifth century after a series of attacks by people living on its frontiers and because of internal disputes.

The European portion of the Roman Empire was fragmented into a large number of estates owned by competing kings, dukes, barons, and other nobles. Victorious nobles seized control of defeated rivals' estates, and after these nobles died, others fought to take possession of their land. Meanwhile, most people were forced to live on an estate, working and fighting for the benefit of the noble.

A handful of powerful kings emerged as rulers over large numbers of estates beginning about the year 1100. The consolidation of neighboring estates under the unified control of a king formed the basis for the development of such modern Western European states as England, France, and Spain. However, much of central Europe—notably present-day Germany and Italy—remained fragmented into a large number of estates that were not consolidated into states until the nineteenth century.

COLONIES

As Europe became organized into states, European countries established colonies in much of the rest of the world. A **colony** is a territory that is legally tied to a sovereign state rather than being completely independent. In some cases, a sovereign state runs only the colony's military and foreign policy. In others, it also controls the colony's internal affairs.

The colonial era began in the 1400s, when European explorers sailed westward for Asia but encountered and settled in the Western Hemisphere instead. The United Kingdom and France assembled the largest colonial empires. The European states eventually lost most of their Western Hemisphere colonies: independence was declared by the United States in 1776 and by most Latin American states between 1800 and 1824. European states then turned their attention to Africa and Asia.

8.2.4 **THE RUINS OF THE ANCIENT CITY OF UR**

STATES OF THE WORLD

8.3 Colonies

- Until the twentieth century, much of the world consisted of colonies of European states.
- Few colonies remain in the world.

European states came to control much of the world through **colonialism**, which is the effort by one country to establish settlements and to impose its political, economic, and cultural principles on such territory.

The European colonization of Africa and Asia is often termed **imperialism**, which is control of territory already occupied and organized by an indigenous society, whereas colonialism is control of previously uninhabited or sparsely inhabited land.

European states established colonies for three basic reasons:

- European missionaries established colonies to promote Christianity.
- Colonies provided resources that helped the economy of European states.
- European states considered the number of colonies to be an indicator of relative power.

The three motives may be summarized as God, gold, and glory.

At one time, colonies were widespread over Earth's surface, but today only a handful remain. Nearly all are islands in the Pacific Ocean or Caribbean Sea. Most African and Asian colonies became independent after World War II. Only 15 African and Asian states were members of the United Nations when it was established in 1945, compared to 106 in 2009. The boundaries of the new states frequently coincide with former colonial provinces, although not always.

The most populous colony is Puerto Rico, which is a Commonwealth of the United States. Its 4 million residents are citizens of the United States, but they do not participate in U.S. elections, nor do they have a voting member of Congress. Puerto Ricans are split between those who want to maintain commonwealth status and those who want to see the island become a U.S. state.

FLAG OF PUERTO RICO FLIES NEXT TO U.S. FLAG OUTSIDE GOVERNMENT BUILDING IN SAN JUAN, CAPITAL OF PUERTO RICO

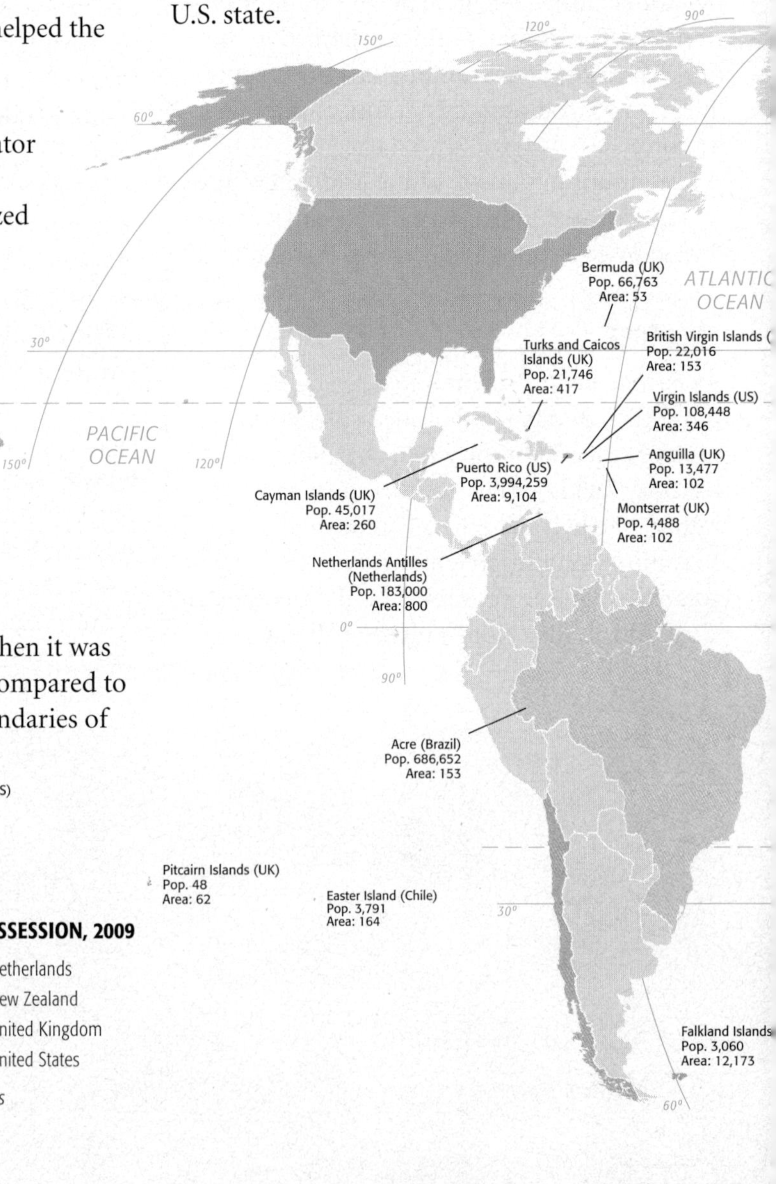

8.3.1 **COLONIAL POSSESSION, 2009**

Australia
Brazil
Chile
France
Netherlands
New Zealand
United Kingdom
United States

Area in square kilometers

Bermuda (UK) Pop. 66,763 Area: 53
Turks and Caicos Islands (UK) Pop. 21,746 Area: 417
British Virgin Islands (UK) Pop. 22,016 Area: 153
Virgin Islands (US) Pop. 108,448 Area: 346
Anguilla (UK) Pop. 13,477 Area: 102
Puerto Rico (US) Pop. 3,994,259 Area: 9,104
Cayman Islands (UK) Pop. 45,017 Area: 260
Montserrat (UK) Pop. 4,488 Area: 102
Netherlands Antilles (Netherlands) Pop. 183,000 Area: 800
Acre (Brazil) Pop. 686,652 Area: 153
American Samoa (US) Pop. 68,200 Area: 199
Pitcairn Islands (UK) Pop. 48 Area: 62
Easter Island (Chile) Pop. 3,791 Area: 164
Falkland Islands Pop. 3,060 Area: 12,173
PACIFIC OCEAN
ATLANTIC OCEAN

Other than Puerto Rico, remaining colonies with populations between 100,000 and 300,000 include France's French Polynesia, Mayotte, and New Caledonia; the Netherlands' Netherlands Antilles; and the United States' Guam and U.S. Virgin Islands.

The world's least-populated colony is Pitcairn Island, possessed by the United Kingdom. Pitcairn, in the South Pacific, has 47 people on an island less than 5 square kilometers (2 square miles). The island was settled in 1790 by British mutineers from the ship *Bounty*, commanded by Captain William Bligh. Today the islanders survive by selling fish and postage stamps.

PACIFIC OCEAN
ATLANTIC OCEAN
ATLANTIC OCEAN
INDIAN OCEAN
PACIFIC OCEAN
Tropic of Cancer
Equator
Tropic of Capricorn

8.3.2 COLONIAL POSSESSIONS, 1914

Colonial possessions, 1914

Belgium
Austria
Denmark
Spain
Germany
Italy
Japan
Netherlands
Ottoman Empire
Portugal
Russia
France
United Kingdom
United States

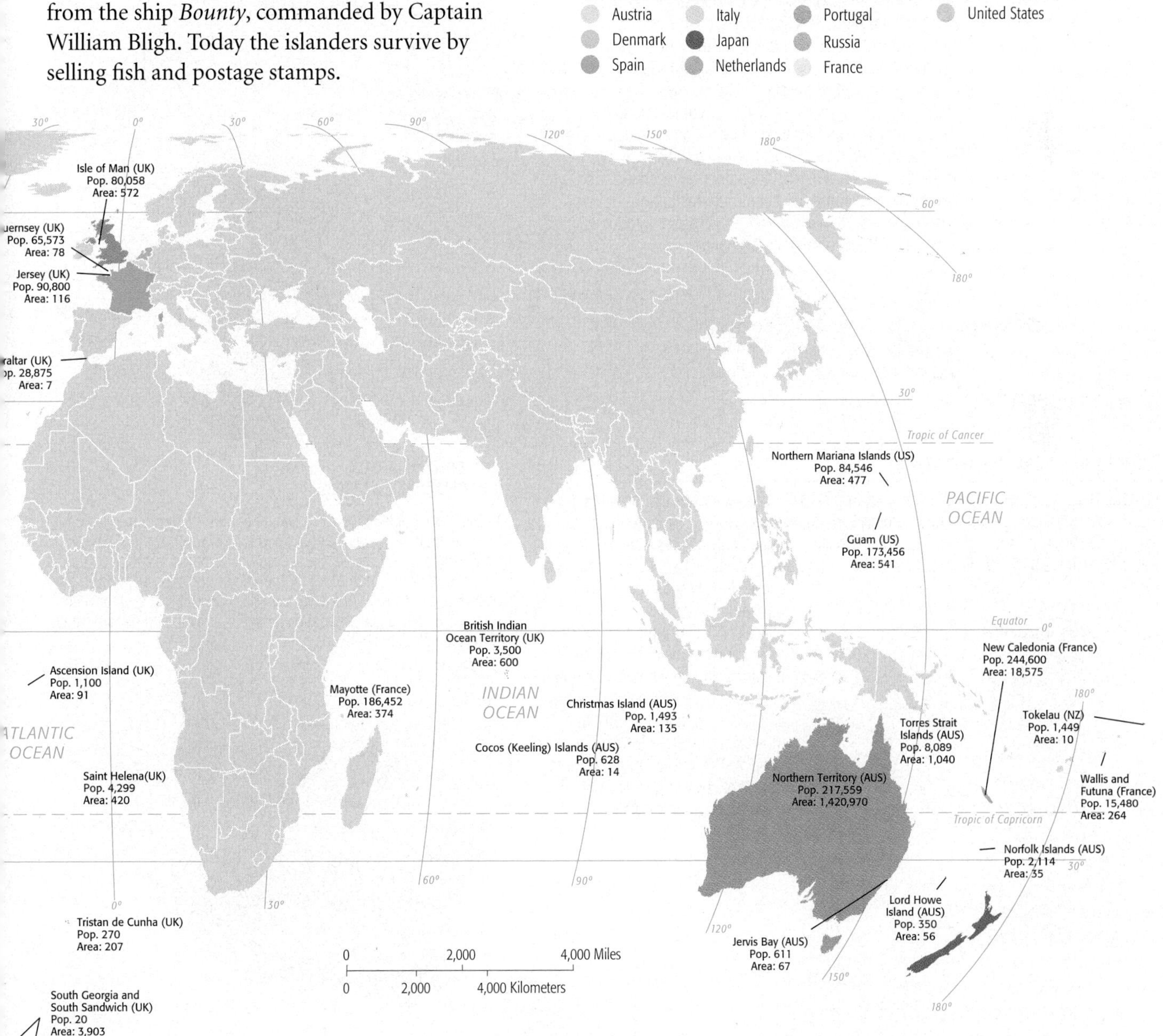

STRUCTURE OF STATES

8.4 Shapes of States

- A state has one of five basic shapes.
- States that have no water boundary are landlocked.

The physical shape of a state, such as the outline of the United States or Canada, is part of its unique identity. The shape of a state can influence the ease or difficulty of internal administration and can affect social unity. The shape affects the potential for communication and conflict with neighbors.

Countries have one of five basic shapes—compact, prorupted, elongated, fragmented, and perforated. Examples of each can be seen in southern Africa. Each shape displays distinctive characteristics and challenges.

8.4.1 GAMBIA: AN ELONGATED STATE

There are a handful of **elongated states,** or states with a long and narrow shape. In West Africa, Gambia is an elongated state extending along the banks of the Gambia River about 500 kilometers (300 miles) east–west but only about 25 kilometers (15 miles) north–south. Except for its short coastline along the Atlantic Ocean, Gambia is otherwise completely surrounded by Senegal. The shape of the two countries is a legacy of competition among European countries to establish colonies during the nineteenth century. Gambia became a British colony, whereas Senegal was French. The border between the two countries divided families and ethnic groups but was never precisely delineated, so people trade and move across the border with little concern for its location.

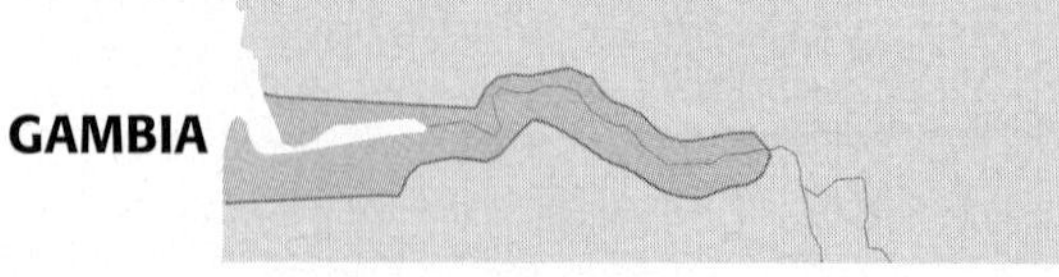

8.4.2 ANGOLA: A FRAGMENTED STATE

A **fragmented state** includes several discontinuous pieces of territory. Technically, all states that have offshore islands as part of their territory are fragmented. However, fragmentation is particularly significant for some states. There are two kinds of fragmented states–those with areas separated by water, and those separated by an intervening state. A difficult type of fragmentation occurs if the two pieces of territory are separated by another state. Picture the difficulty of communicating between Alaska and the lower 48 states if Canada were not a friendly neighbor. All land connections between Alaska and the rest of the United States must pass through a long expanse of Canada. The division of Angola into two pieces by Congo's proruption creates a fragmented state.

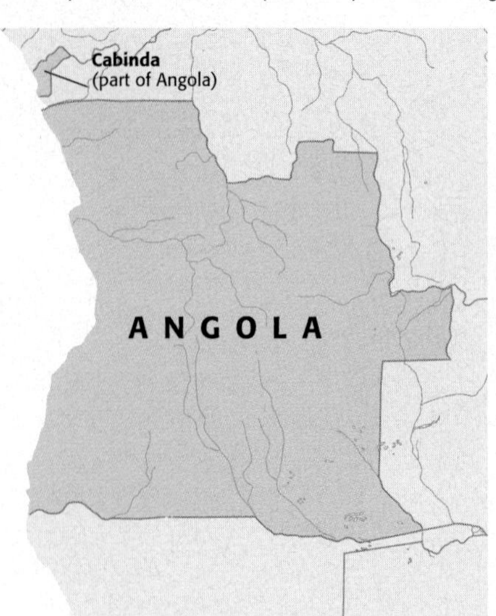

8.4.3 SOUTH AFRICA: A PERFORATED STATE

A state that completely surrounds another one is a **perforated state**. The one good example of a perforated state is South Africa, which completely surrounds the state of Lesotho. Lesotho must depend almost entirely on South Africa for the import and export of goods. Dependency on South Africa was especially difficult for Lesotho when South Africa had a government controlled by whites who discriminated against the black majority population.

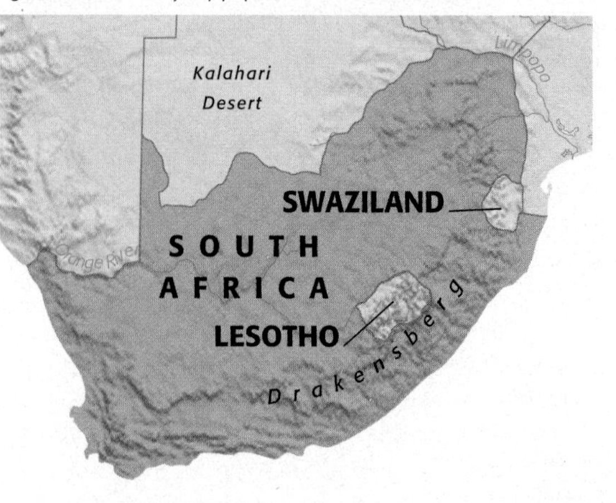

8.4.4 CONGO AND NAMIBIA: PRORUPTED STATES

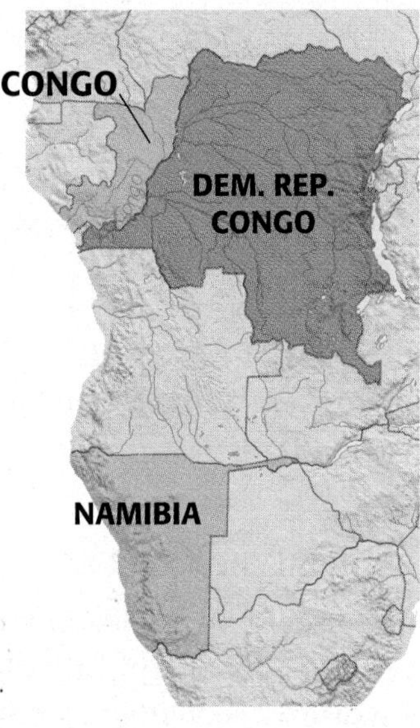

An otherwise compact state with a large projecting extension is a **prorupted state.** Proruptions are created for two principal reasons. First, a proruption can provide a state with access to a resource, such as water. When the Belgians gained control of the Congo, they carved out a westward proruption about 500 kilometers (300 miles) long. The proruption, which followed the Zaire (Congo) River, gave the colony access to the Atlantic. The proruption also divided the Portuguese colony of Angola (now an independent state) into two discontinuous fragments 50 kilometers (30 miles) apart. The northern fragment, called Cabinda, constitutes less than 1 percent of Angola's total land area.

In their former colony of South West Africa (now Namibia), the Germans in 1890 carved out a 500-kilometer (300-mile) proruption to the east. This proruption, known as the Caprivi Strip, provided the Germans with access to one of Africa's most important rivers, the Zambezi. The Caprivi Strip also disrupted communications among the British colonies of southern Africa. South Africa, which controlled Namibia from the 1910s until its independence in 1990, stationed troops in the Caprivi Strip to fight enemies in Angola, Zambia, and Botswana.

8.4.5 SUB-SAHARAN AFRICA: SEVERAL COMPACT STATES

In a **compact state**, the distance from the center to any boundary does not vary significantly. The ideal theoretical compact state would be shaped like a circle, with the capital at the center and the shortest possible boundaries to defend. Compactness is a beneficial characteristic for most smaller states, because good communications can be more easily established to all regions, especially if the capital is located near the center. Examples of compact states in southern African include Burundi, Kenya, Rwanda, and Uganda. However, compactness does not necessarily mean peacefulness, as compact states are just as likely as others to experience civil wars and ethnic rivalries.

LANDLOCKED STATES

A landlocked state lacks a direct outlet to the sea because it is completely surrounded by several other countries. Landlocked states are most common in Africa, where 14 of the continent's 54 states have no direct ocean access. The prevalence of landlocked states in Africa is a remnant of the colonial era, when Britain and France controlled extensive regions.

The European powers built railroads, mostly in the early twentieth century, to connect the interior of Africa with seaports. Railroads moved minerals from interior mines to seaports, and in the opposite direction, rail lines carried mining equipment and supplies from seaports to the interior. Now that the British and French empires are gone and former colonies have become independent states, some important colonial railroad lines pass through several independent countries. This has created new landlocked states, which must cooperate with neighboring states that have seaports.

Direct access to an ocean is critical to states because it facilitates international trade. Bulky goods, such as petroleum, grain, ore, and vehicles, are normally transported long distances by ship. This means that a country needs a seaport where goods can be transferred between land and sea. To send and receive goods by sea, a landlocked state must arrange to use another country's seaport.

STRUCTURE OF STATES

8.5 Boundaries Between States

- Physical boundaries include mountains, deserts, and bodies of water.
- Cultural boundaries include geometric and ethnic boundaries.

A state is separated from its neighbors by a **boundary**, an invisible line marking the extent of a state's territory. Boundaries result from a combination of natural physical features (such as rivers, deserts, and mountains) and cultural features (such as language and religion).

The boundary line, which must be shared by more than one state, is the only location where direct physical contact must take place between two neighboring states. Therefore, the boundary has the potential to become the focal point of conflict between them. Boundaries are of two types—physical and cultural.

8.5.1 THE ANDES: A MOUNTAIN BOUNDARY
Mountains can be effective boundaries because they are permanent and difficult to cross, and mountain regions are often sparsely inhabited. Mountains do not always provide for the amicable separation of neighbors. Argentina and Chile agreed to be divided by the crest of the Andes Mountains but could not decide on the precise location of the crest. The two countries almost fought a war over the boundary line. But with the help of U.S. mediators, they finally decided on the line connecting adjacent mountain peaks.

PHYSICAL BOUNDARIES

Important physical features on Earth's surface can make good boundaries because they are easily seen, both on a map and on the ground. Three types of physical boundaries are mountains, deserts, and bodies of water.

CULTURAL BOUNDARIES

The boundaries between some states coincide with differences in ethnicity, especially language and religion. Other cultural boundaries are drawn according to geometry; they are simply straight lines drawn on a map, although good reasons may exist for where the lines are located. Boundaries between countries have been placed where possible to separate speakers of different languages or followers of different religions. As discussed in Chapter 7, a nation-state exists when the boundaries of a state match the boundaries of the territory inhabited by an ethnic group. Problems exist when the boundaries do not match.

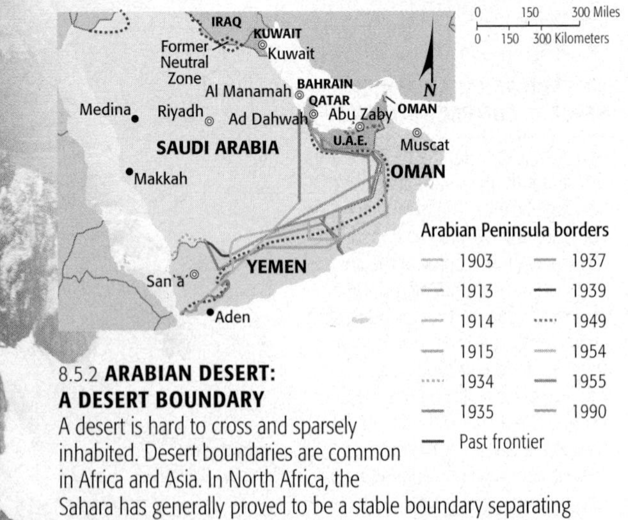

8.5.2 ARABIAN DESERT: A DESERT BOUNDARY
A desert is hard to cross and sparsely inhabited. Desert boundaries are common in Africa and Asia. In North Africa, the Sahara has generally proved to be a stable boundary separating Algeria, Libya, and Egypt at the north from Mauritania, Mali, Niger, Chad, and the Sudan at the south. Rather than boundaries, some states in the Arabian desert are separated by frontiers, which are zones where no state exercises complete political control.

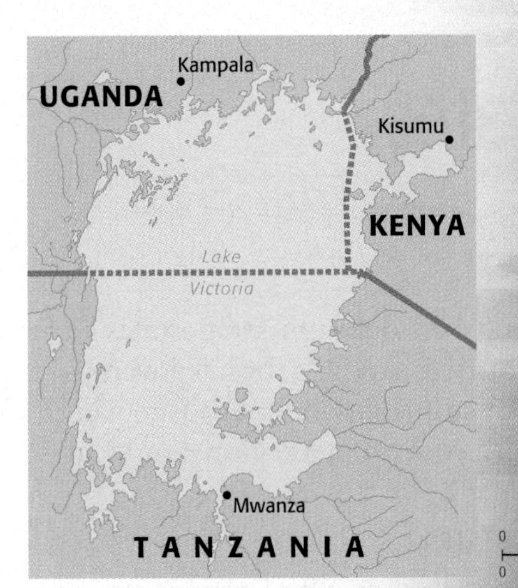

8.5.3 **LAKE VICTORIA: A WATER BOUNDARY**
Rivers, lakes, and oceans are frequently used as boundaries. A water boundary can cause difficulties, because the precise position of the water may change over time. Rivers, in particular, can change their courses. Ocean boundaries also cause problems because states generally claim that the boundary lies not at the coastline, but out at sea. The Law of the Sea, signed by 117 countries in 1983, standardized the territorial limits for most countries at 12 nautical miles (about 22 kilometers or 14 land miles from the coast). States also have exclusive rights to the fish and other marine life within 200 miles (320 kilometers).

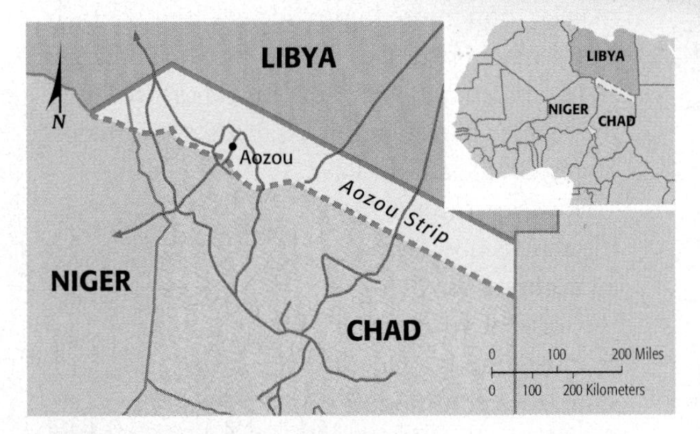

8.5.4 **AOZOU STRIP: A GEOMETRIC BOUNDARY**
Part of the northern U.S. boundary with Canada is a 2,100-kilometer (1,300-mile) straight line (more precisely, an arc) along 49° north latitude, running from Lake of the Woods between Minnesota and Manitoba to the Strait of Georgia between Washington State and British Columbia. This boundary was established in 1846 by a treaty between the United States and Great Britain, which still controlled Canada.

More problematic is the 1,000-kilometer (600-mile) straight line separating Chad and Libya drawn in 1899 by the French and British. In 1912, Italy seized Libya from the Turks and demanded that the French move the boundary with Chad to the south, and the French agreed. The land between the northern and southern boundaries is known as the Aozou Strip. When Libya and Chad both became independent countries, the boundary was set at the original northern location. Claiming that it had been secretly sold the Aozou Strip by the president of Chad, Libya seized the territory in 1973. In 1987, Chad expelled the Libyan army with the help of French forces and regained control of the strip.

8.5.5 **CYPRUS: AN ETHNIC BOUNDARY**
Language is an important cultural characteristic for drawing boundaries, especially in Europe. Religious differences often coincide with boundaries, though in only a few cases, notably in South Asia, has religion been used to select the actual line.

Cyprus is divided between Greek and Turkish ethnicities. Although physically closer to Turkey, the island is 78 percent Greek. Several Greek Cypriot military officers who favored unification of Cyprus with Greece seized control of the government in 1974. Shortly after the coup, Turkey invaded to protect the Turkish Cypriot minority, occupying 37 percent of the island. The Greek coup leaders were removed within a few months, and an elected government was restored, but the Turkish sector declared itself the independent Turkish Republic of Northern Cyprus in 1983, though only Turkey recognizes it as a separate state. A buffer zone patrolled by UN soldiers stretches across the entire island to prevent Greeks and Turks from interacting with the other side. Only one official crossing point exists.

STRUCTURE OF STATES

8.6 Boundaries Inside States

- Two types of government structures are unitary states and federal states.
- States are increasingly adopting federal systems.

In the face of increasing demands by ethnicities for more self-determination, states have restructured their governments to transfer some authority from the national government to local government units. An ethnicity that is not sufficiently numerous to gain control of the national government may be content with control of a regional or local unit of government.

The governments of states are organized according to one of two approaches—the unitary system or the federal system.

- The **unitary state** places most power in the hands of central government officials.
- The **federal state** allocates strong power to units of local government within the country.

A country's cultural and physical characteristics influence the evolution of its governmental system.

UNITARY STATES

In principle, the unitary government system works best in nation-states characterized by few internal cultural differences and a strong sense of national unity. Because the unitary system requires effective communications with all regions of the country, smaller states are more likely to adopt it. Unitary states are especially common in Europe.

In reality, multinational states have often adopted unitary systems, so that the values of one nationality can be imposed on others. In a number of African countries, such as Kenya and Rwanda, for instance, the mechanisms of a unitary state have enabled one ethnic group to extend dominance over weaker groups. When Communist parties controlled the governments, most Eastern European countries had unitary systems so as to promote the diffusion of Communist values.

FEDERAL STATES

In a federal state, such as the United States, local governments possess more authority to adopt their own laws. Multinational states may adopt a federal system of government to empower different nationalities, especially if they live in separate regions of the country. Under a federal system, local government boundaries can be drawn to correspond with regions inhabited by different ethnicities.

The federal system is also more suitable for very large states because the national capital may be too remote to provide effective control over isolated regions. Most of the world's largest states are federal, including Russia (as well as the former Soviet Union), Canada, the United States, Brazil, and India.

However, the size of the state is not always an accurate predictor of the form of government: tiny Belgium is a federal state (to accommodate the two main cultural groups, the Flemish and Waloons, as discussed in Chapter 5), whereas China is a unitary state (to promote Communist values).

8.6.1 **POLES MARCH TO SUPPORT END OF COMMUNIST RULE IN POLAND IN 1988**

TREND TOWARD FEDERAL STATES

Global trends favor federal systems. Unitary systems have been sharply curtailed in a number of countries and scrapped altogether in others.

Eastern European countries such as Poland switched from a unitary to a federal system after control of the national government was wrested from the Communists. The federal system was adopted to dismantle legal structures by which Communists had maintained unchallenged power for more than 40 years.

Under the Communists' unitary system, local governments held no legal authority. The national government appointed local officials and owned public property. This system led to deteriorated buildings, roads, and water systems, because the national government did not allocate sufficient funds to maintain property and no one had clear responsibility for keeping property in good condition.

The transition to a federal system of government proved difficult in Eastern European countries. Local officials elected after the fall of communism had little experience in governing a community. The first task for many newly elected councilors was to attend a training course in how to govern.

Thousands of qualified people had to be found to fill appointed positions, such as directors of education, public works, and planning. Municipalities had the option of hiring some of the 95,000 national government administrators who previously looked after local affairs under the unitary system. However, many of these former officials were rejected by the new local governments because of their close ties to the discredited Communist party.

STRUCTURE OF STATES

8.7 Electoral Geography

- Gerrymandering is the drawing of legislative boundaries to favor the party in power.
- Some U.S. states gerrymandered electoral districts.

The boundaries separating legislative districts within the United States and other countries are redrawn periodically to ensure that each district has approximately the same population. Boundaries must be redrawn because migration inevitably results in some districts gaining population, whereas others are losing. The districts of the 435 U.S. House of Representatives are redrawn every 10 years following the release of official population figures by the Census Bureau.

GERRYMANDERING

The political party in control of the state legislature naturally attempts to redraw boundaries to improve the chances of its supporters to win seats. The process of redrawing legislative boundaries for the purpose of benefiting the party in power is called **gerrymandering**.

The term gerrymandering was named for Elbridge Gerry (1744–1814), governor of Massachusetts (1810–1814) and vice president of the United States (1813–1814). As governor, Gerry signed a bill that redistricted the state to benefit his party. An opponent observed that an oddly shaped new district looked like a "salamander," whereupon another opponent responded that it was a "gerrymander." A newspaper subsequently printed an editorial cartoon of a monster named "gerrymander" with a body shaped like the district.

Gerrymandering works like this: suppose a community has 125 votes to be allocated among five districts of 25 votes each. The Yellow Party has 65 supporters or 52 percent of the community total and the Blue Party 60 supporters or 48 percent. Gerrymandering takes three forms: wasted vote, excess vote, and stacked vote.

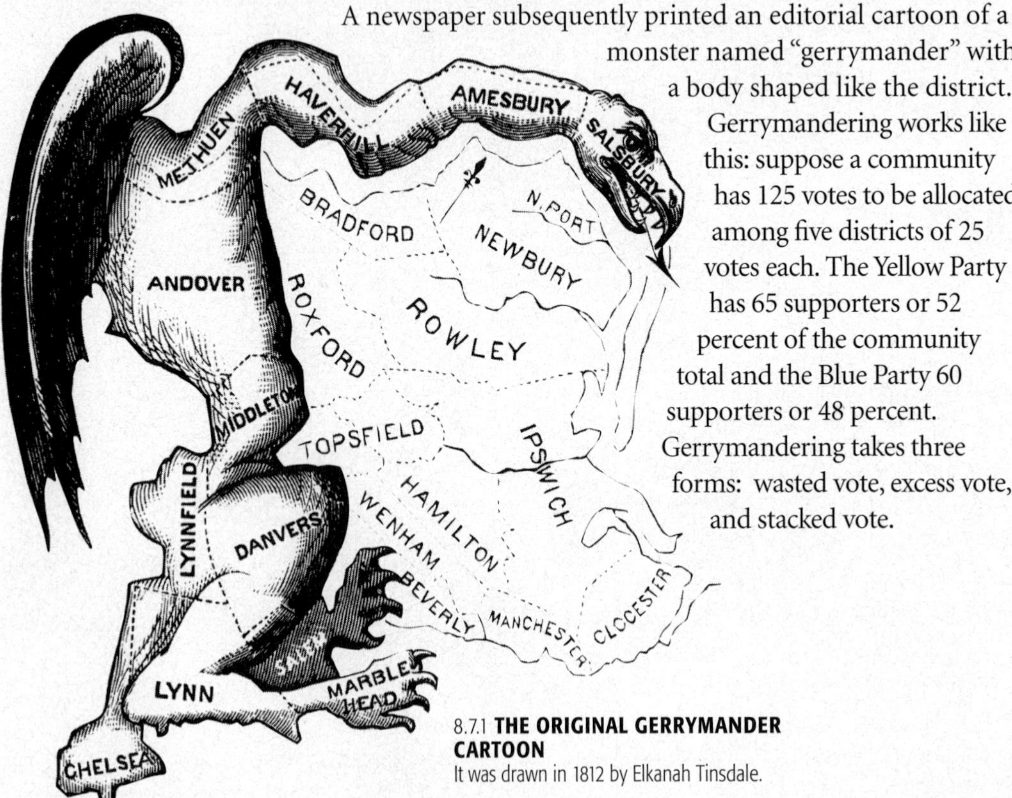

8.7.1 **THE ORIGINAL GERRYMANDER CARTOON**
It was drawn in 1812 by Elkanah Tinsdale.

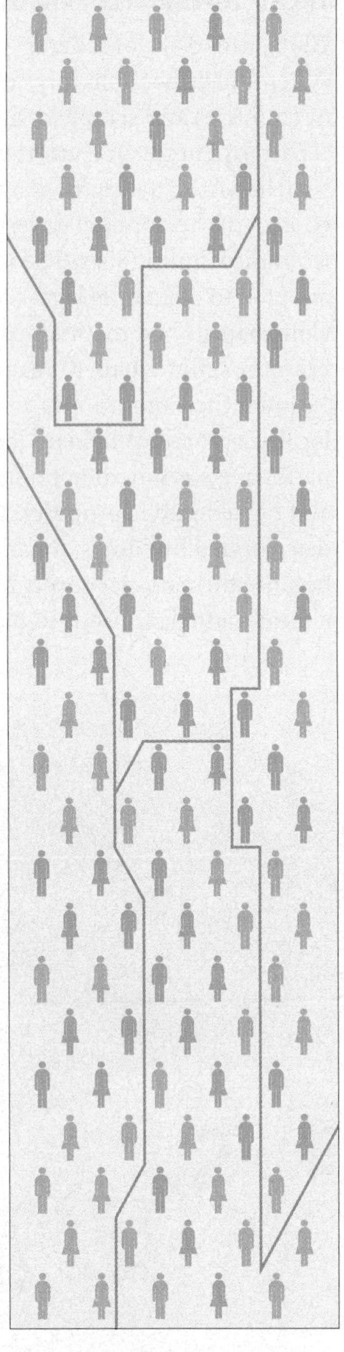

8.7.2 **"WASTED VOTE" GERRYMANDERING**
A "wasted vote" spreads opposition supporters across many districts but always in the minority. If the Yellow Party controls the redistricting process it could do a "wasted vote" gerrymander by putting in each of the five districts 26 of its voters and 24 of the Blue Party voters, thereby giving the Yellow Party the opportunity to win all five districts despite holding an overall edge of only 52–48 percent.

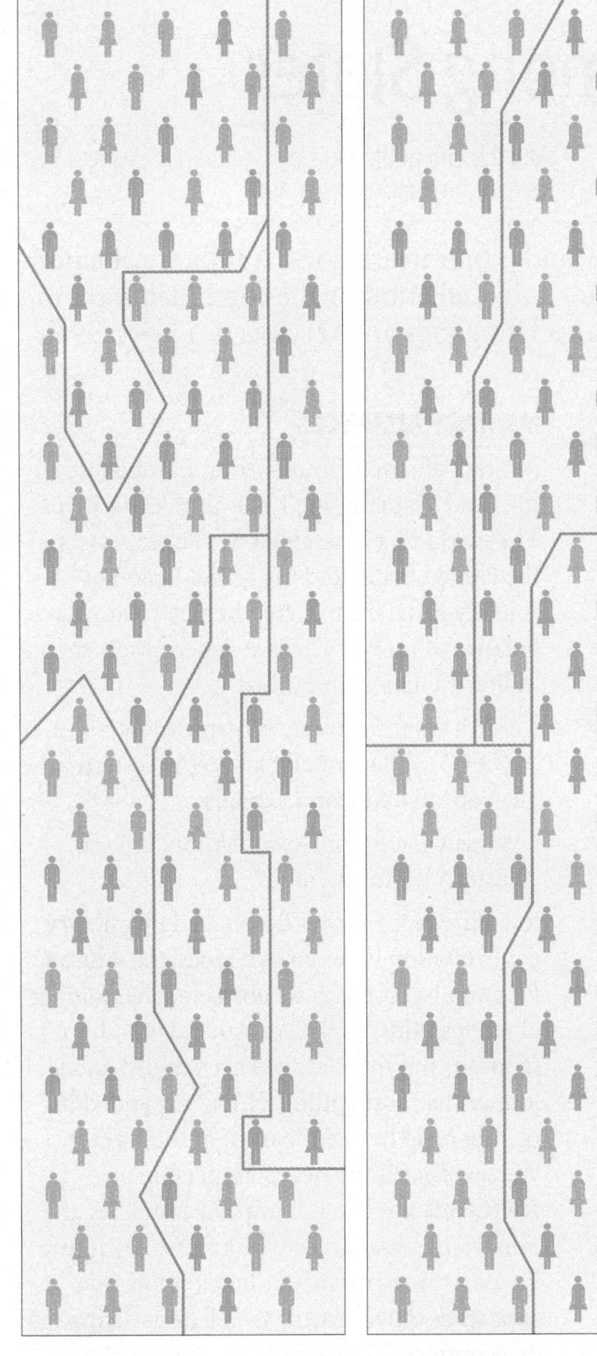

8.7.3 **"EXCESS VOTE" GERRYMANDERING**
An "excess vote" concentrates opposition supporters into a few districts. If the Blue Party controls the redistricting process it could do an "excess vote" gerrymander by putting 26 of its voters and 24 of the Yellow Party voters in four of the five districts and concentrating 34 Yellow Party voters and only 16 Blue Party voters in the fifth district, thereby giving the Blue Party the likelihood of winning four of five districts.

8.7.4 **"STACKED VOTE" GERRYMANDERING**
A "stacked vote" links distant areas of like-minded voters through oddly shaped boundaries. Recent gerrymandering in the United States has been primarily "stacked vote." If the Yellow Party in the example controls the redistricting process, the trend is to create three districts, each with 15 of its voters and 10 of the Blue Party's voters, and two districts, both with 10 of its voters and 15 of the Blue Party's voters. That way, all five districts are safely in possession of one party, with a majority of three for the Yellow Party and two for the Blue Party. When Yellow Party members are especially partisan, they could create four "safe" districts, each with 15 of its voters and 10 of the Blue Party's voters, while conceding the fifth district with 20 of Blue Party supporters and only 5 Yellow Party supporters.

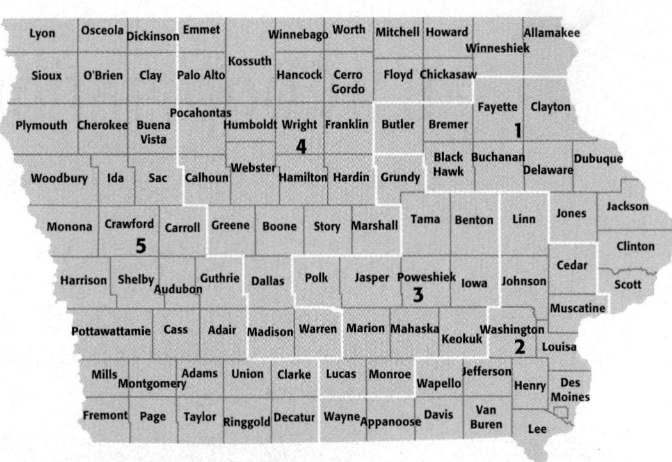

8.7.5 **IOWA: NO GERRYMANDERING**
Iowa is a state where Congressional district boundaries are drawn by an independent non-partisan commission. As a result, the districts are compact and follow county lines.

8.7.6 **GERRYMANDERING IN FLORIDA AND GEORGIA**
Florida's Republican-controlled state legislature gerrymandered Congressional district boundaries to concentrate Democratic voters using a combination of "wasted vote" and "stacked vote" strategies. Although Republican voters outnumbered Democrats slightly in Florida, the gerrymandering resulted in electing 18 Republicans and only seven Democrats in 2002 and 2004. Gerrymandered boundaries are especially clear on the southeast coast, where Republicans were able to win three of six districts in a region that has a substantial Democratic majority. Meanwhile, in Georgia Democrats tried to "stack" and "waste" Republican votes, but Republicans ended up winning 8 of 13 seats in 2002 and 7 in 2004.

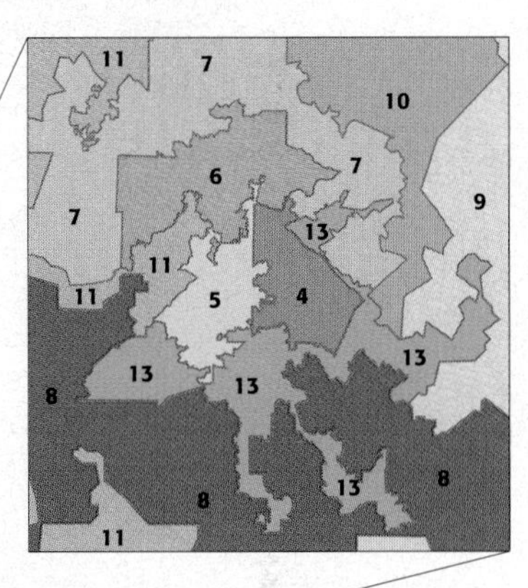

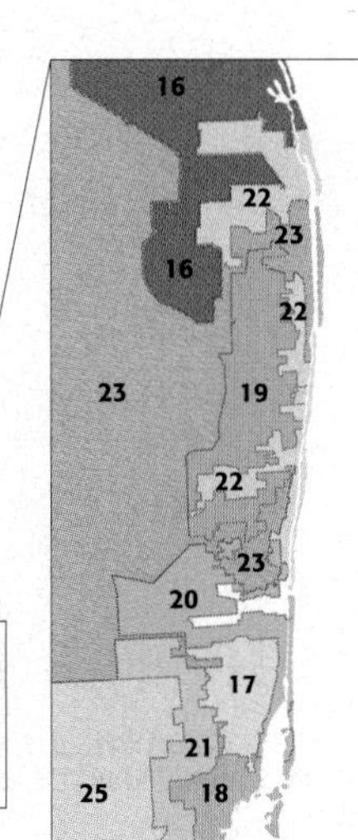

GEOGRAPHY OF WAR AND PEACE

8.8 Cooperation Among States

- During the Cold War, European states joined military alliances.
- With the end of the Cold War, economic alliances have become more important.

States cooperate with each other for military and economic reasons. A military alliance offers protection to one state through the threat of retaliation by the combined force of allies. Economic alliances have enlarged markets for goods and services produced in an individual state.

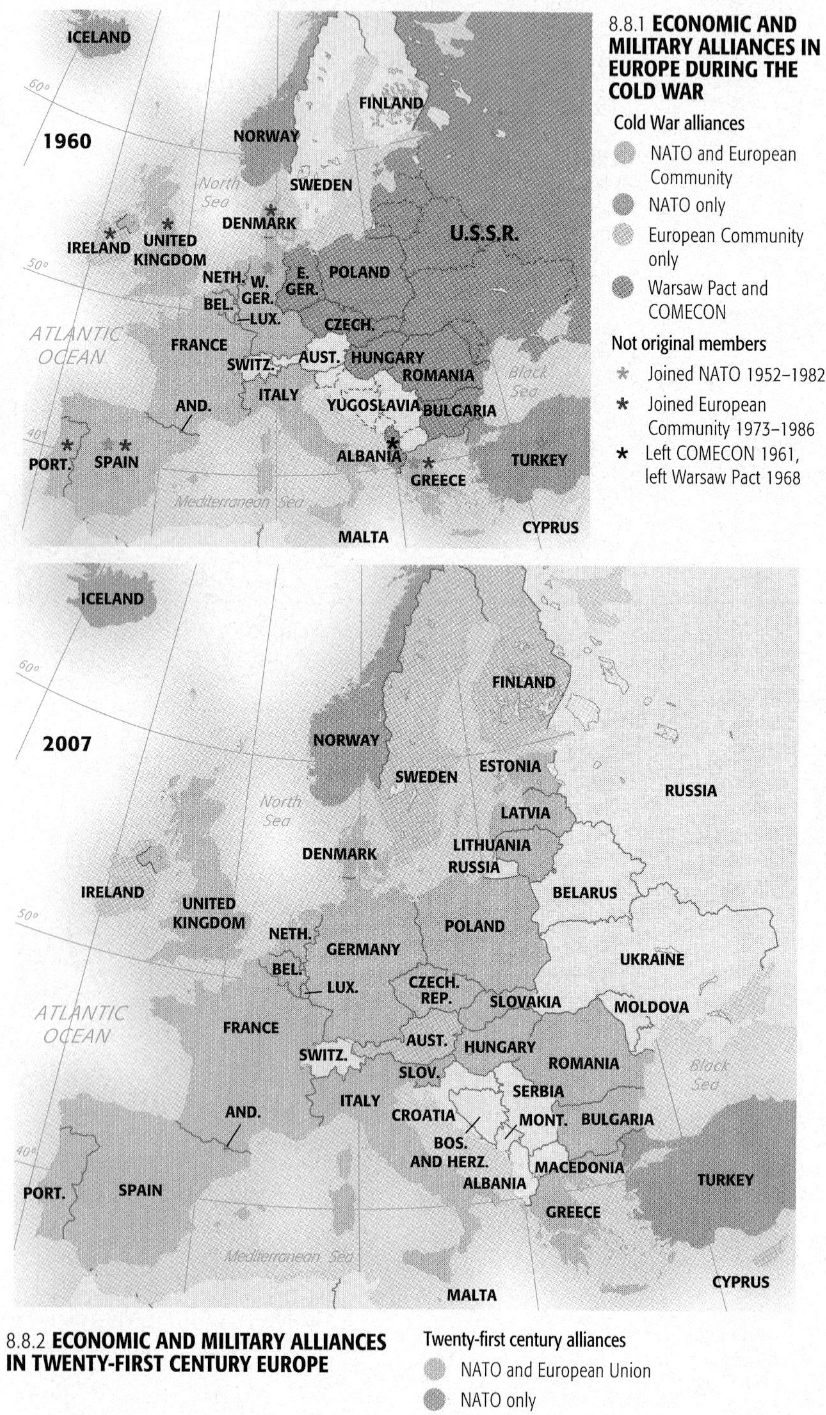

8.8.1 **ECONOMIC AND MILITARY ALLIANCES IN EUROPE DURING THE COLD WAR**

8.8.2 **ECONOMIC AND MILITARY ALLIANCES IN TWENTY-FIRST CENTURY EUROPE**

MILITARY ALLIANCES

Military alliances flourished in Europe during the Cold War era (late 1940s until early 1990s). The world then contained two superpowers—the United States and the Soviet Union—roughly equal in strength, thereby creating a **balance of power**. The two superpowers set up military alliances in Europe:

- North Atlantic Treaty Organization (NATO): The United States, 14 Western European allies, and Canada.
- Warsaw Pact: The Soviet Union and six Eastern European allies.

In a Europe no longer dominated by military confrontation between two blocs, the Warsaw Pact and NATO became obsolete. The number of troops under NATO command was sharply reduced, and the Warsaw Pact was disbanded. Rather than disbanding, NATO expanded its membership to include most of the former Warsaw Pact countries. Membership in NATO offers Eastern European countries an important sense of security against any future Russian threat, no matter how remote that appears at the moment, as well as participation in a common, united Europe.

ECONOMIC COOPERATION

With the decline in the military-oriented alliances, European states increasingly have turned to economic cooperation. Western Europe's most important economic organization is the European Union (formerly known as the European Economic Community, the Common Market, and the European Community).

When it was established in 1958, the predecessor to the European Union included six countries. It expanded to 12 countries during the 1980s and 27 countries during the first decade of the twenty-first century. Others hope to join.

In 1949, during the Cold War, the seven Eastern European Communist states in the Warsaw Pact formed an organization for economic cooperation, the Council for Mutual Economic Assistance (COMECON). Like the Warsaw Pact, COMECON disbanded in the early 1990s after the fall of communism in Eastern Europe.

A European Parliament is elected by the people in each of the member states simultaneously. It has removed most barriers to free trade: with a few exceptions, goods, services, capital, and people can move freely through Europe.

8.8.3 **CUBAN MISSILE CRISIS**
A major confrontation during the Cold War between the United States and Soviet Union came in 1962 when the Soviet Union secretly began to construct missile launching sites in Cuba, less than 150 kilometers (90 miles) from U.S. territory. President Kennedy went on national television to demand that the missiles be removed and ordered a naval blockade to prevent further Soviet material from reaching Cuba.

The U.S. Department of Defense took aerial photographs to show the Soviet buildup in Cuba. (Top) Three Soviet ships with missile equipment are being unloaded at Mariel naval port in Cuba. Within the outline box (enlarged below and rotated 90° clockwise) are Soviet missile transporters, fuel trailers, and oxider trailers (used to support the combustion of missile fuel).

At the United Nations, immediately after Soviet Ambassador Valerian Zorin denied that his country had placed missiles in Cuba, U.S. Ambassador Adlai Stevenson dramatically revealed aerial photographs taken by the U.S. Department of Defense clearly showing them. Faced with irrefutable evidence that the missiles existed, the Soviet Union ended the crisis by dismantling them.

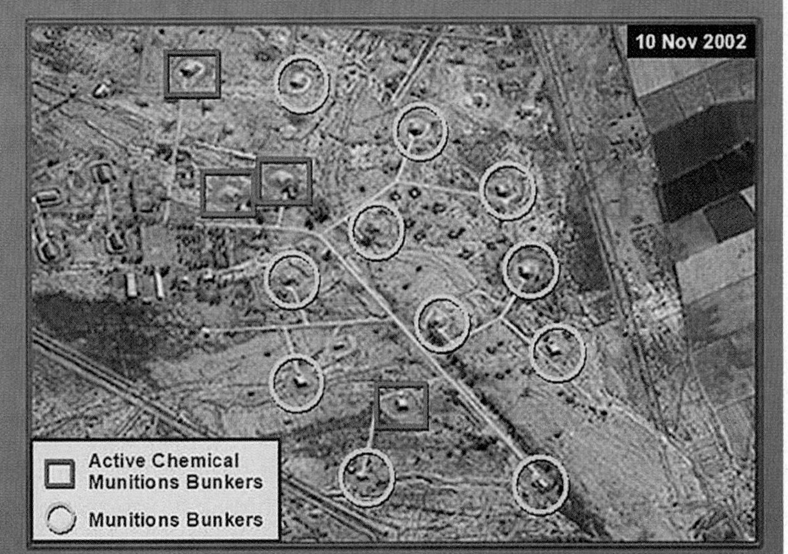

8.8.4 **IRAQ'S ALLEGED WEAPONS**
As the United States moved toward war with Iraq in 2003, Secretary of State Colin Powell scheduled a speech at the United Nations. The speech was supposed to present irrefutable evidence to the world justifying military action against Iraq. Adding credibility to the presentation, Powell was known to be the senior U.S. diplomat most reluctant to go to war.

Recalling the Cuban missile crisis, Powell displayed a series of aerial photos designed to prove that Iraq possessed weapons of mass destruction. Powell first showed an image of fifteen munitions bunkers at Taji, Iraq (top). He also showed close-ups of some of the bunkers (bottom).

Unlike the Cuban missile crisis in 1962, Powell could not make a convincing case at the United Nations for the U.S. position through aerial photos. As a result, the United States went to war with Iraq without the support of the United Nations. A subsequent U.S. State Department analysis found many inaccuracies in the interpretation of aerial photos presented by Powell. For example, the "decontamination vehicle" in the bottom left photo turned out to be a water truck. Two years later, Powell himself said that the 2003 speech had been a "blot" on his record.

GEOGRAPHY OF WAR AND PEACE

8.9 Terrorism

- Al-Qaeda terrorists have attacked the United States several times.
- Al-Qaeda justifies terrorism as a holy war.

Terrorism is the systematic use of violence by a group in order to intimidate a population or coerce a government into granting its demands. Terrorists attempt to achieve their objectives through organized acts such as bombing, kidnapping, hijacking, taking of hostages, and assassination that spread fear and anxiety among the population.

Terrorism differs from other acts of political violence because attacks are aimed at ordinary people rather than at military targets or political leaders. Victims of terrorism are a cross section of citizens who happen to be at the target at the time of the attack. A terrorist considers all citizens responsible for the actions being opposed; therefore, all citizens are equally justified as targets.

Distinguishing terrorism from other acts of political violence may be difficult. For example, if a Palestinian suicide bomber kills several dozen Israeli teenagers in a Jerusalem restaurant, is that an act of terrorism or wartime retaliation against Israeli government policies and army actions? Spokespersons on television make competing claims: Israel's sympathizers denounce the act as a terrorist threat to the country's existence, whereas advocates of the Palestinian cause argue that long-standing injustices and Israeli army attacks on ordinary Palestinian civilians provoked the act. Similarly, Russia claims that Chechen rebels are terrorists (see Chapter 7), and the British have long claimed that Irish Republican Army members are terrorists (see Chapter 6).

8.9.1 **AL-QAEDA ATTACK ON WORLD TRADE CENTER**
September 11, 2001: Tower 1 is burning from the crash of American Flight 11 at 8:45 A.M. as United Flight 175 approaches World Trade Center Tower 2 at 9:03 A.M. (top) and crashes into it (center). Both towers are engulfed (bottom).

ATTACKS ON THE UNITED STATES

The United States has suffered several terrorist attacks, including the most dramatic on September 11, 2001. The tallest buildings in the United States, the 110-story twin towers of the World Trade Center in New York City, were destroyed, and the Pentagon in Washington, D.C., was damaged. The attacks resulted in nearly 3,000 fatalities.

Responsible or implicated in most of the anti-U.S. terrorism has been the al-Qaeda network, founded by Osama bin Laden. One of about 50 children of a billionaire construction firm owner, bin Laden used his several-hundred-million-dollar inheritance to fund al-Qaeda.

Bin Laden recruited militant Muslims from Arab countries to fight against the Soviet Union in Afghanistan during the 1980s. After the Soviet withdrawal in 1989, bin Laden returned to Saudi Arabia, but he was expelled in 1991 for opposing the Saudi government's decision permitting the United States to station troops there during the 1991 war against Iraq. Bin Laden moved to Sudan but was expelled in 1994 for instigating attacks against U.S. troops in Yemen and Somalia; he returned to Afghanistan, where he lived as a "guest" of the Taliban-controlled government.

Bin Laden issued a declaration of war against the United States in 1996 because of U.S. support for Saudi Arabia and Israel. In a 1998 fatwa ("religious decree"), bin Laden argued that Muslims had a duty to wage a holy war against U.S. citizens because the United States was responsible for maintaining the Saud royal family as rulers of Saudi Arabia and a state of Israel dominated by Jews. Destruction of the Saudi monarchy and the Jewish state of Israel would liberate from their control Islam's three holiest sites of Makkah (Mecca), Madinah, and Jerusalem.

AL-QAEDA

Al-Qaeda (an Arabic word meaning "the foundation" or "the base") was created around 1990 to unite jihad fighters in Afghanistan, as well as supporters of bin Laden elsewhere in the Middle East. Membership is estimated at around 20,000, dispersed in as many as 34 countries. Its size has been hard to estimate because the organization consists of a large number of isolated autonomous cells whose members have minimal contact with those in other cells or even others in the same cell.

Al-Qaeda is not a single unified organization. It includes the following:

- The original organization founded by Osama bin Laden, which has been implicated in several bombings since 9/11 in Riyadh and Khobar, Saudi Arabia; Istanbul, Turkey; London, England; Sharm-el-Sheikh, Egypt; and Amman, Jordan.
- Local franchises concerned with country-specific issues, such as Jemaah Islamiyah, which has launched several attacks in Indonesia in support of installing a fundamentalist Islamic government.
- Imitators aligned with al-Qaeda but not financially tied to it, such as a group that blew up trains in Madrid, Spain.

Most al-Qaeda cell members have lived in ordinary society, supporting themselves with jobs, burglary, and credit card fraud. They are examples of "sleepers," so called because they await their cell leader's order to "awake" and perform a job for the network. The cell's planners and attackers typically do not have direct contact with each other, whereas the support members encounter both planners and attackers but do not know the target or attack plan. If arrested, members of one cell are not in a position to identify members of other cells.

Al-Qaeda's use of religion to justify attacks has posed challenges to Muslims and non-Muslims alike. For many Muslims, the challenge has been to express disagreement with the policies of governments in the United States and Europe yet disavow the use of terrorism. For many Americans and Europeans, the challenge has been to distinguish between the peaceful but unfamiliar principles and practice of the world's 1.3 billion Muslims and the misuse and abuse of Islam by a handful of terrorists.

8.9.2 **USS COLE**
Al-Qaeda bombed the ship docked in the port of Aden, Yemen, October 12, 2000, killing 17 U.S. service personnel.

8.9.3 **LONDON**
Al-Qaeda bombed several subway trains and buses, July 7, 2005, killing 56 (including 4 terrorists).

8.9.4 **MADRID**
A local terrorist group loosely associated with al-Qaeda blew up several commuter trains, March 11, 2004, killing 192.

Chapter **Summary**

STATES OF THE WORLD

- The world is divided into nearly 200 states, nearly all of which are members of the United Nations.
- Historically, most lands belonged to empires or were colonies of states.
- Few colonies are left in the world, most with very small populations.

STRUCTURE OF STATES

- States take five types of shapes—compact, prorupted, elongated, fragmented, and perforated.
- A number of states are landlocked, a challenge for international trade.
- Either physical or cultural features can be used to set boundaries between states.
- Boundaries within countries are delineated for elections and can be gerrymandered.
- States may be organized into either unitary or federal systems of local government

GEOGRAPHY OF WAR AND PEACE

- States increasingly cooperate in regional military and economic alliances, especially in Europe.
- During the second half of the twentieth century, the Cold War divided countries into pro-U.S. and pro-Soviet alliances.
- In the twenty-first century, terrorism has replaced the Cold War as the principal challenge to security.

DUST ON VEHICLE THAT WAS PARKED NEAR THE WORLD TRADE CENTER DURING 9/11 ATTACK

Geographic Consequences of Change

Two political trends dominate the early twenty-first century. First, after a half-century dominated by the Cold War between two superpowers—the United States and the former Soviet Union—the world has entered a period characterized by an unprecedented increase in the number of new states created to satisfy the desire of nationalities for self-determination as an expression of cultural distinctiveness. Turmoil has been the result because, in many cases, the boundaries of the new states do not precisely match the territories occupied by distinct nationalities.

At the same time, with the end of the Cold War, military alliances have become less important than patterns of global and regional economic cooperation and competition among states. Economic cooperation has increased among neighboring states in Western Europe and North America, and competition among these two blocs, as well as Japan, has increased.

Both trends have diminished the importance of the nation-state in the twenty-first century. In Western Europe, economic cooperation has been responsible. Western Europeans carry European Union rather than national passports, even though they don't need to show them when traveling within Western Europe. National currencies have been replaced by the euro.

Meanwhile, in Eastern Europe, the nation-state concept has been carried to the other extreme. The desire for self-determination among Eastern Europe's many ethnicities has resulted in the breakup of states into a larger number of smaller ones. The future viability of tiny nation-states such as Kosovo and Macedonia is questionable. Perhaps a country's ability to provide its citizens with food, jobs, economic security, and material wealth, rather than the principle of self-determination, might be a stronger basis for creating and maintaining states.

SHARE YOUR VOICE Student Essay

Jordan Endicott
Illinois Institute of Technology,
Chicago Kent College of Law

Returning to Croatia

Migration is something that is inherent to most creatures that inhabit Earth. My heritage here in America can be traced back to my great-grandfather coming over from Croatia when he was 16. Like many immigrants at the time, he came from a humble background, left by himself, and arrived without much.

In 2006, I had the opportunity to go to Croatia myself to visit my extended family members that are still living there. The side of the family that stayed behind in Croatia has prospered greatly when compared to the meager background that my great-grandfather left behind. Most of my family that I stayed with there I had never met or spoken with before, but it was a feeling of immediate acceptance as soon as I arrived.

While I was there, I had the opportunity to take a drive across the border into Bosnia in order to go and see the home that my great-grandfather was born in. It was in a tiny farming village that had mostly been destroyed and subsequently abandoned during the war there. My grandma's cousin guided us to where the house was, and just seeing the site created such a connection for me to my roots.

The house was just basically rubble, with no roof left and only a couple of outside and interior walls still standing. Most of the area was completely overgrown with weeds and trees, so it was hard to do any sort of exploring of the site. I took my pictures though, and pulled a stone out of the wall to take home with me so I would at least have something physical to remember it by.

Most people in America never get the chance to see where their family migrated from, so the chance to do so really meant quite a bit to me. That is what made my vacation to Croatia such a memorable one. Even though my great-grandfather had migrated over 80 years ago, the connection to my family and my heritage in the land that he came from was not difficult to pick up again and embrace.

Chapter **Resources**

THE IRAQ ATTACK

In the wake of the 9/11 attacks, the U.S.-led war on terrorism initially focused on al-Qaeda but soon switched to what it considered state-supported terrorism.

The war on terrorism began when the United States attacked Afghanistan in 2001 because the Taliban leaders sheltered Osama bin Laden and other al-Qaeda terrorists. Although removed from power, the Taliban were able to regroup and resume an insurgency against the U.S.-backed Afghanistan government.

The United States led an attack against Iraq in 2003 in order to depose the country's longtime president, Saddam Hussein. U.S. officials justified removing Hussein because he had created biological and chemical weapons of mass destruction. These weapons could fall into the hands of terrorists, the U.S. government charged, because close links were said to exist between Iraq's government and al-Qaeda. The United Kingdom and a few other countries joined U.S forces, but most countries did not offer support.

U.S. confrontation with Iraq predated the war on terror. From the time he became president of Iraq in 1979, Hussein's behavior had raised concern around the world:

- A war with neighbor Iran, begun in 1980, ended 8 years later in stalemate.
- Nuclear weapons allegedly were being developed at a reactor near Baghdad; Israeli planes destroyed the nuclear reactor in 1981.
- UN inspectors found evidence of weapons of mass destruction in Iraq in the 1980s, including programs for making weapons from the VX nerve agent and from such toxic agents as botulinum, anthrax, aflatoxin, and clostridium.
- Hussein ordered the use of poison gas in 1988 against Iraqi Kurds, killing 5,000.
- Iraq invaded neighboring Kuwait, which Hussein claimed was part of Iraq, in 1990; the 1991 U.S.-led Operation Desert Storm drove Iraq out of Kuwait.

To justify the invasion of Iraq, U.S. officials claimed that Iraq still had weapons of mass destruction and ties to al-Qaeda. However, UN experts concluded that Iraq had destroyed its weapons of mass destruction in 1991 after its Desert Storm defeat, and the United States never found any. The U.S. assertion that Hussein had close links with al-Qaeda was also refuted by most other countries, as well as ultimately by U.S. intelligence agencies.

The United States argued instead that Iraq needed a "regime change." Hussein's quarter-century record of brutality justified replacing him with a democratically elected government, according to U.S. officials. The U.S. position drew little international support, because sovereign states are reluctant to invade another sovereign state just because they dislike its leader, no matter how odious.

Having invaded Iraq and removed Hussein from power, the United States expected an enthusiastic welcome from the Iraqi people. Instead, the United States became embroiled in a complex and violent struggle among religious sects and tribes that had little to do with fighting terrorists.

At this time Iran was a less immediate threat. However, the United States accused Iran of harboring al-Qaeda members and of trying to gain influence in Iraq, where like Iran, the majority of the people were Shiites. More troubling to the international community was evidence that Iran was developing a nuclear weapons program. Prolonged negotiations are being undertaken to dismantle Iran's nuclear capabilities without resorting to yet another war in the Middle East.

U.S. Marine drapes a flag over a statue of Saddam Hussein in Baghdad, Iraq, in 2003.

KEY TERMS

Balance of power
Condition of roughly equal strength between opposing countries or alliances of countries.

Boundary
Invisible line that marks the extent of a state's territory.

City-state
A sovereign state comprising a city and its immediate hinterland.

Colonialism
Attempt by one country to establish settlements and to impose its political, economic, and cultural principles in another territory.

Colony
A territory that is legally tied to a sovereign state rather than completely independent.

Compact state
A state in which the distance from the center to any boundary does not vary significantly.

Elongated state
A state with a long, narrow shape.

Federal state
An internal organization of a state that allocates most powers to units of local government.

Fragmented state
A state that includes several discontinuous pieces of territory.

Gerrymandering
Process of redrawing legislative boundaries for the purpose of benefiting the party in power.

Landlocked state
A state that does not have a direct outlet to the sea.

Perforated state
A state that completely surrounds another one.

Prorupted state
An otherwise compact state with a large projecting extension.

Sovereignty
Ability of a state to govern its territory free from control of its internal affairs by other states.

State
An area organized into a political unit and ruled by an established government with control over its internal and foreign affairs.

Unitary state
An internal organization of a state that places most power in the hands of central government officials.

ON THE INTERNET

The U.S. Central Intelligence Agency has a World Factbook. Select a country from the drop-down list to find background information, as well as facts and figures about the country's demography, economy, physical geography, government, and military. Maps are also available. On-line at **https://www.cia.gov/library/publications/the-world-factbook/.**

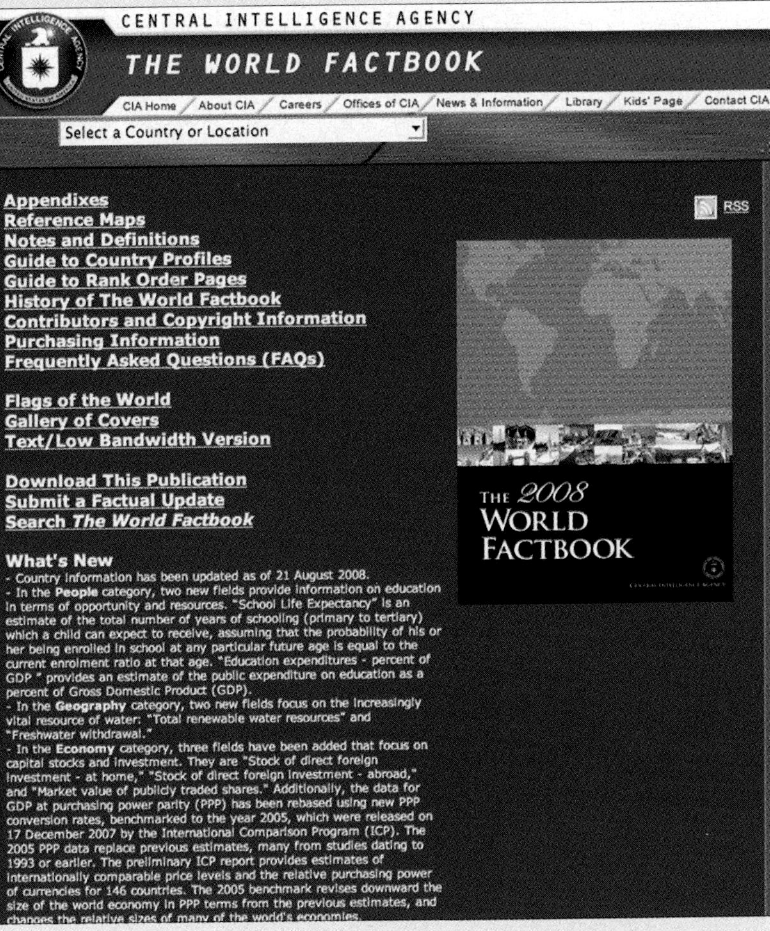

FURTHER READINGS

Agnew, John. "Sovereignty Regimes: Territoriality and State Authority in Contemporary World Politics." *Annals of the Association of American Geographers* 95 (2005): 437–61.

Cohen, Saul B. "Geopolitical Realities and United States Foreign Policy." *Political Geography* 22 (2003): 1–33.

Cutter, Susan L., Douglas B. Richardson, and Thomas J. Wilbanks, eds. *The Geographical Dimensions of Terrorism.* New York: Routledge, 2003.

Dale, E. H. "Some Geographical Aspects of African Land-Locked States." *Annals of the Association of American Geographers* 58 (1968): 485–505.

Johnston, R. J., Peter J. Taylor, and Michael J. Watts. *Geographies of Global Change,* 2d ed. Cambridge, MA: Blackwell, 2002.

Kliot, N., and Y. Mansfield. "The Political Landscape of Partition: The Case of Cyprus." *Political Geography* 16 (1997): 495–521.

Murphy, Alexander B. "Historical Justifications for Territorial Claims." *Annals of the Association of American Geographers* 80 (1990): 531–48.

Nijman, Jan. "The Limits of Superpower: The United States and the Soviet Union Since World War II." *Annals of the Association of American Geographers* 82 (1992): 681–95.

Ó Tuathail, Gearóid. "The Postmodern Geopolitical Condition: States, Statecraft, and Security at the Millennium." *Annals of the Association of American Geographers* 90 (2000): 166–78.

Prescott, J. R.V. *Political Frontiers and Boundaries.* London: Unwin Hyman, 1990.

Taylor, Peter J., and Colin Flint. *Political Geography: World Economy, Nation-State and Locality,* 4th ed. New York: Prentice Hall, 2000.

Chapter 9 DEVELOPMENT

The world is divided into more developed countries (MDCs) and less developed countries (LDCs). The one-fifth of the world's people living in MDCs consume five-sixths of the world's goods, whereas the 14 percent of the world's people who live in Africa consume about 1 percent. The United Nations draws stark contrasts in consumption patterns between the MDCs and the LDCs:

- Americans spend more per year on cosmetics ($8 billion) than the cost of providing schools for the 2 billion children in the world in need of them ($6 billion).
- Europeans spend more on ice cream per year ($11 billion) than the cost of providing a working toilet to the 2 billion people currently without one at home ($9 billion).

To reduce disparities between rich and poor countries, LDCs must develop more rapidly. This means increasing per capita gross domestic product more rapidly and using the additional funds to make more rapid improvements in people's social and economic conditions.

THE KEY ISSUES IN THIS CHAPTER

HUMAN DEVELOPMENT

9.1 Human Development Index
9.2 Gender-related Development

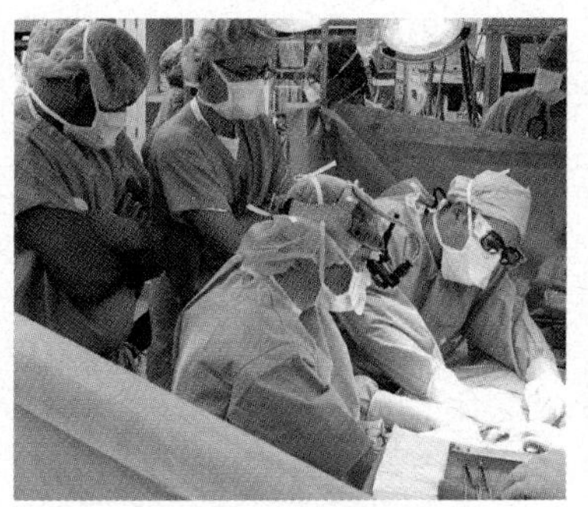

DEVELOPMENT INDICATORS

9.3 Economic Indicators
9.4 Social Indicators
9.5 Health Indicators

DEVELOPMENT THROUGH TRADE

9.6 Two Paths to Development
9.7 World Trade
9.8 Financing Development
9.9 Fair Trade

DISRUPTIONS TO THE DELIVERY OF AID
Locals try to dig a truck out of a usually dry riverbed (called a wadi) in which a truck got stuck during the rainy season, near the border with the Darfur region of Sudan. The rainy season makes road trips almost impossible, which seriously hampers aid to the refugee camps hosting around 300,000 refugees from Darfur.

HUMAN DEVELOPMENT

9.1 Human Development Index

- The Human Development Index (HDI) measures a country's level of development.
- The HDI combines economic, social, and demographic factors.

Earth's nearly 200 countries can be classified according to their level of development, which is the process of improving the material conditions of people through diffusion of knowledge and technology. The development process is continuous, involving never-ending actions to constantly improve the health and prosperity of the people.

A country's level of development can be distinguished according to three factors—economic, social, and demographic. The **Human Development Index (HDI)**, created by the United Nations, recognizes that a country's level of development is a function of all three of these factors.

To create the HDI, the United Nations selects one economic factor, two social factors, and one demographic factor that in the opinion of an international team of analysts best reveal a country's level of development.

- The economic factor is gross domestic product (GDP) per capita.
- The social factors are the literacy rate and amount of education.
- The demographic factor is life expectancy.

The four factors are combined to produce a country's HDI. The UN has computed HDIs for countries every year since 1990, although it has tinkered a few times with the method of computation. The highest HDI possible is 1.0, or 100 percent.

Because many countries cluster at the high or low end of the continuum of development, they can be divided into two groups. A more developed country (MDC) has progressed further along a continuum of development than has a less developed country (LDC).

9.1.1 **HDI ECONOMIC FACTOR: GDP PER CAPITA**

9.1.2 **ANGLO-AMERICA**
Expenditures on health care are high in the region.

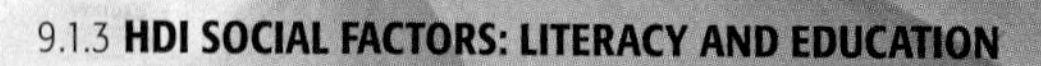

9.1.3 **HDI SOCIAL FACTORS: LITERACY AND EDUCATION**

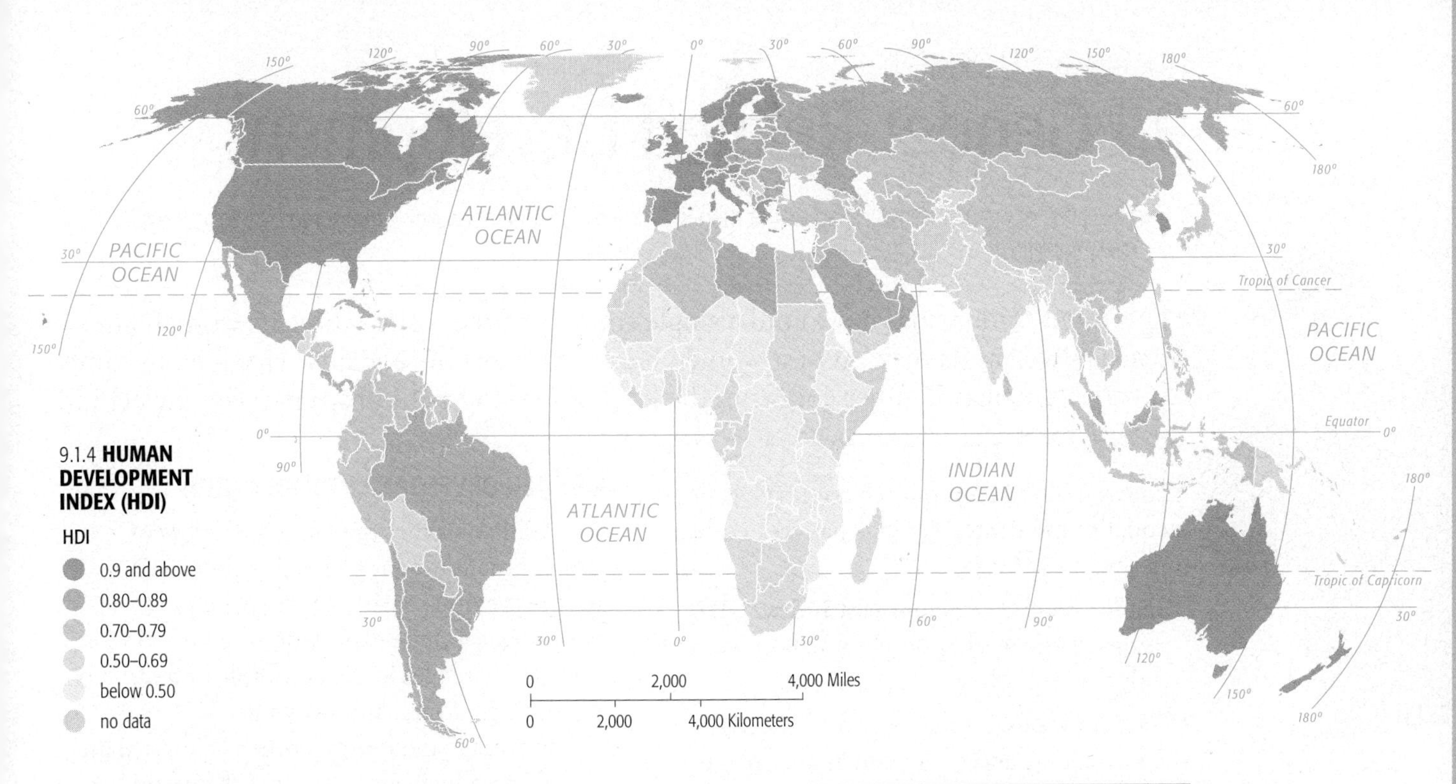

9.1.4 **HUMAN DEVELOPMENT INDEX (HDI)**

Three of the nine major cultural regions—Anglo-America, Western Europe, and Eastern Europe, plus Japan and the South Pacific—are considered more developed. The other six regions—Latin America, the Middle East, East Asia, Southeast Asia, South Asia, and sub-Saharan Africa—are considered less developed.

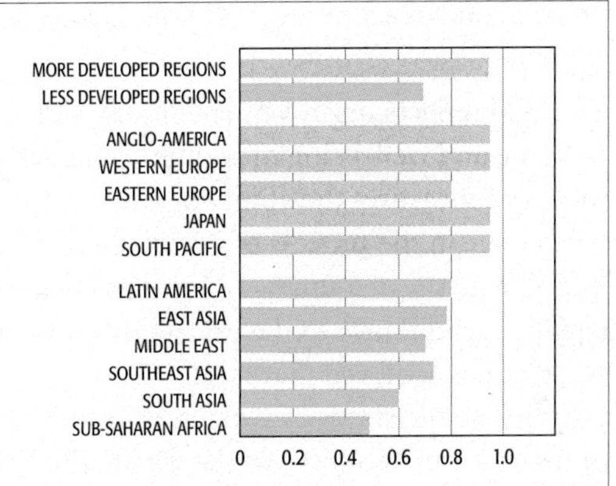

9.1.5 **REGIONAL DIFFERENCES IN HDI**

9.1.6 **HDI DEMOGRAPHIC FACTOR: LIFE EXPECTANCY**

FOCUS ON ANGLO-AMERICA

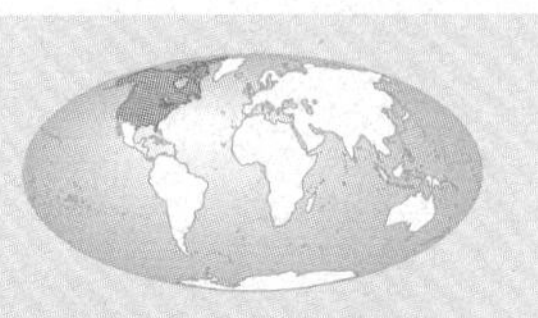

The HDI of the United States is lower than eight European countries, as well as Australia, Canada, and Japan. By most measures of development, the United States ranks within the top handful of countries. However, the HDI is constructed from four measures considered especially important by the United Nations. The United States ranked 6th in GDP per capita, but was tied for 14th in literacy, 17th in education, and 30th in life expectancy. The social indicators reflect a school drop out rate that is relatively high, and life expectancy is lower because of inadequate health care for low-income people.

HUMAN DEVELOPMENT

9.2 Gender-Related Development

- The status of women is lower than that of men in every country.
- Gender inequality is measured by two indexes.

The United Nations has not found a single country in the world where its women are treated as well as its men. At best women have achieved near-equality with men in some countries, whereas in other countries the level of development of women lags far behind the level for men.

To measure the extent of each country's gender inequality, the United Nations has created two indexes:

- **Gender-related Development Index (GDI)** compares the level of development of women with that of both sexes.
- **Gender Empowerment Measure (GEM)** compares the ability of women and men to participate in economic and political decision making.

The GDI reflects improvements in the standard of living and well-being of women, whereas the GEM measures the ability of women to participate in the process of achieving those improvements. A country with complete gender equality would have a GDI and a GEM of 1.0. No country has achieved that level.

Although the status of women is lower than that of men in every country of the world, the United Nations has found that it is improving. Since 1970, the gap has been reduced by two-thirds in LDCs and by one-fourth in MDCs.

GENDER DEVELOPMENT INDEX (GDI)

The GDI combines the same indicators of development used in the HDI (discussed on the previous page)—income, literacy, education, and life expectancy—adjusted to reflect differences in the accomplishments and conditions of men and women.

The average income of women is lower than that of men in every country of the world. Women on average have 60 percent of the income of men in MDCs and 50 percent in LDCs.

Female university students outnumber male students in MDCs, but are less numerous in LDCs. Literacy rates are comparable in MDCs, but women are less literate than men in LDCs. The gender gap in education and literacy is especially high in sub-Saharan Africa.

Women live several years longer than men in MDCs but only a year or two longer in LDCs. In LDCs, women's shorter life span derives primarily from the hazards of childbearing.

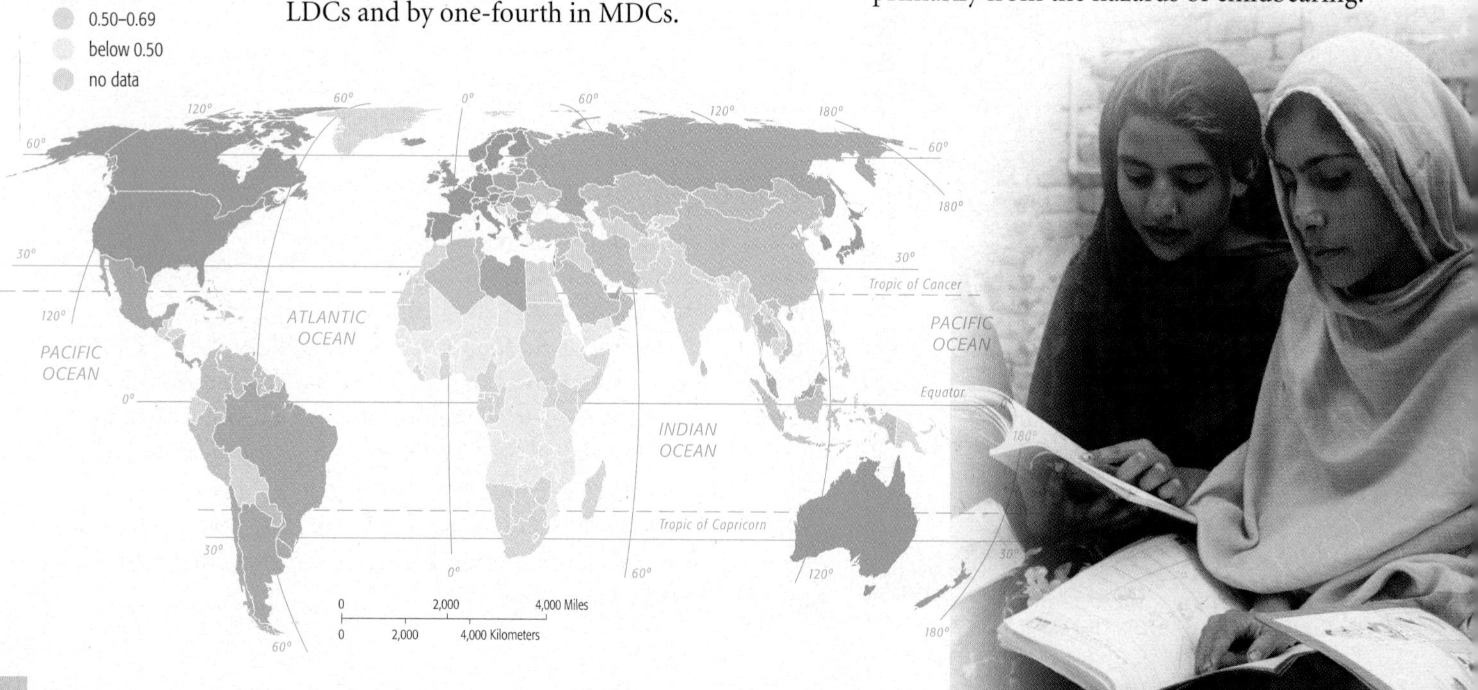

9.2.1 **GENDER-RELATED DEVELOPMENT INDEX (GDI)**

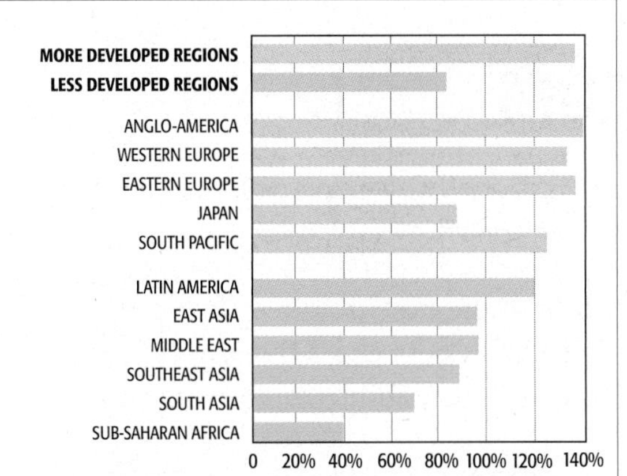

9.2.2 **FEMALE COLLEGE ATTENDANCE AS PERCENTAGE OF MALE COLLEGE ATTENDANCE**

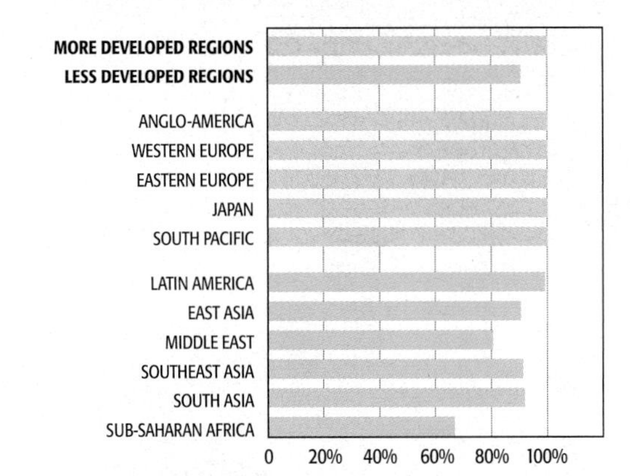

9.2.3 **FEMALE LITERACY RATE AS PERCENTAGE OF MALE LITERACY RATE**

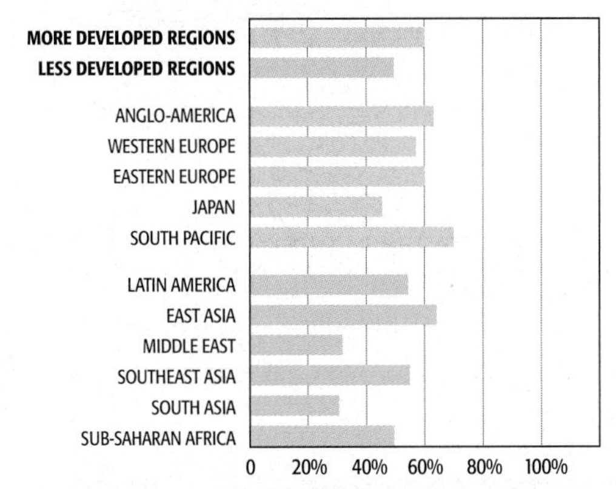

9.2.4 **FEMALE INCOME AS PRECENTAGE OF MALE INCOME**

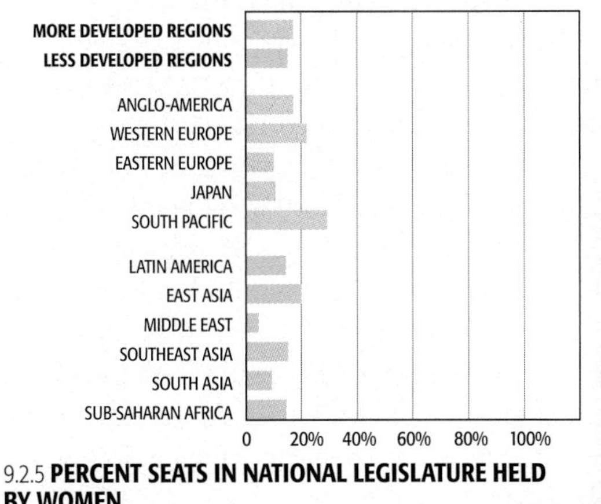

9.2.5 **PERCENT SEATS IN NATIONAL LEGISLATURE HELD BY WOMEN**

9.2.6 **"WHY NOT ME?"** The girl asks as the mother takes the boy to school.

GENDER EMPOWERMENT MEASURE (GEM)

The GEM is calculated by combining two indicators of economic power (income and professional jobs) and two indicators of political power (managerial jobs and elected jobs).

Professional and technical workers are more likely to be female in MDCs and more likely to be male in LDCs.

The percentage of women elected to the national legislature is similar in MDCs and LDCs. Women hold approximately one-sixth of the seats worldwide. Female representation is especially low in the Middle East and South Asia.

FOCUS ON WESTERN EUROPE

Within Western Europe the level of development is the world's highest in a core area that extends from southern Scandinavia to western Germany. To maintain its high level of development, Western Europe must import food, energy, and minerals. To pay for their imports, Western Europeans have provided high-value goods and services, such as insurance, banking, and luxury motor vehicles, including BMW and Mercedes-Benz.

9.2.7 **WESTERN EUROPE** The region has the highest percentage of women in government, with the exception of the South Pacific.

DEVELOPMENT INDICATORS

9.3 Economic Indicators

- Average incomes are higher in MDCs than in LDCs.
- People in MDCs are more productive and possess more goods.

Gross domestic product per capita is the economic indicator included in the HDI calculation. Other economic indicators that distinguish between MDCs and LDCs include economic structure, worker productivity, and availability of consumer goods.

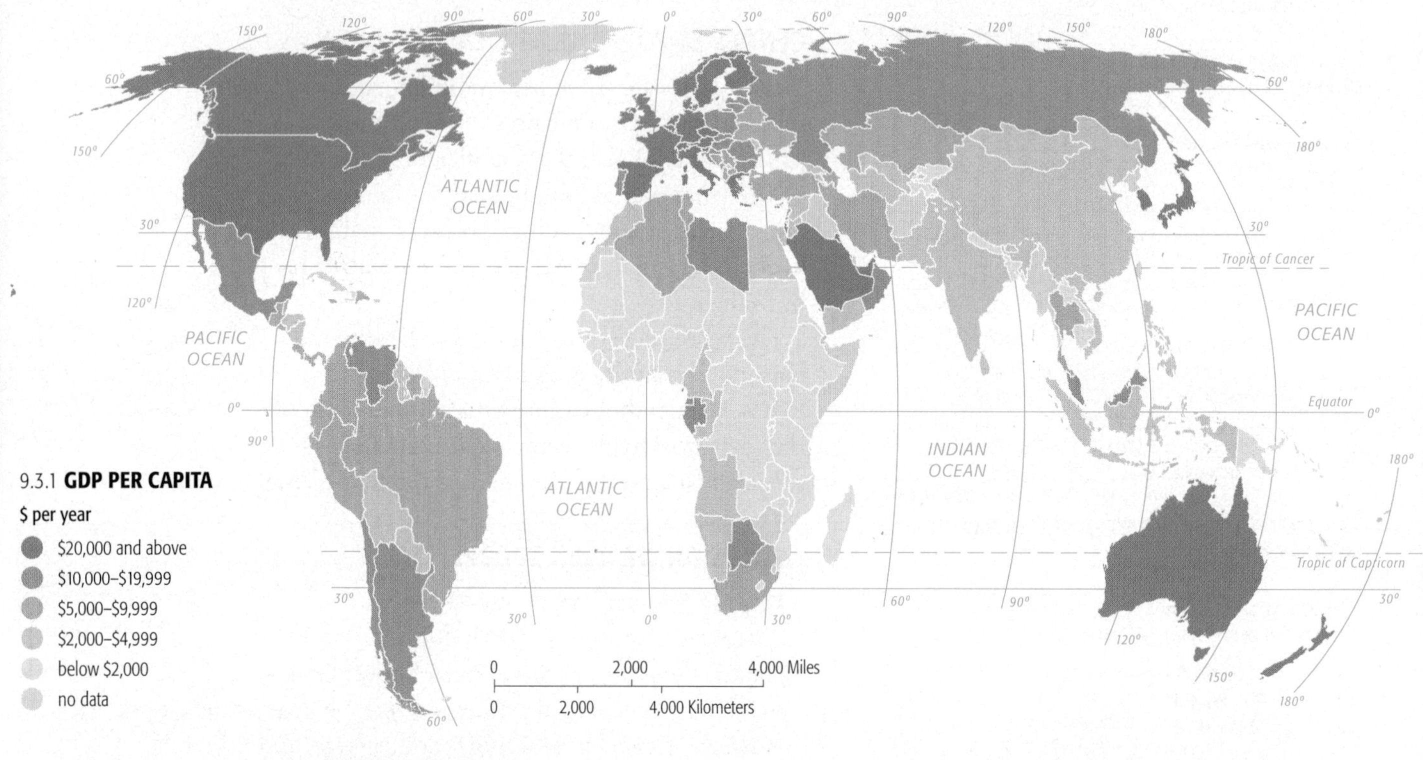

9.3.1 **GDP PER CAPITA**

PER CAPITA INCOME

The average individual earns a much higher income in an MDC than in an LDC. The **gross domestic product (GDP)** is the value of the total output of goods and services produced in a country, normally during a year. Dividing the GDP by total population measures the contribution made by the average individual toward generating a country's wealth in a year.

For example, GDP in the United States is currently about $14 trillion and its population is about 300 million, so GDP per capita is about $46,000.

9.3.2 **ECONOMIC STRUCTURE: SECONDARY SECTOR** Garment workers in Bangladesh factory.

ECONOMIC STRUCTURE

Average per capita income is higher in MDCs because people typically earn their living by different means than in LDCs. Jobs fall into three categories—primary (including agriculture), secondary (including manufacturing), and tertiary (including services). A low percentage of primary-sector workers indicates that a handful of farmers can produce enough food for the rest of society. Freed from the task of growing their own food, most people in an MDC can contribute to an increase in the national wealth by working in the secondary and tertiary sectors.

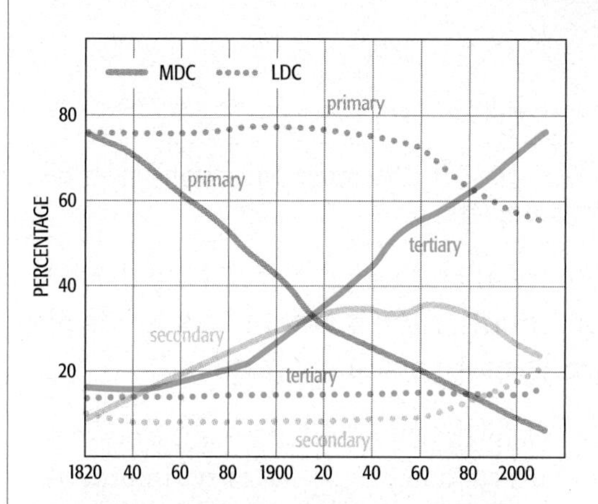

9.3.3 **CHANGES IN PERCENT EMPLOYED IN PRIMARY, SECONDARY, AND TERTIARY SECTORS**

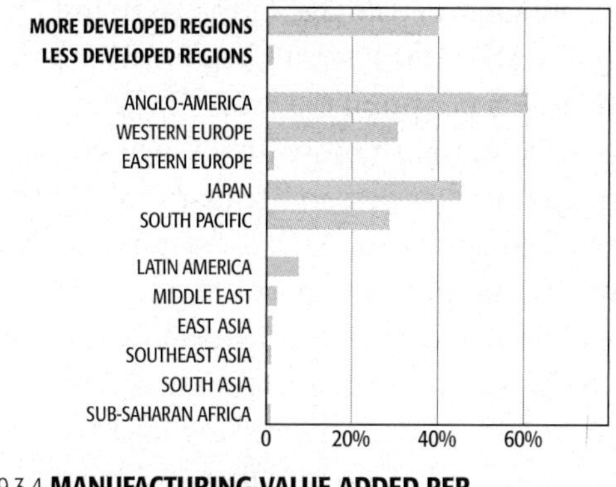

9.3.4 **MANUFACTURING VALUE ADDED PER MANUFACTURING WORKER**

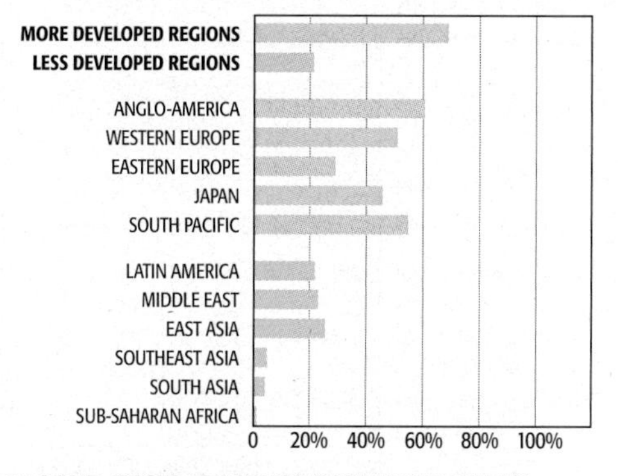

9.3.5 **LAND-BASED TELEPHONE LINES PER CAPITA**

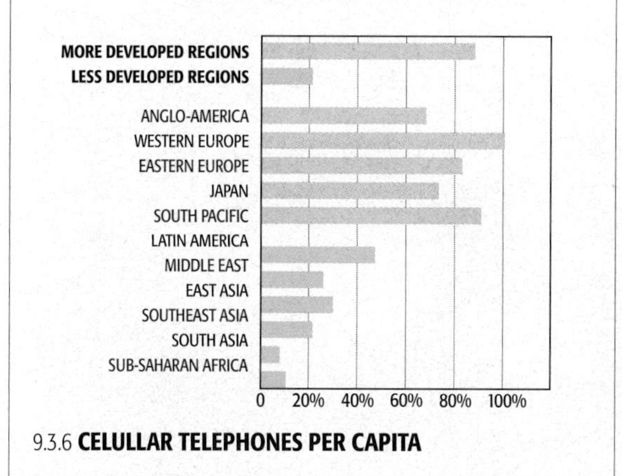

9.3.6 **CELULLAR TELEPHONES PER CAPITA**

PRODUCTIVITY

Workers in MDCs are more productive than those in LDCs. Productivity is the value of a particular product compared to the amount of labor needed to make it. Productivity can be measured by the **value added** per worker. The value added in manufacturing is the gross value of the product minus the costs of raw materials and energy. Workers in MDCs produce more with less effort because they have access to more machines, tools, and equipment to perform much of the work.

9.3.7 **EASTERN EUROPE** Auto production is increasing in the region.

CONSUMER GOODS

Part of the wealth generated in MDCs is used to purchase goods and services. Especially important are goods and services related to transportation and communications, including motor vehicles, telephones, and computers.

Technological change may help to reduce the gap in access to communications between MDCs and LDCs. For example, the distribution of cellular telephone service varies from the pattern for other communications devices. Cell phone ownership is lower than land-line ownership in North America but higher in much of the rest of the world. Cell phones do not require the costly investment of connecting wires to each individual building, and more individuals can obtain service from a single tower or satellite.

FOCUS ON EASTERN EUROPE

Eastern Europe is the only region where the HDI has declined significantly since the United Nations created the index in 1990. The decline is a legacy of the region's history of Communist rule. The Communists promoted development through economies directed by government officials rather than private entrepreneurs. Government policies emphasized heavy industry like mining and steel rather than consumer goods. A lack of consumer products such as motor vehicles, clothing, and refrigerators was a major contributor to the fall of communism in Eastern Europe around 1990.

DEVELOPMENT INDICATORS

9.4 Social Indicators

- People in MDCs attend school for more years and are more likely to be literate.
- MDCs and LDCs differ in key demographic indicators of development.

A higher percentage of people in MDCs are able to read and write. This ability derives in part from the ability to spend more money on schools. MDCs and LDCs also display key differences in demographic features discussed in more detail in Chapter 2.

EDUCATION AND LITERACY

In general, the higher the level of development, the greater are both the quantity and the quality of a country's education. A measure of the quantity of education is the average number of school years attended. The assumption is that no matter how poor the school, the longer the pupils attend, the more likely they are to learn something.

The quality of education is measured in two ways—student/teacher ratio and literacy rate. The fewer pupils a teacher has, the more likely that each student will receive instruction.

The average pupil attends school for about 10 years in MDCs, compared to only a couple of years in LDCs. The student/teacher ratio is twice as high in LDCs as in MDCs.

The **literacy rate** is the percentage of a country's people who can read and write. It exceeds 98 percent in MDCs, compared with less than 60 percent in LDCs.

For many in LDCs, education is the ticket to better jobs and higher social status. Improved education is a major goal of many developing countries, but funds are scarce. Education may receive a higher percentage of the GDP in LDCs, but their GDP is far lower to begin with, so they spend far less per pupil than do MDCs.

The MDCs publish more books, newspapers, and magazines per person because more of their citizens read and write. MDCs dominate scientific and nonfiction publishing worldwide—this textbook is an example. Students in LDCs must learn technical information from books that usually are not in their native language but are printed in English, German, Russian, or French.

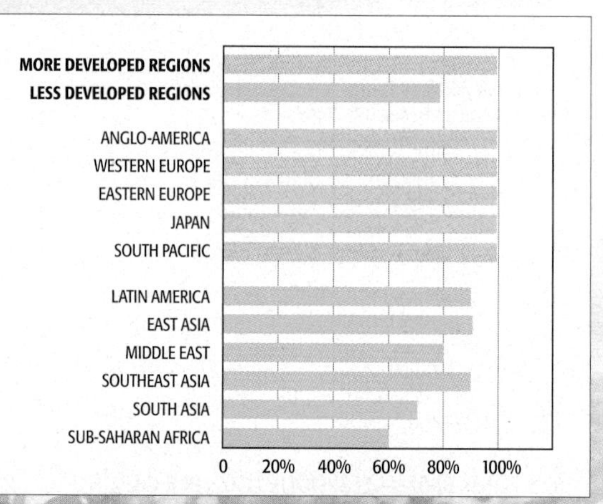

9.4.1 **PERCENT OF ADULTS WHO ARE LITERATE**

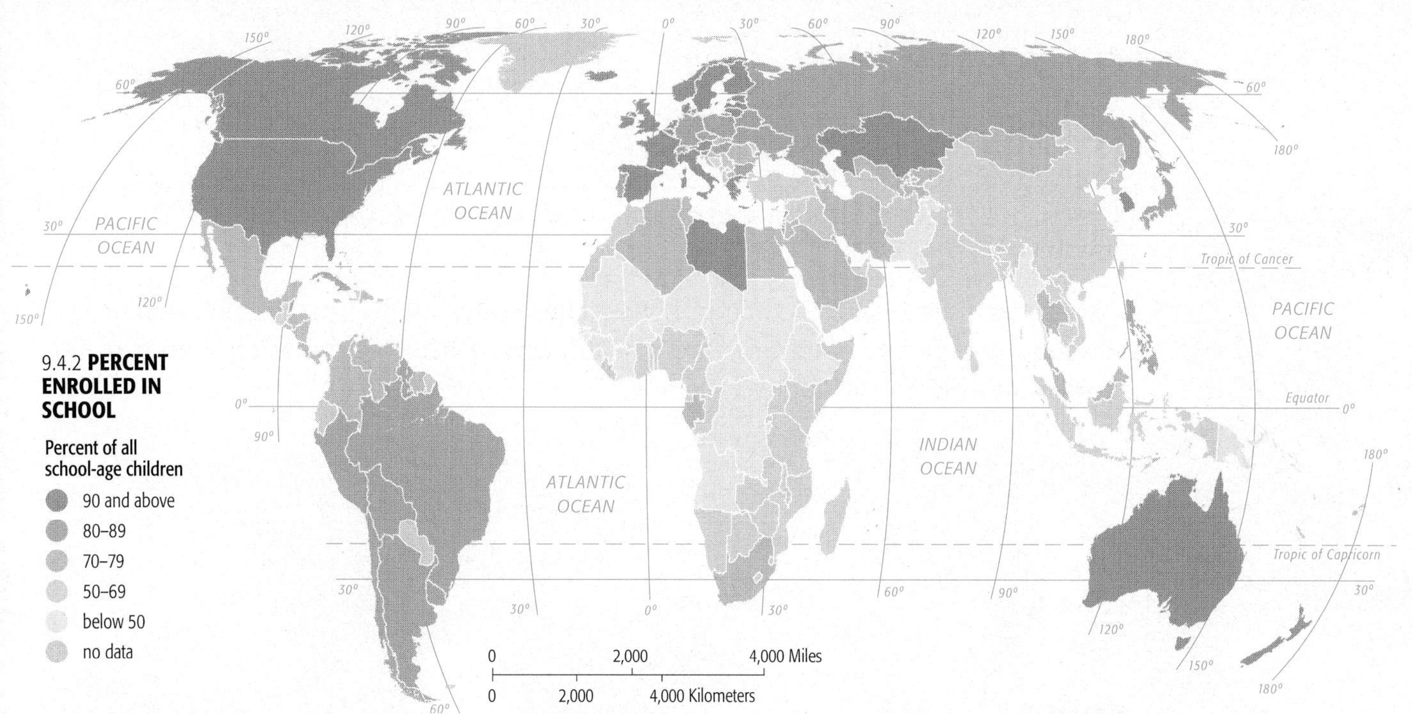

9.4.2 **PERCENT ENROLLED IN SCHOOL**

KEY DEMOGRAPHIC DIFFERENCES

MDCs display many demographic differences compared to LDCs (refer to sections 2.3 and 2.4).

Life Expectancy Babies born today can expect to live into their 60s in LDCs compared to their 70s in MDCs. With longer life expectancies, MDCs have a higher percentage of older people who have retired and receive public support and a lower percentage of children under age 15, who are too young to work and must also be supported by employed adults and government programs.

Infant Mortality The infant mortality rate is greater in LDCs for several reasons. Babies may die from malnutrition or lack of medicine needed to survive illness, such as dehydration from diarrhea. They may also die from poor medical practices that arise from lack of education.

Natural Increase Greater natural increase strains a country's ability to provide hospitals, schools, jobs, and other services that can make its people healthier and more productive. Many LDCs must allocate increasing percentages of their GDPs just to care for the rapidly expanding population rather than to improve care for the current population.

Crude Birth Rate LDCs have higher natural increase rates because they have higher crude birth rates. Women in MDCs choose to have fewer babies for various economic and social reasons, and they have access to varied birthcontrol devices to achieve this goal.

9.4.3 **LATIN AMERICA** Literacy rates are higher in the region than other developing regions.

FOCUS ON LATIN AMERICA

The level of development is relatively high along the South Atlantic Coast between Curitiba, Brazil, and Buenos Aires, Argentina.
The area has high agricultural productivity, and it ranks among the world's leaders in production and export of wheat and corn (maize). Mexico's development has been aided by proximity to the United States. Development is lower in Central America, several Caribbean islands, and the interior of South America. Overall development in Latin America is hindered by inequitable income distribution.

DEVELOPMENT INDICATORS

9.5 Health Indicators

- One-third of Africans and one-sixth of all people are undernourished.
- MDCs spend more on health care.

People are healthier in MDCs than in LDCs. The greater wealth that is generated in MDCs is used in part to obtain food and health care. A healthier population in turn can be more economically productive.

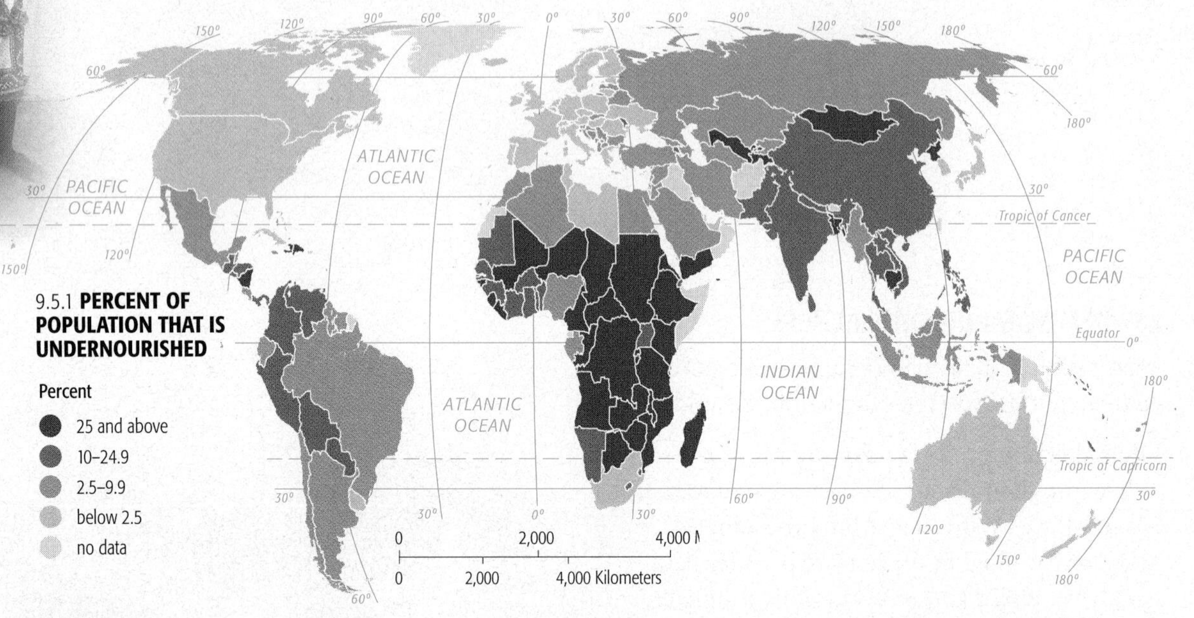

9.5.1 **PERCENT OF POPULATION THAT IS UNDERNOURISHED**

NUTRITION

The health of a population is influenced by diet. The United Nations Food and Agriculture Organization (FAO) has determined that in the twenty-first century 14 percent of the population worldwide is undernourished.

Considerable progress was made in providing food security during the 1970s and 1980s. The number of undernourished people declined from 1 billion or 27 percent in 1970, to 800 million or 16 percent in 1990. But little further progress in reducing undernourishment has occurred since 1990, according to the FAO. Approximately 200 million people lack food security in sub-Saharan Africa and in India, 150 million each in China and the rest of Asia, and 100 million in the rest of the world. In sub-Saharan Africa, more than one-third of the population is undernourished.

On average, people in MDCs receive more calories and proteins daily than they need. But in the LDCs of Africa and Asia, most people receive less than the daily minimum allowance of calories and proteins recommended by the United Nations.

9.5.2 **EAST ASIA**
Food production has increased rapidly in the region.

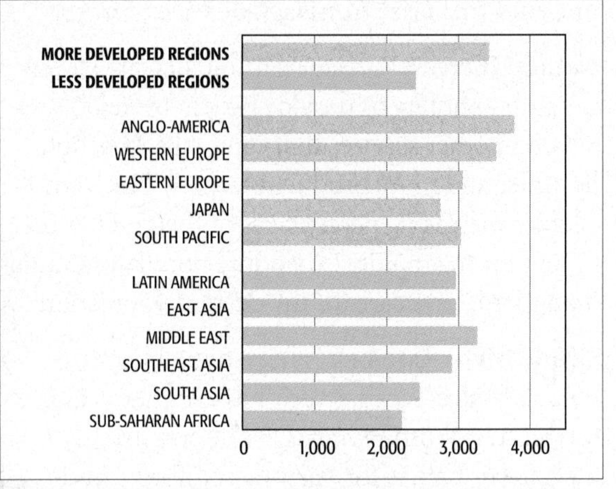

9.5.3 **CALORIE SUPPLY PER CAPITA (KILOCALORIES PER DAY)**

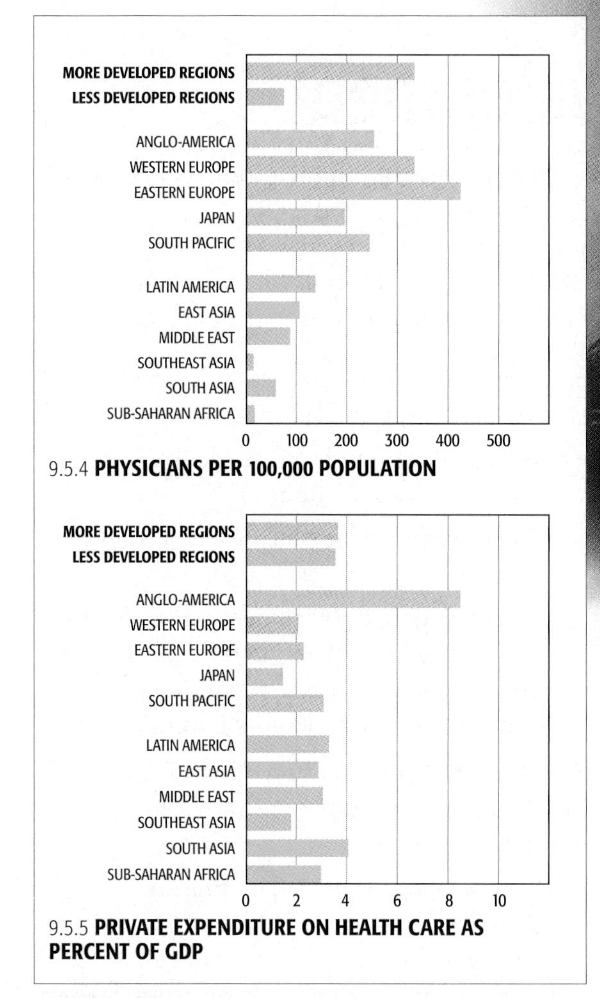

9.5.4 **PHYSICIANS PER 100,000 POPULATION**

9.5.5 **PRIVATE EXPENDITURE ON HEALTH CARE AS PERCENT OF GDP**

HEALTH CARE

When people get sick, MDCs possess the resources to care for them. Total expenditures on health care exceed 8 percent of GDP in MDCs, compared to less than 6 percent in LDCs. So not only do MDCs have much higher GDP per capita than LDCs, they spend a higher percentage of that GDP on health care. Some of that additional expenditure on health is reflected in more hospitals, doctors, and nurses per capita in MDCs.

In most MDCs, health care is a public service that is available at little or no cost. The government programs pay more than 70 percent of health-care costs in most European countries, and private individuals pay less than 30 percent. In comparison, private individuals must pay more than half of the cost of health care in LDCs. An exception is the United States, where private individuals are required to pay 55 percent of health care, more closely resembling the pattern in LDCs.

The MDCs use part of their wealth to protect people who, for various reasons, are unable to work. In these states some public assistance is offered to those who are sick, elderly, poor, disabled, orphaned, veterans of wars, widows, unemployed, or single parents. Western European countries such as Denmark, Norway, and Sweden typically provide the highest level of public-assistance payments.

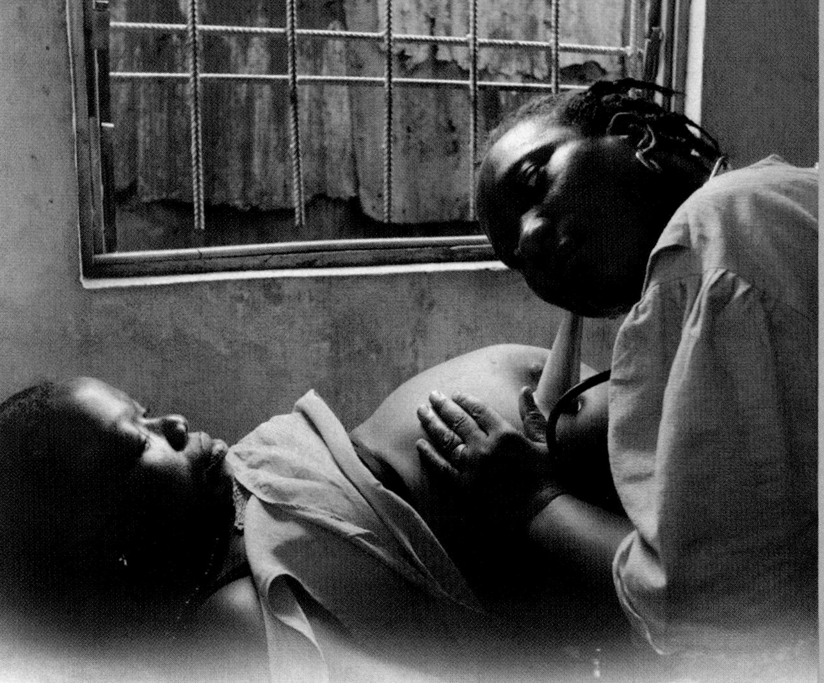

9.5.6 **HEALTH-CARE CLINIC IN SIERRA LEONE**
The clinic lacks electricity and clean drinking water.

MDCs are hard-pressed to maintain their current levels of public assistance. In the past, rapid economic growth permitted these states to finance generous programs with little hardship. But in recent years economic growth has slowed, whereas the percentage of people needing public assistance has increased. Governments have faced a choice between reducing benefits or increasing taxes to pay for them.

FOCUS ON EAST ASIA

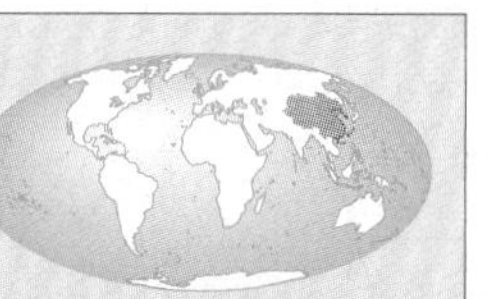

China accounted for much of the substantial reduction in undernourishment in the world during the 1970s and 1980s. The number of undernourished Chinese declined from 400 million or 46 percent of the total population in 1970 to 150 million or 16 percent in 1990. Now the world's second largest economy, behind only the United States, China accounts for one-third of total world economic growth, and GDP per capita has risen faster there than in any other country. China's watershed year was 1949, when the Communist Party won a civil war and created the People's Republic of China. Under communism, the government took control of most agricultural land. The system assured the production and distribution of enough food to support China's one-billion-plus population. In recent years, farmers have been permitted to own land and control their own production, but the number of undernourished in China has remained around 150 million.

DEVELOPMENT THROUGH TRADE

9.6 Two Paths to Development

- The self-sufficiency development path erects barriers to trade.
- The international trade path allocates scarce resources to a few activities.

To reduce disparities with the MDCs, the LDCs must promote development. Two development models are typical: one emphasizes self-sufficiency, the other international trade.

DEVELOPMENT THROUGH SELF-SUFFICIENCY

Self-sufficiency, or balanced growth, was the more popular of the development alternatives for most of the twentieth century. According to the balanced growth approach:

- Investment is spread as equally as possible across all sectors of a country's economy and in all regions.
- The pace of development may be modest, but the system is fair because residents and enterprises throughout the country share the benefits of development.
- Reducing poverty takes precedence over encouraging a few people to become wealthy consumers.
- Fledgling businesses are isolated from competition with large international corporations.
- The import of goods from other places is limited by barriers such as tariffs, quotas, and licenses.

SELF-SUFFICIENCY IN INDIA

To promote self-sufficiency, India once imposed many barriers to trade:

- To import goods into India, most foreign companies had to secure a license that had to be approved by several dozen government agencies.
- An importer with a license was severely restricted in the quantity it could sell in India.
- Heavy taxes on imported goods doubled or tripled the price to consumers.
- Indian money could not be converted to other currencies.
- Businesses required government permission to sell a new product, modernize a factory, expand production, set prices, hire or fire workers, and change the job classification of existing workers.

The experience of India and other LDCs revealed two major problems with self-sufficiency.

- **Self-sufficiency protected inefficient industries.** Businesses could sell all they made, at high government-controlled prices, to customers culled from long waiting lists. They had little incentive to improve quality, lower production costs, reduce prices, or increase production. Nor did they keep abreast of rapid technological changes elsewhere.
- **A large bureaucracy was needed to administer the controls.** A complex administrative system encouraged abuse and corruption. Aspiring entrepreneurs found that struggling to produce goods or offer services was less rewarding financially than advising others how to get around the complex regulations.

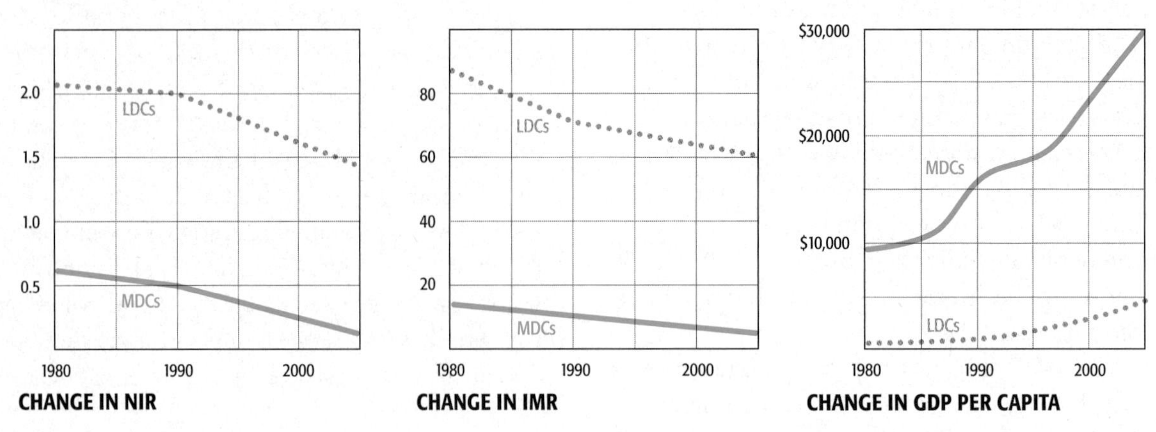

9.6.1 **DIFFERENCES BETWEEN LDCs AND MDCS. CHANGE IN NATURAL INCREASE RATE (left), INFANT MORTALITY RATE (center), AND GDP PER CAPITA (right).** The gap in key development indicators between LDCs and MDCs remains wide. Differences in natural increase and infant mortality are getting smaller, primarily because the rates can't get much lower in MDCs. On the other hand, the gap in GDP per capita has grown much wider.

9.6.2 **MIDDLE EAST** Moroccan women wearing traditional chador head covering peel shrimp destined for export to the Netherlands.

DEVELOPMENT THROUGH INTERNATIONAL TRADE

According to the international trade approach, a country can develop economically by concentrating scarce resources on expansion of its distinctive local industries. The sale of these products in the world market brings funds into the country that can be used to finance other development. W. W. Rostow proposed a five-stage model of development in the 1950s:

- **The traditional society.** A very high percentage of people engaged in agriculture and a high percentage of national wealth allocated to what Rostow called "nonproductive" activities, such as the military and religion.
- **The preconditions for takeoff.** An elite group of well-educated leaders initiates investment in technology and infrastructure, such as water supplies and transportation systems, designed increase productivity.
- **The takeoff.** Rapid growth is generated in a limited number of economic activities, such as textiles or food products.
- **The drive to maturity.** Modern technology, previously confined to a few takeoff industries, diffuses to a wide variety of industries.
- **The age of mass consumption.** The economy shifts from production of heavy industry, such as steel and energy, to consumer goods, such as motor vehicles and refrigerators.

Among the first countries to adopt the international trade alternative were the following.

- **The "Four Dragons."** South Korea, Singapore, Taiwan, and the then-British colony of Hong Kong (also known as the "four little tigers" and "the gang of four") developed by producing a handful of manufactured goods, especially clothing and electronics, that depended on low labor costs.
- **Petroleum-rich Arabian Peninsula countries.** Once among the world's least developed countries, they were transformed overnight into some of the wealthiest thanks to escalating petroleum prices during the 1970s.

Rostow's optimistic development model was based on two factors:

- If Japan, as well as countries in Southern and Eastern Europe, could develop by following this model, why couldn't other countries?
- Many LDCs contain an abundant supply of raw materials sought by manufacturers and producers in MDCs; the sale of these raw materials could generate funds for LDCs to promote development.

FOCUS ON MIDDLE EAST

Middle East countries that are oil-rich have used petroleum revenues to finance large-scale projects, such as housing, highways, airports, universities, and telecommunications networks. Imported consumer goods are readily available. Some business practices typical of international trade are difficult to reconcile with Islamic religious principles. Women are excluded from holding most jobs and visiting some public places. All business halts several times a day when Muslims are called to prayers.

DEVELOPMENT THROUGH TRADE

9.7 World Trade

- Most LDCs have adopted the international trade development path.
- The World Trade Organization has facilitated adoption of the international trade path.

9.7.1 **THE WORLD TRADE ORGANIZATION GENERATES STRONG SUPPORT (above) AND OPPOSITION (opposite page)**

In the twenty-first century, the international trade approach has been embraced by most countries as the preferred alternative for stimulating development. Longtime advocates of the self-sufficiency approach quickly converted to international trade during the 1990s for one simple reason: overwhelming evidence that international trade promoted development more rapidly.

Three problems have hindered countries outside Asia's four dragons and the Middle East from developing through the international trade approach:

- International trade worked for Middle East countries because the price of petroleum has escalated so rapidly. Other LDCs have not been so fortunate with the price of their commodities.
- LDCs trying to take advantage of their low-cost labor find that markets in MDCs are growing more slowly than when the four dragons used this strategy a generation ago.
- Building up a handful of takeoff industries has forced some LDCs to cut back on production of food, clothing, and other necessities for their own people.

9.7.2 **SOUTHEAST ASIA** Containers stacked together in Singapore, one of the world's largest ports.

FOREIGN DIRECT INVESTMENT

International trade requires corporations based in a particular country to invest in other countries. Investment in the economy of another country is known as **foreign direct investment** (FDI). FDI has increased from $13 billion in 1970 to $1 trillion in 2005. Only one-third of FDI goes from a MDC to a LDC, whereas the other two-thirds goes from one MDC to another MDC. Nearly one-half of all FDI destined for LDCs goes to China, nearly one-fourth to other Asian countries, and one-fourth to Latin America; less than 10 percent goes to Africa.

India, for example, dismantled its formidable collection of barriers to international trade during the 1990s:

- Foreign companies are allowed to set up factories and sell in India.
- Tariffs and restrictions on the import and export of goods have been reduced or eliminated. Monopolies in communications, insurance, and other industries have been eliminated.
- With increased competition, Indian companies have improved the quality of their products.

9.7.3 **FOREIGN DIRECT INVESTMENT**

WORLD TRADE ORGANIZATION

To promote the international trade development path, countries representing 97 percent of world trade established the World Trade Organization (WTO) in 1995. The WTO works to reduce barriers to international trade in two principal ways:

- Negotiations to reduce or eliminate restrictions on trade and movement of money.
- Rulings on charges brought by countries, corporations, or individuals that other countries, corporations, or individuals have violated trade agreements.

Protesters routinely gather in the streets outside high-level meetings of the WTO. Critics charge that the WTO makes decisions behind closed doors and compromises the sovereignty of individual countries.

FOCUS ON SOUTHEAST ASIA

Southeast Asia has become a major manufacturer of textiles and clothing, taking advantage of cheap labor. Thailand has become the region's center for the manufacturing of automobiles and other consumer goods. Indonesia, the world's fourth most populous country, is a major producer of petroleum. Development has slowed because of painful reforms to restore confidence among international investors shaken by unwise and corrupt investments. These investments were made possible by lax regulations and excessively close cooperation among manufacturers, financial institutions, and government agencies.

DEVELOPMENT THROUGH TRADE

9.8 Financing Development

- LDCs finance some development through loans.
- To qualify for loans, a country may need to enact economic reforms.

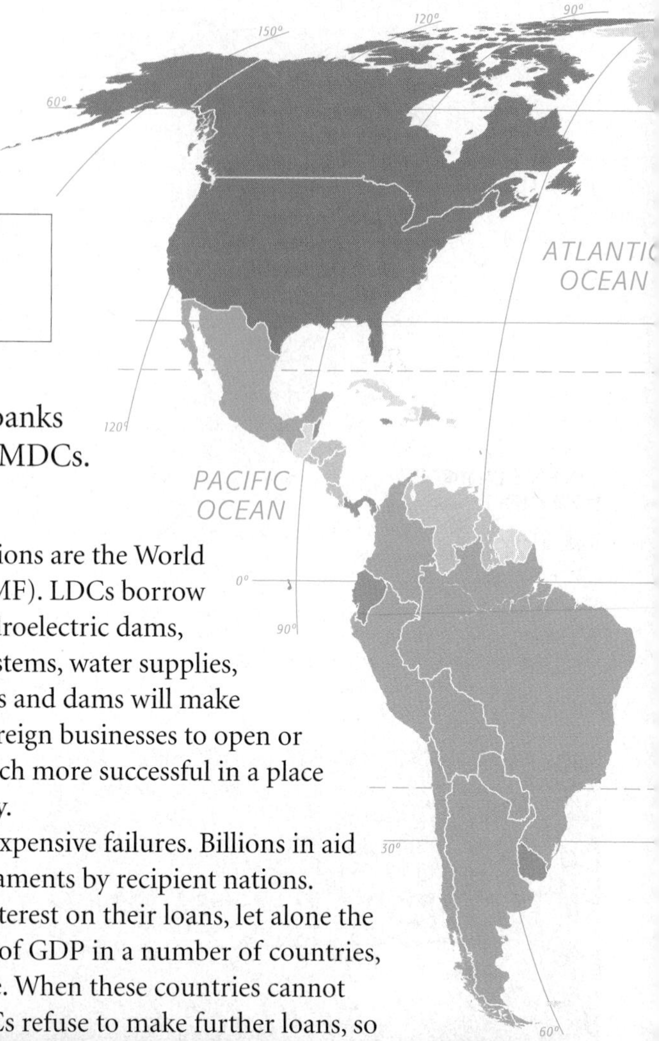

LDCs lack the money needed to finance development, so they obtain loans from banks and international organizations based in MDCs.

LOANS

The two major international lending organizations are the World Bank and the International Monetary Fund (IMF). LDCs borrow money to build new infrastructure, such as hydroelectric dams, electric transmission lines, flood-protection systems, water supplies, roads, and hotels. In theory, building new roads and dams will make conditions more favorable for domestic and foreign businesses to open or expand. After all, businesses are likely to be much more successful in a place with paved roads, running water, and electricity.

Some of the new infrastructure projects are expensive failures. Billions in aid have been squandered, stolen, or spent on armaments by recipient nations.

Many LDCs have been unable to repay the interest on their loans, let alone the principal. Interest payments exceed 10 percent of GDP in a number of countries, especially in Latin America and Eastern Europe. When these countries cannot repay their debts, financial institutions in MDCs refuse to make further loans, so construction of needed infrastructure stops. The inability of many LDCs to repay loans also damages the financial stability of banks in the MDCs.

9.8.1 **SOUTH ASIA** Sections of a water line are being welded in Bihar State, India.

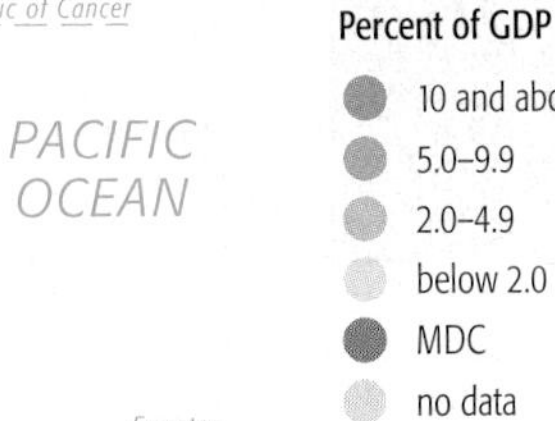

9.8.2 **DEBT SERVICE AS PERCENTAGE OF GDP**

9.8.3 **SOUTH ASIA'S GRAMEEN BANK**
Bangladeshi women meet with Grameen Bank officials to discuss loan opportunities.

STRUCTURAL ADJUSTMENT PROGRAMS

The IMF, World Bank, and governments in MDCs fear that granting, canceling, or refinancing debts without strings attached will perpetuate bad habits in LDCs. Therefore before granting debt relief, an LDC may be required to prepare a Policy Framework Paper outlining a **structural adjustment program**.

A structural adjustment program typically includes economic reforms, such as reduction of government spending and curbing of inflation. Critics charge that structural adjustment programs cause a rise in unemployment and cuts in health, education, and social services that were originally implemented to benefit the poor. They argue that structural reforms punish Earth's poorest people for actions they did not commit—waste, corruption, misappropriation, military buildup.

International organizations respond that the poor suffer more when a country does not undertake reforms. Economic growth is what benefits the poor the most in the long run.

FOCUS ON SOUTH ASIA

An alternative source of loans for development, the Grameen Bank, based in Bangladesh, has made several hundred thousand loans to women in South Asia. Only 1 percent of the borrowers have failed to make their weekly loan repayments, an extraordinarily low percentage for a bank. Several million loans have also been provided to women by the Bangladesh Rural Advancement Committee. For founding the bank, Muhammad Yunus was awarded the Nobel Peace Prize in 2006.

DEVELOPMENT THROUGH TRADE

9.9 Fair Trade

- Fair trade is a model of development that protects small businesses and workers.
- With fair trade, a higher percentage of the sales price goes back to the producers.

A variation of the international trade model of development is **fair trade**, in which products are made and traded according to standards that protect workers and small businesses in LDCs.

Fairtrade Labelling Organisations International (FLO) sets international standards for fair trade.

TransFair USA, a nonprofit organization, certifies the products sold in the United States that are fair trade.

In North America, fair trade products have been primarily craft products such as decorative home accessories, jewelry, textiles, and ceramics. Ten Thousand Villages is the largest fair trade organization in North America specializing in handicrafts. In Europe, most fair trade sales are in food, including coffee, tea, banana, chocolate, cocoa, juice, sugar, and honey products.

Two sets of standards distinguish fair trade: one set applies to producers and the other to workers on farms and in factories.

FAIR TRADE PRODUCER STANDARDS

Fair trade advocates work with small businesses, especially worker owned and democratically run cooperatives. Small-scale farmers and artisans in LDCs are unable to borrow from banks the money they need to invest in their businesses. By banding together, they can get credit, reduce their raw material costs, and maintain higher and fairer prices for their products.

Cooperatives thus benefit the local farmers and artisans who are members, rather than absentee corporate owners interested only in maximizing profits. Because cooperatives are managed democratically, farmers and artisans practice leadership and organizational skills. The people who grew or made the products thereby have a say in how local resources are utilized and sold. Safe and healthy working conditions can be protected.

Consumers pay higher prices for fair trade coffee than for grocery store brands, but prices are comparable to those charged by gourmet brands. However, fair trade coffee producers receive a significantly higher price per pound than traditional coffee producers.

North American consumers pay $4 to $11 a pound for coffee bought from growers for about 80 cents a pound.

9.9.1 **SUB-SAHARAN AFRICA**
Fair trade bananas are being harvested and sorted in Ghana.

Growers who sell to fair trade organizations earn $1.12 to $1.26 a pound.

In some cases the quality is higher because fair traders factor in the environmental cost of production. For instance, in the case of coffee, fairly traded coffee is usually organic and shade grown, which results in a higher-quality coffee.

FAIR TRADE WORKER STANDARDS

Only a tiny percentage of the price a consumer pays for a good reaches the individual in the LDC responsible for making or growing it, charge critics of international trade. A Haitian sewing clothing for the U.S. market, for example, earns less than 1 percent of the retail price, according to the National Labor Committee. In contrast, fair trade returns on average one-third of the price back to the producer in the LDC. The rest goes to the wholesaler who imports the item and for the retailer's rent, wages, and other expenses.

Protection of workers' rights is not a high priority in the international trade development path, according to its critics:

- With minimal oversight by governments and international lending agencies, people in LDCs allegedly work long hours in poor conditions for low pay. The workforce may include children or forced labor.
- Health problems may result from poor sanitation and injuries from inadequate safety precautions.
- Injured, ill, or laid-off workers are not compensated.

In contrast, fair trade requires employers to

- pay workers fair wages (at least the country's minimum wage).
- permit union organizing.
- comply with minimum environmental and safety standards.

Two-thirds of the artisans providing fair trade handcrafted products are women. Often these women are mothers and the sole wage earners in the home.

Because the minimum wage is often not enough for basic survival, whenever feasible, workers are paid more than that—enough to cover food, shelter, education, health care, and other basic needs. Cooperatives are encouraged to reinvest profits back into the community, such as by providing health clinics, child care, and training.

Paying fair wages does not necessarily mean that products cost the consumer more. Because fair trade organizations bypass exploitative intermediaries and work directly with producers, they are able to cut costs and return a greater percentage of the retail price to the producers. The cost remains the same as traditionally traded goods, but the distribution of the cost of the product is different, because the large percentage taken by intermediaries is removed from the equation.

FOCUS ON SUB-SAHARAN AFRICA

Sub-Saharan Africa has the least favorable prospect for development. The region has the world's highest percentage of people living in poverty and suffering from poor health and low education levels. And conditions are getting worse: the average African consumes less today than a quarter-century ago. The fundamental problem in many countries of sub-Saharan Africa is a dramatic imbalance between the number of inhabitants and the capacity of the land to feed the population.

Chapter **Summary**

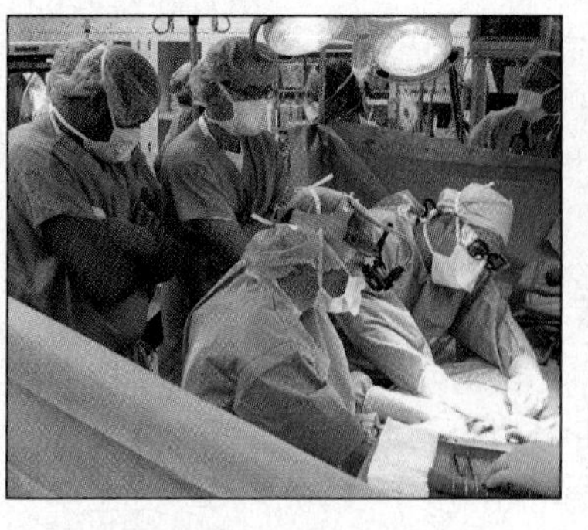

HUMAN DEVELOPMENT

- The world is divided into more developed countries (MDCs) and less developed countries (LDCs).
- The United Nations has created the Human Development Index to measure the relative level of development of every country.
- A Gender-related Development Index compares the level of development of women and men in every country.

DEVELOPMENT INDICATORS

- MDCs and LDCs can be compared according to a number of indicators.
- The most widely used economic indicator is gross domestic product (GDP) per capita.
- MDCs display higher levels of education and literacy.
- Key demographic differences include longer life expectancy and lower rates of births, natural increase, and infant mortality in MDCs.
- People in MDCs have better nutrition and health care.

DEVELOPMENT THROUGH TRADE

- The two principal paths to development are self-sufficiency and international trade.
- Self-sufficiency was the most commonly used path in the past, but most countries now follow international trade.
- LDCs finance trade through loans, but may be required to undertake economic reforms.
- Fair trade is an alternative approach to development through trade that provides greater benefits to the producers in LDCs.

CORE AND PERIPHERY
Viewed from this North Polar projection, MDCs appear clustered in an inner core, whereas LDCs are generally relegated to a peripheral or outer-ring location.

Geographic Consequences of Change

The division of the world into MDCs and LDCs is often described as a north–south split. Most of the countries that have achieved relatively high levels of development are located above 30° north latitude.

The relationship between MDCs and LDCs appears somewhat different on a North Polar projection. Viewed in this way, MDCs form an inner-core area, whereas LDCs are generally relegated to a peripheral or outer-ring location. This unorthodox world map projection emphasizes the central role played by MDCs in the world economy and the secondary role of LDCs.

In an increasingly unified world economy, MDCs play a dominant role in forming the economies of the LDCs on the periphery. Anglo-America, Western Europe, and Japan account for a high percentage of the world's economic activity and wealth. LDCs have less access to the world centers of consumption, communications, wealth, and power, which are clustered in the core. Development prospects of Latin America are tied to governments and businesses in Anglo-America, those of Africa, the Middle East, and Eastern Europe to Western Europe, and those of Asia to Japan and to a lesser extent Western Europe and Anglo-America.

Yet many people in MDCs oppose increased trade with LDCs, because of alleged unfair labor practices, inadequate environmental safeguards, and unfair pricing of products. Many people in LDCs also oppose increased trade, believing that MDCs are exploiting the people and resources of less-developed peripheral regions. But from the perspective of others in less-developed regions, integration into a world economy through trade with MDCs may be a small price to pay to receive material benefits of development, such as a steady job and a television.

On the other hand, rapid development has come to many LDCs, led by China and India. China will have the world's largest economy by the year 2020, ahead of the United States, and India will closely follow. As LDCs are transformed economically, political and cultural changes will follow.

SHARE YOUR VOICE Student Essay

Megan Hunt
Arizona State University

Differences in the Health-care Field

I have always had a great interest in medicine, and my experiences working in very different health-care settings has made a huge impact on how I view health care in the United States. Beginning my freshmen year in college I volunteered at a high-end medical clinic and hospital. Although it specialized in the treatment of rare diseases, some patients came for routine medical needs and general hospital visits, despite the higher expense compared to other area hospitals.

The staff there was excellent, and the hospital and clinic were both amazing facilities with high-tech equipment and large single-occupant rooms. The patients were usually very happy during their stay, which only made it easier for doctors and nurses to work. Although I enjoyed working with the staff, I learned little about the realities of health care for most Americans.

Two years later, I took an internship at a county hospital. Being at a hospital that works primarily with underserved populations, particularly patients without health insurance, was a huge change. The staff was very dedicated, but the conditions were often out of their control. The hospital was crowded and old, and for many patients English was not their first language, creating a problematic language barrier.

Rooms were often shared by four patients, and halls were noisy and crowded. Many patients were stressed and found it difficult to relax, and a staff shortage often meant that doctors and nurses felt the same way. It was a world apart from the private hospital at which I was still volunteering. Those who could afford to pay for the private hospital were often not those in need of the most serious medical treatment, primarily because those without health insurance would wait until their conditions were extremely serious before going to the hospital.

I used to think that I would like to work at the private hospital, but after my experiences I know that I want to work at the county hospital to help provide excellent medical care to everyone who needs it, not just those who can afford it.

Chapter **Resources**

WAL-MART AND CHINA

No corporation exposes the effects of globalization on the world's economy more effectively than Wal-Mart.

Wal-Mart Stores, Inc., founded in 1962 by Sam Walton in Arkansas, was the world's largest corporation in 2007, with revenues of $388 billion. As a result of its prominence, Wal-Mart has become a lightning rod for both defenders and critics of the impacts of globalization.

- As the world's largest retailer, Wal-Mart is a good example of a **tertiary sector** activity. The company operated 7,390 stores worldwide in 2008, and employed 2.1 million. Wal-Mart accounts for 9 percent of all retail sales in North America.
- In the **secondary sector,** the clothes, appliances, and other goods sold in the stores have to be manufactured somewhere in the world, and Wal-Mart is largely responsible for deciding where.
- In the **primary sector**, as the second-largest grocer in the United States (behind Kroger), with 15 percent of the market, Wal-Mart is one of the world's largest purchasers of agricultural products and influences what is grown where.

Wal-Mart's distinctive relationship with suppliers of goods has been a key element in the company's success. Gaining or losing a contract with the world's largest retailer can make or break a manufacturer. But producing for Wal-Mart is hard work: suppliers must manufacture and deliver goods to Wal-Mart on short notice in order to eliminate the high cost of storing extra inventory in warehouses.

Wal-Mart's impact on the secondary sector of China has been especially strong. Supplying Wal-Mart has been a central element in the rise of China to become the world's second largest manufacturer. Three-fourths of products sold in Wal-Mart are at least partially manufactured in China. Most of the remainder comes from elsewhere in Asia. The company accounts for around 10 percent of the U.S. trade deficit with China.

For Wal-Mart, obtaining most of its goods in China is a simple matter of economic geography: because of much lower labor costs, producing in China costs less than in the United States, even with the added cost of shipping across the Pacific Ocean. Critics charge that the United States has lost 1 million manufacturing jobs because Wal-Mart obtains almost no merchandise from domestic suppliers. Wal-Mart responds that it still has 68,000 U.S. suppliers, employing 3.5 million people.

Wal-Mart has also been subject to local-scale criticism. Wal-Mart employees are paid lower than those at other retailers, and many lack health-care benefits. The company is on record as strongly opposing efforts to unionize the workforce in order to bargain for higher wages and benefits. Wal-Mart's low-wage low-benefit employment policy, as well as its preference for locating on the outskirts of cities, has been satirized in episodes of *The Simpsons* TV show set in a store called Sprawl-Mart.

KEY TERMS

Development
A process of improvement in the material conditions of people through diffusion of knowledge and technology.

Fair trade
Alternative to international trade that emphasizes small businesses and worker-owned and democratically run cooperatives and requires employers to pay workers fair wages, permit union organizing, and comply with minimum environmental and safety standards.

Foreign direct investment
Investment made by a foreign company in the economy of another country.

Gender Empowerment Measure (GEM)
Compares the ability of women and men to participate in economic and political decision making.

Gender-related Development Index (GDI)
Compares the level of development of women with that of both sexes.

Gross domestic product (GDP)
The value of the total output of goods and services produced in a country in a given time period (normally 1 year).

Human Development Index (HDI)
Indicator of level of development for each country, constructed by United Nations, combining income, literacy, education, and life expectancy.

Literacy rate
The percentage of a country's people who can read and write.

Primary sector
The portion of the economy concerned with the direct extraction of materials from Earth's surface, generally through agriculture, although sometimes by mining, fishing, and forestry.

Productivity
The value of a particular product compared to the amount of labor needed to make it.

Secondary sector
The portion of the economy concerned with manufacturing useful products through processing, transforming, and assembling raw materials.

Structural adjustment program
Economic policies imposed on LDCs by international agencies to create conditions encouraging international trade, such as raising taxes, reducing government spending, controlling inflation, selling publicly owned utilities to private corporations, and charging citizens more for services.

Tertiary sector
The portion of the economy concerned with transportation, communications, and utilities, sometimes extended to the provision of all goods and services to people in exchange for payment.

Value added
The gross value of the product minus the costs of raw materials and energy.

ON THE INTERNET

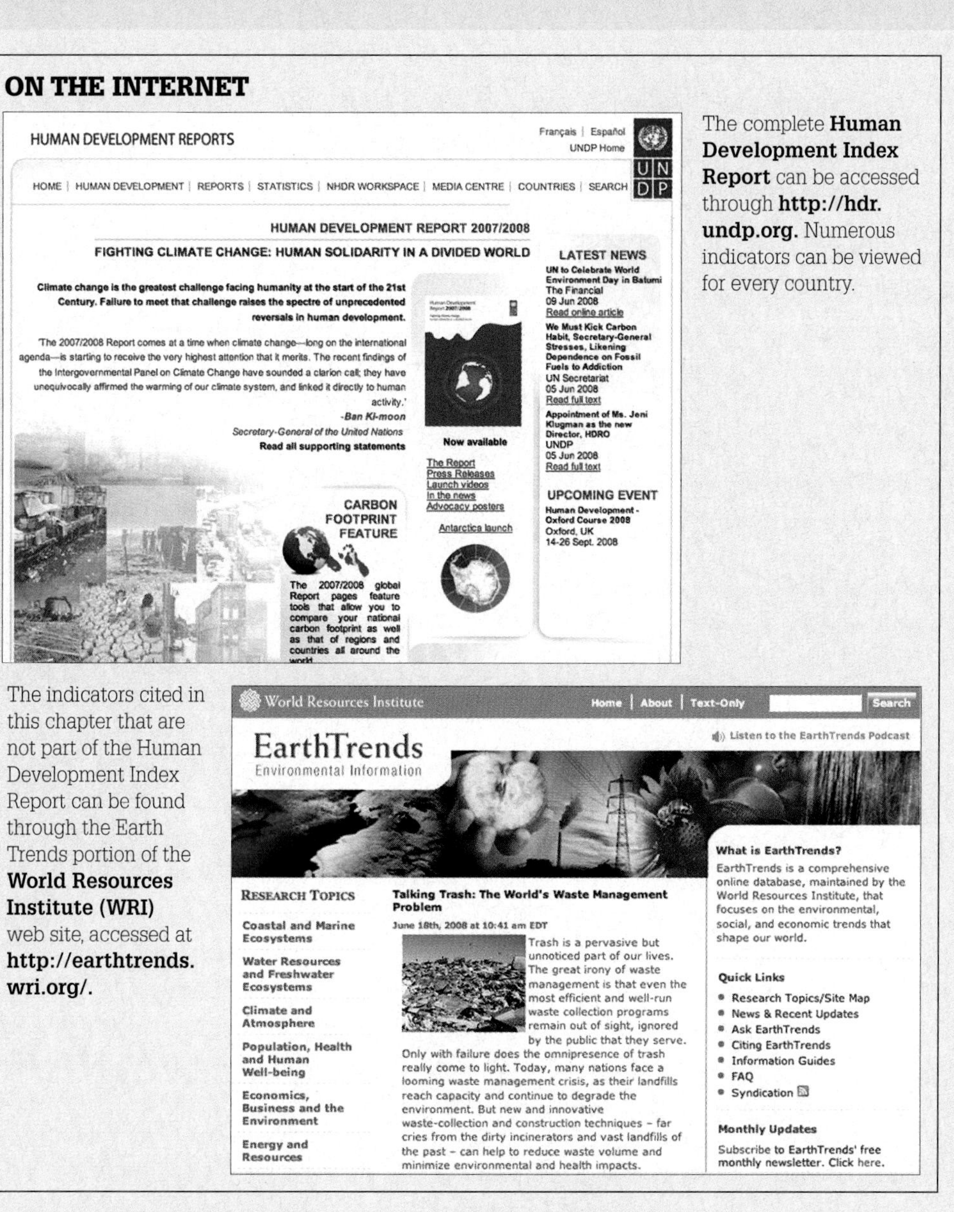

The complete **Human Development Index Report** can be accessed through **http://hdr.undp.org.** Numerous indicators can be viewed for every country.

The indicators cited in this chapter that are not part of the Human Development Index Report can be found through the Earth Trends portion of the **World Resources Institute (WRI)** web site, accessed at **http://earthtrends.wri.org/.**

FURTHER READINGS

Clark, Gordon L., Maryann P. Feldman, and Meric S. Gertler, eds. *The Oxford Handbook of Economic Geography.* New York: Oxford University Press, 2000.

Dicken, Peter. *Global Shift*, 4th ed. New York: Guilford, 2003.

Food and Agriculture Organization of the United Nations. *The State of Food Insecurity in the World.* Rome, Italy: Viale delle Terme di Caracalla, 2006.

Khan, A. U. "A Decade of Indian Economic Reforms and the Inflow of Foreign Investment." *Regional Studies* 21 (2003): 63–85.

Mabogunje, Akin L. *The Development Process: A Spatial Perspective*, 2d ed. London: Unwin Hyman, 1989.

Meier, Gerald M., and James E. Rauch. *Leading Issues in Economic Development*, 8th ed. New York: Oxford University Press, 2005.

Peet, Richard. *Theories of Development.* New York: Guilford, 1999.

Rostow, Walter W. *The Stages of Economic Growth.* Cambridge: Cambridge University Press, 1960.

Sharp, Joanne, John Briggs, Hoda Yacoub, and Nabila Hamed. "Doing Gender and Development: Understanding Empowerment and Local Gender Relations." *Transactions of the Institute of British Geographers New Series* 28 (2003): 281–95.

Storper, Michael. *The Regional World: Territorial Development in a Global Economy.* New York: Guilford Press, 1997.

Wallerstein, Immanuel. *The Capitalist World-Economy.* Cambridge: Cambridge University Press, 1979.

Chapter 10 AGRICULTURE

When you buy food in the supermarket, are you reminded of a farm? Not likely. The meat is carved into pieces that no longer resemble an animal and is wrapped in paper or plastic film. Often the vegetables are canned or frozen. The milk and eggs are in cartons.

Providing food in the United States and Canada is a vast industry. Only a few people are full-time farmers, and they may be more familiar with the operation of computers and advanced machinery than the typical factory or office worker.

The mechanized, highly productive American or Canadian farm contrasts with the subsistence farm found in much of the world. The most "typical" human—if there is such a person—is an Asian farmer who grows enough food to survive, with little surplus. This sharp contrast in agricultural practices constitutes one of the most fundamental differences between the more developed and less developed countries of the world.

More than 40 percent of the people in the world are farmers. The overwhelming majority of them are growing enough food to feed themselves, but little more. In most African and Asian countries one-half of the people are farmers. In contrast, fewer than 2 percent of the people in the United States and Canada are farmers. Yet the advanced technology used by these farmers allows them to produce enough food for people in the United States and Canada at a very high standard, plus food for many people elsewhere in the world.

RICE TERRACES IN LONGJI, CHINA

THE KEY ISSUES IN THIS CHAPTER

DISTRIBUTION OF AGRICULTURE

10.1 Origin of Agriculture

10.2 Agricultural Regions

10.3 Rural Settlements

TYPES OF AGRICULTURE

10.4 Comparing Subsistence and Commercial Agriculture

10.5 Subsistence Agriculture Regions

10.6 Commercial Agriculture Regions

CHALLENGES FOR AGRICULTURE

10.7 Subsistence Agriculture and Population Growth

10.8 Commercial Agriculture and Market Forces

10.9 Sustainable Agriculture

DISTRIBUTION OF AGRICULTURE

10.1 Origin of Agriculture

- Early humans obtained food through hunting and gathering.
- Agriculture originated in multiple hearths and diffused in many directions.

Agriculture is deliberate modification of Earth's surface through cultivation of plants and rearing of animals to obtain sustenance or economic gain. Agriculture thus originated when humans domesticated plants and animals for their use.

HUNTING AND GATHERING

Before the invention of agriculture, all humans probably obtained the food they needed for survival through hunting for animals, fishing, or gathering plants (including berries, nuts, fruits, and roots). Hunters and gatherers lived in small groups, with usually fewer than 50 persons, because a larger number would quickly exhaust the available resources within walking distance. They survived by collecting food often, perhaps daily. The food search might take only a short time or much of the day, depending on local conditions. The men hunted game or fished, and the women collected berries, nuts, and roots. This division of labor sounds like a stereotype but is based on evidence from archaeology and anthropology.

The group traveled frequently, establishing new home bases or camps. The direction and frequency of migration depended on the movement of game and the seasonal growth of plants at various locations. We can assume that groups communicated with each other concerning hunting rights, intermarriage, and other specific subjects. For the most part, they kept the peace by steering clear of each other's territory.

Why did nomadic groups convert from hunting, gathering, and fishing to agriculture? In gathering wild vegetation, people inevitably cut plants and dropped berries, fruits, and seeds. These hunters probably observed that, over time, damaged or discarded food produced new plants. They may have deliberately cut plants or dropped berries on the ground to see if they would produce new plants. Subsequent generations learned to pour water over the site and to introduce manure and other soil improvements. Over thousands of years, plant cultivation apparently evolved from a combination of accident and deliberate experiment.

Prehistoric people may have originally domesticated animals for noneconomic reasons, such as for sacrifices and other religious ceremonies. Other animals probably were domesticated as household pets, surviving on the group's food scraps.

Today perhaps a quarter-million people, or less than 0.005 percent of the world's population, still survive by hunting and gathering rather than by agriculture. These people live in isolated locations, including the Arctic and the interior of Africa, Australia, and South America. Examples include African Bushmen of Namibia and Botswana and Aborigines in Australia.

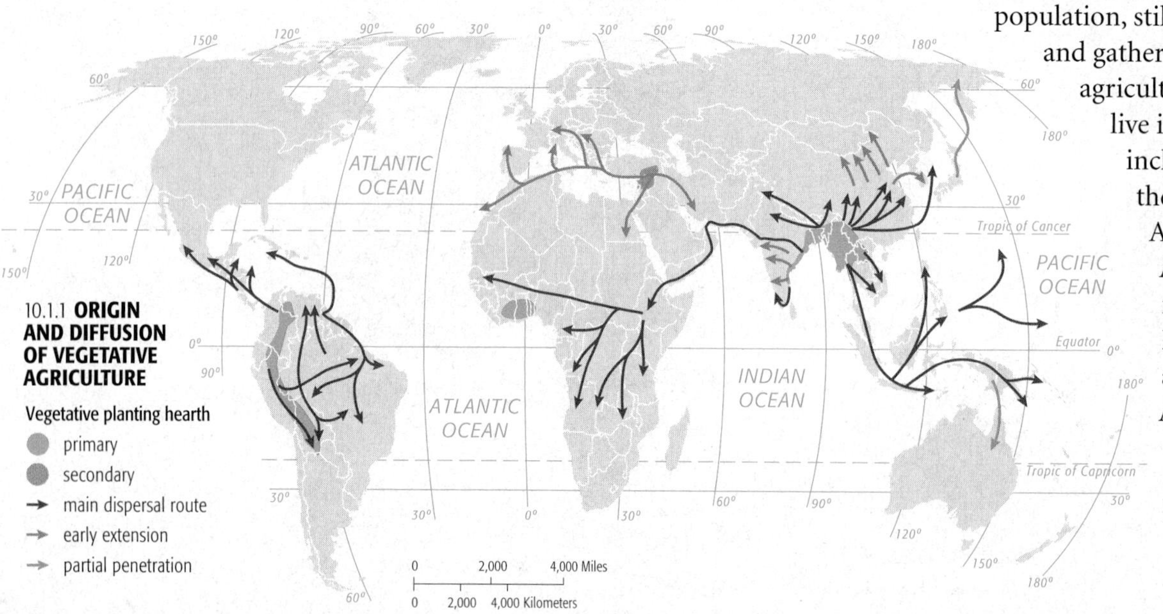

10.1.1 **ORIGIN AND DIFFUSION OF VEGETATIVE AGRICULTURE**

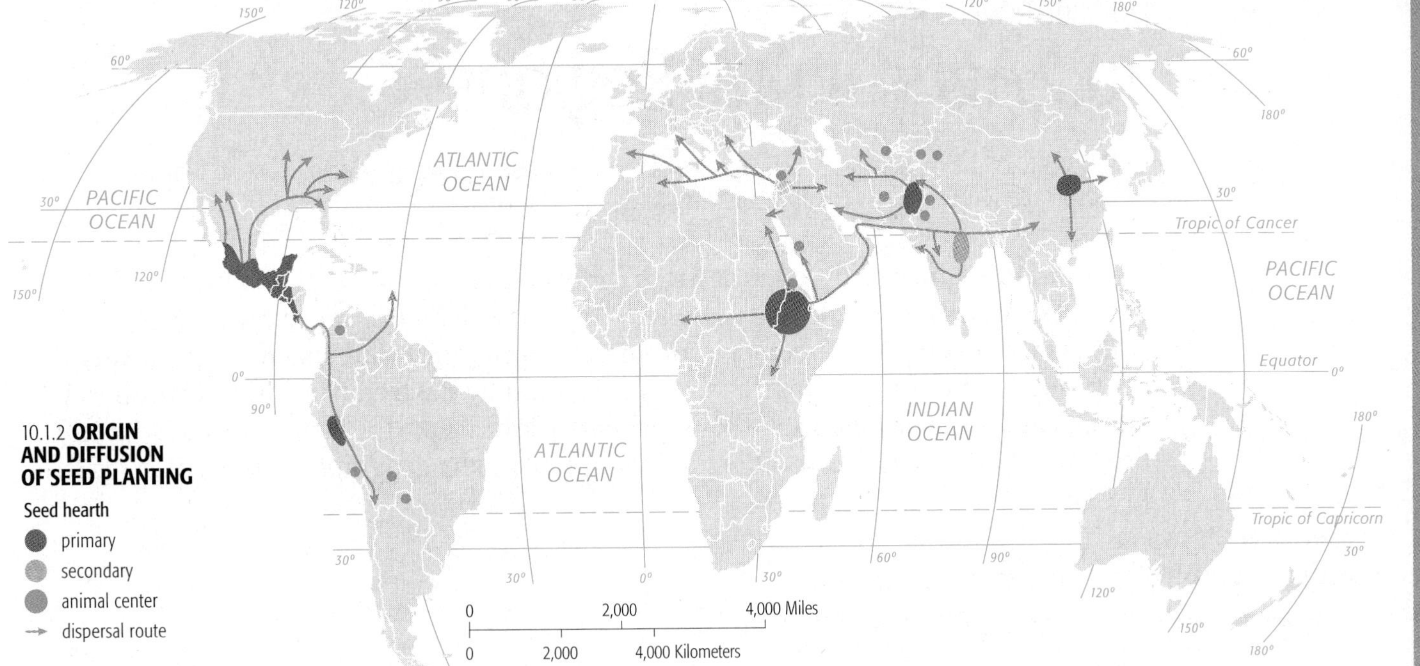

10.1.2 **ORIGIN AND DIFFUSION OF SEED PLANTING**

INVENTION OF AGRICULTURE

The origins of agriculture cannot be documented with certainty, because it began before recorded history. Scholars try to reconstruct a logical sequence of events based on fragments of information about ancient agricultural practices and historical environmental conditions.

Agriculture probably did not originate in one location but began in multiple, independent hearths, or points of origin. From these hearths, agricultural practices diffused across Earth's surface. That agriculture had multiple origins means that, from earliest times, people have produced food in distinctive ways in different regions. This diversity derives from a unique legacy of wild plants, climatic conditions, and cultural preferences in each region.

Improved communications in recent centuries have encouraged the diffusion of some plants to varied locations around the world. Many plants and animals thrive across a wide portion of Earth's surface, not just in their place of original domestication. Only after 1500, for example, were wheat, oats, and barley introduced to the Western Hemisphere, and maize (corn) to the Eastern Hemisphere.

Prominent cultural geographer Carl Sauer documented the origin and diffusion of two types of plant cultivation:

- **Vegetative planting.** The reproduction of plants by direct cloning from existing plants, such as cutting stems and dividing roots. Plants found growing wild were deliberately divided and transplanted. Sauer believed **vegetative planting** originated in Southeast Asia, because its diverse landscape encouraged growth of a wide variety of plants suitable for dividing and transplanting. Also, the people obtained food primarily by fishing rather than by hunting, so they may have been more sedentary and therefore able to devote more attention to growing plants. The first plants domesticated in Southeast Asia through vegetative planting probably included roots such as the taro and yam, and tree crops such as the banana and palm.
- **Seed agriculture.** The reproduction of plants through annual planting of seeds that result from sexual fertilization. **Seed agriculture** is practiced by most farmers today. Seed agriculture originated in three hearths in the Eastern Hemisphere and two in the Western Hemisphere. Important early advances were made in Southwest Asia, including the domestication of wheat and barley, two grains that became particularly important thousands of years later in European and American civilizations. Southwest Asia may have domesticated herd animals such as cattle, sheep, and goats to plow the land before planting seeds and, in turn, they were fed part of the harvested crop. This integration of plants and animals is a fundamental element of modern agriculture.

A KALAHARI BUSHMAN USING TRADITIONAL TOOLS

DISTRIBUTION OF AGRICULTURE

10.2 Agricultural Regions

- The world can be divided into several regions of subsistence agriculture and commercial agriculture.
- These regions are related in part to climate conditions.

The most fundamental differences in agricultural practices are between those in LDCs and those in MDCs. Farmers in LDCs generally practice subsistence agriculture, whereas farmers in MDCs practice commercial agriculture. **Subsistence agriculture** is the production of food primarily for consumption by the farmer's family. **Commercial agriculture** is the production of food primarily for sale off the farm. The most widely used map of world agricultural regions was prepared by geographer Derwent Whittlesey in 1936.

10.2.1 **AGRICULTURAL REGIONS**

SUBSISTENCE AGRICULTURE

Shifting cultivation
Primarily the tropical regions of South America, Africa, and Southeast Asia.

Pastoral nomadism
Primarily the drylands of North Africa and Asia.

Intensive subsistence, wet rice dominant
Primarily the large population concentrations of East and South Asia.

Intensive subsistence, crops other than rice dominant
Primarily the large population concentrations of East and South Asia where growing rice is difficult.

Plantation
Primarily the tropical and subtropical regions of Latin America, Africa, and Asia.

COMMERCIAL AGRICULTURE

Mixed crop and livestock
Primarily U.S. Midwest and central Europe.

Dairying
Primarily near population clusters in northeastern United States, southeastern Canada, and northwestern Europe.

Grain
Primarily north-central United States and Eastern Europe.

Ranching
Primarily the drylands of western United States, southeastern South America, Central Asia, southern Africa, and Australia.

Mediterranean
Primarily lands surrounding the Mediterranean Sea, western United States, and Chile.

Commercial gardening
Primarily southeastern United States and southeastern Australia.

10.2.2 **CLIMATE REGIONS**
Similarities between the maps on these two pages are striking. Climate influences the crop that is grown, or whether animals are raised instead of growing any crop. For example, pastoral nomadism is the predominant type of agriculture in the Middle East, which has a dry climate, whereas **shifting cultivation** is the predominant type of agriculture in central Africa, which has a humid low-latitude climate. The correlation between agriculture and climate is by no means perfect, but clearly some relationship exists between climate and agriculture. Cultural preferences account for some of the agricultural differences in areas of similar climate.

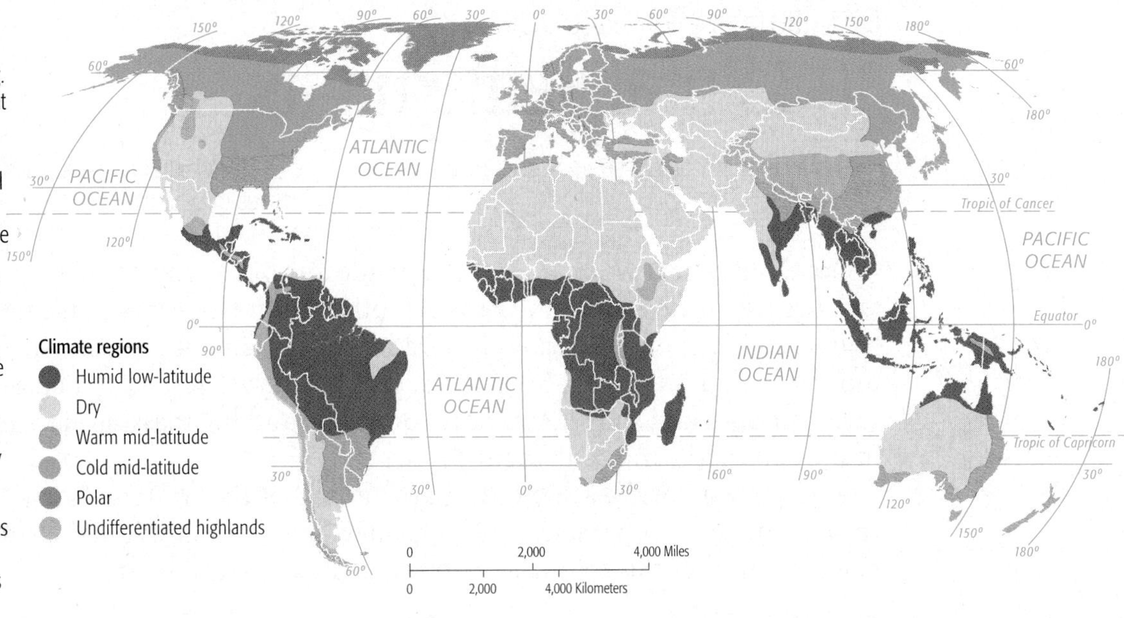

ATLANTIC OCEAN

INDIAN OCEAN

PACIFIC OCEAN

Tropic of Cancer

Equator

Tropic of Capricorn

0 1,000 2,000 Miles
0 1,000 2,000 Kilometers

DISTRIBUTION OF AGRICULTURE

10.3 Rural Settlements

- Settlements can be clustered or dispersed.
- Clustered rural settlements are laid out in many types of patterns.

A **clustered rural settlement** is an agricultural-based community in which a number of families live in close proximity to each other, with fields surrounding the collection of houses and farm buildings. A clustered rural settlement typically includes homes, barns, toolsheds, and other farm structures, plus consumer services, such as religious structures, schools, and shops. A handful of public and business services may also be present in the clustered rural settlement.

A **dispersed rural settlement**, characteristic of the contemporary North American rural landscape, is characterized by farmers living on individual farms isolated from neighbors rather than alongside other farmers in settlements.

CLUSTERED NEW ENGLAND SETTLEMENTS

New England colonists built clustered settlements centered on an open area called a common. Settlers grouped their homes and public buildings, such as the church and school, around the common. Beyond the fields, the town held pastures and woodland for the common use of all residents.

Clustered settlements were favored by New England colonists for a number of reasons. Typically, those who traveled to the American colonies together wanted to settle as a group. New England settlements were also clustered to reinforce common cultural and religious values.

Outsiders could obtain land in the settlement only by permission of the town's residents. Land was not sold, but rather was awarded to an individual after the town's residents felt confident that the recipient would work hard. Settlements accommodated a growing population by establishing new settlements nearby.

Dispersed rural settlements were more common in other colonies, as well as the interior of the United States. Land was plentiful and cheap, and people bought as much as they could manage.

10.3.1 **NEWFANE, VERMONT, A CLUSTERED NEW ENGLAND SETTLEMENT**
Public buildings are grouped around a common, including Windham County Courthouse (foreground), Congregational Church (right), and Newfane Village Union Hall (background, opposite courthouse).

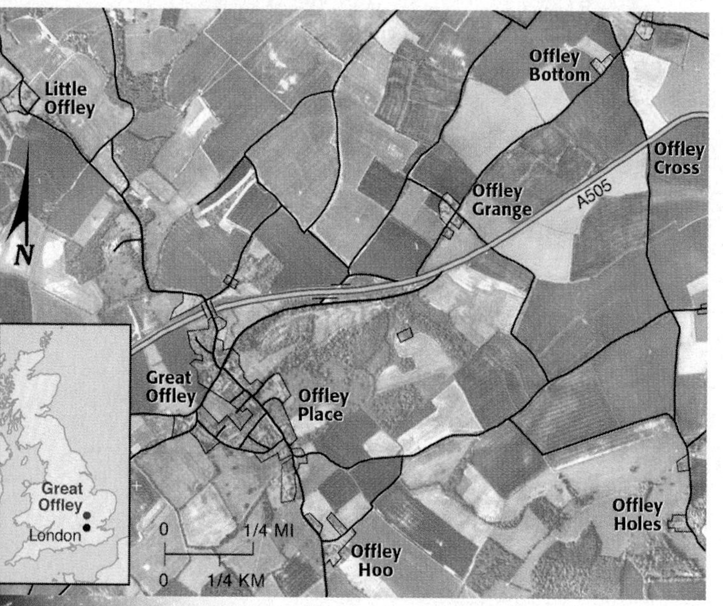

10.3.2 **CLUSTERED RURAL SETTLEMENTS IN ENGLAND**
Traditionally, when the population of a settlement grew too large for the capacity of the surrounding fields, new settlements were established nearby. For example, the parish of Offley, in Hertfordshire, England, contains the rural settlements of Great Offley (the largest), Little Offley, Offley Grange (barn), Offley Cross, Offley Bottom, Offley Place, Offley Hoo (house), and Offley Hole. All are within a few kilometers of each other. The name "Offley" means the wooded clearing of Offa, who was a ruler of Mercia during the eighth century and is said to have died at the site of the settlement.

CLUSTERED LINEAR RURAL SETTLEMENTS

Clustered linear rural settlements feature buildings clustered along a road, river, or dike to facilitate communications. The fields extend behind the buildings in long, narrow strips.

In the French long-lot, or seigneurial system, houses were erected along a river, which was the principal water source and means of communication. Narrow lots from 5 to 100 kilometers deep (3 to 60 miles) were established perpendicular to the river, so that each original settler had river access. This created a linear settlement along the river.

These long, narrow lots eventually were subdivided. French law required that each son inherit an equal portion of an estate, so the heirs established separate farms in each division. Roads were constructed parallel to the river for access to inland farms. In this way, a new linear settlement emerged along each road, parallel to the original riverfront settlement.

Today, in North America, linear rural settlements exist in areas settled by the French. The French settlement pattern was commonly used along the St. Lawrence River in Québec and the lower Mississippi River.

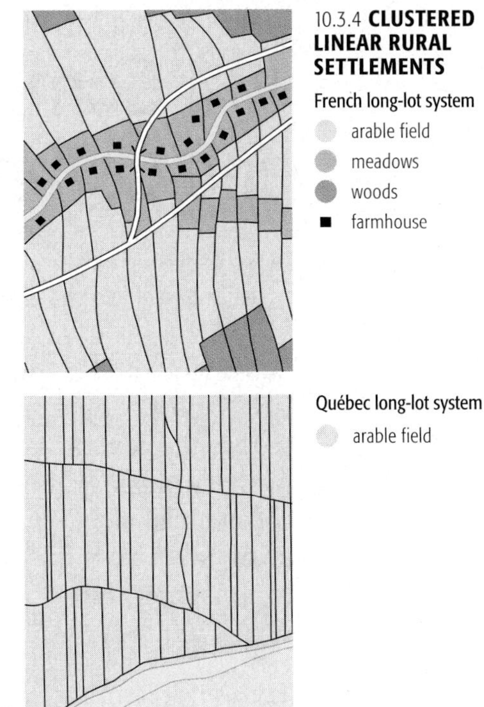

10.3.4 **CLUSTERED LINEAR RURAL SETTLEMENTS**

CLUSTERED CIRCULAR RURAL SETTLEMENTS

The clustered circular rural settlement consists of a central open space surrounded by structures. The kraal villages in southern Africa have enclosures for livestock in the center, surrounded by a ring of houses (compare our English word *corral*). In East Africa the Masai people, who are pastoral nomads, built kraal settlements as camps; women had principal responsibility for constructing them.

The German Gewandorf settlement consisted of a core of houses, barns, and churches, encircled by different types of agricultural activities. Small garden plots were located in the first ring surrounding the village, with cultivated land, pastures, and woodlands in successive rings. Von Thünen observed this circular rural pattern in his landmark agricultural studies in the early nineteenth century (refer to page 234, Figure 10.8.1).

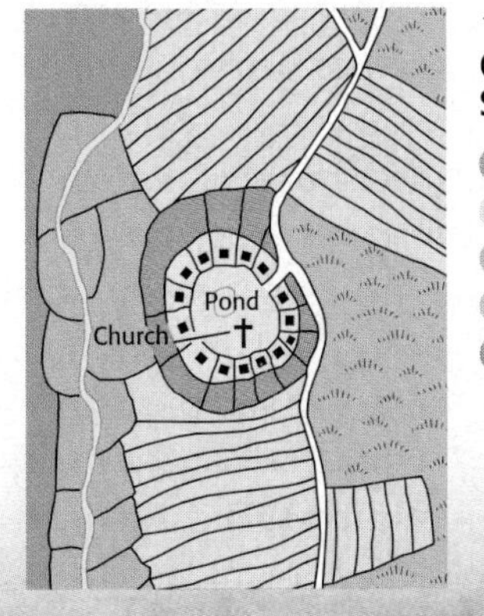

10.3.5 **CLUSTERED CIRCULAR RURAL SETTLEMENT**

10.3.3 **A CIRCULAR SETTLEMENT**
An aerial view of a traditional Maasi homestead with a number of huts around the periphery of a central cattle kraal.

TYPES OF AGRICULTURE

10.4 Comparing Subsistence and Commercial Agriculture

- Subsistence agriculture is characterized by a high percentage of farmers, small farms, and few machines.
- Commercial agriculture has a small percentage of farmers, large farms, and many machines.

Subsistence and commercial agriculture are undertaken for different purposes. In LDCs most people produce food for their own consumption. Some surplus may be sold to the government or to private firms, but the surplus product is not the farmer's primary purpose and may not even exist some years because of growing conditions. In commercial farming, farmers grow crops and raise animals primarily for sale off the farm rather than for their own consumption. Agricultural products are not sold directly to consumers but to food-processing companies.

FARM SIZE

The average farm size is relatively large in commercial agriculture, especially in the United States and Canada. U.S. farms average about 171 hectares (444 acres), whereas in LDCs the average farmer has only 0.7 hectares (1.7 acres).

Large size is partly a consequence of mechanization. Combines, pickers, and other machinery perform most efficiently at very large scales, and their considerable expense cannot be justified on a small farm. As a result of the large size and the high level of mechanization, commercial agriculture is an expensive business. Farmers spend hundreds of thousands of dollars to buy or rent land and machinery before beginning operations. This money is frequently borrowed from a bank and repaid after the output is sold.

Although the United States currently has fewer farms and farmers than in 1900, the amount of land devoted to agriculture has increased. The United States had 60 percent fewer farms and 85 percent fewer farmers in 2000 than in 1900, but 13 percent more farmland, primarily through irrigation and reclamation.

Most commercial farms in MDCs are family owned and operated—98 percent in the United States. Commercial farmers frequently expand their holdings by renting nearby fields. Nonetheless, commercial agriculture is increasingly dominated by a handful of large farms. In the United States, the 29,000 largest farms comprise only 1.4 percent of all U.S. farms, but they account for 48 percent of all agricultural sales. At the other extreme, one-half of the 2 million U.S. farms generate less than $5,000 per year in sales.

PERCENTAGE OF FARMERS IN SOCIETY

In MDCs less than one-tenth of the workers are engaged directly in farming, compared to more than one-half in LDCs. The percentage of farmers is even lower in the United States and Canada, at only 2 percent. Yet the small percentage of farmers in the United States and Canada produces enough food not only for themselves and the rest of the region, but also a surplus to feed people elsewhere.

The number of farmers declined dramatically in MDCs during the twentieth century. The United States had about 6 million farms in 1940 and 4 million in 1960; the number has stabilized during the past two decades at around 2 million. Both push and pull migration factors have been responsible for the decline: people were pushed away from farms by lack of opportunity to earn a decent income, and at the same time they were pulled to higher-paying jobs in urban areas.

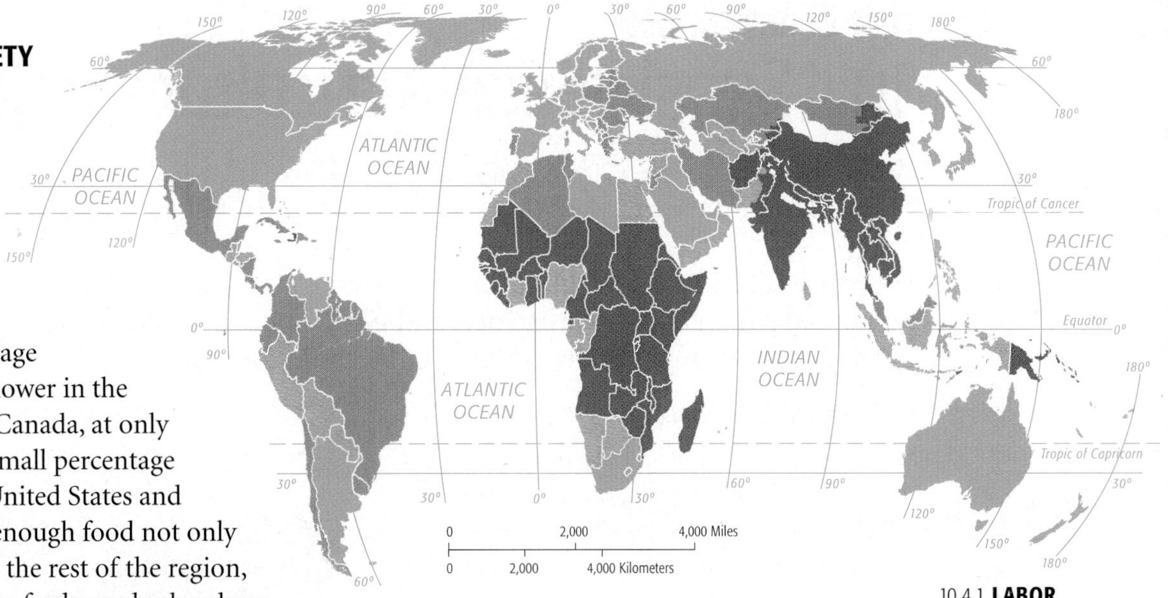

10.4.1 **LABOR FORCE ENGAGED IN AGRICULTURE**

Percent
- 50 and above
- 25.0–49.9
- 10.0–24.9
- below 10

USE OF MACHINERY

A small number of farmers in more developed societies can feed many people because they rely on machinery to perform work, rather than relying on people or animals. The first all-iron plow was made in the 1770s and was followed in the nineteenth and twentieth centuries by such machines as tractors, combines, corn pickers, and planters. In LDCs, farmers still do much of the work with hand tools and animal power.

Transportation improvements have also aided commercial farmers. The building of railroads in the nineteenth century, and highways and trucks in the twentieth century, have enabled farmers to transport crops and livestock farther and faster. Cattle arrive at market heavier and in better condition when transported by truck or train than when driven on hoof. Crops reach markets without spoiling.

Commercial farmers use scientific advances to increase productivity. Experiments conducted in university laboratories, industry, and research organizations generate new fertilizers, herbicides, hybrid plants, animal breeds, and farming practices, which can produce higher crop yields and healthier animals. Global positioning system (GPS) units determine the precise coordinates for spreading different types and amounts of fertilizers. On large ranches, GPS is also used to monitor the location of cattle.

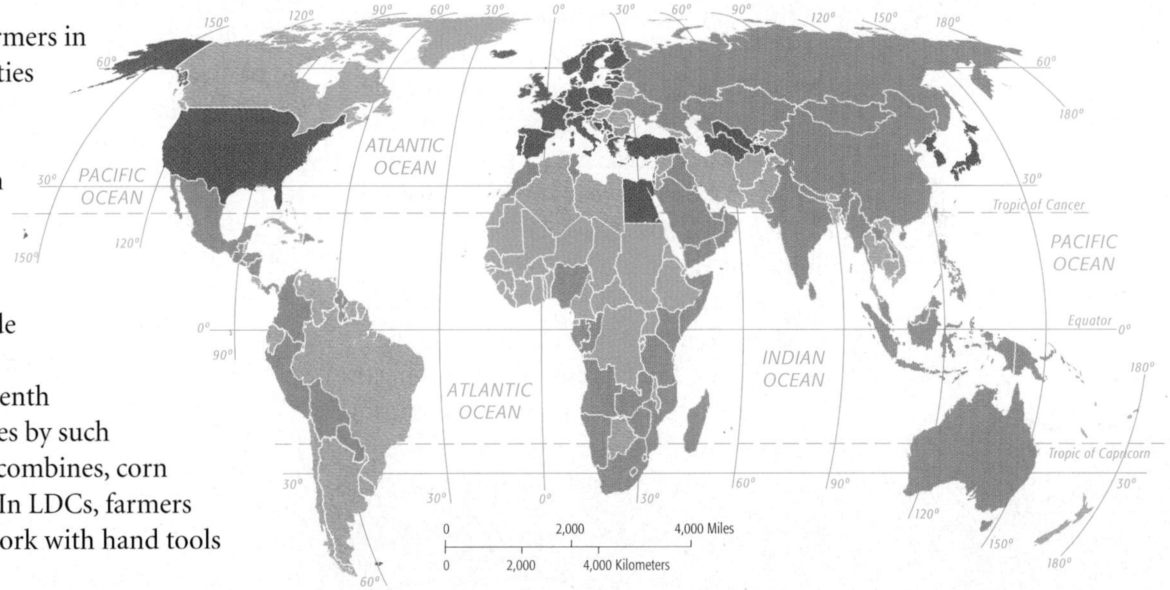

10.4.2 **TRACTORS PER 1,000 HECTARES OF CROPLAND**

- 25 and above
- 10–24
- 1–9
- below 1

TYPES OF AGRICULTURE

10.5 Subsistence Agriculture Regions

- Shifting cultivation is practiced in wet lands and pastoral nomadism in dry lands.
- Asia's large population concentrations practice intensive subsistence agriculture.

Four types of subsistence agriculture predominate in LDCs—shifting cultivation, pastoral nomadism, intensive subsistence, and plantation. Intensive subsistence agriculture is divided into two regions, depending on the choice of crop.

SHIFTING CULTIVATION

Shifting cultivation is practiced in much of the world's Humid Low-Latitude regions, which have relatively high temperatures and abundant rainfall. Each year villagers designate for planting an area surrounding the settlement. Before planting, they remove the dense vegetation using axes and machetes. On a windless day the debris is burned under carefully controlled conditions; consequently, shifting cultivation is sometimes called **slash-and-burn agriculture.** The rains wash the fresh ashes into the soil, providing needed nutrients.

The cleared area is known by a variety of names, including *swidden, ladang, milpa, chena,* and *kaingin*. The **swidden** can support crops only briefly, usually 3 years or less, before soil nutrients are depleted. Villagers then identify a new site and begin clearing it, leaving the old swidden uncropped for many years, so that it is again overrun by natural vegetation.

Shifting cultivation is being replaced by logging, cattle ranching, and cultivation of cash crops. Selling timber to builders or raising beef cattle for fast-food restaurants is a more effective development strategy than maintaining shifting cultivation. Defenders of shifting cultivation consider it a more environmentally sound approach for tropical agriculture.

10.5.1 **SHIFTING CULTIVATION IN BRAZIL**
Burning the debris.

10.5.2 **PASTORAL NOMAD, KENYA**

PASTORAL NOMADISM

Pastoral nomadism is a form of subsistence agriculture based on the herding of domesticated animals. It is adapted to dry climates, where planting crops is impossible. Pastoral nomads live primarily in the large belt of arid and semiarid land that includes North Africa, the Middle East, and parts of Central Asia. The Bedouins of Saudi Arabia and North Africa and the Masai of East Africa are examples of nomadic groups.

Pastoral nomads depend primarily on animals rather than crops for survival. The animals provide milk, and their skins and hair are used for clothing and tents. Like other subsistence farmers, though, pastoral nomads consume mostly **grain** rather than meat. Their animals are usually not slaughtered, although dead ones may be consumed. To nomads, the size of their herd is both an important measure of power and prestige and their main security during adverse environmental conditions.

Only about 15 million people are pastoral nomads, but they sparsely occupy about 20 percent of Earth's land area. Nomads used to be the most powerful inhabitants of the drylands. Today, national governments control the nomadic population using force, if necessary.

INTENSIVE SUBSISTENCE

In densely populated East, South, and Southeast Asia, most farmers practice **intensive subsistence agriculture**. Because the agricultural density—the ratio of farmers to arable land—is so high in parts of East and South Asia, families must produce enough food for their survival from a very small area of land. Most of the work is done by hand or with animals rather than with machines, in part due to abundant labor, but largely from lack of funds to buy equipment.

The intensive agriculture region of Asia can be divided between areas where wet rice dominates and areas where it does not. The term **wet rice** refers to the practice of planting rice on dry land in a nursery and then moving the seedlings to a flooded field to promote growth.

Wet rice is most easily grown on flat land, because the plants are submerged in water much of the time. The pressure of population growth in parts of East Asia has forced expansion of areas under rice cultivation. One method of developing additional land suitable for growing rice is to terrace the hillsides of river valleys.

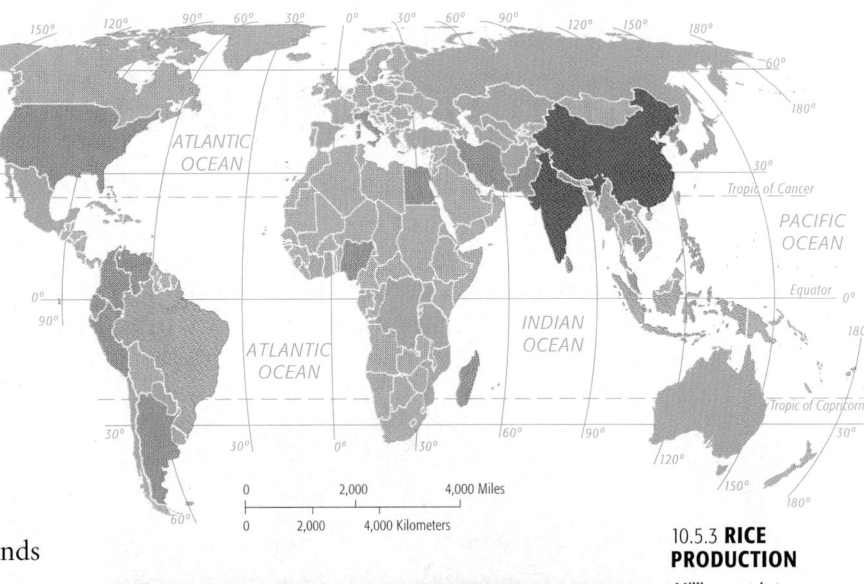

10.5.3 **RICE PRODUCTION**

10.5.4 **WET RICE TERRACES, GUANGXI ZHUANGZU, CHINA**

PLANTATION

A **plantation** is a large farm that specializes in one or two crops. Among the most important crops grown on plantations are cotton, sugarcane, coffee, rubber, and tobacco. The plantation is a form of commercial agriculture found in the tropics and subtropics, especially in Latin America, Africa, and Asia. Although generally situated in LDCs, plantations are often owned or operated by Europeans or North Americans and grow crops for sale primarily in MDCs.

Until the Civil War, plantations were important in the U.S. South, where the principal crop was cotton, followed by tobacco and sugarcane. Slaves brought from Africa performed most of the labor until the abolition of slavery and the defeat of the South in the Civil War. Thereafter, plantations declined in the United States; they were subdivided and either sold to individual farmers or worked by tenant farmers.

10.5.5 **TEA PLANTATION, MALAYSIA**

TYPES OF AGRICULTURE

10.6 Commercial Agriculture Regions

- Six main types of commercial agriculture are found in MDCs.
- The type of agriculture is influenced by physical geography conditions.

Commercial agriculture in MDCs can be divided into six main types. Each type is predominant in distinctive regions within MDCs, depending largely on climate.

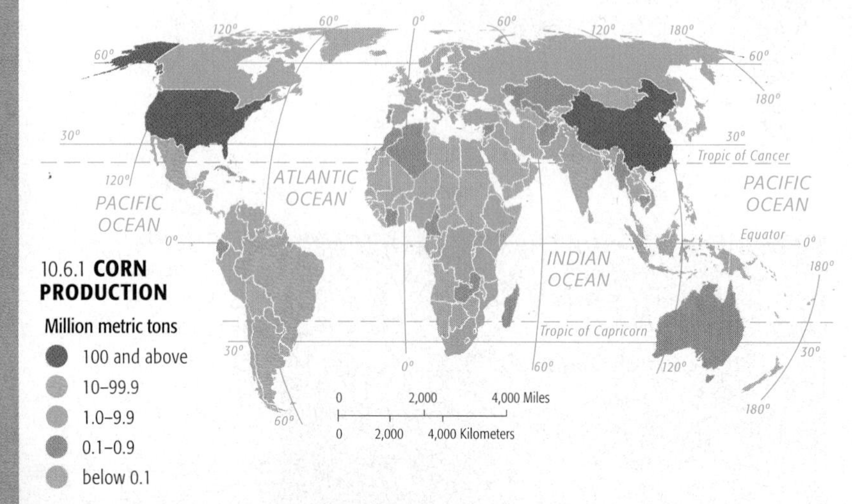

10.6.1 **CORN PRODUCTION**

MIXED CROP AND LIVESTOCK

The most distinctive characteristic of mixed crop and livestock farming is its integration of crops and livestock. Corn is the most commonly grown crop, followed by soybeans. Most of the crops are fed to animals rather than consumed directly by humans. A typical mixed commercial farm devotes nearly all land area to growing crops but derives more than three-fourths of its income from the sale of animal products, such as beef, milk, and eggs.

Mixed crop and livestock farming typically involves **crop rotation**. The farm is divided into a number of fields, and each field is planted on a planned cycle, often of several years duration.

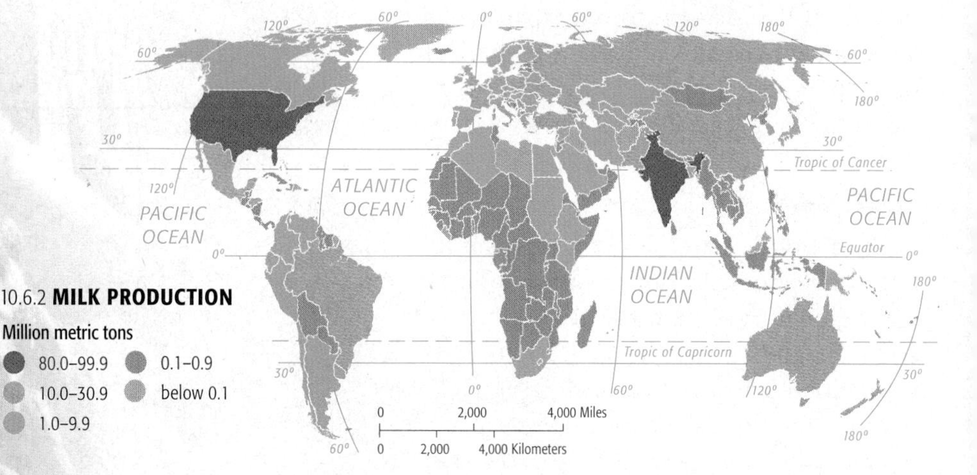

10.6.2 **MILK PRODUCTION**

DAIRY FARMING

Dairy farming is the most important agriculture practiced near large urban areas in MDCs. Dairy farms must be closer to their markets than other products because milk is highly perishable.

The share of the world's dairy farming conducted in LDCs has risen dramatically in recent years. Rising incomes permit urban residents to buy more milk products.

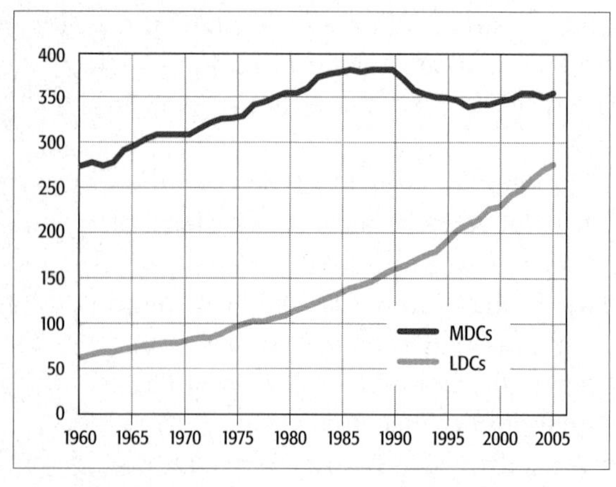

10.6.3 **WORLD MILK PRODUCTION CHART (billion tons)**

GRAIN FARMING

Commercial grain farms are generally located in regions that are too dry for mixed crop and livestock farming. Unlike mixed crop and livestock farming, crops on a grain farm are grown primarily for consumption by humans rather than by livestock.

The most important crop grown is wheat, used to make flour. It can be stored relatively easily without spoiling and can be transported a long distance. Because wheat has a relatively high value per unit weight, it can be shipped profitably from remote farms to markets. The United States is by far the largest commercial producer of grain, primarily in the Plains states.

LIVESTOCK RANCHING

Ranching is the commercial grazing of livestock over an extensive area. It is practiced primarily on semiarid or arid land where the vegetation is too sparse and the soil too poor to support crops.

Ranching has been glamorized in novels and films, although the cattle drives and "Wild West" features of this type of farming actually lasted only a few years in the mid-nineteenth century. Contemporary ranching has become part of the meat-processing industry, rather than carried out on isolated farms.

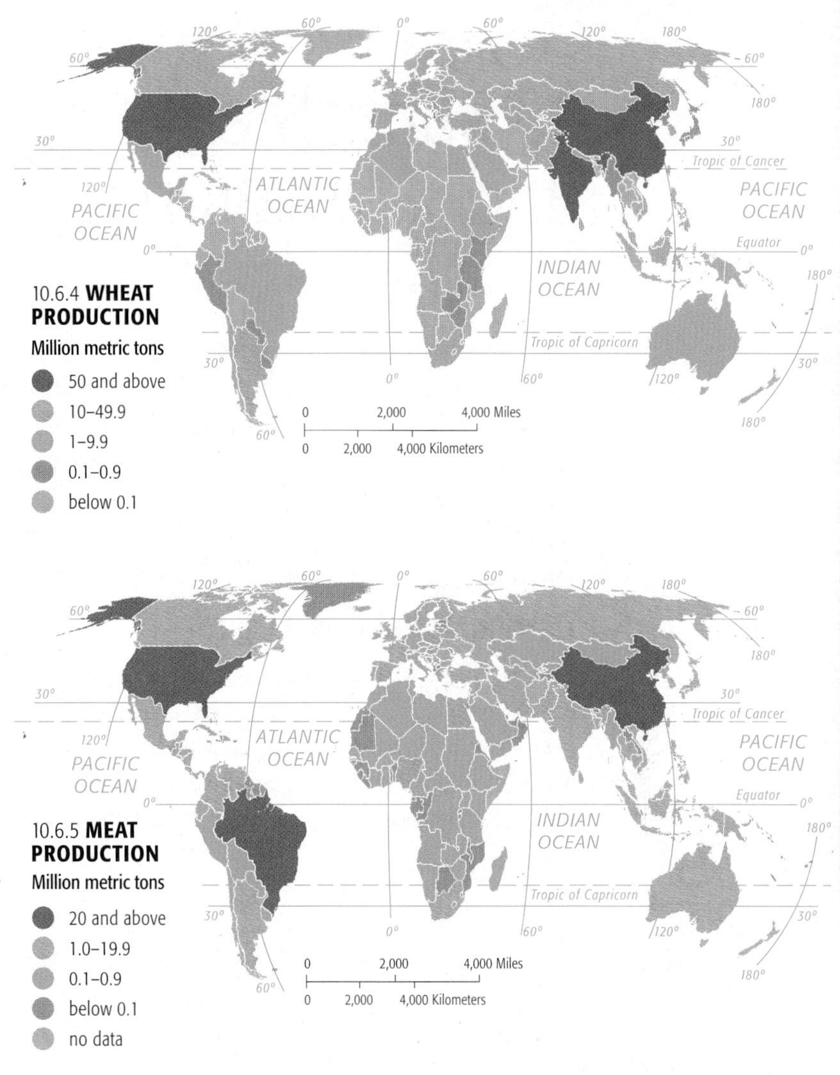

10.6.4 **WHEAT PRODUCTION**
Million metric tons

10.6.5 **MEAT PRODUCTION**
Million metric tons

COMMERCIAL GARDENING AND FRUIT FARMING

Commercial gardening and fruit farming are the predominant types of agriculture in the U.S. Southeast. The region has a long growing season and humid climate and is accessible to the large markets in the big cities along the East Coast. It is frequently called **truck farming**, because "truck" was a Middle English word meaning bartering or the exchange of commodities.

Truck farms grow many of the fruits and vegetables that consumers demand in MDCs, such as apples, cherries, lettuce, and tomatoes. A form of truck farming called specialty farming has spread to New England. Farmers are profitably growing crops that have limited but increasing demand among affluent consumers, such as asparagus, mushrooms, peppers, and strawberries.

MEDITERRANEAN AGRICULTURE

Mediterranean agriculture exists primarily on lands that border the Mediterranean Sea and other places that share a similar physical geography. Winters are moist and mild, summers hot and dry. The land is very hilly, and mountains frequently plunge directly to the sea, leaving very little flat land. The two most important crops are olives (primarily for cooking oil) and grapes (primarily for wine).

FARMER INSPECTING INVERTED PEANUT CROP DURING HARVEST, GEORGIA

CHALLENGES FOR AGRICULTURE

10.7 Subsistence Agriculture and Population Growth

- Four strategies can increase food supply in LDCs.
- Africa faces strong challenges in producing enough food.

Two issues discussed in earlier chapters influence the challenges faced by subsistence farmers. First, because of rapid population growth in LDCs (discussed in Chapter 2), subsistence farmers must feed an increasing number of people. Second, because of adopting the international trade approach to development (discussed in Chapter 9), subsistence farmers must grow food for export instead of for direct consumption. Four strategies have been identified to increase food supply.

1. EXPAND AGRICULTURAL LAND

Historically, world food production increased primarily by expanding the amount of land devoted to agriculture. When the world's population increased more rapidly during the Industrial Revolution beginning in the eighteenth century, pioneers could migrate to uninhabited territory and cultivate the land. New land might appear to be available, because only 11 percent of the world's land area is currently used for agriculture. But excessive or inadequate water makes expansion difficult. The rate of expansion of agricultural land has slowed considerably since the 1990s, and expansion of agricultural land has been slower than the increase of the human population since around 1950.

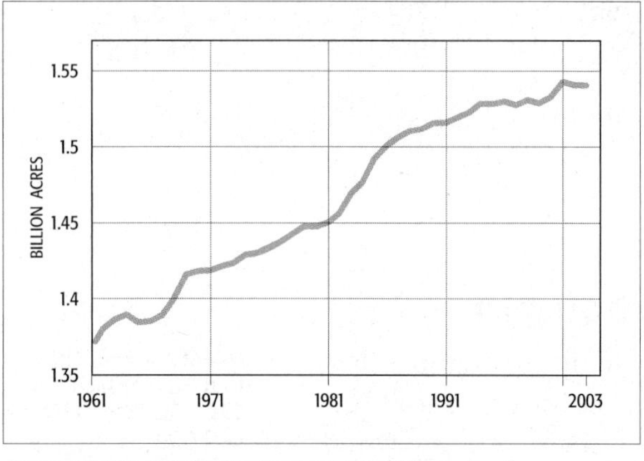

10.7.1 **WORLD AGRICULTURAL LAND (billion acres)**

10.7.2 **INTERNATIONAL RICE RESEARCH INSTITUTE, HOME OF THE "GREEN REVOLUTION"**

2. INCREASE AGRICULTURAL PRODUCTIVITY

New agricultural practices have permitted farmers worldwide to achieve much greater yields from the same amount of land. The invention and rapid diffusion of more productive agricultural techniques during the 1970s and 1980s is called the **green revolution**.

Scientists began experiments during the 1950s to develop a higher-yield form of wheat. A decade later, the International Rice Research Institute created a miracle rice seed. The Rockefeller and Ford Foundations sponsored many of the studies, and the program's director, Dr. Norman Borlaug, won the Nobel Peace Prize in 1970. More recently, scientists have developed new high-yield maize (corn). Scientists have continued to create higher-yield hybrids that are adapted to environmental conditions in specific regions.

The green revolution was largely responsible for preventing a food crisis in LDCs during the 1970s and 1980s. Will these scientific breakthroughs continue in the twenty-first century? To take full advantage of the new miracle seeds, farmers must use more fertilizer and machinery, both of which depend on increasingly expensive fossil fuels. To maintain the green revolution, governments in LDCs must allocate scarce funds to subsidize the cost of seeds, fertilizers, and machinery.

The new miracle seeds were diffused rapidly around the world. India's wheat production, for example, more than doubled in 5 years. After importing 10 million tons of wheat annually in the mid-1960s, India by 1971 had a surplus of several million tons.

3. IDENTIFY NEW FOOD SOURCES

New food sources could come from:

- Increased cultivation of the oceans. Oceans supply only a small percentage of the world's food, but overfishing has reduced the population of some fish species—by 90 percent in the case of tuna and swordfish.
- Higher protein **cereal grains**. People in LDCs depend on grains that lack certain proteins. Hybrids with higher protein content could achieve better nutrition without changing food-consumption habits.
- Improved palatability of rarely consumed foods. Some foods are rarely consumed because of taboos, religious values, and social customs. In MDCs, consumers avoid consuming recognizable soybean products like tofu and sprouts, but could be induced to eat soybeans shaped like burgers and franks. Krill—a small shrimplike crustacean—could be an important source of food from the oceans. But Krill does not taste very good.

10.7.3 **KRILL**

4. EXPAND EXPORTS

The three top export grains are wheat, maize (corn), and rice. The United States is the world's leading exporter of grain by a wide margin. Countries in South Asia and Southeast Asia were net importers of grain in the twentieth century but have become net exporters. Japan is by far the world's leading grain importer, followed by China. On a regional scale, the Middle East has become the leading net importer of all three major grains.

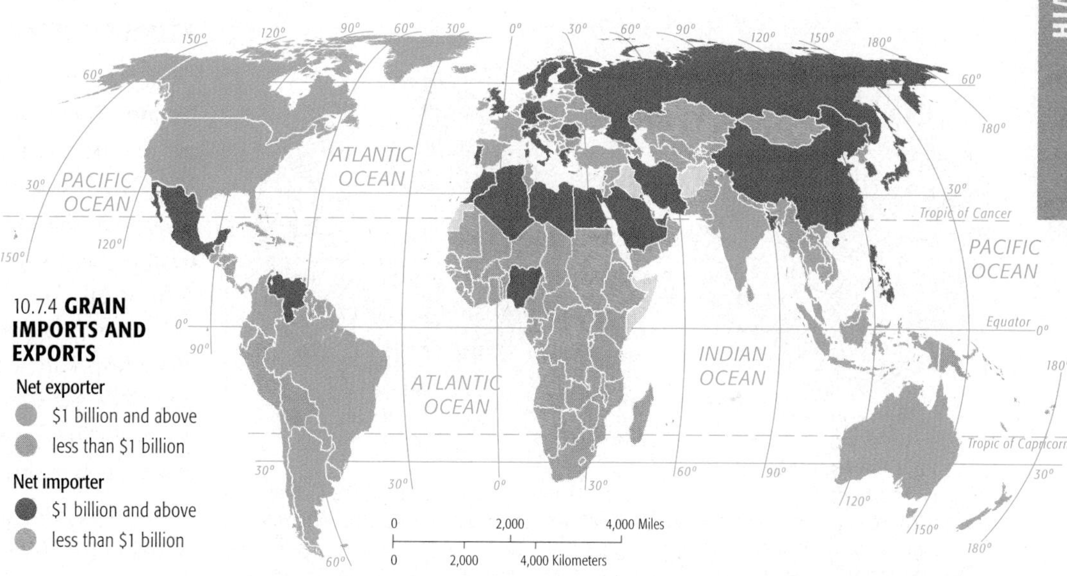

10.7.4 **GRAIN IMPORTS AND EXPORTS**

Net exporter
- $1 billion and above
- less than $1 billion

Net importer
- $1 billion and above
- less than $1 billion

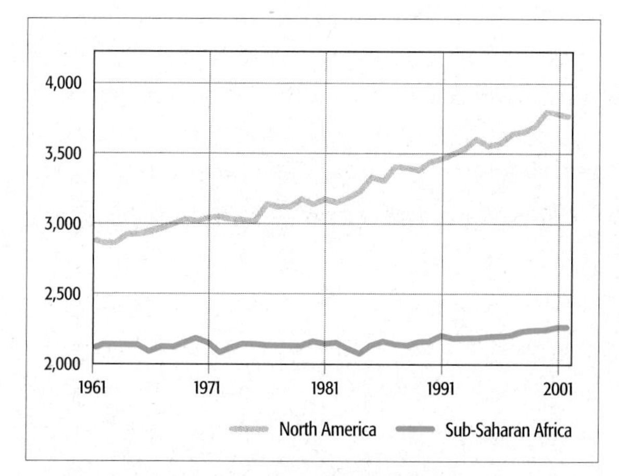

10.7.5 **DAILY CALORIES CONSUMED PER CAPITA**

AFRICA'S FOOD-SUPPLY CRISIS

Sub-Saharan Africa is struggling to keep food production ahead of population growth. Since 1961, per capita caloric intake has increased substantially in North America, while remaining virtually unchanged in Sub-Saharan Africa. Production of the three main grains tripled in Sub-Saharan Africa during the period, whereas population increased more than sixfold. Forty million Africans face famine, according to the World Food Program.

The threat of famine is particularly severe in the Horn of Africa and the Sahel. Traditionally, this region supported limited agriculture. Pastoral nomads moved their herds frequently, permitting vegetation to regenerate. Farmers grew groundnuts for export and used the receipts to import rice. With rapid population growth, farmers overplanted, and herd size increased beyond the capacity of the land to support the animals. Animals overgrazed the limited vegetation and clustered at scarce water sources. Many of the animals died of hunger.

Government policies have aggravated the food-shortage crisis. To make food affordable for urban residents, governments keep agricultural prices low. Constrained by price controls, farmers are unable to sell their commodities at a profit and therefore have little incentive to increase productivity.

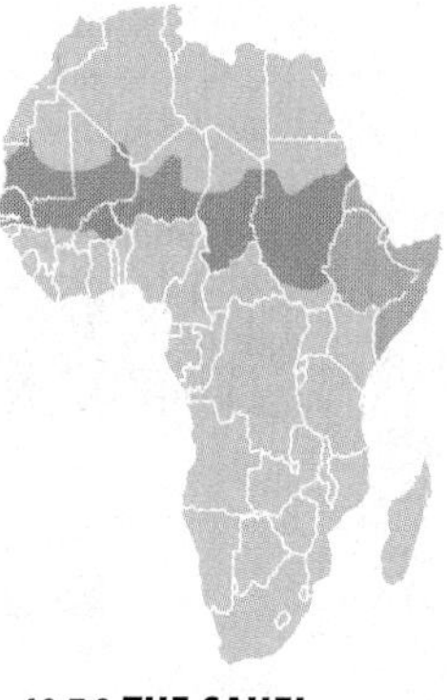

10.7.6 **THE SAHEL**

- Horn of Africa
- Sahel

CHALLENGES FOR AGRICULTURE

10.8 Commercial Agriculture and Market Forces

- Farming is part of agribusiness in MDCs.
- Because of overproduction, farmers in MDCs may receive government subsidies to reduce output.

The system of commercial farming found in MDCs has been called **agribusiness**, because the family farm is not an isolated activity, but rather is integrated into a large food-production industry. Agribusiness encompasses such diverse enterprises as tractor manufacturing, fertilizer production, and seed distribution.

Farmers are less than 2 percent of the U.S. labor force, but around 20 percent of U.S. labor works in food production and service related to agribusiness—food processing, packaging, storing, distributing, and retailing. Although most farms are owned by individual families, many other aspects of agribusiness are controlled by large corporations.

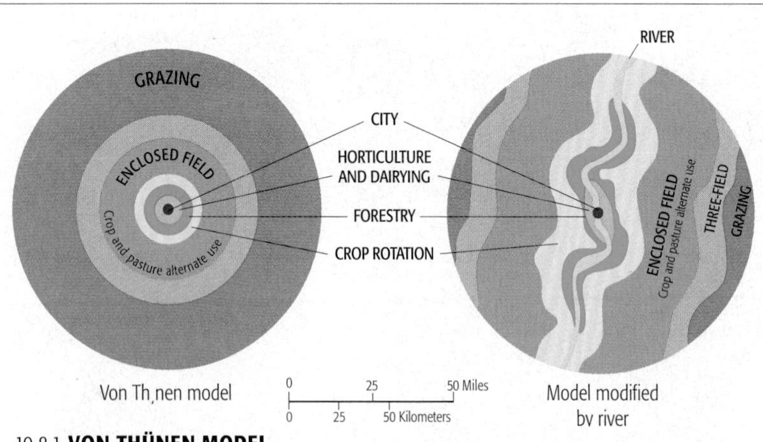

10.8.1 **VON THÜNEN MODEL**
Johann Heinrich von Thünen, a farmer in northern Germany, proposed a model to explain the importance of proximity to market in the choice of crops on commercial farms. The von Thünen model was first proposed in 1826 in a book titled *The Isolated State*.

According to the model, which was later modified by geographers, a commercial farmer initially considers which crops to cultivate and which animals to raise based on market location. Von Thünen based his general model of the spatial arrangement of different crops on his experiences as owner of a large estate in northern Germany. He found that specific crops were grown in different rings around the cities in the area.

PRODUCTIVITY CHALLENGES

Commercial farmers suffer from low incomes because they are capable of producing much more food than is demanded by consumers in MDCs. Although the food supply has increased in MDCs, demand has remained constant, because of low population growth and market saturation.

A surplus of food can be produced because of widespread adoption of efficient agricultural practices. New seeds, fertilizers, pesticides, mechanical equipment, and management practices have enabled farmers to obtain greatly increased yields per area of land.

The experience of dairy farming in the United States demonstrates the growth in productivity. The United States had 20 million dairy cows producing 57 million metric tons (63 million tons) of milk a year during the 1960s. The number of dairy cows in the United States declined to 9 million in 2005, but production increased to 80 million metric tons (90 million tons). In other words, yield per cow tripled.

GOVERNMENT SUBSIDIES

The U.S. government has three policies that are supposed to address the problem of excess productive capacity:

1. Farmers are encouraged to avoid producing crops that are in excess supply. Because soil erosion is a constant threat, the government encourages planting fallow crops, such as clover, to restore nutrients to the soil and to help hold the soil in place. These crops can be used for hay, forage for pigs, or to produce seeds for sale.
2. The government pays farmers when certain commodity prices are low. The government sets a target price for the commodity and pays farmers the difference between the price they receive in the market and a target price set by the government as a fair level for the commodity. The target prices are calculated to give farmers the same price for the commodity today as in the past, when compared to other consumer goods and services.
3. The government buys surplus production and sells or donates it to foreign governments. In addition, low-income Americans receive food stamps in part to stimulate their purchase of additional food.

Farming in Europe is subsidized even more than in the United States. Government policies in MDCs point out a fundamental irony in worldwide agricultural patterns. In MDCs, farmers are encouraged to grow less food, whereas LDCs struggle to increase food production to match the rate of growth in the population.

10.8.2 **AGRIBUSINESS IN DAIRY FARMING**
(Top) Large-scale milking. (Below) Transportation from farm to processing; (below left) processing; (below right) bottling; and (bottom) retailing.

CHALLENGES FOR AGRICULTURE

10.9 Sustainable Agriculture

- Sustainable agriculture and organic farming rely on sensitive land management.
- Sustainable agriculture also limits use of chemicals and integrates crops and livestock.

Some commercial farmers are converting their operations to **sustainable agriculture**, an agricultural practice that preserves and enhances environmental quality. An increasingly popular form of sustainable agriculture is organic farming. Farmers practicing sustainable agriculture typically generate lower revenues than do conventional farmers, but they also have lower costs.

There are three principal practices that distinguish sustainable agriculture (and at its best, organic farming) from conventional agriculture.

1. SENSITIVE LAND MANAGEMENT

Sustainable agriculture protects soil in part through **ridge tillage**, which is a system of planting crops on ridge tops. Crops are planted on 10- to 20-centimeter (4- to 8-inch) ridges that are formed during cultivation or after harvest. The crop is planted on the same ridges, in the same rows, year after year. Ridge tillage is attractive for two main reasons—lower production costs and greater soil conservation.

Production costs are lower with ridge tillage in part because it requires less investment in tractors and other machinery than conventional planting. An area that would be prepared for planting under conventional farming with three to five tractors can be prepared for ridge tillage with only one or two tractors. The primary tillage tool is a row-crop cultivator that can form ridges. There is no need for a plow, or field cultivator, or a 300-horsepower four-wheel-drive tractor.

With ridge tillage, the space between rows needs to match the distance between wheels of the machinery. If 75 centimeters (30 inches) are left between rows, tractor tires will typically be on 150-centimeter (60-inch) centers and combine wheels on 300-centimeter (120-inch) centers. Wheel spacers are available from most manufacturers to fit the required spacing.

Ridge tillage features a minimum of soil disturbance from harvest to the next planting. A compaction-free zone is created under each ridge and in some row middles. Keeping the trafficked area separate from the crop-growing area improves soil properties. Over several years the soil will tend to have increased organic

10.9.1 **RIDGE TILLAGE**
An important sustainable agriculture practice.

matter, greater water holding capacity, and more earthworms. The channels left by earthworms and decaying roots enhance drainage.

Ridge tillage compares favorably with conventional farming for yields while lowering the cost of production. Although more labor intensive than other systems, it is profitable on a per-acre basis. In Iowa, for example, ridge tillage has gained favor for production of organic and herbicide-free soybeans, which sell for more than regular soybeans.

How much farming is organic?

The U.S. Department of Agriculture sets national standards for what constitutes organic. About 0.22 percent of U.S. farmland was certified organic in 2006, including 4 percent of apple orchards and 4 percent of lettuce fields. Percentages have been lower for the leading field crops: only 0.4 percent of wheat fields, 0.2 percent of soybean fields, and 0.1 percent of corn fields were certified organic in 2003.

The UN Food and Agriculture Organization estimates the share of agricultural land farmed through organic practices at 0.23 percent worldwide (1/4 of 1 percent), including 0.4 percent in MDCs and 0.2 percent in LDCs. Australia was the world leader, with organic farming practiced on 2.7 percent of its farmland.

2. LIMITED USE OF CHEMICALS

Sustainable agriculture involves application of limited, if any, herbicides to control weeds. In principle, farmers can control weeds without chemicals, although it requires additional time and expense that few farmers can afford. Researchers have found that combining mechanical weed control with some chemicals yields higher returns per acre than relying solely on one of the two methods.

Ridge tilling also promotes decreased use of chemicals, which can be applied only to the ridges and not the entire field. Combining herbicide banding—which applies chemicals in narrow bands over crop rows—with cultivating may be the best option for many farmers.

3. INTEGRATED CROP AND LIVESTOCK

Sustainable agriculture attempts to integrate the growing of crops and the raising of livestock as much as possible at the level of the individual farm. Animals consume crops grown on the farm and are not confined to small pens.

In conventional farming, integration between crops and livestock generally takes place through intermediaries rather than inside an individual farm. That is, many farmers in the mixed crop and livestock region actually choose to only grow crops or only raise animals. They sell their crops off the farm or purchase feed for their animals from outside suppliers.

Sustainable agriculture is sensitive to the complexities of interdependencies between crops and livestock:

- Finding the correct number and distribution of livestock for the area based on the landscape and forage sources. Prolonged concentration of livestock in a specific location can result in permanent loss of vegetative cover, so the farmer needs to move the animals.
- Animal confinement. The moral and ethical debate regarding the welfare of confined livestock is particularly intense. Manure from non-confined animals can contribute to soil fertility.
- Management in extreme weather. Herd size may need to be reduced during periods of short- and long-term droughts.
- Flexible feeding and marketing. Feed costs are the largest single variable cost in livestock operation. Feed costs can be kept to a minimum by monitoring animal condition and performance and understanding seasonal variations in feed and forage quality on the farm.

Chapter **Summary**

DISTRIBUTION OF AGRICULTURE

- Prior to the development of agriculture, people survived by hunting animals, gathering wild vegetation, or fishing.
- Agriculture originated in multiple hearths and diffused to numerous places simultaneously.
- Several agricultural regions can be identified based on specific farming practices.
- Most farm communities are clustered settlements, where farm families live near each other surrounded by fields.

TYPES OF AGRICULTURE

- Subsistence agriculture, typical of LDCs, involves growing food for one's own consumption.
- Commercial agriculture, typical of MDCs, involves growing food to sell off the farm.
- Commercial agriculture involves larger farms, fewer farmers, and more mechanization than does subsistence agriculture.

CHALLENGES FOR AGRICULTURE

- Subsistence agriculture faces distinctive economic challenges resulting from rapid population growth and pressure to adopt international trade strategies to promote development.
- Commercial agriculture faces distinct challenges resulting from access to markets and overproduction.
- Sustainable farming can preserve and enhance environmental quality.

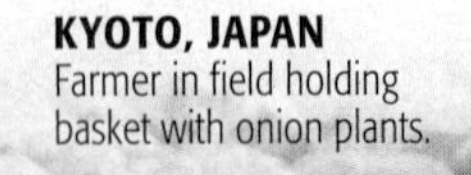

KYOTO, JAPAN
Farmer in field holding basket with onion plants.

Geographic Consequences of Change

A country's agriculture remains one of the best measures of its level of development and standard of material comfort. Despite major changes, agriculture in LDCs still employs the majority of the population, and producing food for local survival is still paramount.

Farming in MDCs directly employs few people, but all of agribusiness combined is actually the largest employment sector. However, most agribusiness is classified in the industrial and service sectors.

The future is uncertain for both subsistence farmers in LDCs and commercial farmers in MDCs. In one respect, the uncertainty stems from a similar problem: farming in neither location produces sufficient income to support the standard of living farm families desire. However, the underlying cause of low incomes differs significantly between MDCs and LDCs.

In LDCs people migrate from the farms to the cities in search of higher-paying jobs and a better life. Given the high natural increase rate and pressure to produce more for international trade, the migrants are not missed on the farms. The need for more food will not be met by adding more workers on the farms, but from more intensive use of existing farms and purchase of food from abroad.

In MDCs people also migrate from the farms to the cities in search of higher-paying jobs. However, these migrants are missed. Small farming communities in the United States are dying, and their death causes a loss of rural-based culture and values. Farming is the backbone of many small-town economies. Without farmers, banks and shops lose their main sources of income. For every five people that give up farming, one business closes in a small town. People still live on farms, but work in factories, offices, or businesses in the nearest big city. And they shop at the big-city Wal-Mart instead of the small-town Main Street.

SHARE YOUR VOICE Student Essay

LAURA CRONIN

Laura Cronin
University of Oklahoma

Moving to the City

I was born and raised in a small town and never understood how anyone would want to live any other way. The thought of moving away from my farm town in Oklahoma was the craziest idea to me. If someone told me that I would be moving to a large city while I was growing up, I would have thought that they were crazy. Now that I have moved to Chicago, I am beginning to understand why so many people choose to live in an urban setting.

Growing up, I was completely content with my small town world. I was taught to love the outdoors and to respect nature but I was also taught that people from the city were stuck up and rude. That is perhaps the reason that I had so many apprehensions about big cities. I have encountered many preconceived notions about urban living that had a large influence on me, so it was understandable that I had my doubts about moving to Chicago.

Eventually, I decided to take the plunge and make the move. All of my biases and preconceived notions soon melted away as I developed an understanding for the urban world that I had been taught to avoid for so long. I learned to appreciate the culture and way of life here. Things are so much easier in a large city: everything is closer, there is so much to do, and best of all, there is a tremendous amount of diversity. The biological and cultural diversity is unlike anything I have ever experienced.

After living in both places, I feel that I have gotten an opportunity to experience the best of both worlds. I have my small town background that has taught me to love the outdoors and the simple things in life, but I also have an understanding for the urban world and see why it is such a popular place to live. I have had my issues with adjusting to life in an urban landscape, but through my experiences, I have come to love and appreciate it.

Chapter Resources

GENETICALLY MODIFIED FOODS AND AFRICA'S FOOD NEEDS

LDCs have been urged by the United States to increase their food supply in part through increased use of genetic modification (GM) of crops and livestock.

Farmers have been manipulating crops and livestock for thousands of years: the very nature of agriculture is to deliberately manipulate nature. The science of genetics beginning in the nineteenth century expanded understanding of how to manipulate plants and animals to secure dominance of the most favorable traits. GM is different: the genetic composition of an organism is not merely studied, it is actually altered. GM involves mixing genetic material among two or more species that would not otherwise mix in nature.

The positives of GM are higher yields, increased nutrition, and more resistance to pests. GM foods are better tasting, at least to some palates.

GM is widespread in the United States, especially in the processed food that Americans consume in restaurants and at home heated in microwave ovens. The United States was responsible for 63 percent of the world's GM crops in 2003.

Outside the United States, opposition to GM is strong. GM may cause safety problems, such as lowered resistance to antibiotics, and could destroy long-standing ecological balances in local agriculture. Food is regarded as less nutritious than that from traditionally bred crops and livestock.

In Africa, opposition to GM stems in part from practical economics. European countries, the main markets for Africa's agricultural exports, require GM foods to be labeled. Because European consumers shun GM food, African farmers fear that if they are no longer able to certify their exports as GM-free, European customers will stop buying them.

Africans are especially uneasy with GM primarily because it would increase dependence on the U.S.-based transnational corporations responsible for manufacturing most of the GM seeds. Every country in Africa except South Africa rejected an offer of GM seeds made by Monsanto and other U.S.-based biotech corporations at a 1998 UN meeting. After the 1998 UN meeting, African countries released this statement: "We strongly object that the image of the poor and hungry from our countries is being used by giant multinational corporations to push a technology that is neither safe, environmentally friendly, nor economically beneficial to us."

Africans fear that the biotech companies could—and would—introduce a so-called "terminator" gene in the GM seeds, to prevent farmers from replanting them after harvest and require them to continue to purchase seeds year after year from the transnational corporations. Mozambique's Prime Minister said, "We don't want to create a habit of using genetically modified maize that the country cannot maintain." If agriculture is regarded as a way of life, not just a food production business, GM represents for many Africans an unhealthy level of dependency on MDCs.

When a drought threatened millions in southern Africa with starvation in 2006, the United States offered one-half million tons of GM grain, but most countries rejected the offer. Zambia's president called GM food "poison." Ultimately, some countries accepted the offer, but only after the United States agreed to several safeguards, such as milling the grain before delivery rather than sending seeds that risked cross-breeding with local strains.

Zimbabwe subsistence farmer holds corn (maize) devastated by drought.

KEY TERMS

Agribusiness
Commercial agriculture characterized by the integration of different steps in the food-processing industry, usually through ownership by large corporations.

Agriculture
The deliberate effort to modify a portion of Earth's surface through the cultivation of crops and the raising of livestock for sustenance or economic gain.

Clustered rural settlement
A rural settlement in which the houses and farm buildings of each family are situated close to each other and fields surround the settlement.

Cereal grain
A grass yielding grain for food.

Commercial agriculture
Agriculture undertaken primarily to generate products for sale off the farm.

Crop rotation
The practice of rotating use of different fields from crop to crop each year, to avoid exhausting the soil.

Dispersed rural settlement
A rural settlement pattern characterized by isolated farms rather than clustered villages.

Grain
Seed of a cereal grass.

Green revolution
Rapid diffusion of new agricultural technology, especially new high-yield seeds and fertilizers.

Intensive subsistence agriculture
A form of subsistence agriculture in which farmers must expend a relatively large amount of effort to produce the maximum feasible yield from a parcel of land.

Pastoral nomadism
A form of subsistence agriculture based on herding domesticated animals.

Plantation
A large farm in tropical and subtropical climates that specializes in the production of one or two crops for sale, usually to a more developed country.

Ranching
A form of commercial agriculture in which livestock graze over an extensive area.

Ridge tillage
System of planting crops on ridge tops in order to reduce farm production costs and promote greater soil conservation.

Seed agriculture
Reproduction of plants through annual introduction of seeds, which result from sexual fertilization.

Shifting cultivation
A form of subsistence agriculture in which people shift activity from one field to another; each field is used for crops for a relatively few years and left fallow for a relatively long period.

Slash-and-burn agriculture
Another name for shifting cultivation, so named because fields are cleared by slashing the vegetation and burning the debris.

Subsistence agriculture
Agriculture designed primarily to provide food for direct consumption by the farmer and the farmer's family.

Sustainable agriculture
Farming methods that preserve long-term productivity of land and minimize pollution, typically by rotating soil-restoring crops with cash crops and reducing inputs of fertilizer and pesticides.

Swidden
A patch of land cleared for planting through slashing and burning.

Truck farming
Commercial gardening and fruit farming, so named because truck was a Middle English word meaning bartering or the exchange of commodities.

Vegetative planting
Reproduction of plants by direct cloning from existing plants.

Wet rice
Rice planted on dryland in a nursery and then moved to a deliberately flooded field to promote growth.

ON THE INTERNET

Agricultural statistics can be found on the Internet at the United Nation's Food and Agriculture Organization's website **www.fao.org**. The FAO maintains a data base known as FAOSTAT, with information on crops, food, and land use. Information about sustainable agriculture can be found through the Sustainable Agriculture Research and Education website **www.sare.org.**

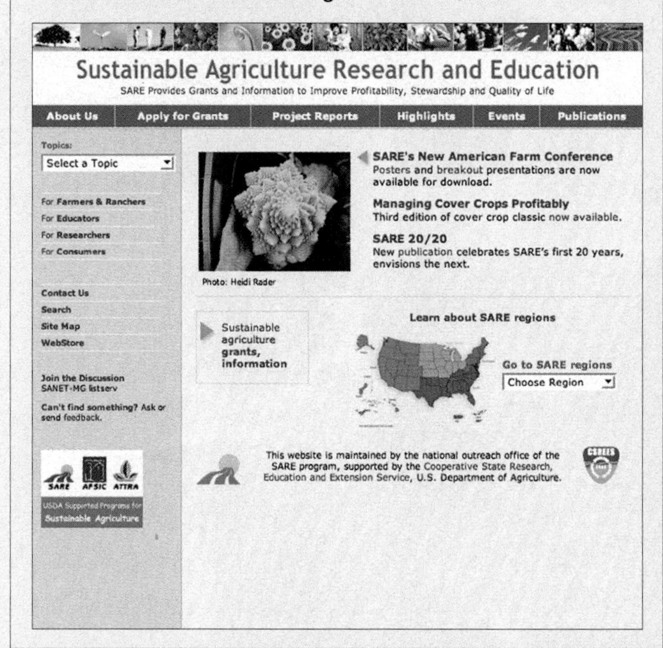

FURTHER READINGS

Chakravarti, A. K. "Green Revolution in India." *Annals of the Association of American Geographers* 63 (1973): 319–30.

Durand, Loyal, Jr. "The Major Milksheds of the Northeastern Quarter of the United States." *Economic Geography* 40 (1964): 9–33.

Grigg, David B. *The Agricultural Systems of the World: An Evolutionary Approach*. London: Cambridge University Press, 1974.

———. "Food Imports, Food Exports and Their Role in National Food Consumption." *Geography* 86 (2001): 171–76.

Hart, John Fraser. *The Changing Scale of American Agriculture*. Charlottesville: University of Virginia Press, 2003.

Ilbery, Brian W. *Agricultural Geography: A Social and Economic Analysis*. New York: Oxford University Press, 1985.

Morgan, W. B. *Agriculture in the Third World: A Spatial Analysis*. Boulder, CO: Westview Press, 1978.

Newbury, Paul A. R. *A Geography of Agriculture*. Estover: Macdonald and Evans, 1980.

Pacione, Michael, ed. *Progress in Agricultural Geography*. London and Dover, NH: Croom Healm, 1986.

Sauer, Carl O. *Agricultural Origins and Dispersals*, 2d ed. Cambridge, MA: M.I.T. Press, 1969.

Symons, Leslie. *Agricultural Geography*, rev. ed. London: G. Bell, 1979.

Tarrant, John R. *Agricultural Geography*. New York: John Wiley, 1974.

Turner, B. L., II, and Stephen B. Brush, eds. *Comparative Farming Systems*. New York: Guilford Press, 1987.

von Thünen, Johann Heinrich. *Von Thünen's Isolated State: An English Edition of "Der Isolierte Staat."* Trans. by Carla M. Wartenberg. Elmsford, NY: Pergamon Press, 1966.

Whittlesey, Derwent. "Major Agricultural Regions of the Earth." *Annals of the Association of American Geographers* 26 (1936): 199–240.

Chapter 11 INDUSTRY

Manufacturing jobs are viewed as a special asset by communities around the world. They are seen as the "engine" of economic growth and prosperity. Different communities possess distinctive assets for attracting particular types of industries as well as challenges in retaining them.

A generation ago, industry was highly clustered in a handful of communities within more developed countries, but industry has diffused to more communities, including some in less developed countries. Meanwhile, loss of manufacturing jobs in MDCs such as the United States has caused economic problems for communities traditionally dependent on them.

SOCK FACTORY, YIWU, CHINA

THE KEY ISSUES IN THIS CHAPTER

ORIGIN AND DISTRIBUTION OF INDUSTRY

11.1 The Industrial Revolution

11.2 Distribution of Industry

SITUATION FACTORS IN INDUSTRIAL LOCATION

11.3 Situation Factors in Locating Industry

11.4 Changes in Location of Steel Production

11.5 Changing Distribution of Auto Production

11.6 Ship by Boat, Rail, Truck, or Air?

SITE FACTORS IN INDUSTRIAL LOCATION

11.7 Site Factors in Locating Industry

11.8 Distribution of Textile and Apparel Production

11.9 Labor-intensive Industries

ORIGIN AND DISTRIBUTION OF INDUSTRY

11.1 The Industrial Revolution

- The Industrial Revolution transformed how goods are produced for society.
- The United Kingdom was home to key events in the birth of the Industrial Revolution.

The root of the **Industrial Revolution** was technology, involving several inventions that transformed the way in which goods were manufactured. The revolution in industrial technology created an unprecedented expansion in productivity, resulting in substantially higher standards of living.

ORIGINS OF THE INDUSTRIAL REVOLUTION

The Industrial Revolution originated in northern England and southern Scotland, in part because the region was home to a remarkable concentration of innovative engineers and mechanics during the late eighteenth century. The term *Industrial Revolution* is somewhat misleading because it was far more than industrial and it didn't happen overnight.

- The Industrial Revolution resulted in new social, economic, and political inventions, not just industrial ones.
- The changes involved a gradual diffusion of new ideas and techniques over decades, rather than an instantaneous revolution.

Nonetheless, the term is commonly used to define the process that began in the United Kingdom in the late 1700s.

Prior to the Industrial Revolution, industry was geographically dispersed across the landscape. People made household tools and agricultural equipment in their own homes or obtained them in the local village. Home-based manufacturing was known as the **cottage industry** system. The large supply of steam power available from James Watt's steam engines induced firms to concentrate all manufacturing steps in a process in one building attached to a single power source.

11.1.1 **THE CRYSTAL PALACE**
The 1851 London World's Fair, more formally known as the "Great Exhibition of the Works of Industry of All Nations," celebrated the industrial accomplishments of the United Kingdom, which was the world's dominant industrial power. It was responsible for more than half of the world's cotton fabric and iron and mined two-thirds of its coal. The World's Fair was housed in a glass and iron building resembling a very large greenhouse called the Crystal Palace that became the symbol of the Industrial Revolution.

TRANSFORMATION OF KEY INDUSTRIES

Industries impacted by the Industrial Revolution included the following.

- **Iron.** The first industry to benefit from Watt's steam engine. The usefulness of iron had been known for centuries, but the scale of production was small. The process demanded constant heating and cooling of the iron, a time-consuming and skilled operation, because energy could not be generated to keep the ovens hot for a sufficiently long period of time. The Watt steam engine provided a practical way to keep the ovens constantly heated.
- **Coal.** The source of energy to operate the ovens and the steam engines. Wood, the main source prior to the Industrial Revolution, became increasingly scarce in England because it was needed for construction of ships, buildings, and furniture, as well as for heat. The solution was to use high-energy coal. To minimize the cost of transporting bulky, heavy coal, iron production was consolidated into large factories near the coal mines.
- **Textiles.** Transformed from a dispersed cottage industry to a concentrated factory system during the late eighteenth century. In 1746, John Roebuck and Samuel Garbett established a factory to bleach cotton with sulfuric acid obtained from burning coal. In 1768, Richard Arkwright, a barber and wigmaker in Preston, England, invented machines to untangle cotton prior to spinning.
- **Food processing.** Essential to feed the factory workers no longer living on farms. In 1810, French confectioner Nicholas Appert started canning food in glass bottles sterilized in boiling water.
- **Transportation.** Critical for diffusing the Industrial Revolution. First canals and then railroads enabled factories to attract large numbers of workers, bring in bulky raw materials such as iron ore and coal, and ship finished goods to consumers.

11.1.2 **JAMES WATT'S STEAM ENGINE**
The invention most important to the development of factories was the steam engine, patented in 1769 by James Watt, a maker of mathematical instruments in Glasgow, Scotland. Watt built the first useful steam engine that could pump water far more efficiently than the watermills then in common use, let alone human or animal power. Steam injected in a cylinder (left side of engine) pushes a piston attached to a crankshaft that drives machinery (right side of engine).

11.1.3 **TRANSFORMATION OF AN INDUSTRY**
In the early nineteenth century, the textile industry was a cottage industry based on people working in their homes. By the middle of the century, the industry had become based in factories and mills. In this interior view of a cotton mill in 1835 girls and women tend carding, drawing, and roving machinery.

ORIGIN AND DISTRIBUTION OF INDUSTRY

11.2 Distribution of Industry

- Approximately three-fourths of the world's manufacturing is clustered in four regions.
- The major industrial areas are divided into subareas.

11.2.1 WESTERN EUROPE

The first region to industrialize during the nineteenth century. Numerous industrial centers developed in Europe as countries competed with each other for supremacy.

• United Kingdom
Dominated world production of steel and textiles during the nineteenth century. These industries have declined, but the country has attracted international investment through new high-tech industries that serve the European market.

• Rhine-Ruhr Valley
Has a concentration of iron and steel manufacturing because of proximity to large coalfields. Rotterdam, the world's largest port, lies at the mouth of several branches of the Rhine River as it flows into the North Sea.

• Mid-Rhine
Western Europe's most centrally located industrial area. Frankfurt is a financial and commercial center and the hub of Germany's transport network. Stuttgart specializes in high-value goods that require skilled labor. Mannheim, an inland port along the Rhine, has a large chemical industry that manufactures synthetic fibers, dyes, and pharmaceuticals.

• Po Basin
Has attracted textiles and other industries because of two key assets, compared to Europe's other industrial regions: numerous workers willing to accept lower wages and inexpensive hydroelectricity from the nearby Alps.

• Northeastern Spain
Western Europe's fastest growing manufacturing area in recent years. Spain's leading industrial area, Catalonia, centered on the city of Barcelona, is the center of Spain's textile industry and the country's largest motor-vehicle plant.

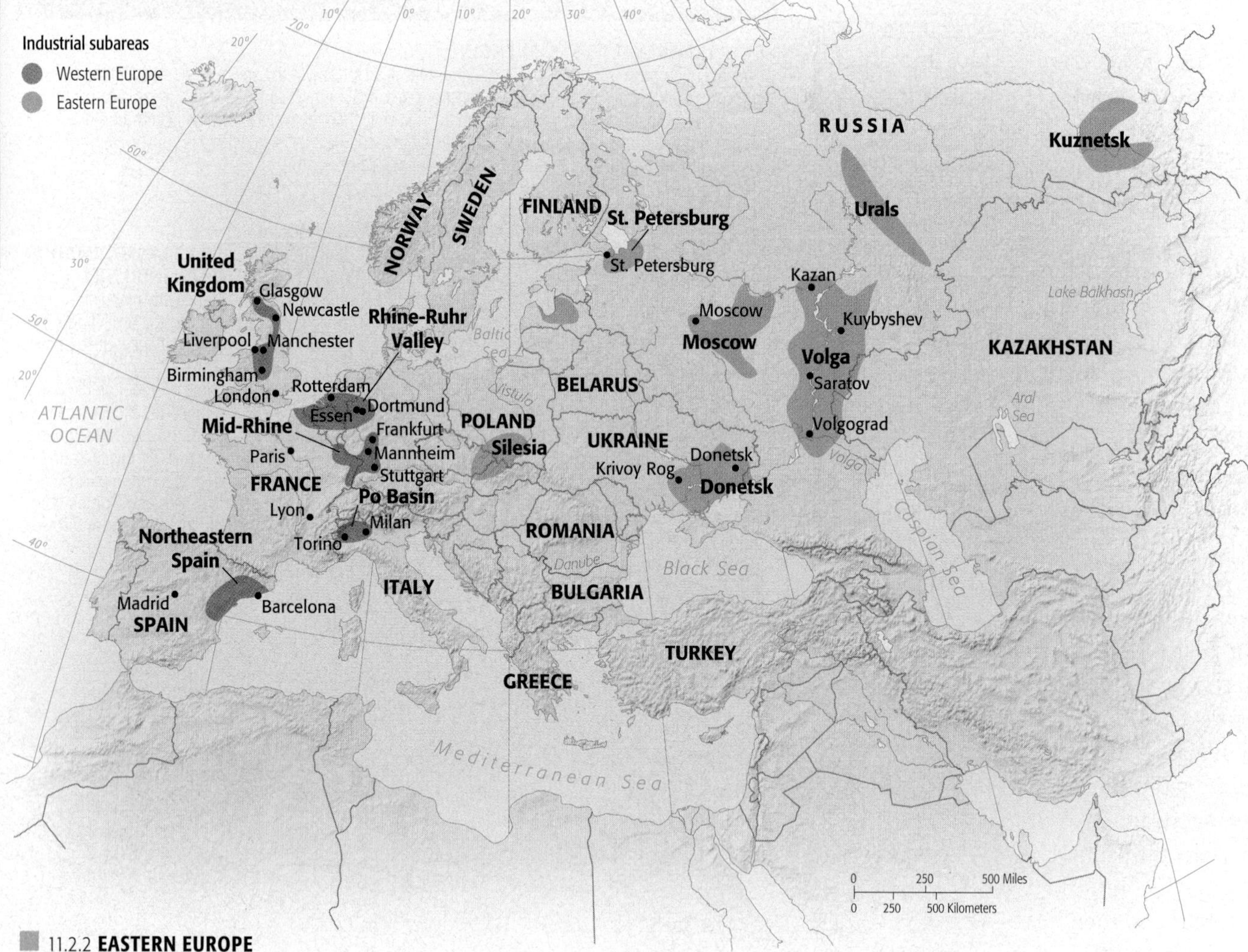

11.2.2 EASTERN EUROPE

Russia's industries developed initially around its two largest cities, Moscow and St Petersburg. Under the Soviet Union, industrial areas were established in other regions of the country.

• Moscow
Russia's oldest industrial region, centered around the country's capital and largest city.

• St. Petersburg
Eastern Europe's second largest city, specializing in shipbuilding and other industries serving Russia's navy and ports in the Baltic Sea.

• Volga
Russia's largest petroleum and natural gas fields.

• Urals
The Ural mountain range contains more than 1,000 types of minerals, the most varied collection found in any mining region in the world.

• Kuznetsk
Russia's most important manufacturing district east of the Ural Mountains, with the country's largest reserves of coal and an abundant supply of iron ore.

• Donetsk
Eastern Ukraine's coalfield, with one of the world's largest coal reserves.

11.2.3 EAST ASIA

Centered on Japan and China.

- **Japan**
Became an industrial power in the 1950s and 1960s initially by producing goods that could be sold in large quantity at cut-rate prices to consumers in other countries Manufacturing is concentrated in the central region, especially in the two large urban areas of Tokyo-Yokohama and Osaka-Kobe-Kyoto.

- **China**
Has become the world's second-largest manufacturer measured in output and has the largest labor force employed in manufacturing. Manufacturers cluster in three areas along the east coast: near Guangdong and Hong Kong, the Yangtze River valley between Shanghai and Wuhan, and along the Gulf of Bo Hai from Tianjin and Beijing to Shenyang.

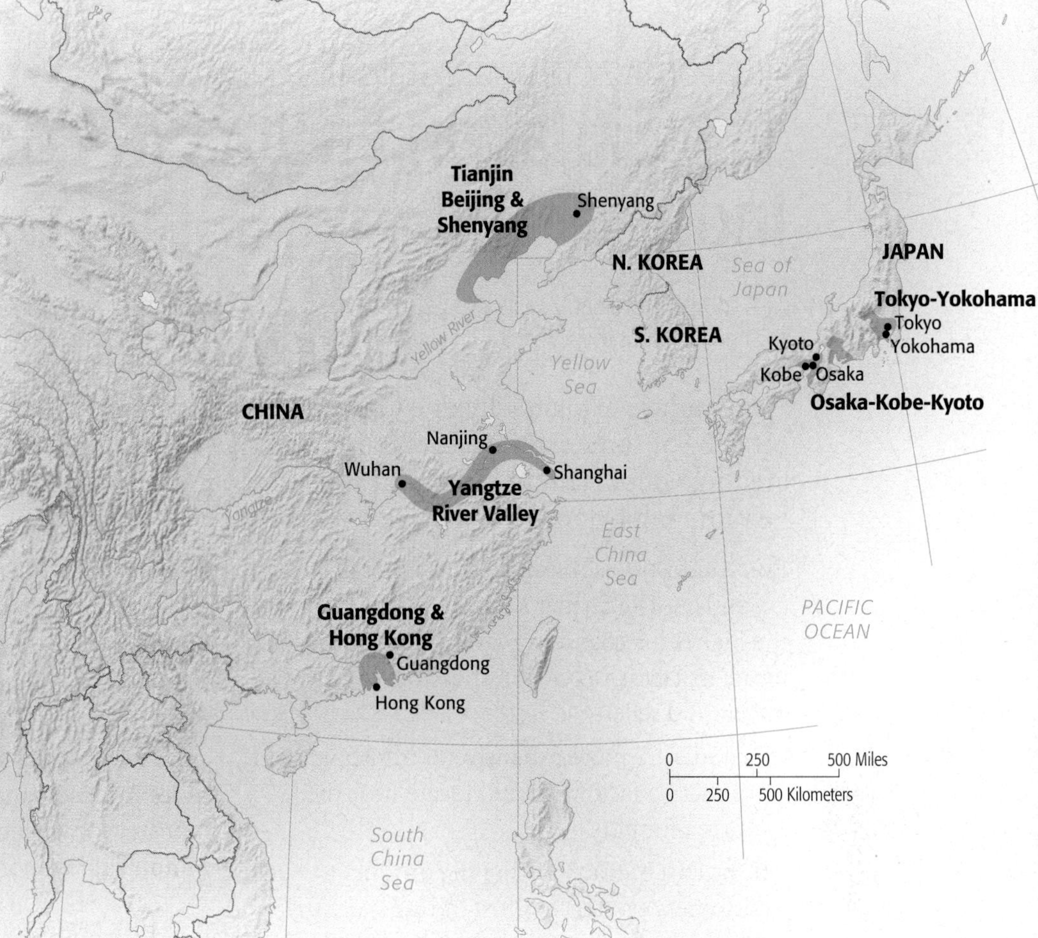

11.2.4 NORTH AMERICA

Traditionally, highly concentrated in northeastern United States and southeastern Canada. In recent years, manufacturing has relocated to the South, lured by lower wages and legislation that has made it difficult for unions to organize factory workers.

- **New England**
A cotton textile center in the early nineteenth century. Cotton was imported from southern states and finished cotton products were shipped to Europe.

- **Middle Atlantic**
The largest U.S. market, so the region attracts industries that need proximity to a large number of consumers and depend on foreign trade through one of this region's large ports.

- **Mohawk Valley**
A linear industrial belt in upper New York state, taking advantage of inexpensive electricity generated at nearby Niagara Falls.

- **Pittsburgh-Lake Erie**
The leading steel-producing area in the nineteenth century because of proximity to Appalachian coal and iron ore.

- **Western Great Lakes**
Centered on Chicago, the hub of the nation's transportation network, now the center of steel production.

- **Southern California**
Now the country's largest area of clothing and textile production, the second-largest furniture producer, and a major food-processing center.

- **Southeastern Ontario**
Canada's most important industrial area, central to the Canadian and U.S. markets and near the Great Lakes and Niagara Falls.

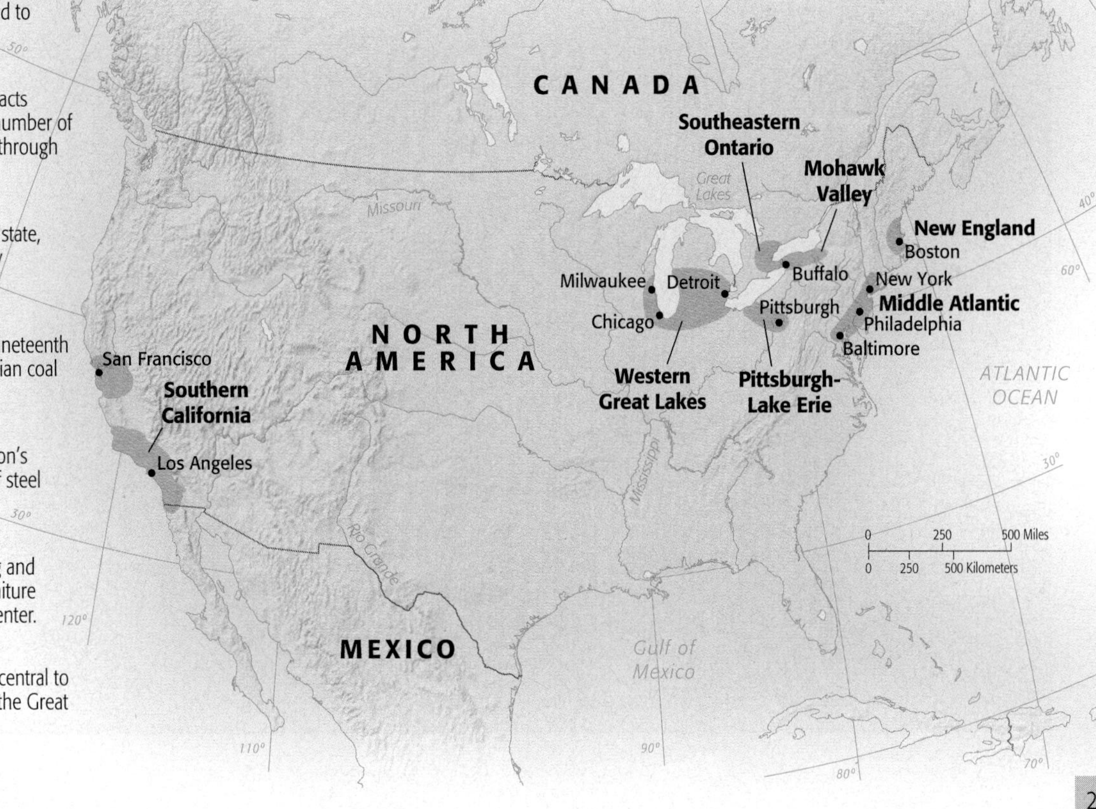

SITUATION FACTORS IN INDUSTRIAL LOCATION

11.3 Situation Factors in Locating Industry

- A company ordinarily faces two geographical costs: situation and site.
- Situation factors involve transporting materials to and from a factory.

Geographers explain why one location may prove more profitable for a factory than others. For some manufacturers minimizing **situation factors** (that is transport costs) are most important in determining where to locate a factory. For others, site factors may be more important, as discussed in Section 11.7.

Every industry uses some inputs and sells to customers. The farther something is transported, the higher the cost, so a manufacturer tries to locate its factory as close as possible to both buyers and sellers.

- If inputs are more expensive to transport than products, a factory should locate near the source of inputs.
- If the cost of transporting the product to customers exceeds the cost of transporting inputs, then the optimal plant location is as close as possible to the customer.

PROXIMITY TO INPUTS

A **bulk-reducing industry** is an economic activity in which the inputs weigh more than the final product. Inputs may be resources from the physical environment (minerals, wood, or animals), or they may be parts or materials made by other companies. A bulk-reducing industry locates near the source of its inputs to minimize transportation cost.

11.3.1 **BULK-REDUCING INDUSTRIES: COPPER**
Copper mining, concentrating, smelting, and refining are bulk-reducing industries performed near the principal source of inputs, including Mount Isa, Australias largest copper mine.

PROXIMITY TO MARKETS

For many firms, the optimal location is close to markets where the product is sold. The cost of transporting goods to consumers is a critical location factor for three types of industries:

- **Bulk-gaining industries** make something that gains volume or weight during production. A prominent example is the fabrication of parts and machinery from steel and other metals. A fabricated-metal factory brings together metals such as steel and previously manufactured parts as the main inputs and transforms them into a more complex product. Common fabricated goods include televisions, refrigerators, and air conditioners. Because fabricated and machined products typically occupy a larger volume than the sum of their individual parts and metals, the cost of shipping the final product to consumers is usually the most critical factor.

- Single-market manufacturers make products sold primarily in one location, so they also cluster near their markets. For example, the manufacturers of parts for motor vehicles are specialized manufacturers often with only one or two customers—the major carmakers such as General Motors and Toyota.

- Perishable products must be located near their markets so their products can reach consumers as rapidly as possible.

11.3.2 **BULK-GAINING INDUSTRY: BEER BOTTLING**
A brewery is a good example of a bulk-gaining industry that needs to be located near consumers, because water, the principal ingredient in beer, is heavy. Consequently, the two best-selling brewing companies locate their plants near major population concentrations. Most breweries are clustered in the heavily populated Northeast.

11.3.3 **SINGLE-MARKET MANUFACTURER: SEAT MAKER**
Most seat-making plants ship to only one customer, a single carmaker. Consequently, a seat-making plant is typically located within an hour of its customer's assembly plant.

11.3.4 **PERISHABLE PRODUCT: MILK PRODUCTION**
Food producers such as bakers and milk bottlers must locate near their customers to assure rapid delivery, because few people want stale bread or sour milk.

SITUATION FACTORS IN INDUSTRIAL LOCATION

11.4 Changes in Location of Steel Production

- Steel production has traditionally been a prominent example of a bulk-reducing industry.
- Restructuring has made steel production more sensitive to market locations.

Steelmaking is a bulk-reducing industry that traditionally located to minimize the cost of transporting inputs. Globally, the shift to new industrial regions can be seen clearly in steel production.

CHANGING U.S. DISTRIBUTION

In the United States, the steel industry concentrated during the mid-nineteenth century around Pittsburgh in southwestern Pennsylvania, where iron ore and coal were both mined. Steel mills were built during the late 1800s around Lake Erie, in the Ohio cities of Cleveland, Youngstown, and Toledo, and around Detroit, Michigan.

The westward shift was largely influenced by the discovery of rich iron ore in the Mesabi Range, a series of low mountains in northern Minnesota. This area soon became the source for virtually all iron ore used in the U.S. steel industry. The ore was transported by way of Lake Superior, Lake Huron, and Lake Erie. Coal was shipped from Appalachia by train.

New steel mills were located farther west around 1900, near the southern end of Lake Michigan—in Gary, Indiana, Chicago, and other communities. The main raw materials

11.4.1 **INTEGRATED STEEL MILL**
An integrated steel mill processes iron ore, converts coal into coke, converts the iron into steel, and forms the steel into sheets, beams, rods, or other shapes.

11.4.2 **CLOSED STEEL MILL**
A large percentage of integrated steel mills in North America and Europe have closed in recent years. Steel production has moved to less developed countries and to minimills.

continued to be iron ore and coal, but changes in steelmaking required more iron ore in proportion to coal. Thus, new steel mills were built closer to the Mesabi Range to minimize transportation cost. Coal was available from nearby southern Illinois, as well as from Appalachia.

Most large U.S. steel mills built during the first half of the twentieth century were located in communities near the East and West coasts, including Baltimore, Los Angeles, and Trenton, New Jersey. These coastal locations partly reflected further changes in transportation cost.

Iron ore increasingly came from other countries, especially Canada and Venezuela, and locations near the Atlantic and Pacific oceans were more accessible to those foreign sources. Further, scrap iron and steel—widely available in the large metropolitan areas of the East and West coasts—had become an important input in the steel-production process.

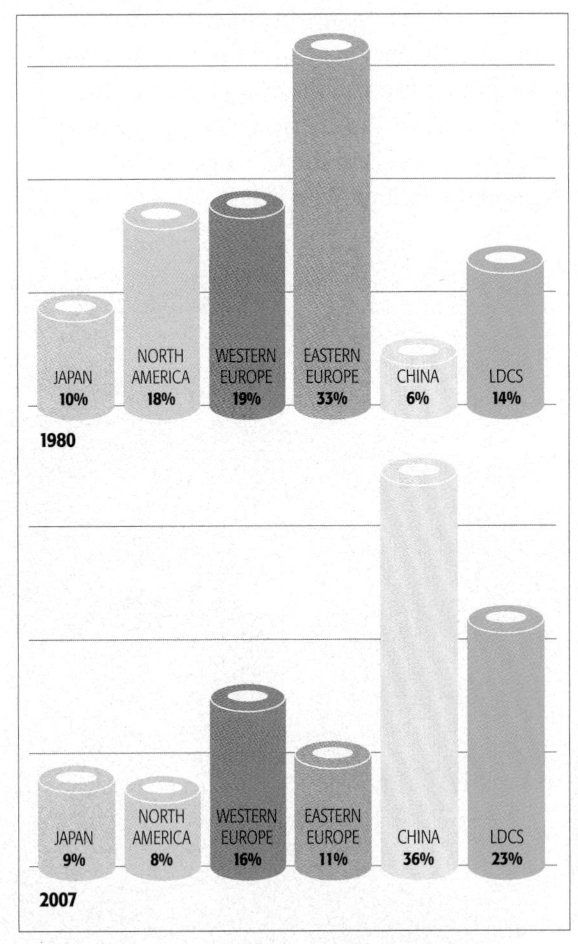

11.4.3 **CHANGE IN WORLDWIDE STEEL PRODUCTION**. Worldwide in 1980, 80 percent of world steel was produced in MDCs and 20 percent in LDCs. Between 1980 and 2007, the share of world steel production declined to 44 percent in MDCs and increased to 56 percent in LDCs. China, now the world's largest steel producer, accounted for 36 percent of world steel output in 2007.

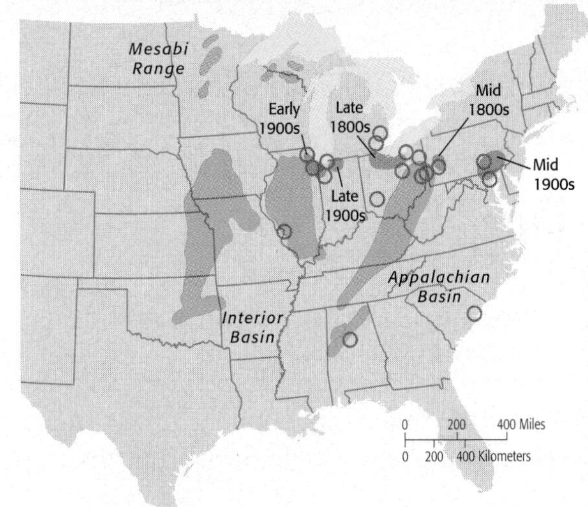

11.4.4 **INTEGRATED STEEL MILLS IN THE UNITED STATES**
Integrated steel mills are densely clustered near the southern Great Lakes, especially Lake Erie and Lake Michigan. Historically, the most critical factor in siting a steel mill was to minimize transportation cost for raw materials, especially heavy, bulky iron ore and coal. In recent years many integrated steel mills have closed. Most surviving mills are in the Midwest to maximize access to consumers.

RESTRUCTURING THE STEEL INDUSTRY

Recently, more steel plants have closed than opened in the United States. Among the survivors, plants around southern Lake Michigan and along the East Coast have significantly increased their share of national production. This success derives primarily from access to markets rather than inputs.

In contrast to the main historical location factor—transportation cost of raw materials—successful steel mills today are located increasingly near major markets. Coastal plants provide steel to large East Coast population centers, and southern Lake Michigan plants are centrally located to distribute their products countrywide.

The growth of steel minimills also demonstrates the increasing importance of access to markets rather than to inputs. Traditionally, most steel was produced at large, integrated mills. Minimills, generally limited to one step in the process—steel production—have captured one-fourth of the U.S. steel market.

11.4.5 **MINIMILL**
A minimill makes steel from scrap metal. It is less expensive than an integrated mill to build and operate, and can locate near its market because its main input—scrap metal—is widely available.

SITUATION FACTORS IN INDUSTRIAL LOCATION

11.5 Changing Distribution of Auto Production

- Motor vehicles are bulk-gaining products that are made near their markets.
- In the United States, most carmaking operations have clustered in auto alley.

A prominent example of a fabricated metal product is a motor vehicle. The motor vehicle industry also demonstrates the importance of globalization of manufacturing.

TWO TYPES OF AUTO PLANTS

Carmakers put together vehicles at final assembly plants, using thousands of parts supplied by independent companies. Carmakers account for only around 30 percent of the value of the vehicles that bear their names. As a result of outsourcing, independent companies supply the other 70 percent of the value.

Three-fourths of the world's final assembly plants are controlled by seven carmakers, all with global reach. These include U.S.-based General Motors, Ford, and Chrysler; France-based Renault (which controls Japanese-based Nissan) and Peugeot; and Japan-based Toyota and Honda. Nationality matters to these companies in terms of location of corporate headquarters, top managers, research facilities, and shareholders.

As bulk-gaining industries, the leading carmakers make most of their cars in the regions where they are to be sold. Thus, three-fourths of vehicles sold in the United States are assembled in the United States, and most of the remainder are assembled in Canada and Mexico.

In the United States, vehicles are fabricated at about 60 final assembly plants, from parts made at several thousand other plants. Most of the assembly and parts plants are located in the interior of the country, between Michigan and Alabama, centered in a corridor known as "auto alley," formed by north-south interstate highways 65 and 75.

For a fabricated product like a motor vehicle, the critical location factor is minimizing transportation to the market, in this case the 15 million North Americans who buy new vehicles each year. If a company has a product that is made at only one plant, and the critical location factor is to minimize the cost of distribution throughout North America, then the optimal factory location is in the U.S. interior rather than on the East or West Coast.

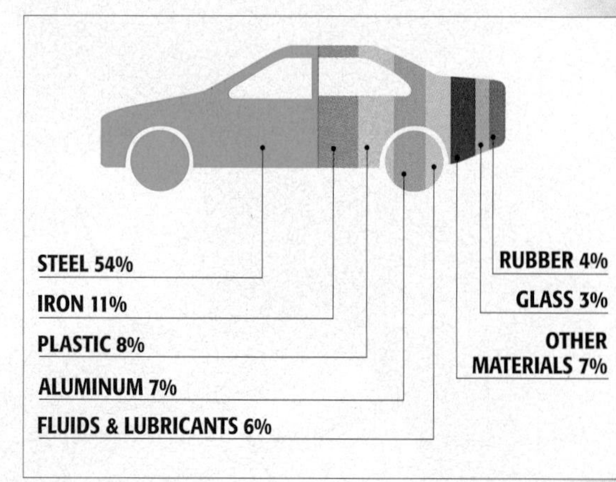

11.5.1 **COMPOSITION OF A CAR**

LOCATION OF PARTS MAKERS

In the United States, one-half of the parts are made in the United States by U.S.-owned companies, one-fourth are made in the United States by foreign-owned transnational corporations, and one-fourth are made overseas and imported into the United States.

Some parts makers build their factories as close as possible to their customers, the final assembly plants. Seats, for example, are invariably manufactured within an hour of the final assembly plant. A seat is an especially large and bulky object, and carmakers do not want to waste valuable space in their assembly plants by piling up an inventory of them. Most engines, transmissions, and metal body parts are also produced within a couple hours of an assembly plant.

On the other hand, many parts do not need to be manufactured close to the customer. For them, site factors are more important. Some locate in Mexico and China to take advantage of lower labor costs. For example, most wiring is made in Mexico, whereas China is the leading source of aluminum wheels.

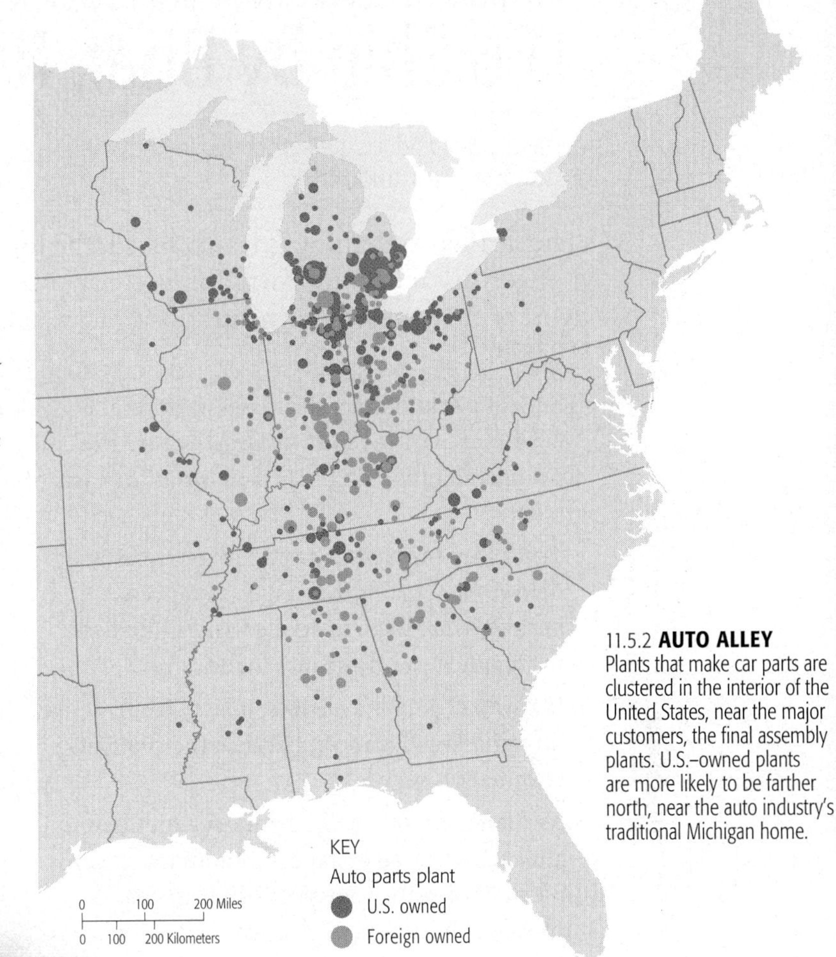

11.5.2 **AUTO ALLEY**
Plants that make car parts are clustered in the interior of the United States, near the major customers, the final assembly plants. U.S.-owned plants are more likely to be farther north, near the auto industry's traditional Michigan home.

Oshawa
Flint
Tarrytown
Janesville
Oakland/ Fremont
Baltimore
Kansas City/ Leeds
St. Louis
Cincinnati/ Norwood
Los Angeles/ Van Nuys
Atlanta/ Lakewood
0 400 Miles
0 400 Kilometers

11.5.3 **CHEVROLET ASSEMBLY PLANTS 1955**
In 1955, General Motors assembled Chevrolets at 10 plants around the country located near major population centers.

Chevrolet assembly plants 1955

0 400 Miles
0 400 Kilometers

11.5.4 **CHEVROLET ASSEMBLY PLANTS 2008**
In 2008, General Motors and other carmakers clustered production in the interior of the country because they were producing a wider variety of vehicles for national distribution.

Assembly plants 2008
Car
Truck

SITUATION FACTORS IN INDUSTRIAL LOCATION

11.6 Ship by Boat, Rail, Truck, or Air?

- Inputs and products are transported in one of four ways: ship, rail, truck, or air.
- The cheapest of the four alternatives will change with the distance that goods are being sent.

The farther something is transported, the lower is the cost per kilometer (or mile). Longer-distance transportation is cheaper per kilometer in part because firms must pay workers to load goods on and off vehicles, whether the material travels 10 kilometers or 10,000.

The cost per kilometer decreases at different rates for each of the four modes, because the loading and unloading expenses differ for each mode.

- Trucks are most often used for short-distance delivery and trains for longer distances, because trucks can be loaded and unloaded more quickly and cheaply than trains.
- If a water route is available, boats are attractive for very long distances, because the cost per kilometer is very low.
- Airplanes are normally the most expensive alternative for all distances, but an increasing number of firms transport by air to ensure speedy delivery of small-bulk, high-value packages.

Air-transport companies such as FedEx, Airborne, and UPS promise overnight delivery for most packages. They pick up packages in the afternoon and transport them by truck to the nearest airport. Late at night, planes filled with packages are flown to a central hub airport in the interior of the country, such as Memphis, Tennessee, and Dayton, Ohio. The packages are then transferred to other planes, flown to airports nearest their destination, transferred to trucks, and delivered the next morning.

11.6.1 **SHIP BY AIR**
Air transport is used to ship packages long distances in a hurry.

BREAK-OF-BULK POINTS

Regardless of transportation mode, cost rises each time that inputs or products are transferred from one mode to another. For example, workers must unload goods from a truck and then reload them onto a plane. The company may need to build or rent a warehouse to store goods temporarily after unloading from one mode and before loading to another mode.

Some companies may calculate that the cost of one mode is lower for some inputs and products, whereas another mode may be cheaper for other goods. Many companies that use multiple transport modes locate at a break-of-bulk point. A **break-of-bulk point** is a location where transfer among transportation modes is possible.

Important break-of-bulk points include seaports and airports. For example, a steel mill near the port of Baltimore receives iron ore by ship from South America and coal by train from Appalachia.

JUST-IN-TIME DELIVERY

Proximity to market has long been important for many types of manufacturers, as discussed earlier in this chapter. The factor has become even more important in recent years because of the rise of just-in-time delivery.

As the name implies, **just-in-time** is shipment of parts and materials to arrive at a factory moments before they are needed. Just-in-time delivery is especially important for delivery of inputs, such as parts and raw materials, to manufacturers of fabricated products, such as cars and computers.

With just-in-time, parts and materials arrive at a factory frequently, in many cases daily if not hourly. Suppliers of the parts and materials are told a few days in advance how much will be needed over the next week or two and told first thing each morning exactly what will be needed at precisely what time that day.

11.6.2 **BREAK-OF-BULK AT A SEAPORT**
Many goods that are shipped long distances are packed in uniformly sized containers, which can be quickly transferred between ships and trucks or trains.

11.6.3 **ELIMINATING INVENTORY**
Leading computer manufacturers have cut costs in part through eliminating the need to store inventory in warehouses. Computers are built only after the buyer has placed the order.

Just-in-time delivery reduces the money that a manufacturer must tie up in wasteful inventory. The percentage of the U.S. economy tied up in inventory has been cut in half during the past quarter-century. Manufacturers also save money through just-in-time delivery by reducing the size of the factory, because space does not have to be wasted by piling up a mountain of inventory.

To meet a tight timetable, a supplier of parts and materials must locate factories near its customers. If only an hour or two notice is given, a supplier has no choice but to locate a factory within 50 miles or so of the customer.

Just-in-time delivery sometimes merely shifts the burden of maintaining inventory to suppliers. Wal-Mart, for example, holds low inventories but tells its suppliers to hold high inventories "just in case" a sudden surge in demand requires restocking on short notice.

JUST-IN-TIME DELIVERY DISRUPTIONS

Just-in-time delivery means that producers have less inventory to cushion against disruptions in the arrival of needed parts. Two kinds of disruptions can result from reliance on just-in-delivery

- **Labor unrest.** A strike at one supplier plant can shut down the entire production within a couple of days. Also disrupting deliveries could be a strike in the logistics industry, such as truckers or dockworkers.
- **Unpredictable disruptions.** Most common disruptions are weather-related, such as blizzards that close highways or floods that damage factories. The most notable non-weather-related disruption in recent years followed the September 11, 2001, terrorist attacks on the United States. The grounding of all civilian aircraft for several days after the attacks prevented delivery of compact high-value parts. Although trucks and trains could still move across the United States after the attacks, suppliers in Canada and Mexico were unable to maintain just-in-time deliveries to manufacturers in the United States because the border crossings were closed.

11.6.4 **SHIP BY TRUCK**
Movement of goods by truck can be slowed by accidents, construction, and congestion on the highways.

11.6.5 **SHIP BY RAIL**

SITE FACTORS IN INDUSTRIAL LOCATION

11.7 Site Factors in Locating Industry

- Site factors result from the unique characteristics of a location.
- The three main site factors are labor, land, and capital.

Site factors are industrial location factors related to the costs of production inside the plant, notably labor, land, and capital.

LABOR

A **labor-intensive industry** is one in which wages and other compensation paid to employees constitute a high percentage of expenses. Labor costs an average of 11 percent of overall manufacturing costs in the United States, so a labor-intensive industry would have a much higher percentage than that.

The average wage paid to manufacturing workers exceeds $20 per hour in MDCs. Health care, retirement pensions, and other benefits add substantially to the compensation. In LDCs such as China, average wages are less than $5 per hour with limited additional benefits. For some manufacturers—but not all—the difference between paying workers $5 and $20 per hour is critical.

LAND

In the early years of the Industrial Revolution, multistory factories were constructed in the heart of the city. Now, they are more likely to be built in suburban or rural areas, in part to provide enough space for one-story buildings.

Raw materials are typically delivered at one end and moved through the factory on conveyors or forklift trucks. Products are assembled in logical order and shipped out at the other end.

Locations on the urban periphery are also attractive for factories to facilitate delivery of inputs and shipment of products. In the past, when most material moved in and out of a factory by rail, a central location was attractive because rail lines converged there.

With trucks now responsible for transporting most inputs and products, proximity to major highways is more important for a factory. Especially attractive is the proximity to the junction of a long-distance route and the beltway or ring road that encircles most cities. Factories cluster in industrial parks located near suburban highway junctions.

Also, land is much cheaper in suburban or rural locations than near the center city. A hectare (or an acre) of land in the United States may cost only a few thousand dollars in a rural area, tens of thousands in a suburban location, and hundreds of thousands near a center city.

11.7.1 **LABOR: CHINESE CLOTHING WORKERS** Around the world, approximately 150 million people are employed in manufacturing, according to the U.N. International Labor Organization (ILO). China has around 20 percent of the world's manufacturing workers and the United States around 10 percent.

CAPITAL

Manufacturers typically borrow funds to establish new factories or expand existing ones. The ability to borrow money has become a critical factor in the distribution of industry in LDCs. Financial institutions in many LDCs are short of funds, so new industries must seek loans from banks in MDCs. But enterprises may not get loans if they are located in a country that is perceived to have an unstable political system, a high debt level, or ill-advised economic policies.

11.7.3 **CAPITAL: SILICON VALLEY**
The most important factor in the clustering in California's Silicon Valley of high-tech industries, such as Oracle Industries headquarters, has been availability of capital. Banks in Silicon Valley have long been willing to provide money for new software and communications firms even though lenders elsewhere have hesitated. High-tech industries have been risky propositions—roughly two-thirds of them fail—but Silicon Valley financial institutions have continued to lend money to engineers with good ideas so that they can buy the software, communications, and networks they need to get started. One-fourth of all capital in the United States is spent on new industries in the Silicon Valley.

11.7.2 **LAND: PETROCHEMICAL PLANT**
Contemporary factories, such as this petrochemical plant in New Plymouth, New Zealand, often require large tracts of land, because they can operate more efficiently when laid out in one-story buildings.

SITE FACTORS IN INDUSTRIAL LOCATION

11.8 Distribution of Textile and Apparel Production

- Textile and apparel production is a prominent example of a labor-intensive industry.
- Textile and apparel production generally requires less skilled, low-wage workers.

Textile and apparel production generally requires less skilled, low-cost workers.

The textile and apparel industry accounts for 6 percent of the dollar value of world manufacturing, but a much higher 14 percent of world manufacturing employment, an indicator that it is a labor-intensive industry. The industry accounts for an even higher percentage of the world's women employed in manufacturing.

Textile and apparel production involves three principal steps:

- **Spinning of fibers to make yarn.** Fibers can be spun from natural or synthetic elements. Cotton is the principal natural fiber—three-fourth of the total—followed by wool. Historically, natural fibers were the sole source, but today synthetics account for three-fourths and natural fibers only one-fourth of world thread production. Because it is a labor-intensive industry, spinning is done primarily in low-wage countries.

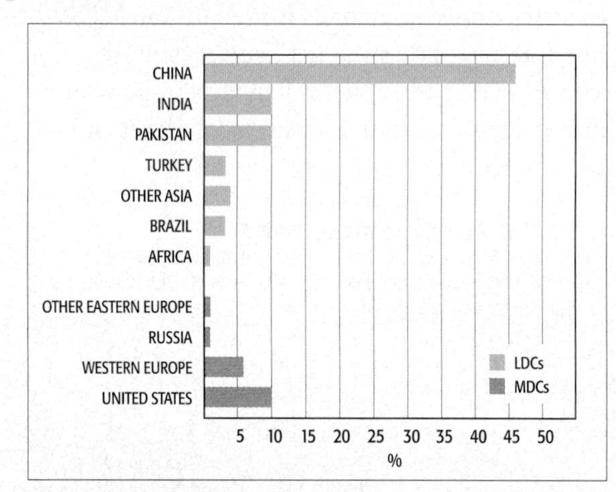

11.8.1 **TEXTILE AND APPAREL SPINNING**
LDCs account for around three-fourths of world spinning production. Natural fiber spinning is even more highly clustered in LDCs–more than 80 percent of world total–synthetics less so. Spinning of cotton fiber is concentrated in the LDCs where the principal input, cotton, is grown. China produces 40 percent of the world's cotton thread, India and Pakistan about 10 percent each. Among MDCs, the United States, the only major cotton grower, is also the only major cotton thread producer. Synthetic fibers were once spun primarily in a handful of MDCs where the chemical industry was concentrated. LDCs now account for about half of the world's spinning of synthetics, with especially rapid expansion in China and Indonesia.

- **Weaving or knitting yarn into fabric.** For thousands of years, fabric has been woven or laced together by hand on a loom, which is a frame on which two sets of threads are placed at right angles to each other. One set of threads, called a warp, is strung lengthwise. A second set of threads, called a weft, is carried in a shuttle that is inserted over and under the warp. Due to the hard, labor-intensive work, total production costs are more for weaving than for spinning and assembly steps. As the process of weaving was physically hard work, weavers were traditionally men.
- **Cutting and sewing of fabric for assembling into clothing and other products.** Textiles are assembled into four main types of products: garments, carpets, home products such as bed linens and curtains, and industrial uses such as headliners inside motor vehicles. Most of the 80 billion articles of clothing sold worldwide in a year are produced in Asia. The percentage varies according to the article of clothing. More than three-fourths of shirts and other tops are manufactured in Asia, compared to less than one-half of suits and dresses. Most underwear, lingerie, and nightwear are also made in Asia. European and North American countries produce more woolens. The percentage has been increasing in Asia and decreasing in North America and Europe.

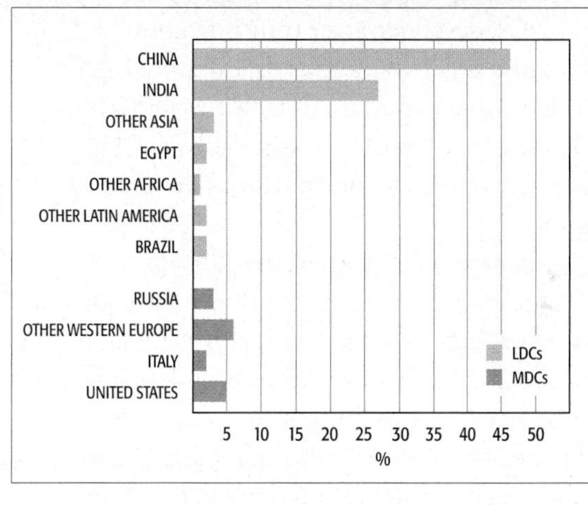

11.8.3 **WOVEN COTTON FABRIC PRODUCTION**
Woven cotton fabric is likely to be produced in LDCs, because the process is more labor intensive than the other major processes in textile and clothing manufacturing. Five-sixths of the world's woven cotton fabric is produced in LDCs. China alone accounts for one-half of the world's woven cotton fabric production, and India another one-fourth. Despite their remoteness from European and North American markets, China and India have become the dominant fabric producers because lower labor cost offsets the expense of shipping inputs and products long distances.

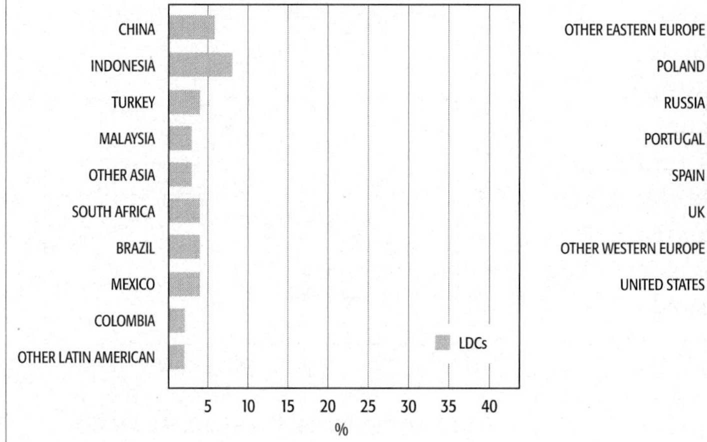

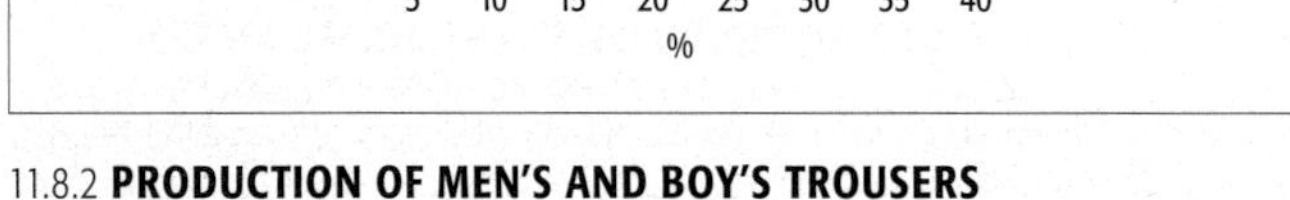

11.8.2 **PRODUCTION OF MEN'S AND BOY'S TROUSERS**
Sewing of cotton fabric into men's and boys' trousers is more likely to take place in MDCs, although some production has moved to LDCs in recent years. Clothing producers must balance the need for low-wage workers with the need for proximity to customers.

SITE FACTORS IN INDUSTRIAL LOCATION

11.9 Labor-intensive Industries

- Labor is the site factor that is changing most dramatically in the twenty-first century.
- Depending on the industry, a low-skilled or high-skilled work force may be critical.

To minimize labor costs, some manufacturers are locating in places where prevailing wage rates are lower than in traditional industrial regions. At the same time, some industries are remaining in traditional industrial regions because of availability of skilled labor.

LOOKING FOR LOW-COST LABOR

Transnational corporations have been especially aggressive in using low-cost labor in LDCs. To remain competitive in the global economy, they carefully review their production processes to identify steps that can be performed by low-paid, low-skilled workers in LDCs.

Given the substantial difference in wages between MDCs and LDCs, transnational corporations can profitably transfer some work to LDCs, despite greater transportation cost. At the same time, operations that require highly skilled workers remain in factories in MDCs. This selective transfer of some jobs to LDCs is known as the **new international division of labor.**

Transnational corporations allocate production to low-wage countries through **outsourcing**, which is turning over much of the responsibility for production to independent suppliers. Outsourcing contrasts with the approach typical of traditional mass production, called vertical integration, in which a company would control all phases of a highly complex production process.

Outsourcing has had a major impact on the distribution of manufacturing, because each step in the production process is now scrutinized closely in order to determine the optimal location. The leading destination for outsourcing production from the United States to low-wage countries has been Mexico.

Under U.S. and Mexican laws, companies receive tax breaks if they ship materials from the United States, assemble components at a so-called ***maquiladora*** plant in Mexico, and export the finished product back to the United States. Altogether, 1.3 million Mexicans are employed at over 3,000 *maquiladoras*.

Competition from Asia has forced some *maquiladora* plants to close. Although the average wage rate at *maquiladoras* of around $400 per month is considerably cheaper than in the United States, it is much higher than the $100 per month being paid in China and other Asian countries.

11.9.1. **CLOTHING MANUFACTURING IN GUATEMALA**
Clothing is manufactured in Latin America to take advantage of low-cost labor, as well as proximity to the U.S. market.

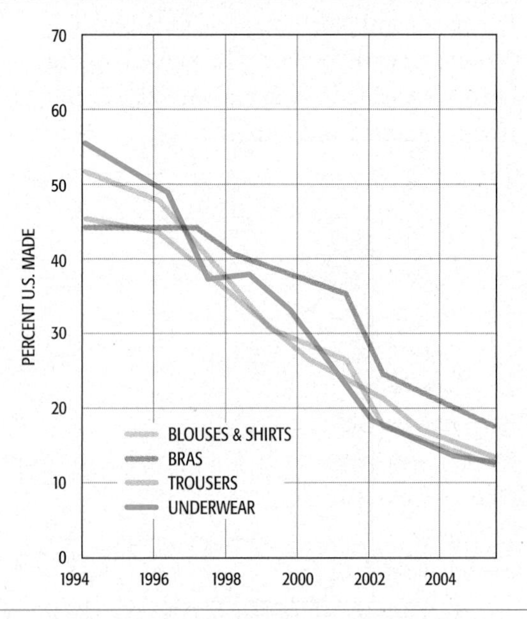

11.9.2 **CLOTHING PRODUCED IN THE UNITED STATES**
The percentage of everyday clothing accounted for by domestic production has decreased sharply in the United States since the 1990s. Conversely, the percentage accounted for by imports has increased sharply. The reason is wages: labor costs per hour are under $1 in many LDCs compared to more than $10 in MDCs.

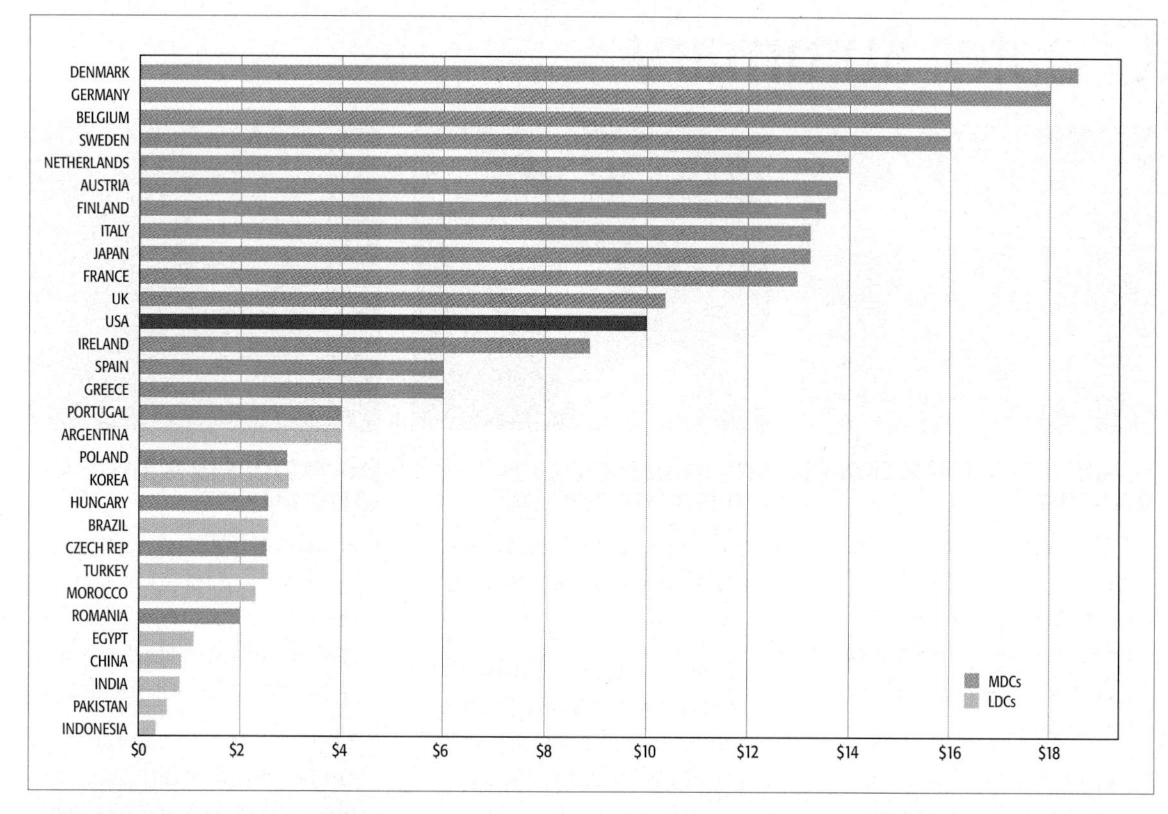

11.9.3 **CLOTHING MANUFACTURING: COST OF LABOR PER HOUR.**
The textile and apparel industry has been especially prominent in opening production in lower wage locations while closing production in higher wage locations. Wages are in the teens in MDCs, compared to under a dollar in some LDCs.

LOOKING FOR SKILLED LABOR

Henry Ford boasted that he could take people off the street and put them to work with only a few minutes of training. That has changed for many industries, including motor vehicle assembly, which now want skilled workers instead. The search for skilled labor has important geographic implications, because it is an asset found principally in the traditional industrial regions.

Factories traditionally assigned each worker one specific task to perform repeatedly. Some geographers call this approach **Fordist** or mass production, because the Ford Motor Company was one of the first to organize its production this way early in the twentieth century.

At its peak, Ford's factory complex along the River Rouge in Dearborn, Michigan, near Detroit, employed more than 100,000. Most of these workers did not need education or skills to do their jobs, and many were immigrants from Europe or the southern United States.

In recent years, some factories have adopted new work rules, known as lean or flexible production. The term **post-Fordist** production is sometimes used to describe flexible production, as a contrast with Fordist production. Again, a carmaker is best known for pioneering lean production, in this case Toyota.

A high-wage industry is not the same as a labor-intensive industry. "Labor-intensive" is measured as a percentage, whereas "high-wage" is measured in dollars or other currencies. For example, auto workers are paid much higher hourly wages than textile workers, yet the textile industry is labor-intensive whereas the auto industry is not.

Although auto workers earn relatively high wages, most of the value of a car is accounted for by the parts and the machinery needed to put the parts together. On the other hand, labor accounts for a large percentage of the cost of producing a towel or shirt when compared with materials and machinery.

Three types of work rules distinguish post-Fordist lean production.

1. **Teams.** Workers are placed in teams and told to figure out for themselves how to perform a variety of tasks.
2. **Problem-solving.** A problem is addressed through consensus after consulting with all affected parties rather than through filing a complaint or grievance.
3. **Leveling.** Factory workers are treated alike and managers and veterans do not get special treatment; they wear the same uniform, eat in the same cafeteria, park in the same lot, and participate in the same athletic and social activities.

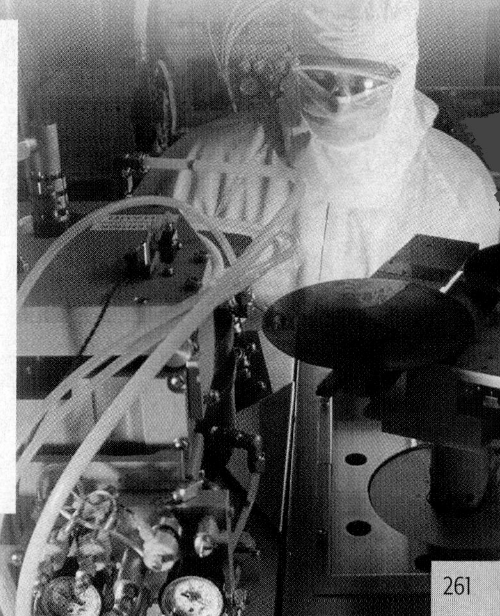

11.9.4 **COMPUTER-CHIP MANUFACTURING.**
Computer manufacturing is an example of an industry that has concentrated in relatively high-wage high-skilled regions of the United States, especially California, along with Massachusetts, New York, and Texas. These states have concentrations of skilled workers in association with proximity to major university centers.

Chapter **Summary**

ORIGIN AND DISTRIBUTION OF INDUSTRY

- The Industrial Revolution dates from the late 1700s in the United Kingdom, when a series of inventions transformed industrial production.
- Approximately three-fourths of the world's industrial output concentrated during the twentieth century in four regions: Western Europe, Eastern Europe, Northeastern North America, and East Asia.
- New industrial regions have emerged in recent years, led by China, which is now the world's largest manufacturing region by some measures.

SITUATION FACTORS IN INDUSTRIAL LOCATION

- Factories try to identify a location where production cost is minimized.
- Situation factors involve the cost of transporting both inputs into the factory and products from the factory to consumers.
- Steel and motor vehicle production are two prominent examples of industries that locate primarily because of situation factors.

SITE FACTORS IN INDUSTRIAL LOCATION

- For some industries, the most critical location factors are related to site factors that result from the unique characteristics of a location.
- Land, labor, and capital are the three traditional production factors that may vary among locations.
- Production of textiles and apparel is a prominent example of an industry that has traditionally located because of site factors, especially proximity to an abundant supply of low-cost labor.

WORKERS IN HAMBURG, GERMANY, FIX BOLTS TO A PROPELLER TO BE INSTALLED IN A SOUTH KOREAN CONTAINER SHIP

Geographic Consequences of Change

Three recent changes in the structure of manufacturing have geographic consequences:

1. Factories have become more productive through introduction of new machinery and processes. A factory may continue to operate at the same location but require fewer workers to produce the same output. Faced with meager prospects of getting another job in the same community, workers laid off at these factories migrate to other regions.
2. Companies are locating production in communities where workers are willing to adopt more flexible work rules. Firms are especially attracted to smaller towns where low levels of union membership and high visibility reduce vulnerability to work stoppages, even if wages are kept low and layoffs become necessary.
3. By spreading production among many countries, or among many communities within one country, large corporations have increased their bargaining power with local governments and labor forces. Production can be allocated to locations where the local government is especially helpful and generous in subsidizing the costs of expansion and the local residents are especially eager to work in the plant.

SHARE YOUR VOICE Student Essay

Matthew Earley
Miami University (Ohio)

The Motor City

For one hundred years, General Motors has been making automobiles for both the United States and the world, setting up facilities abroad and domestically. With sales lagging and Toyota taking over the top spot in terms of sales, GM has been forced to make structural changes, in terms of the geography of both its people and its facilities. As GM continues to go through rough times, the question remains as to what its presence will be in Michigan by the time it is able to fully rebound from their current slump.

GM is based in Detroit and has always been a very prevalent part of the Michigan economy. Headquarters is in the Renaissance Center in downtown Detroit. Rising 73 stories above the Detroit River, the GM Renaissance Center dominates the glittering downtown Detroit skyline. With the closing of assembly plants and the laying off of more and more blue collar jobs, the company is showing that it is not able to sustain itself within the state of Michigan at a time when the state's economy is in poor condition.

Since moving to Detroit in 1994, I've been able to see the decline of the Big 3 first hand. It's made some of the rough economic times more real to me because I've been able to see the impact that it has on the city, the state, and the people living in the area. Some of my friend's parents who once worked for either Chrysler, Ford, or GM have either been laid off or have had to take a pay cut or a reduced role in since the early 2000's.

My friends and I have had many conversations on the future of the city, and whether or not we would return to make our own lives in the area. Every conversation boils down to the fact that if the autos cannot right themselves and return to the success they once had, then we will be forced to go elsewhere. As the Big 3 goes, so will the city of Detroit and the state of Michigan.

Chapter Resources

THE WORLD'S LARGEST BICYCLE FACTORY

Huffy bicycles were manufactured in Ohio throughout the twentieth century. But in the twenty-first century they are no longer made in Ohio.

Huffy bicycles were manufactured in Ohio for more than a century, first in Dayton, where George P. Huffman founded the company's predecessor Davis Sewing Machine Co. in 1892, and beginning in 1954 at the world's largest bicycle factory in Celina.

Huffy closed the Celina plant in 1998 and moved production to Farmington, Missouri. Less than a year later, Huffy stopped making bicycles altogether in the United States.

After it closed its U.S. bicycle factory, Huffy Corp. contracted with a Mexican company Elamex to manufacture bicycles in Nuevo Laredo. The arrangement lasted only 3 years. In 2001, Huffy moved production from Mexico to the small town of Sha Jiang in Shenzhen province China.

Huffy's story is not isolated. The United States lost one-fifth of its manufacturing jobs during the first 8 years of the twenty-first century. At the same time, China's manufacturing growth is highly visible to consumers when they open boxes stamped "made in China" or read about a factory closure.

Americans' fears of manufacturing job losses were echoed elsewhere in the world. A former president of the European Union warned against the "deindustrialization of Europe." Japan's loss of manufacturing jobs to overseas locations was called a "hollowing out" by Japanese politicians. In Mexico, the loss of manufacturing jobs during the early twenty-first century led to "a wave of soul-searching."

Manufacturing jobs are viewed as a special asset by communities around the world. Communities mourn when factories close and rejoice when they open. To attract and retain them, officials offer financial support considered excessive when scrutinized by independent analysts.

For companies like Huffy, the changing geography of production was clearly justified by its corporate balance sheet. Huffy lost $2.2 million during the last year it made bicycles in Celina and $33.3 million during the year it made them in Farmington.

In 2000, the first year of overseas production, Huffy earned $35 million. Reducing labor costs was critical to the turnaround: Instead of paying workers an average of about $10.50 an hour in Celina, or $8 an hour in Farmington, or even $5 an hour in Nuevo Laredo, Huffy was able to find workers for only 25¢ an hour in China.

Ultimately responsible for the changing geography of manufacturing is the American consumer. When Huffy bicycles were made in Celina, they sold for $80. After production was shifted overseas, the price came down to $40. For nearly all consumers, low price is much more important than place of origin.

KEY TERMS

Break-of-bulk point
A location where transfer is possible from one mode of transportation to another.

Bulk-gaining industry
An industry in which the final product weighs more or comprises a greater volume than the inputs.

Bulk-reducing industry
An industry in which the final product weighs less or comprises a lower volume than the inputs.

Cottage industry
Manufacturing based in homes rather than in a factory, commonly found prior to the Industrial Revolution.

Fordist production
Form of mass production in which each worker is assigned one specific task to perform repeatedly.

Industrial Revolution
A series of improvements in industrial technology that transformed the process of manufacturing goods.

Just-in-time delivery
Shipment of parts and materials to arrive at a factory moments before they are needed.

Labor-intensive industry
An industry for which labor costs comprise a high percentage of total expenses.

Maquiladora
Factories built by U.S. companies in Mexico near the U.S. border to take advantage of much lower labor costs in Mexico.

New international division of labor
Transfer of some types of jobs, especially those requiring low-paid, less-skilled workers, from more developed to less developed countries.

Outsourcing
A decision by a corporation to turn over much of the responsibility for production to independent suppliers.

Post-Fordist production
Adoption by companies of flexible work rules, such as the allocation of workers to teams that perform a variety of tasks.

Site factors
Location factors related to the costs of factors of production inside the plant, such as land, labor, and capital.

Situation factors
Location factors related to the transportation of materials into and from a factory.

FURTHER READINGS

Bluestone, Barry, and Bennett Harrison. *The Deindustrialization of America: Plant Closings, Community Abandonment, and the Dismantling of Basic Industry.* New York: Basic Books, 1982.

Essletzbichler, Jürgen. "The Geography of Job Creation and Destruction in the U.S. Manufacturing Sector, 1967–1997." *Annals of the Association of American Geographers* 94 (2004): 602–19.

Hughes, Alex, and Suzanne Reimer, eds. *Geographies of Commodity Chains.* New York: Routledge, 2004.

Klier, Thomas, and James M. Rubenstein. *Who Really Made Your Car? Restructuring and Geographic Change in the Auto Industry.* Kalamazoo, Mich.: W. E. Upjohn Institute for Employment Research, 2008.

Ó hUallacháin, Breandan, and Richard A. Matthews. "Economic Restructuring in Primary Industries: Transaction Costs and Corporate Vertical Integration in the Arizona Copper Industry, 1980–1991." *Annals of the Association of American Geographers* 84 (1994): 399–417.

Rubenstein, James M. *The Changing U.S. Auto Industry.* London: Routledge, 1992.

———. Making and Selling Cars: *Innovation and Change in the U.S. Automotive Industry.* Baltimore: Johns Hopkins University Press, 2001.

Scott, Allen J., and Michael Storper, eds. *Production, Work, Territory.* Boston: Allen and Unwin, 1986.

Storper, Michael, and Richard Walker. *The Capitalist Imperative: Territory, Technology, and Industrial Growth.* New York: Basil Blackwell, 1989.

ON THE INTERNET

Statistics on employment in manufacturing, as well as other sectors of the U.S. economy, are compiled by the U.S. Department of Labor Bureau of Labor Statistics at **www.bls.gov.** Some foreign labor statistics are also supplied.

Chapter 12 SETTLEMENTS AND SERVICES

Flying across the United States on a clear night, you look down on the lights of settlements, large and small. You see small clusters of lights from villages and towns, and large, brightly lit metropolitan areas. It may appear that the light clusters are random, but geographers discern a regular pattern in them. These regularities have been documented, and concepts from economic geography can be applied to understand why this pattern exists.

The regular distribution observed over North America and over other MDCs is not seen in LDCs. Geographers explain this difference and why the absence of a regular pattern is significant.

The regular pattern of settlement in MDCs reflects where services are provided. In some MDCs, more than three-fourths of the workers are employed in the tertiary sector of the economy, providing goods and services to people in exchange for payment. These services are provided primarily in cities. In contrast, less than one-third of the labor force in some LDCs provide services.

Everyone needs food for survival. In LDCs, most people work in the primary sector, growing food. In MDCs, people purchase food at supermarkets or restaurants. The people employed at the supermarkets and restaurants are examples of service-sector workers, and the customers pay for the food with money earned in other service-sector jobs, such as retailing, banking, law, education, and government.

THE KEY ISSUES IN THIS CHAPTER

CITIES OF THE WORLD

12.1 Distribution of Cities

12.2 Cities in History

12.3 Urbanization

BUSINESS SERVICES

12.4 Types of Services

12.5 Hierachy of Business Services

12.6 Coolness and Innovation

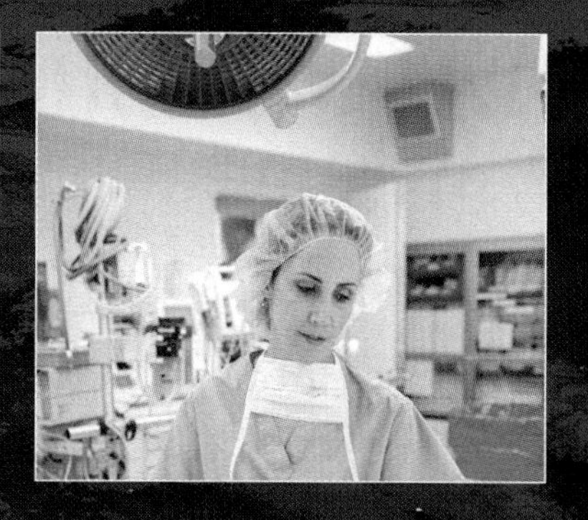

CONSUMER SERVICES

12.7 Central Place Theory

12.8 Hierarchy of Consumer Services

12.9 Market Area Analysis

THE UNITED STATES AT NIGHT
This is a composite of hundreds of images produced by the Defense Meteorological Satellite Program.

CITIES OF THE WORLD

12.1 Distribution of Cities

- The world has 22 cities with at least 10 million inhabitants.
- Seven of the 10 largest cities are in LDCs.

Identifying the world's largest cities is difficult, because each country defines *city* in a unique manner. The Demographia website (see chapter resources) uses maps and satellite imagery to delineate urban areas consistently regardless of country. According to Demographia, 177 urban areas have at least 2 million inhabitants, 100 at least 3 million, 55 at least 5 million, 22 at least 10 million, and 3 (Tokyo, New York, and Seoul) at least 20 million.

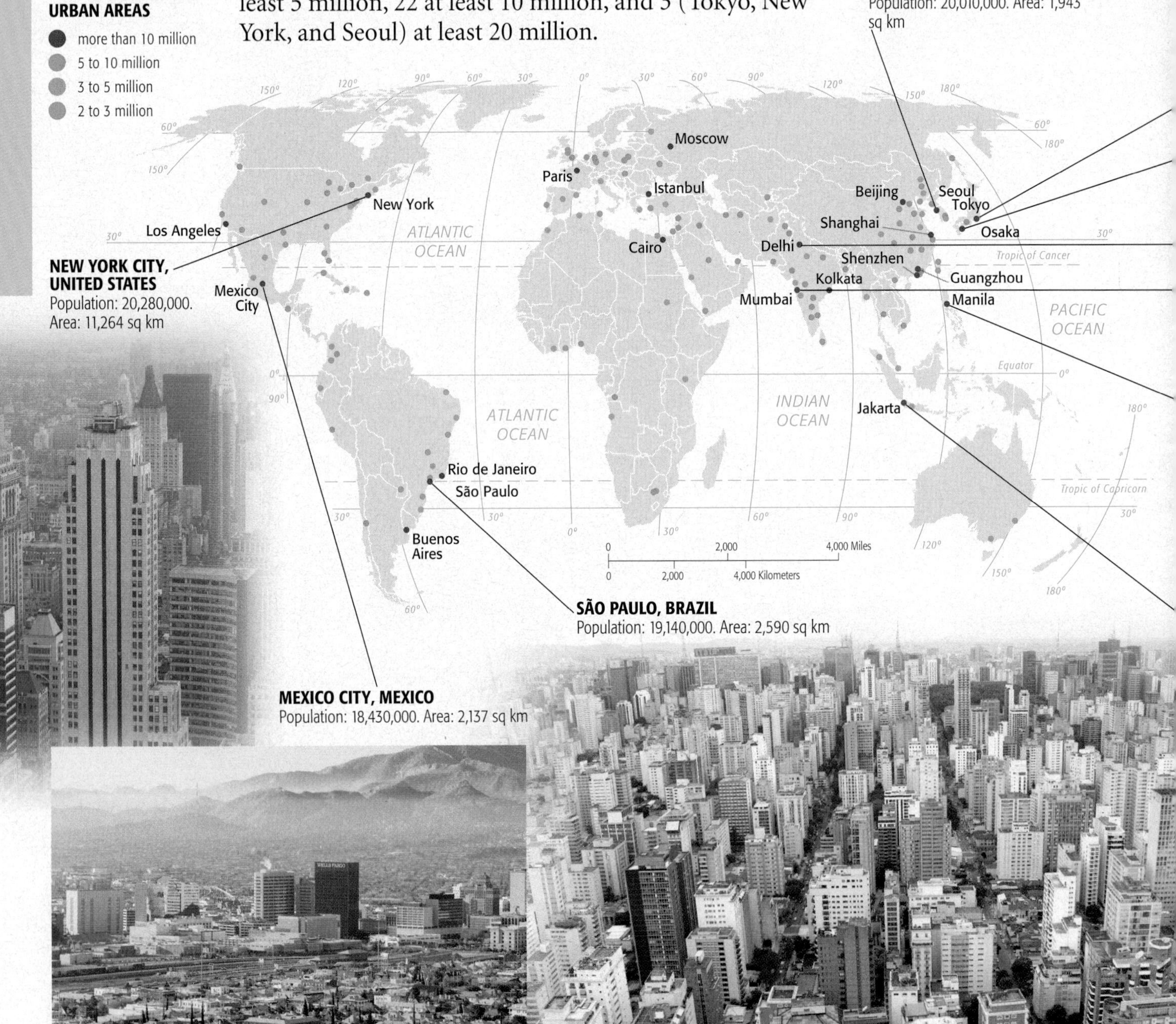

TOKYO–YOKOHAMA, JAPAN
Population: 34,400,000. Area: 7,835 sq km

OSAKA–KOBE–KYOTO, JAPAN
Population: 17,270,000. Area: 2,720 sq km

DELHI, INDIA
Population: 17,640,000. Area: 1,425 sq km

MUMBAI (BOMBAY), INDIA
Population: 19,360,000. Area: 777 sq km

MANILA, PHILIPPINES
Population: 17,080,000. Area: 1,347 sq km

JAKARTA, INDONESIA
Population: 19,880,000. Area: 2,849 sq km

CITIES OF THE WORLD

12.2 Cities in History

- Cities originated in multiple hearths and diffused in multiple directions.
- Through history, the world's largest city has usually been in the Middle East or East Asia.

Urban settlements date from the beginning of documented history in the Middle East and Asia. They may have originated in Mesopotomia, part of the Fertile Crescent of the Middle East, and diffused at an early date to Egypt, China, and South Asia's Indus Valley. Or they may have originated independently in each of the four hearths. In any case, from these four hearths, the concept of urban settlements diffused to the rest of the world.

MEMPHIS, EGYPT
The Alabaster Sphinx was constructed around 3,500 years ago near Memphis, Egypt, which at the time was probably the world's largest city.

LAGASH, BABYLONIA (Iraq)
THEBES, Egypt
BABYLON, BABYLONIA (Iraq)
MEMPHIS, Egypt
NINEVEH, ASSYRIA (Iraq)

3000 B.C.
2750 B.C.
2500 B.C.
2250 B.C.
2000 B.C.
1750 B.C.
1500 B.C.
1250 B.C.
1000 B.C.
750 B.C.

MEMPHIS, Egypt
population over 30,000
AKKAD, BABYLONIA (Iraq)
UR, BABYLONIA (Iraq)
population 65,000
AVARIS, Egypt
THEBES, Egypt
BABYLON, BABYLONIA (Iraq)
FIRST ABOVE 200,000

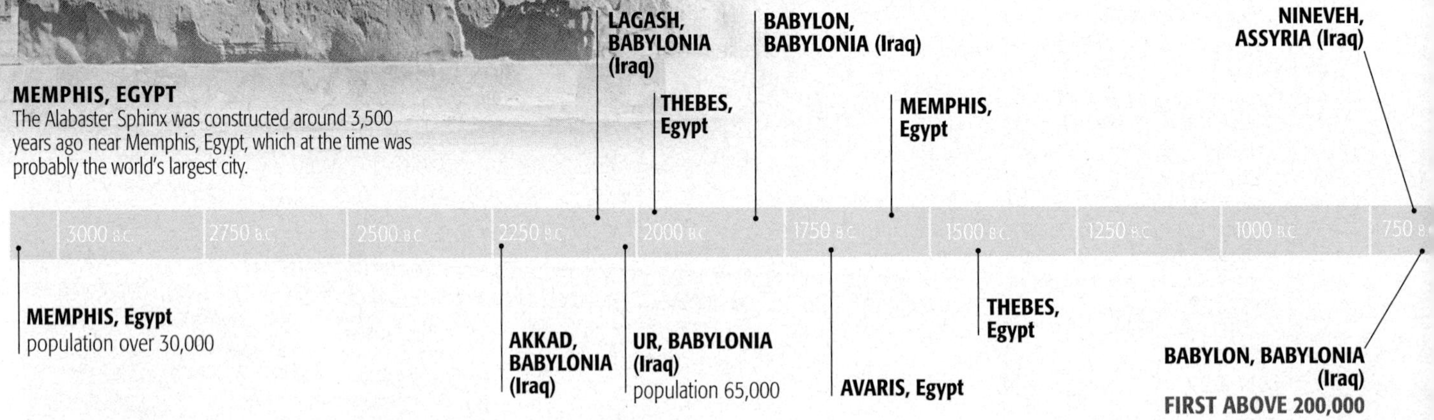

12.2.1 **LARGEST CITIES THROUGH HISTORY**
The chart shows the largest cities at various points in history and estimated population if known.

Rome
Nineveh
Avaris
Akkad
Alexandria
Lagash
Ch'ang-an
Memphis
Pataliputra
Ur
Thebes
Babylon

UR
Ur, which means "fire," was where Abraham lived prior to his journey to Canaan in approximately 1900 B.C., according to the Bible. Archaeologists have unearthed ruins in Ur that date from approximately 3000 B.C.

CONSTANTINOPLE
Originally called Byzantium, and now known as Istanbul, Constantinople became the world's largest city after the decline of Rome, and remained Europe's largest city for most of the next 1,000 years. St Sophia, built during the sixth century was the largest cathedral in Europe until converted to a mosque in the fifteenth century.

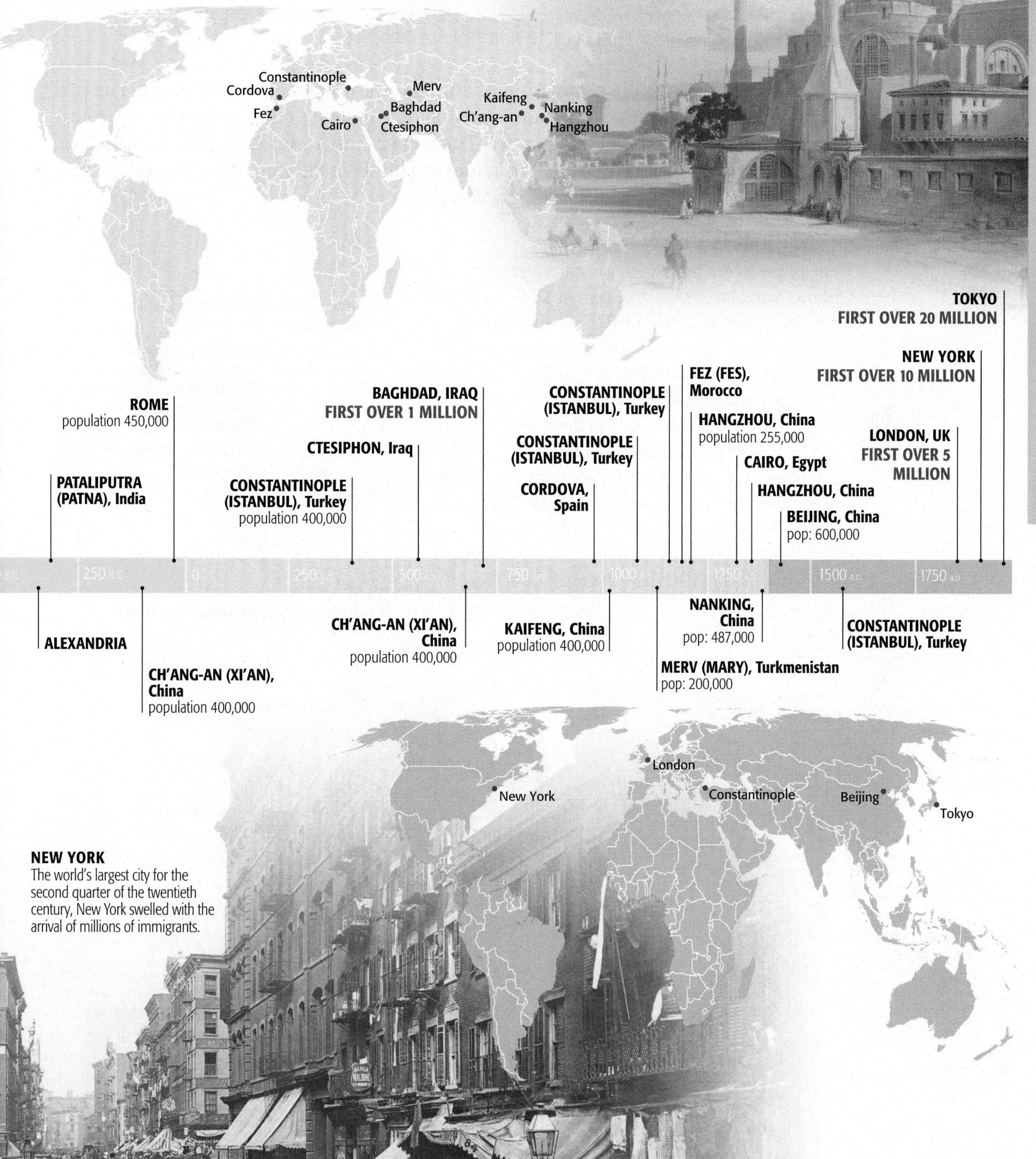

NEW YORK
The world's largest city for the second quarter of the twentieth century, New York swelled with the arrival of millions of immigrants.

CITIES OF THE WORLD

12.3 Urbanization

- MDCs have a higher percentage of people living in urban areas, a consequence of economic restructuring.
- The distinctive economic structure of a city is its economic base.

The process by which the population of cities grows, known as **urbanization**, has two dimensions—an increase in the number of people living in cities and an increase in the *percentage* of people living in cities. The large number of people in cities was demonstrated in Figure 12.1.1. Most large cities are in LDCs, because high natural increase rates have helped to swell their populations.

PERCENTAGE OF PEOPLE IN CITIES

The percentage of people living in cities increased from 3 percent in 1800 to 6 percent in 1850, 14 percent in 1900, and 30 percent in 1950. The population of urban settlements exceeded that of rural settlements for the first time in human history in 2008.

A large percentage of people living in urban areas is a measure of a country's level of development. The higher percentage of urban residents in MDCs is a consequence of changes in economic structure during the past two centuries—first the Industrial Revolution in the nineteenth century and then the growth of services in the twentieth. During the past 200 years, rural residents in MDCs have migrated from the countryside to work in the factories and services that are concentrated in cities.

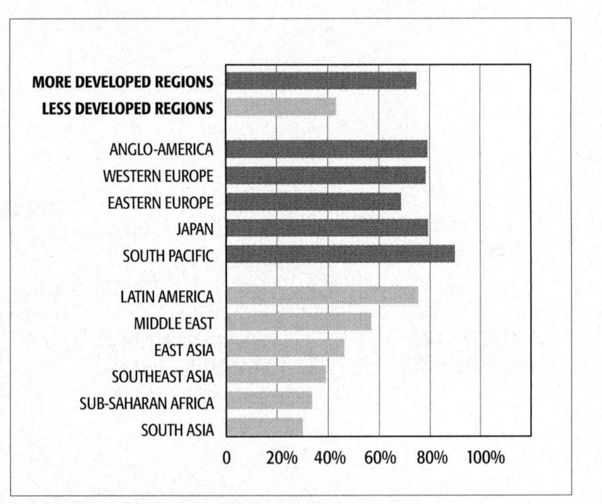

12.3.1 **PERCENT URBAN BY WORLD REGION**

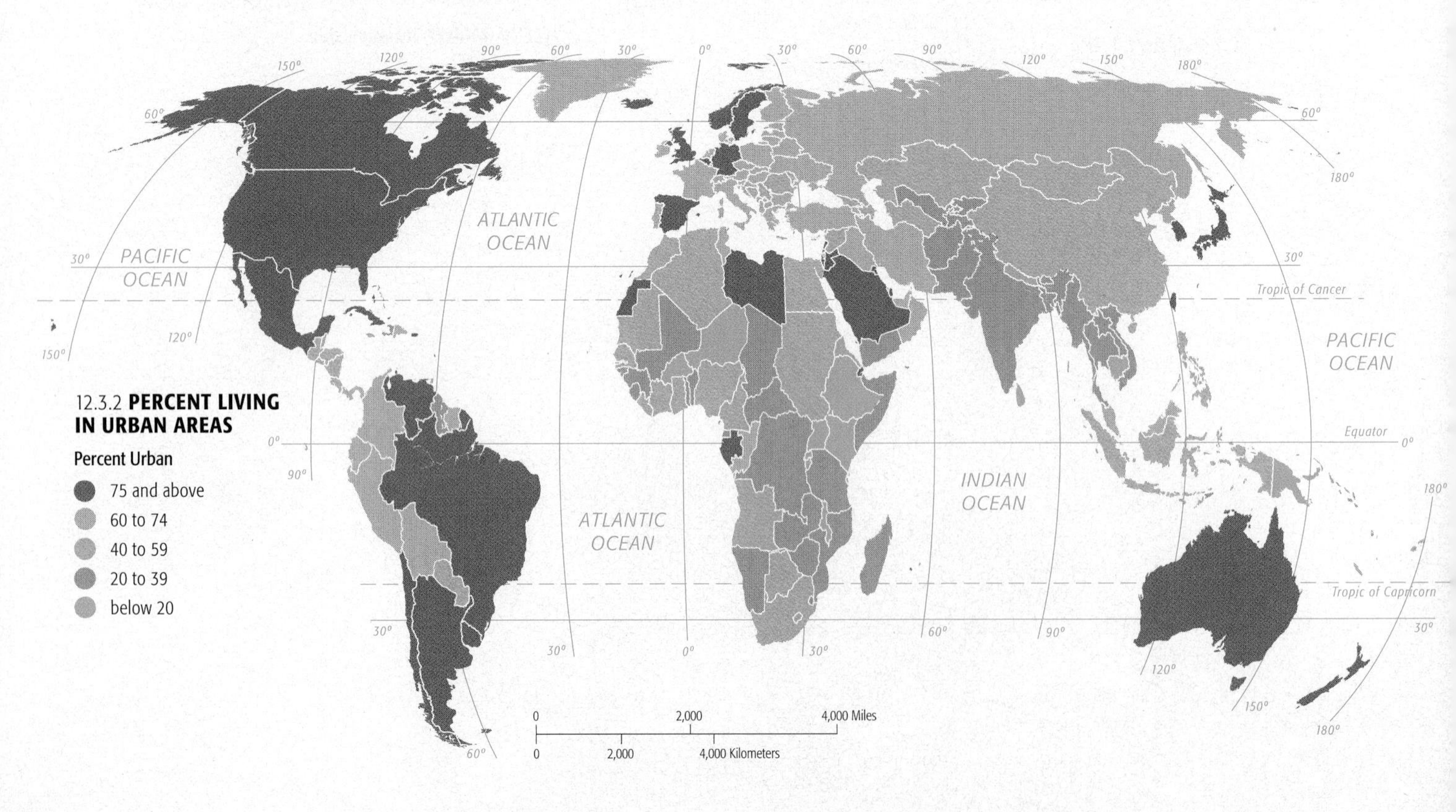

12.3.2 **PERCENT LIVING IN URBAN AREAS**

ECONOMIC BASE OF CITIES

A settlement's distinctive economic structure derives from its **basic industries**, which export primarily to consumers outside the settlement. **Nonbasic industries** are enterprises whose customers live in the same community, essentially consumer services. A community's unique collection of basic industries defines its **economic base.**

A settlement's economic base is important because exporting by the basic industries brings money into the local economy, thus stimulating the provision of more nonbasic consumer services for the settlement. New basic industries attract new workers to a settlement, and they bring their families with them. The settlement attracts additional consumer services to meet the needs of the new workers and their families. Thus a new basic industry stimulates establishment of new supermarkets, laundromats, restaurants, and other consumer services. But a new nonbasic service, such as a supermarket, will not induce construction of new basic industries.

A community's basic industries can be identified by computing the percentage of the community's workers employed in different types of businesses. The percentage of workers employed in a particular industry in a settlement is then compared to the percentage of all workers in the country employed in that industry. If the percentage is much higher in the local community, then that type of business is a basic economic activity.

Settlements in the United States can be classified by their type of basic activity. Each type of basic activity has a different spatial distribution. Compared to the national average, some settlements have a very high percentage of workers employed in the primary sector, notably mining. Mining settlements are located near reserves of coal, petroleum, and other resources.

The economic base of some settlements is in the secondary sector. Some of these specialize in durable manufactured goods, such as steel and automobiles, others in nondurable manufactured goods, such as textiles, apparel, food, chemicals, and paper. Most communities that have an economic base of manufacturing durable goods are clustered between northern Ohio and southeastern Wisconsin, near the southern Great Lakes. Detroit, Michigan specializes in manufacturing motor vehicles; Gary, Indiana, specializes in steel. Nondurable manufacturing industries, such as textiles, are clustered in the Southeast, especially in the Carolinas.

The concept of basic industries originally referred to manufacturing. But in a postindustrial society, such as the United States, increasingly the basic economic activities are in services. Steel was once the most important basic industry of Cleveland and Pittsburgh, but now health services such as hospitals and clinics and medical high-technology research are more important.

Northern and eastern cities that were once major manufacturing centers have been transformed into business service centers. These cities have moved more aggressively to restructure their economic bases to offset sharp declines in manufacturing jobs.

THE GRAPES PUBLIC HOUSE IN THE SKELDERGATE AREA OF YORK, UK

BUSINESS SERVICES

12.4 Types of Services

- Three types of services are consumer, business, and public.
- Employment has grown more rapidly in some services than in others.

The service sector of the economy is subdivided into three types—consumer services, business services, and public services. Each of these sectors is divided into several major subsectors.

CONSUMER SERVICES

The principal purpose of **consumer services** is to provide services to individual consumers who desire them and can afford to pay for them. Nearly one-half of all jobs in the United States are in consumer services. Four main types of consumer services are retail, education, health, and leisure.

12.4.2 **CONSUMER SERVICES**

RETAIL AND WHOLESALE
About 11 percent of all U.S. jobs. One-fifth each in department stores, grocers, and motor vehicle sales and service. In addition, 4 percent of all jobs are in wholesale services that provide retailers their merchandise.

EDUCATION
About 11 percent of all U.S. jobs. Two-thirds in public schools, one-third in private schools.

HEALTH CARE
About 10 percent of all U.S. jobs. Primarily hospitals, doctors' offices, and nursing homes.

LEISURE AND HOSPITALITY
About 10 percent of all U.S. jobs. Three-fourths in restaurants and bars, the other one-fourth divided evenly among lodging and entertainment.

OTHER U.S. JOBS

12.4.1 **PERCENTAGE OF GDP FROM SERVICES**
Services are provided in all societies, but in MDCs a majority of workers are engaged in the provision of services. The percentage of service workers varies widely in LDCs but is typically less than one-fourth. Services generate more than two-thirds of GDP in most MDCs, compared to less than one-half in most LDCs.

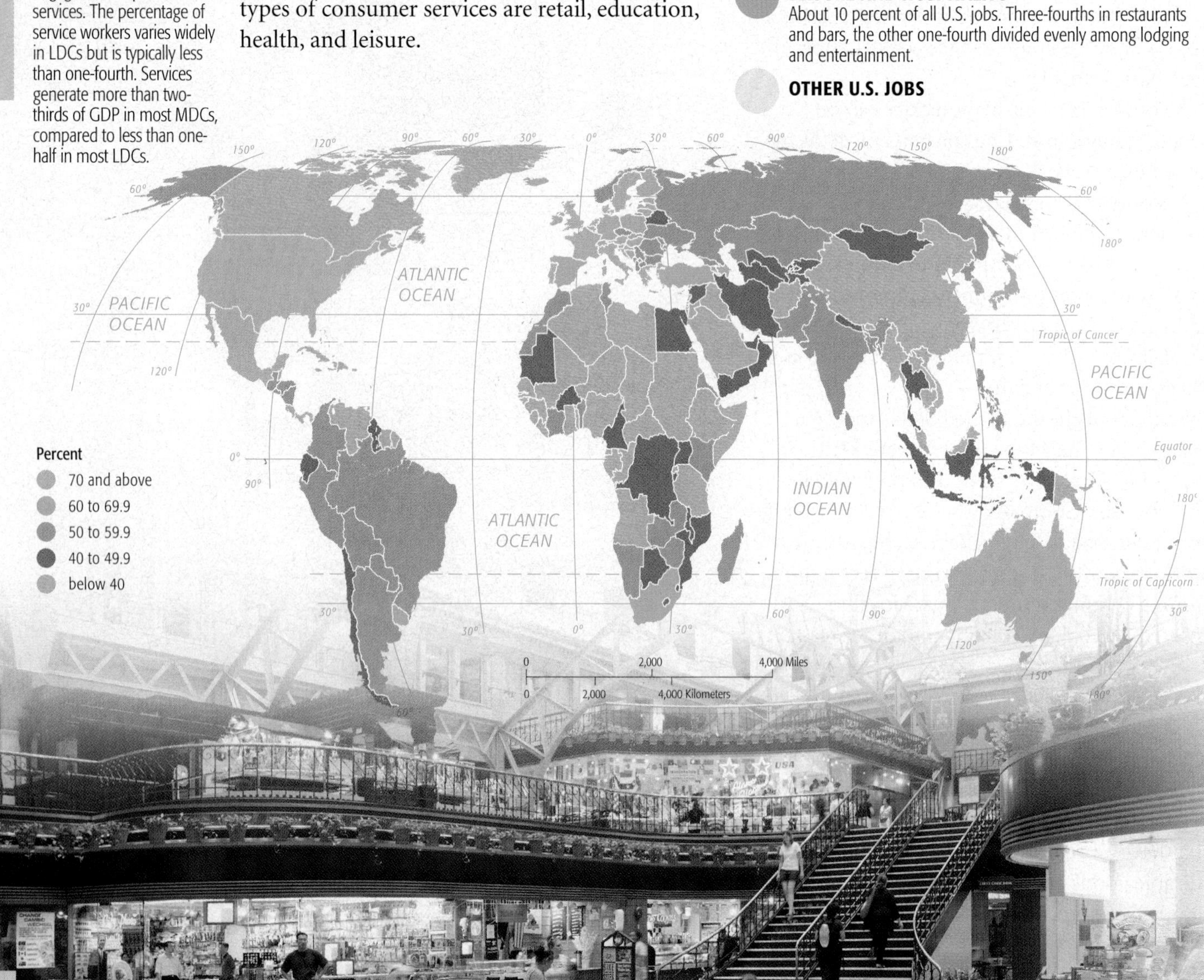

12.4.3 **EMPLOYMENT CHANGE IN THE UNITED STATES BY SECTOR**
The growth in employment in the United States has been in services, whereas employment in primary- and secondary-sector activities has declined.

- Within business services, jobs expanded in professional services (such as engineering, management, and law), data processing, advertising and temporary employment agencies. Jobs grew more slowly in finance and transportation services, because of improved efficiency—fewer workers are needed to run trains and answer phones, for example.
- Within consumer services, the most rapid increase has been in the provision of health care, including hospital staff, clinics, nursing homes, and home health-care programs. Other large increases have been recorded in recreation and entertainment. The share of jobs in retailing has not increased—more stores are opening all the time, but they don't need as many employees as in the past.
- The share of employment in public services has declined during the past two decades

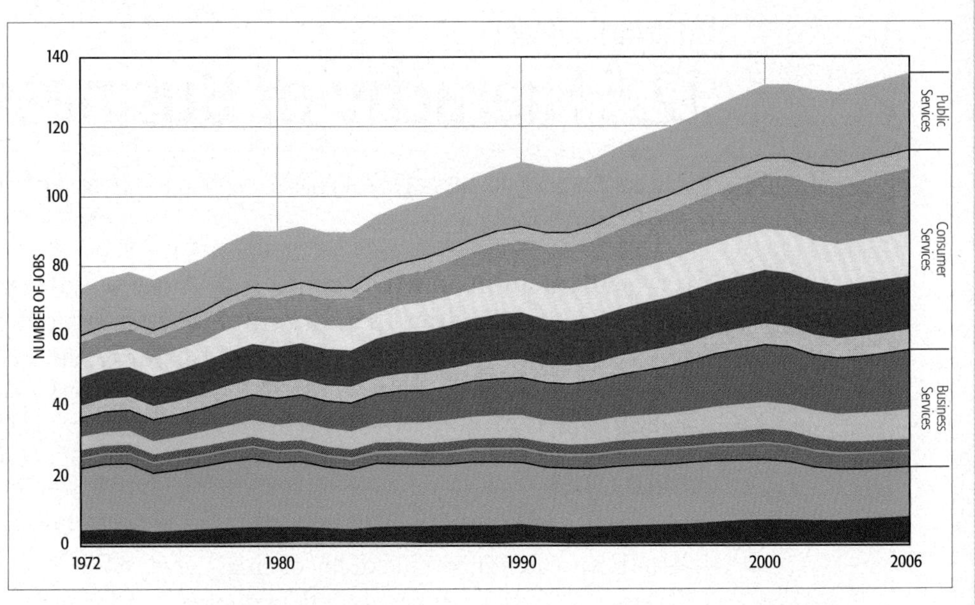

JOB SECTOR
TERTIARY SECTOR
Public Services
- government

Consumer Services
- other services
- education and health
- leisure and hospitality
- retail
- wholesale

Business Services
- professional and business
- finance
- information
- utilities
- transportation and warehousing

SECONDARY SECTOR
- manufacturing
- construction

PRIMARY SECTOR
- agriculture and mining

BUSINESS SERVICES

The principal purpose of **business services** is to facilitate other businesses. One-fourth of all jobs in the United States are in business services. Professional services, financial services, and transportation services are the three main types of business services.

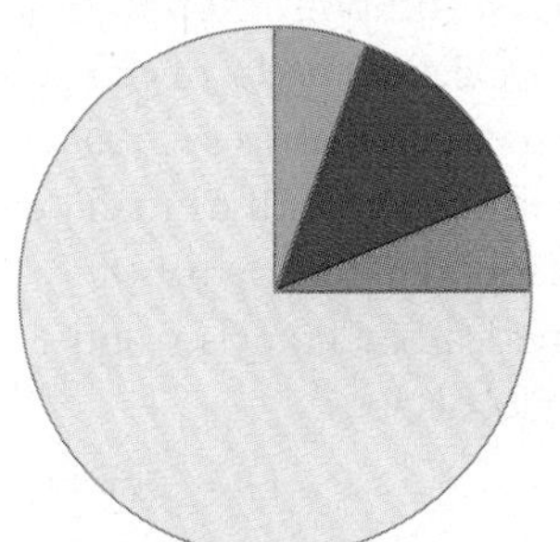

12.4.4 **BUSINESS SERVICES**

FINANCIAL SERVICES
About 6 percent of all U.S. jobs. One-half in banks and other financial institutions, one-third in insurance companies, the remainder in real estate.

PROFESSIONAL SERVICES
About 13 percent of all U.S. jobs. One-tenth in management positions; two-fifths in technical services, including law, accounting, architecture, engineering, design, and consulting; one-half in support services, such as clerical, secretarial, and custodial work.

TRANSPORTATION AND INFORMATION SERVICES
About 6 percent of U.S. jobs. One-half in transportation, primarily trucking, one-half in information services such as publishing and broadcasting, and utilities such as water and electricity.

OTHER U.S. JOBS

PUBLIC SERVICES

The purpose of **public services** is to provide security and protection for citizens and businesses. About 16 percent of all U.S. jobs are in the public sector, 9 percent of public school employees were excluded from the total and counted instead under education (consumer) services. Excluding educators, one-sixth of public-sector employees work for the federal government, one-fourth for one of the 50 state governments, and three-fifths for one of the tens of thousands of local governments. When educators are counted, the percentages for state and local governments would be higher.

The distinction among services is not absolute. For example, individual consumers use business services, such as consulting lawyers and keeping money in banks, and businesses use consumer services, such as purchasing stationery and staying in hotels. A public service worker at a national park may provide the same service as a consumer service worker at Disneyland. Geographers find the classification useful because the various types of services have different distributions and different factors influence locational decisions.

12.4.5 **PUBLIC SERVICES**

PUBLIC SECTOR

OTHER U.S. JOBS

BUSINESS SERVICES

12.5 Hierachy of Business Services

- A hierarchy of world cities can be identified based on business services.
- A hierarchy of cities also exists inside the United States

Every settlement in an MDC such as the United States provides consumer services to people in a surrounding area. But not every settlement of a given size has the same number and types of business services. Business services disproportionately cluster in a handful of settlements, and individual settlements specialize in particular business services.

WORLD CITIES

Geographers J. V. Beaverstock, P. Taylor, and J. G. Smith have identified a hierarchy of cities. Each city received a score based on the extent and type of its business services. At the top of the hierarchy are a handful of world cities.

- London, New York, Paris, and Tokyo are the four leading world cities.
- Chicago, Frankfurt, Hong Kong, Los Angeles, Milan, and Singapore are world cities of somewhat lower importance than the Big Four.

The list of top 10 world cities does not match the list of 10 largest cities on page 268. Only New York and Tokyo are on both lists. Seven of the 10 largest cities are in LDCs, compared to only two of the world cities (Hong Kong and Singapore). World cities are predominantly in MDCs because they are home to a large percentage of the world's global-scale business services, including:

- Financial services such as banking and insurance
- Information-gathering services such as publishing and media
- Professional services such as law, medicine, science, and education

World cities also contain a disproportionately high share of the world's arts, culture, consumer spending on luxury goods, and political power.

A second level of sub-global cities perform global-scale business services in specialized areas such as finance or information, rather than for the broad array of business services found in the top tier of world cities. In North America, San Francisco and Toronto are the two sub-global cities.

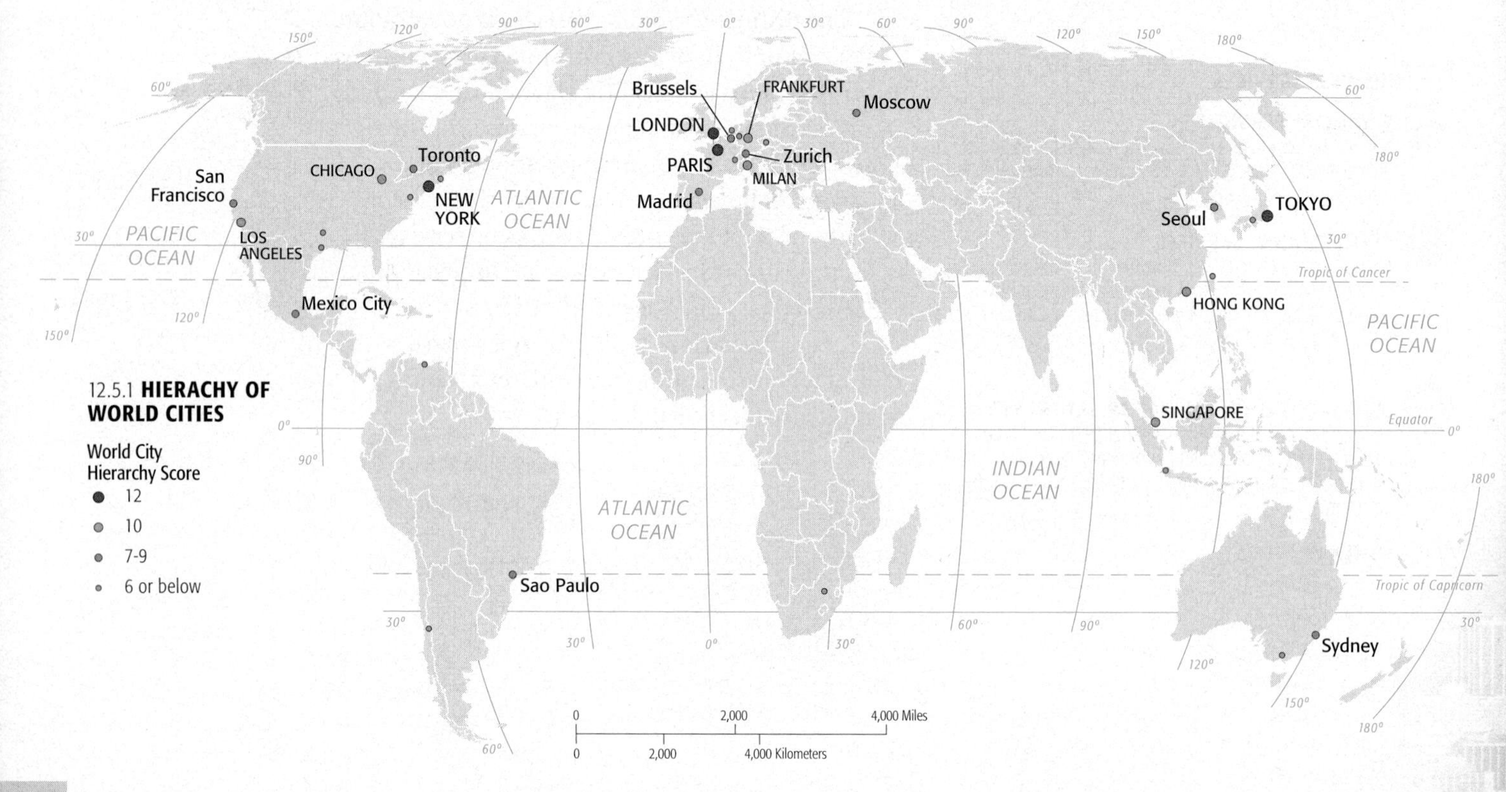

12.5.1 **HIERACHY OF WORLD CITIES**

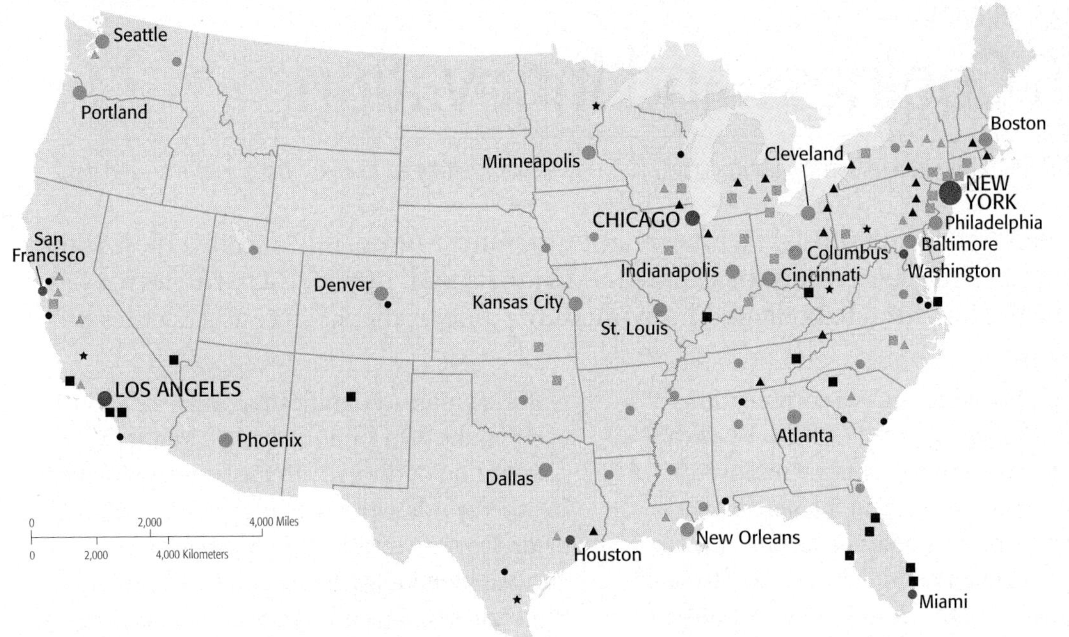

12.5.2 **HIERARCHY OF BUSINESS SERVICE CITIES IN THE UNITED STATES.**

World city
- dominant city
- major city
- secondary city

Command and control center
- regional command & control center
- subregional command & control center

Specialized producer-service center
- functional service center
- government/Education service center

Dependent center
- resort, retirement, and residential centers
- manufacturing centers
- military centers
- mining centers

HIERARCHY OF U.S. CITIES

Below the world cities, three tiers of cities can be identified in the United States.

- **Second tier: command and control centers.** Contain the headquarters of many large corporations, well-developed banking facilities, and concentrations of other business services, including insurance, accounting, advertising, law, and public relations. Important educational, medical, and public institutions can be found in these command and control centers.
- **Third tier: specialized producer-service centers.** Offer narrower and more highly specialized services. One group of these cities specializes in the management and R&D (research and development) activities related to specific industries, such as motor vehicles in Detroit; steel in Pittsburgh; office equipment in Rochester, New York; and semiconductors in San Jose, California. A second group of these cities specializes as centers of government and education, notably state capitals that also have a major university, such as Albany, Lansing, Madison, and Raleigh-Durham.
- **Fourth tier: dependent centers.** Provide relatively unskilled jobs. Four subtypes include
  - **Resort, retirement, and residential centers,** such as Albuquerque, Fort Lauderdale, Las Vegas, and Orlando, clustered in the South and West.
  - **Manufacturing centers,** such as Buffalo, Chattanooga, Erie, and Rockford, clustered mostly in the old northeastern manufacturing belt.
  - **Industrial and military centers,** such as Huntsville, Newport News, and San Diego, clustered mostly in the South and West.
  - **Mining and industrial centers,** such as Charleston (West Virginia) and Duluth, located in mining areas.

12.5.3 **THE U.S. WORLD CITIES: NEW YORK (top), CHICAGO (middle), LOS ANGELES (bottom).**

BUSINESS SERVICES

12.6 Coolness and Innovation

- Talent is not distributed uniformly among cities.
- Talent is attracted to cities that are diverse and cool.

Individuals possessing special talents are not distributed uniformly among cities. Some cities have a higher percentage of talented individuals than others. Richard Florida has studied what has attracted talented individuals to some of the 50 largest U.S. cities but not to others.

Florida measured talent as a combination of the percentage of people in the city with college degrees, the percentage employed as scientists or engineers, and the percentage employed as professionals or technicians. To some extent, talented individuals are attracted to the cities with the most job opportunities and financial incentives. But Florida determined that economic factors were less important than cultural diversity in attracting talented people to a city. Individuals with special talents gravitate toward cities that offer more cultural diversity.

Florida found three measures of cultural diversity to be significant—the number of cultural facilities per capita, the percentage of gay men, and a "coolness" index. A city's gay population was based on census figures for the percentage of households consisting of two adult men. Two adult men who share a house may not be gay, but Florida assumed that the percentage of adult men living together who were gay did not vary from one city to another. The "coolness" index, developed by POV Magazine, combined the percentage of population in their 20s, the number of bars and other nightlife places per capita, and the number of art galleries per capita.

Florida also found that talented individuals were not attracted to a city by its climate or proximity to sports.

Florida found a significant positive relationship between the distribution of talent and the distribution of diversity in the largest U.S. cities. In other words, cities with high cultural diversity tended to have relatively high percentages of talented individuals. For example, Washington, San Francisco, Boston, and Seattle ranked among the top in both talent and diversity, whereas Las Vegas was near the bottom in both.

Attracting talented individuals is important for a city, because these individuals are responsible for promoting economic innovation. They are likely to start new businesses and infuse the local economy with fresh ideas. High-tech industries and high-paying jobs locate in cities with concentrations of talented people.

Florida concludes that promoting cultural diversity is more than a social justice goal—it is good for a city's economy. Providing subsidies to businesses does less to promote a city's economic health than measures to make a city more welcoming to a diverse population. In other words, attract diverse talented people, and the businesses will follow. Florida's work has been criticized for not explaining why creative individuals congregate in particular cities, nor why a clustering of creative individuals necessarily produces economic growth for that city.

12.6.1 **RELATIONSHIP BETWEEN DIVERSITY AND ECONOMIC INNOVATION**

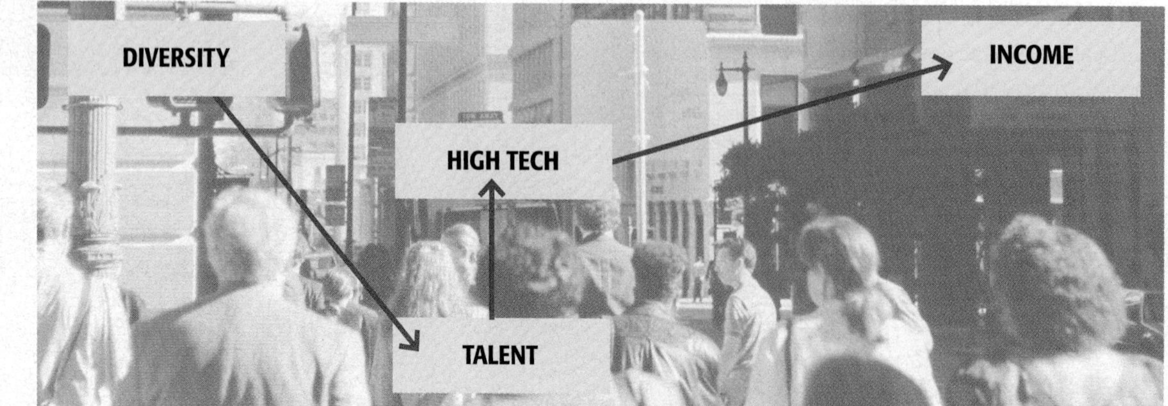

12.6.2 **GEOGRAPHY OF TALENT**
Scientists and engineers (top), professional and technical workers (middle), university graduates (bottom).

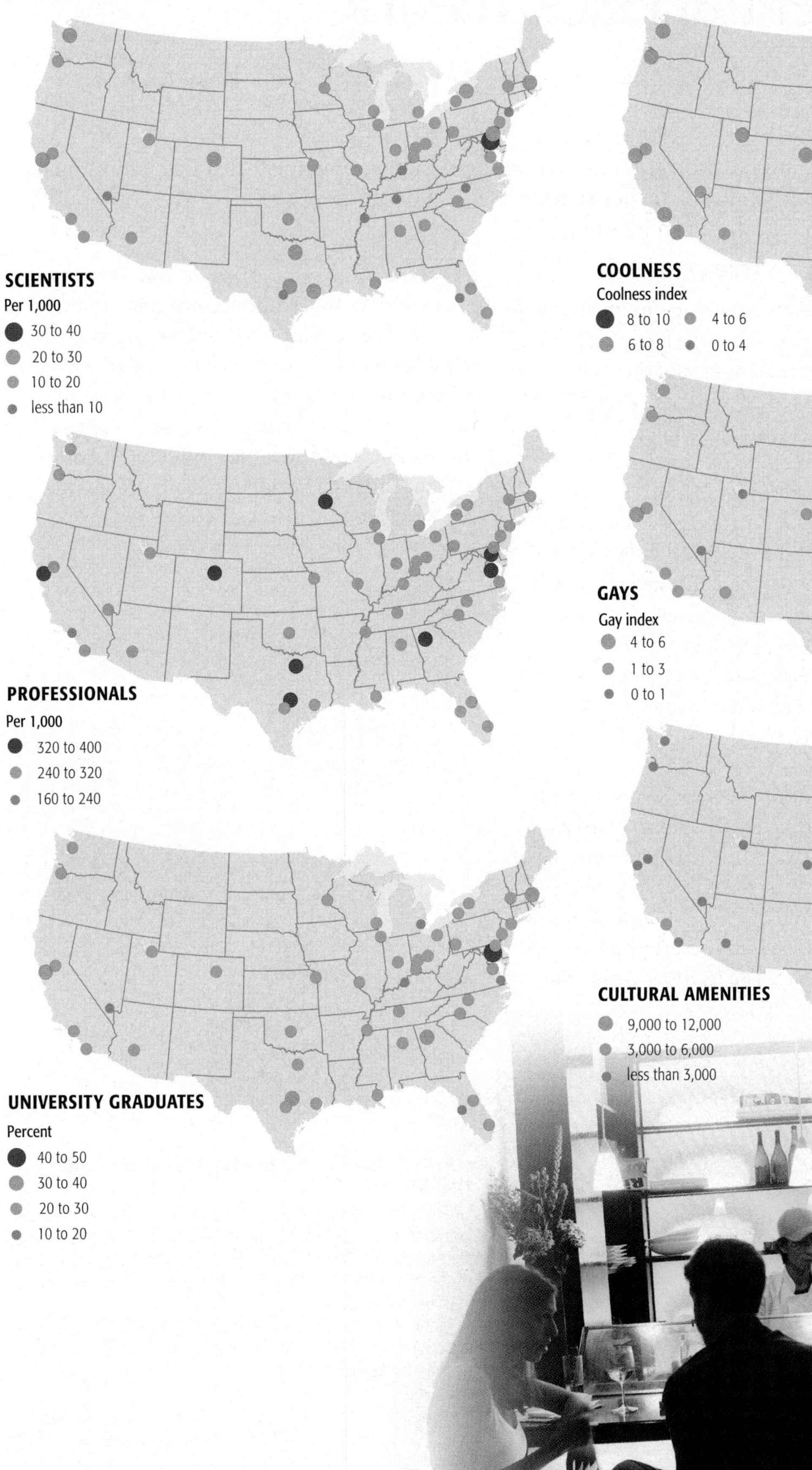

12.6.3 **GEOGRAPHY OF CULTURAL DIVERSITY**
Coolness index (top), gays (middle), cultural amenities (bottom).

COOLNESS
Coolness index
8 to 10
4 to 6
6 to 8
0 to 4

GAYS
Gay index
4 to 6
1 to 3
0 to 1

CULTURAL AMENITIES
9,000 to 12,000
3,000 to 6,000
less than 3,000

CONSUMER SERVICES

12.7 Central Place Theory

- Central place theory explains the location of consumer services.
- A central place has a market area, range, and threshold.

Selecting the right location for a new shop is probably the single most important factor in the profitability of a consumer service. Central place theory helps to explain how the most profitable location can be identified.

MARKET AREA OF A SERVICE

A **central place** is a market center for the exchange of goods and services by people attracted from the surrounding area. The central place is so called because it is centrally located to maximize accessibility from the surrounding region.

Central places compete against each other to serve as markets for goods and services. This competition creates a regular pattern of settlements, according to central place theory. **Central place theory** explains how services are distributed and why a regular pattern of settlements exists—at least in MDCs such as the United States. Central place theory was first proposed in the 1930s by German geographer Walter Christaller, based on his studies of southern Germany.

The area surrounding a service from which customers are attracted is the **market area** or **hinterland**. A market area is a good example of a nodal region—a region with a core where the characteristic is most intense. To establish the market area, a circle is drawn around the node of service on a map. The territory inside the circle is its market area.

Because most people prefer to get services from the nearest location, consumers near the center of the circle obtain services from local establishments. The closer to the periphery of the circle, the greater is the percentage of consumers who will choose to obtain services from other nodes. People on the circumference of the market-area circle are equally likely to use the service, or go elsewhere.

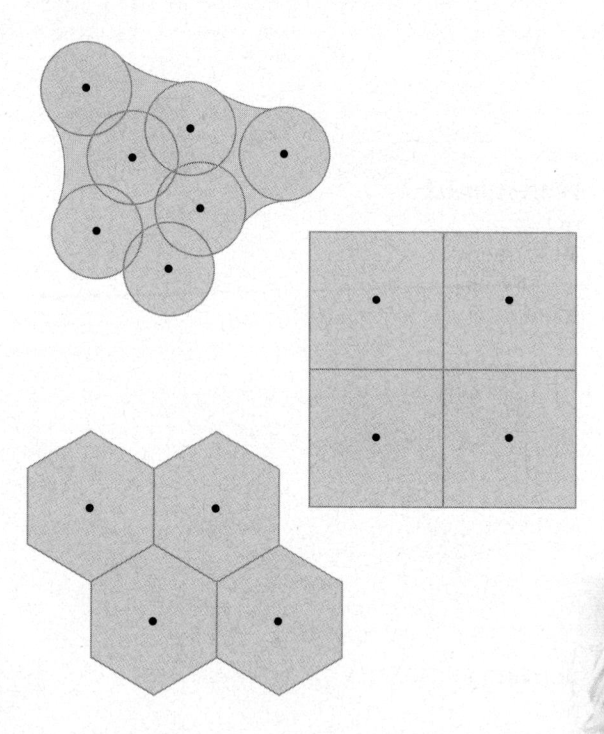

12.7.1 **MARKET AREA, RANGE, AND THRESHOLD OF A SERVICE**

12.7.2 **WHY HEXAGONS ARE USED IN CENTRAL PLACE THEORY.**

Circles can be drawn to designate market areas. They are equidistant from center to edge, but they overlap or leave gaps. People living in the gaps absurdly would be outside of any market area. Overlapping circles are unsatisfactory because one service or another logically must be the closest.

Central place theory requires a geometric shape without gaps or overlaps, so circles are out. Squares fit without gaps, but their sides are not equidistant from the center so the range cannot be calculated.

The hexagon is the best compromise between circles and squares to represent a market area. All points along the hexagon are not the same distance from the center, but the variation is less than with a square, and hexagons, like squares, nest without gaps.

RANGE OF A SERVICE

The market area of every service varies. To determine the extent of a market area, geographers need two pieces of information about a service—its range and its threshold.

How far are you willing to drive for a pizza? To see a doctor for a serious problem? To watch a ballgame? The **range** is the maximum distance people are willing to travel to use a service. The range is the radius of the circle drawn to delineate a service's market area.

People are willing to go only a short distance for everyday consumer services, like groceries, laundromats, or video rentals. But they will travel a long distance for other services, such as a major league baseball game or concert. Thus a convenience store has a small range, whereas a stadium has a large range.

If firms at other locations compete by providing the service, the range must be modified. As a rule, people tend to go to the nearest available service: someone in the mood for a McDonald's hamburger is likely to go to the nearest McDonald's. Therefore, the range of a service must be determined from the radius of a circle that is irregularly shaped rather than perfectly round. The irregularly shaped circle takes in the territory for which the proposed site is closer than the competitors' sites.

The range must be modified further because most people think of distance in terms of time, rather than in terms of a linear measure like kilometers or miles. If you ask people how far they are willing to travel to a restaurant or a baseball game, they are more likely to answer in minutes or hours than in distance.

THRESHOLD OF A SERVICE

The second piece of geographic information needed to compute a market area is the **threshold**, which is the minimum number of people needed to support the service. Every enterprise has a minimum number of customers required to generate enough sales to make a profit. Once the range has been determined, a service provider must determine whether a location is suitable by counting the potential customers inside the irregularly shaped circle.

How potential consumers inside the range are counted depends on the product. Convenience stores and fast-food restaurants appeal to nearly everyone, whereas other goods and services appeal primarily to certain consumer groups. Movie theaters attract younger people; chiropractors attract older folks. Poorer people are drawn to thrift stores; wealthier ones might frequent upscale department stores. Amusement parks attract families with children, but nightclubs appeal to singles. If a good or service appeals to certain customers, then only the type of good or service that appeals to them should be counted inside the range.

12.7.3 **TYPES OF CONSUMER SERVICES**
Clockwise from far left:
HEALTH CARE, such as a hospital.
WHOLESALE, such as a distribution center.
RETAIL, such as a department store.
HOSPITAL!TY, such as a restaurant.
EDUCATION, such as a school.

CONSUMER SERVICES

12.8 Hierarchy of Consumer Services

- Small settlements provide services with small thresholds, ranges, and market areas.
- The size of settlements follows the rank-size rule more closely in MDCs than in LDCs.

Small settlements are limited to consumer services that have small thresholds, short ranges, and small market areas, because too few people live in small settlements to support many services. A large department store or specialty store cannot survive in a small settlement because the minimum number of people needed exceeds the population within range of the settlement.

Larger settlements provide consumer services having larger thresholds, ranges, and market areas. In addition, neighborhoods within large settlements also provide services having small thresholds and ranges. Services patronized by a small number of locals can coexist in a neighborhood ("mom-and-pop stores") along with services that attract many from throughout the settlement.

This difference is vividly demonstrated by comparing the yellow pages for a small settlement with those for a major city. The major city's yellow pages are thick with more services, and diverse headings show widely varied services unavailable in small settlements.

We spend as little time and effort as possible in obtaining consumer services and thus go to the nearest place that fulfills our needs. There is no point in traveling to a distant department store if the same merchandise is available at a nearby one. We travel greater distances only if the price is much lower or if the item is unavailable locally.

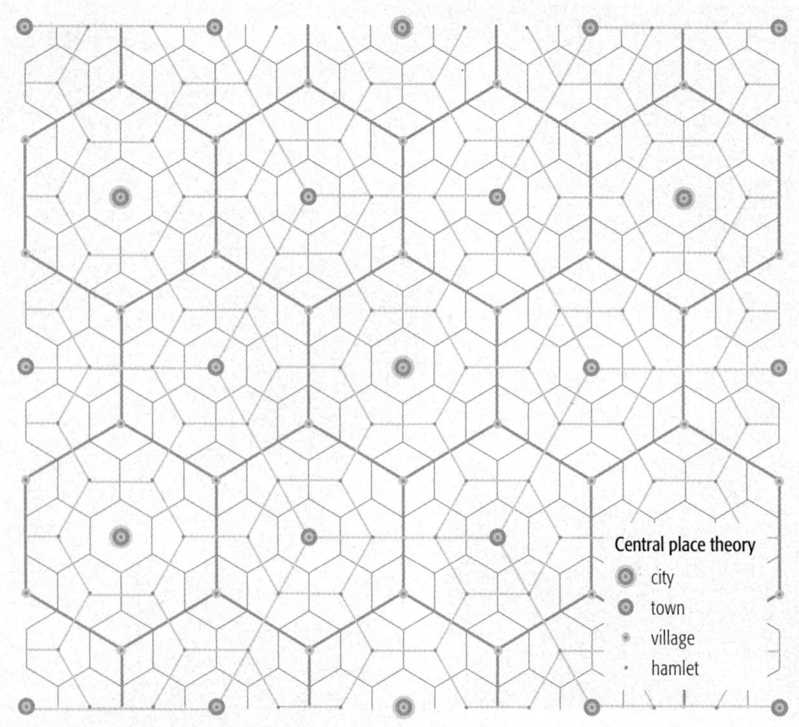

12.8.1 **NESTING OF SETTLEMENTS AND SERVICES.**
According to central place theory, market areas are arranged in a regular pattern. Larger market areas, based in larger settlements, are fewer in number and farther apart from each other than are smaller market areas and settlements. However, larger settlements also provide goods and services for smaller market areas; consequently, larger settlements have both larger and smaller market areas drawn around them.

NESTING OF SERVICES AND SETTLEMENTS

According to central place theory, market areas across an MDC would be a series of hexagons of various sizes, unless interrupted by physical features such as mountains and bodies of water. MDCs have numerous small settlements with small thresholds and ranges, and far fewer large settlements with large thresholds and ranges.

The nesting pattern can be illustrated with overlapping hexagons of different sizes. Four different levels of market area—for hamlet, village, town, and city—are shown in Figure 12.8.1. Hamlets with very small market areas are represented by the smallest contiguous hexagons. Larger hexagons represent the market areas of larger settlements and are overlaid on the smaller hexagons, because consumers from smaller settlements shop for some goods and services in larger settlements.

In his original study, Walter Christaller showed that the distances between settlements in southern Germany followed a regular pattern. He identified seven sizes of settlements (market hamlet, township center, county seat, district city, small state capital, provincial head capital, and regional capital city). For example, the smallest (market hamlet) had an average population of 800 and a market area of 45 square kilometers (17 square miles). The average distance between market hamlets was 7 kilometers (4.4 miles). The figures were higher for the average settlement at each increasing level in the hierarchy. Brian Berry has documented a similar hierarchy of settlements in parts of the U.S. Midwest.

RANK-SIZE DISTRIBUTION OF SETTLEMENTS

In many MDCs, geographers observe that ranking settlements from largest to smallest (population) produces a regular pattern or hierarchy. This is the **rank-size rule,** in which the country's nth-largest settlement is 1/n the population of the largest settlement. In other words, the second-largest city is one-half the size of the largest, the fourth-largest city is one-fourth the size of the largest, and so on. When plotted on logarithmic paper, the rank-size distribution forms a fairly straight line. The distribution of settlements closely follows the rank-size rule in the United States and a handful of other countries.

If the settlement hierarchy does not graph as a straight line, then the society does not have a rank-size distribution of settlements. Several MDCs in Europe follow the rank-size distribution among smaller settlements, but not among the largest ones. Instead, the largest settlement in these countries follows the **primate city rule.** According to the primate city rule, the largest settlement has more than twice as many people as the second-ranking settlement. In this distribution, the country's largest city is called the **primate city.**

The existence of a rank-size distribution of settlements is not merely a mathematical curiosity. It has a real impact on the quality of life for a country's inhabitants. A regular hierarchy—as in the United States—indicates that the society is sufficiently wealthy to justify the provision of goods and services to consumers throughout the country. Conversely, the absence of the rank-size distribution in an LDC indicates that there is not enough wealth in the society to pay for a full variety of services.

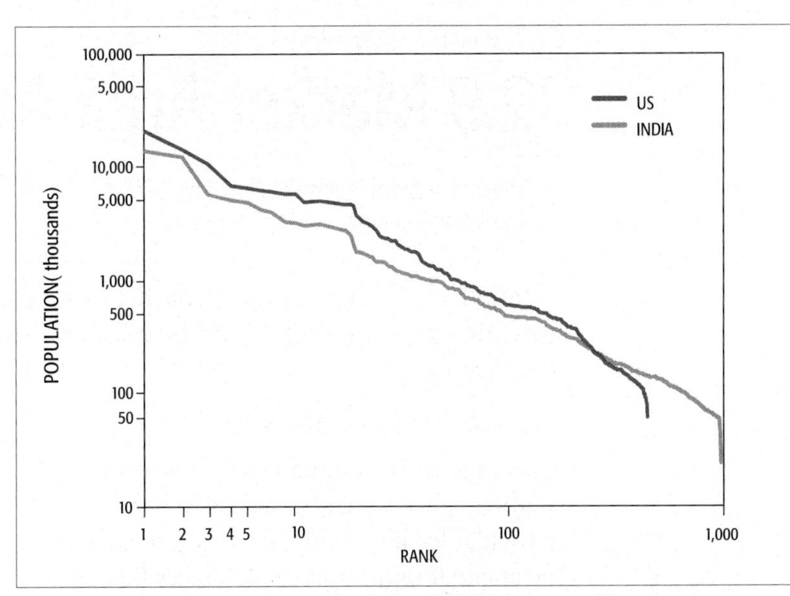

12.8.2 **RANK-SIZE DISTRIBUTION OF SETTLEMENTS IN THE UNITED STATES AND INDIA**

12.8.3 **TWO URBAN AREAS WITH APPROXIMATELY 2.5 MILLION INHABITANTS**
Jaipur in India, and Baltimore in the United Staes (inset).

CONSUMER SERVICES

12.9 Market Area Analysis

- Retailers determine profitability of a site by calculating the range and threshold.
- Site selection is done by placing consumer behavior data on GIS.

Retailers and other service providers make use of market-area studies to determine if locating in the market would be profitable and where within the market area the best location would be.

PROFITABILITY OF A LOCATION

The range and threshold together determine whether a service can be profitable in a particular location. The best location will be the one that minimizes the distances that all potential customers must travel to reach the service.

To illustrate, consider this: Would a convenience store be profitable in your community? First, compute the range—the maximum distance people are willing to travel. You might survey local residents and determine that people are generally willing to travel up to 15 minutes to reach a convenience store.

Then, compute the threshold. Suppose a convenience store must sell at least $10,000 worth of goods per week to make a profit, and the average customer spends $2 a week. The store needs at least 5,000 customers each week, spending $2 each, to achieve the break-even sales level of $10,000. If the average customer goes to a convenience store once a week, the threshold in this example would be 5,000.

Finally, on a map, draw a circle around your community with a 15-minute travel radius, adjusting the boundaries to account for any competitors. Count the number of people within the irregularly shaped circle. If more than 5,000 people are within the radius, then the threshold may be high enough to justify locating the new convenience store in your community. However, your store may need a larger threshold and range to attract some of the available customers if competitors are located nearby.

The threshold must also be adjusted to the fact that the further customers are from the service the less likely they are to patronize it. Geographers have adapted the gravity model from physics. The **gravity model** predicts that the optimal location of a service is directly related to the number of people in the area and inversely related to the distance people must travel to access it.

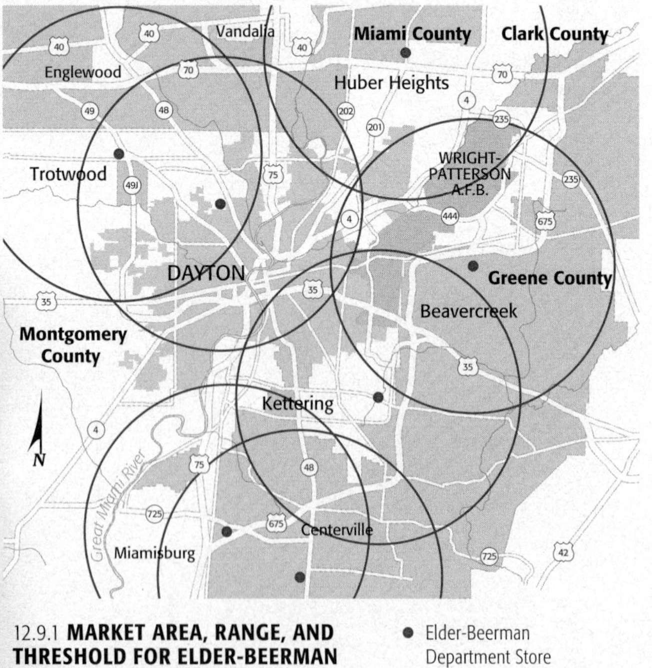

12.9.1 **MARKET AREA, RANGE, AND THRESHOLD FOR ELDER-BEERMAN DEPARTMENT STORES IN THE DAYTON, OHIO, METROPOLITAN AREA.**

- ● Elder-Beerman Department Store
- ○ market area of Elder-Beerman Department Store
- household income of $40,000 and above

LOCATING A NEW DEPARTMENT STORE

Major U.S. department store chains, mall developers, and other large retailers employ geographers to determine the best locations to build new stores. A large retailer has many locations to choose from when deciding to build new stores. A suitable site is one with the potential for generating enough sales to justify using the company's scarce capital to build it. The role of the geographer is to forecast the sales expected at a proposed new store.

The first step in forecasting sales for a proposed new retail outlet is to define the market or trade area where the store would derive most of its sales. Analysis relies heavily on the company's records of their customers' credit card transactions at existing stores. What are the zip codes of customers who paid by credit card? The market area of a department store is typically defined as the zip codes where two-thirds to three-fourths of the customers live.

Based on the zip codes of credit card customers, geographers estimate that the range for a typical midpriced department store is about a 15-minute driving time. Upscale department stores, such as Nordstrom's and Lord & Taylor, define a larger range, because customers are willing to travel farther to shop there.

In terms of population, the threshold for a typical department store is about 250,000. In other words, a typical department store needs about 250,000 people living within the 15-minute range. For threshold, the amount of money available in the area to spend in department stores is more important than simply the number of people.

If a potential site has enough customers with enough money within the market area, then the geographic analysis proceeds to the second step in estimating sales—market share. The proposed new department store will have to share customers with competitors' department stores. Geographers typically predict market share through the so-called analog method. One or more existing stores are identified in locations that the geographer judges to be comparable to the location of the proposed store. The market share of the comparable stores is applied to the proposed new store.

Information about the viability of a proposed new store is depicted through GIS. One layer of the GIS depicts the trade area of the proposed store. Other layers display characteristics of the people living in the area, such as distribution of households, average income, and competitors' stores. A simplified example in Figure 12.9.1 shows the location of Elder-Beerman department stores in the Dayton, Ohio, metropolitan area, compared to average income. The market areas are smaller in higher-income areas and larger in lower-income areas.

The ability of the retail geographer is judged on the accuracy of the forecasts. After a new store is open for several years, how close to the actual sales were the forecasts that the geographer made several years earlier?

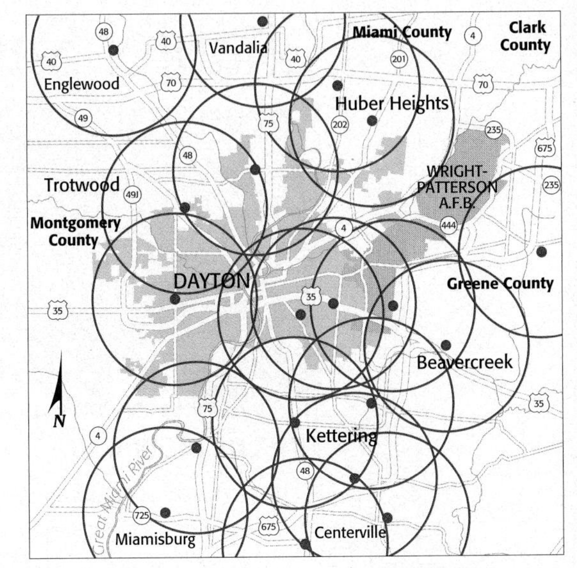

12.9.3 **MARKET AREA, RANGE, AND THRESHOLD FOR KROGER SUPERMARKETS IN THE DAYTON, OHIO, METROPOLITAN AREA.** Fewer stores are in the southwest and northeast, which are predominantly industrial areas, and in the west, which contains lower-income residents

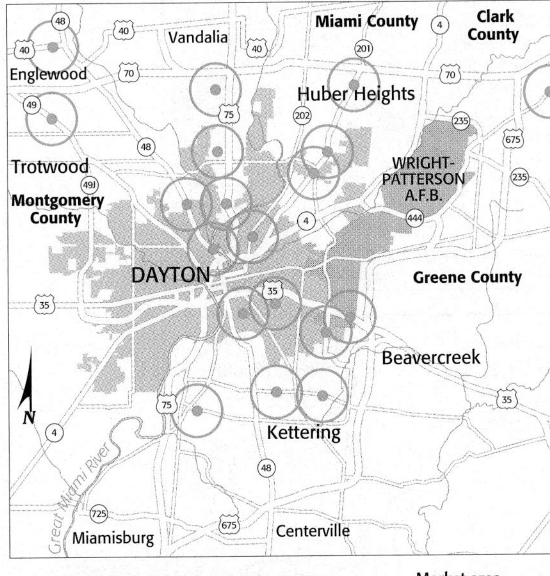

12.9.2 **MARKET AREA, RANGE, AND THRESHOLD FOR UDF CONVENIENCE STORES IN THE DAYTON, OHIO, METROPOLITAN AREA.** Compared to Kroger supermarkets, UDF stores are more numerous and have smaller market areas, ranges, and thresholds.

Chapter **Summary**

CITIES OF THE WORLD

- The first urban settlements predate recorded history.
- LDCs have most of the world's largest cities, but MDCs have higher percentages of urban dwellers.
- The high percentage of urban residents in MDCs is a result of distinctive economic base.

BUSINESS SERVICES

- Job growth in MDCs is in the service sector.
- Three types of services are consumer, business, and public.
- Business services are disproportionately clustered in cities.
- Talented people are attracted to world cities by cultural diversity.

CONSUMER SERVICES

- In MDCs, the distribution of consumer services follows a regular pattern, explained through central place theory.
- Services have market areas, ranges, and thresholds that can be measured.
- Smaller settlements offer services with smaller market areas, ranges, and thresholds.
- Geographers apply central place theory to identify profitable locations for services.

SHARE YOUR VOICE Cosumer Services

Geographic Consequences of Change

Geographers do not merely observe the distribution of services, they play a major role in creating it. Shopping center developers, large department store and supermarket chains, and other retailers employ geographers to identify new sites for stores and assess the performance of existing stores.

Many service providers make location decisions on the basis of instinct, intuition, and tradition. In an increasingly competitive market, retailers and other services that locate in the optimal location secure a critical advantage.

Geographers conduct statistical analyses based on the gravity model to delineate underserved market areas where new stores could be profitable, as well as to identify overserved market areas where poorly performing stores are candidates for closure.

Developers of new retail services obtain loans from banks and financial institutions to construct new stores and malls. Lending institutions want assurance that the proposed retailing has a market area with potential to generate sufficient profits to repay the loan. They employ geographers to make objective market-area analyses often independent of the excessively optimistic forecasts submitted by the retailer.

SHARE YOUR VOICE Student Essay

CHRISTIAN TONEY

Christian Toney
University of Oklahoma

Brief History of the Oklahoma City Economy

"A city set on a hill cannot be hidden" (Matthew 5:14b) depicts a city that is alive and thriving. For a city to flourish, it must be made of successful businesses that attract people of determination to work together and be "the light of the world" (Matthew 5:14a).

Oklahoma City, Oklahoma, was established on April 22, 1889 after the land run. Ten thousand people settled in the area. Not 10 years later, the city more than doubled. Oklahoma City was not the official capital until 1910, after Oklahoma became a state in 1907.

During World War II, most of the people of the United States were struggling to make ends meet. In Oklahoma City, however, some people discovered oil and made a fortune. Unfortunately, most of the Oklahoma population struggled to make a living, but the oil boom aided the financial status of the state of Oklahoma, especially for the present-day Oklahoma City.

It was not until the 1960s that Oklahoma City's oil boom began to plummet, resulting in a loss of jobs and reduced population.. Therefore many recovery attempts were encouraged, but none were successful until the early 1990s when the Metropolitan Area Projects was established. This created a new ball park, library and a remodeled Brick Town, which is an attraction of the city today.

In 1995, some progress was interrupted by the dreadful Oklahoma City bombing. This memory will live on in the hearts and lives of each Oklahoman. In 2000, the Oklahoma City National Memorial was established in remembrance of the lives lost on the tragic April day.

Today Oklahoma City continues to improve its economy with new health agencies, biotechnology, information technology and services. This has expanded many job opportunities for Oklahoma City and the suburban areas. Many shopping centers and parks are becoming attractions for the regenerated area. Oklahoma City officially adopted the OKC Thunder NBA Basketball Team, the team born from the New Orleans Hornets after the 2005 Hurricane Katrina.

Chapter Resources

ECOTOURISM SERVICES AND ECONOMIC DEVELOPMENT

A rapidly growing sector of consumer services is tourism, especially in MDCs where people have enough money and time off work to take holidays in other places.

Global tourism is a multitrillion dollar industry that generates several hundred million jobs.

Tourism is especially important in Latin America, especially for the Caribbean islands close to the United States, which is the source of many of the world's tourists. Caribbean islands attract tourists because of their natural features—wide sandy beaches, balmy weather, and exotic vegetation. Caribbean islands also have a long tradition of serving the needs of consumers in MDCs—most were colonies until recently, and some still are.

Tourism brings consumer service jobs to local residents. These include personal service jobs such as cleaning rooms and leading tour groups, as well as retail service jobs such as operating restaurants and souvenir shops. The economic benefit of tourism is unclear, because wages are low and most of the profits are exported to the transnational corporations that own the resorts. Tourism is also vulnerable to uncertainties such as bad weather and terrorism threats.

Geographers are especially concerned with the adverse impact of tourism on local culture and environment. The arrival of a large number of foreigners from MDCs can overwhelm the capacity of the local environment to provide water and handle waste. Foreign tourists may demand familiar foods and standards of comfort that cannot be accommodated in the local environment. Environmentally sensitive sites may be damaged by trampling visitors.

Local cultural traditions such as dances and clothing become tourist attractions, and once-sacred rites are performed for the amusement of tourists. Local women may be exploited as prostitutes for Western tourists.

Ecotourism is promoted as a way to bring economic benefits while not causing social and environmental damage. Under ecotourism, facilities are constructed to minimize environmental damage and are marketed to tourists seeking to learn about and support environmental sensitivity.

KEY TERMS

Basic industries
Industries that sell their products or services primarily to consumers outside the settlement.

Business services
Services that primarily meet the needs of other businesses, including professional, financial, and transportation services.

Central place
A market center for the exchange of services by people attracted from the surrounding area.

Central place theory
A theory that explains the distribution of services, based on the fact that settlements serve as centers of market areas for services; larger settlements are fewer and farther apart than smaller settlements and provide services for a larger number of people who are willing to travel farther.

Consumer services
Businesses that provide services primarily to individual consumers, including retail services and education, health, and leisure services.

Economic base
A community's collection of basic industries.

Gravity model
A model that holds that the potential use of a service at a particular location is directly related to the number of people in a location and inversely related to the distance people must travel to reach the service.

Market area (or hinterland)
The area surrounding a central place, from which people are attracted to use the place's goods and services.

Nonbasic industries
Industries that sell their products primarily to consumers in the community.

Primate city
The largest settlement in a country, if it has more than twice as many people as the second-ranking settlement.

Primate city rule
A pattern of settlements in a country, such that the largest settlement has more than twice as many people as the second-ranking settlement.

Public services
Services offered by the government to provide security and protection for citizens and businesses.

Range (of a service)
The maximum distance people are willing to travel to use a service.

Rank-size rule
Pattern of settlements in a country, such that the nth largest settlement is 1/n the population of the largest settlement.

Threshold
The minimum number of people needed to support the service.

Urbanization
An increase in the percentage and in the number of people living in urban settlements.

ON THE INTERNET

A number of organizations publish lists of the population of cities, but these are all based on officially published national statistics. One such site is created by Thomas Brinkoff, a professor at the University of Oldenburg (Carl von Ossietzky Universität Oldenburg) in Germany, accessed at **http://www.citypopulation.de/World.html**.

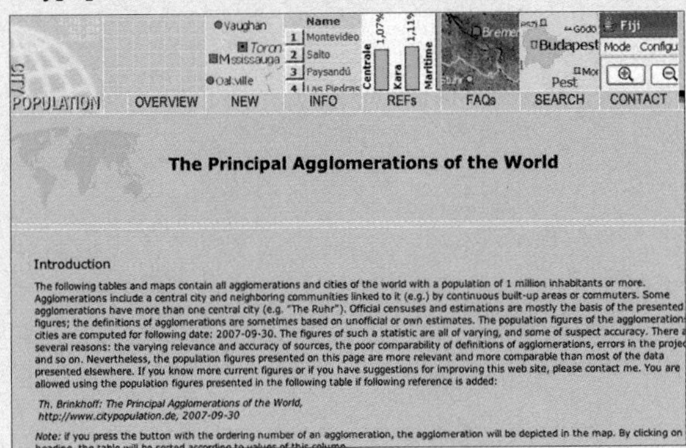

An attempt to apply a uniform definition to measuring the population of cities everywhere has been published by *Demographia*, accessed at **www.demographia.com**. Wendell Cox, a private consultant, is the sole owner of *Demographia*.

FURTHER READINGS

Archer, Clark J., and Ellen R. White. "A Service Classification of American Metropolitan Areas." *Urban Geography* 6 (1985): 122–51.

Bagchi-Sen, Sharmistha. "Service Employment in Large, Medium, and Small Metropolitan Areas in the United States." *Urban Geography* 18 (1997): 264–81.

Beaverstock, J.V., P. Taylor and R.G. Smith, "A Roster of World Cities," *Cities* 16 (1999): 445-58.

Berry, Brian J. L. *The Geography of Market Centers and Retail Distribution.* Englewood Cliffs, NJ: Prentice Hall, 1967.

Christaller, Walter. *The Central Places of Southern Germany.* Englewood Cliffs, NJ: Prentice Hall, 1966.

Crewe, Louise. "Geographies of Retailing and Consumption." *Progress in Human Geography* 24 (2000): 275–90.

Daniels, P. W. *Service Industries: A Geographical Appraisal.* London: Methuen, 1986.

Florida, Richard. "The Economic Geography of Talent." *Annals of the Association of American Geographers* 92 (2002): 743–55.

Keeble, D., and L. Nachum. "Why Do Business Service Firms Cluster? Small Consultancies, Clustering and Decentralization in London and Southern England." *Transactions of the Institute of British Geographers, New Series* 27 (2002): 67–90.

King, Leslie J. *Central Place Theory.* Beverly Hills, CA: Sage, 1984.

Kirn, Thomas J. "Growth and Change in the Service Sector of the U.S.: A Spatial Perspective." *Annals of the Association of American Geographers* 77 (1987): 353–72.

Lord, J. Dennis. "Retail Saturation: Inevitable or Irrelevant?" *Urban Geography* 21 (2000): 342–60.

Lösch, August. *The Economics of Location.* New Haven, CT: Yale University Press, 1954.

Marshall, John U. "Beyond the Rank-Size Rule: A New Descriptive Model of City Sizes." *Urban Geography* 18 (1997): 36–55.

Ó hUallacháin, Breandan, and Neil Reid. "The Location and Growth of Business and Professional Services in American Metropolitan Areas, 1976–1986." *Annals of the Association of American Geographers* 81 (1991): 254–70.

Scott, Peter. *Geography and Retailing.* London: Hutchinson University Press, 1970.

Taylor, P. J., G. Catalano, and D. R. F. Walker. "Measurement of the World City Network." *Urban Studies* 39 (2002): 2367–76.

Chapter 13 URBAN PATTERNS

When you stand at the corner of Fifth Avenue and 34th Street in New York City, staring up at the Empire State Building, you know that you are in a city. When you are standing in an Iowa cornfield, you have no doubt that you are in the country. Geographers help explain what makes city and countryside different places.

A large city is stimulating and agitating, entertaining and frightening, welcoming and cold. A city has something for everyone, but a lot of those things are for people different from you. Urban geography helps to sort out the complexities of familiar and unfamiliar patterns in urban areas. Models help to explain where different people and activities are distributed within urban areas, and why those differences occur.

EVENING RUSH HOUR ON I-405 IN LOS ANGELES

THE KEY ISSUES IN THIS CHAPTER

THE URBAN AREA

13.1 The Central Business District

13.2 Defining Urban Settlements

13.3 Fragmented Government

PATTERNS WITHIN URBAN AREAS

13.4 Models of Internal Structure

13.5 Social Area Analysis

13.6 Urban Patterns Outside the United States

CITY AND SUBURB

13.7 Suburbanization

13.8 Inner-City Decline and Renewal

13.9 Urban Transportation

THE URBAN AREA

13.1 The Central Business District

- Downtown is known as the central business district (CBD).
- The CBD contains consumer, business, and public services.

The best-known and most visually distinctive area of most cities is the central area, commonly called downtown and known to geographers by the more precise term **central business district (CBD)**. The CBD is usually one of the oldest districts in a city, often the original site of the settlement.

Consumer, business, and public services are attracted to the CBD because of its accessibility. The center is the easiest part of the city to reach from the rest of the region and is the focal point of the region's transportation network.

CONSUMER SERVICES

Consumer services in the CBD serve the many people who work in the center and shop during lunch or working hours. These businesses sell office supplies, computers, and clothing, or offer shoe repair, rapid photocopying, dry cleaning, and so on.

Large department stores once clustered in the CBD, often across the street from one another, but they have relocated to suburban malls. In several CBDs, new shopping areas attract suburban shoppers as well as out-of-town tourists with unique recreation and entertainment experiences.

BUSINESS SERVICES

Even with modern telecommunications, many professionals still exchange information primarily through face-to-face contact. Business services, such as advertising, banking, finance, journalism, and law, are centrally located to facilitate rapid communication of fast-breaking news. Face-to-face contact also helps to establish a relationship of trust based on shared professional values.

People in some business services depend on proximity to professional colleagues. Lawyers, for example, locate to be near government offices and courts. Services such as temporary secretarial agencies and instant printers locate downtown to be near lawyers, forming a chain of interdependency that continues to draw offices to the center city.

Extreme competition for limited building sites results in very high land values in the CBD. Because of its high value, land is used more intensively in the center than elsewhere in the city.

Compared to other parts of the city, the central area uses more space below and above ground level. Beneath most central cities runs a vast underground network of garages, loading docks, utilities, walkways, and transit lines. Demand for space in the central city has also made high-rise structures economically feasible.

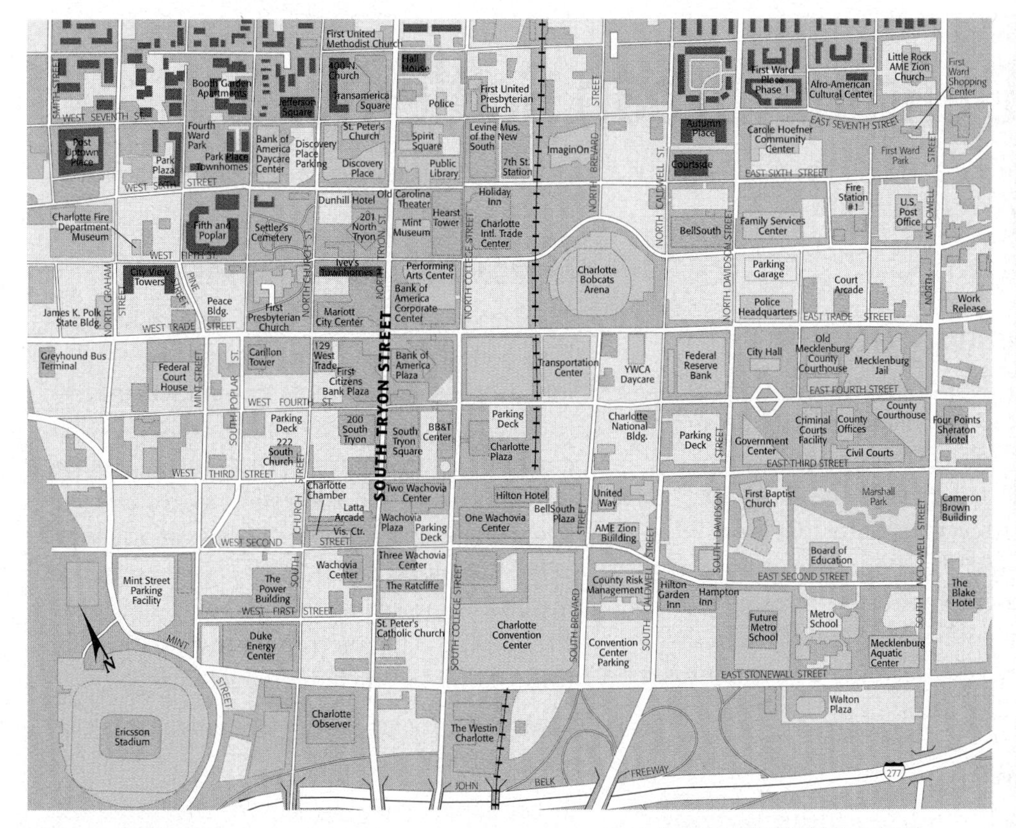

BUSINESS SERVICES

PUBLIC SERVICES

CONSUMER SERVICES

PUBLIC SERVICES

Public services typically located downtown include City Hall, courts, and libraries. These facilities cluster in the CBD to facilitate access for people living in all parts of town.

Sports facilities and convention centers have been constructed or expanded downtown in many cities. These structures attract a large number of people, including many suburbanites and out-of-towners. Cities place these facilities in the CBD because they hope to stimulate more business for downtown restaurants, bars, and hotels.

13.1.1 **CBD OF CHARLOTTE, NORTH CAROLINA** (pictured below)
Charlotte's CBD is dominated by public services, such as City Hall, government offices, and post office. Consumer and business services are clustered along South Tryon Street.

- public and semipublic
- commercial
- parking and other paved areas
- industry and warehouse
- residential
- open areas

0 250 500 Feet

THE URBAN AREA

13.2 Defining Urban Settlements

- Urban places can be defined legally as cities, or as urbanized or metropolitan areas.
- Urban growth has caused adjacent metropolitan areas to overlap.

Urban places can be defined in three ways.

LEGAL DEFINITION OF CITY

The term *city* defines an urban settlement that has been legally incorporated into an independent, self-governing unit. A city has locally elected officials, the ability to raise taxes, and responsibility for providing essential services. The boundaries of the city define the geographic area within which the local government has legal authority. In the United States, a city surrounded by suburbs is sometimes called a *central city.*

URBANIZED AREA

With the rapid growth of settlements, many residents live in suburbs, beyond the boundaries of the central city. In the United States, the central city and the surrounding built-up suburbs are called an **urbanized area**. More precisely, an urbanized area consists of a central city plus its contiguous built-up suburbs where population density exceeds 1,000 persons per square mile (400 persons per square kilometer). Approximately 70 percent of Americans live in urbanized areas, including about 30 percent in central cities and 40 percent in surrounding jurisdictions.

Working with urbanized areas is difficult because few statistics are available about them. Most data in the United States and other countries are collected for cities, counties, and other local government units, but urbanized areas do not correspond to government boundaries.

13.2.1 **ST. LOUIS CITY, URBANIZED AREA, AND METROPOLITAN STATISTICAL AREA**

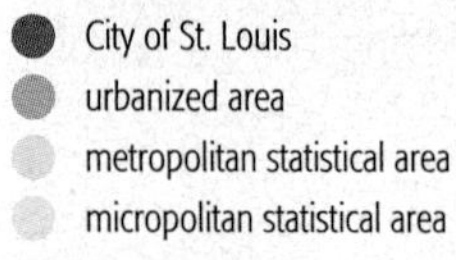

City of St. Louis
urbanized area
metropolitan statistical area

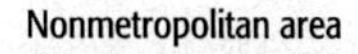

micropolitan statistical area

Nonmetropolitan area

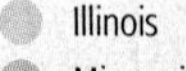

Illinois
Missouri

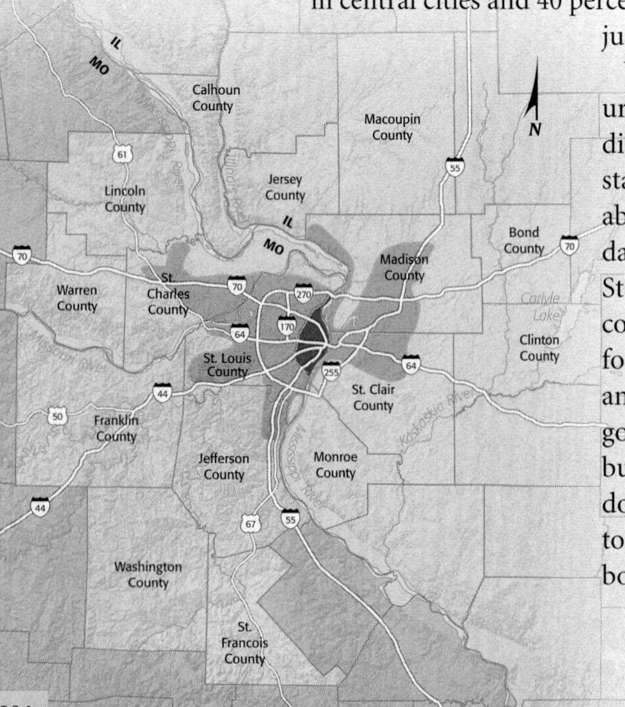

METROPOLITAN STATISTICAL AREA

The area of influence of a city extends beyond legal boundaries and adjacent built-up jurisdictions. For example, commuters may travel a long distance to work and shop in the city or built-up suburbs. People in a wide area watch the city's television stations, read the city's newspapers, and support the city's sports teams.

The U.S. Bureau of the Census has created a method of measuring the functional area of a city, known as the **metropolitan statistical area (MSA)**. An MSA includes the following:

1. An urbanized area with a population of at least 50,000.
2. The county within which the city is located.
3. Adjacent counties with a high population density and a large percentage of residents working in the central city's county (e.g., a county with a density of 25 persons per square mile and at least 50 percent working in the central city's county).

The census has also designated smaller urban areas as **micropolitan statistical areas.** These include an urbanized area of between 10,000 and 50,000 inhabitants, the county in which it is found, and adjacent counties tied to the city. The United States had 560 micropolitan statistical areas in 2003, for the most part found around southern and western communities previously considered rural in character. About 10 percent of Americans live in a micropolitan statistical area.

OVERLAPPING METROPOLITAN AREAS

Some adjacent MSAs overlap. A county between two central cities may send a large number of commuters to jobs in each. In the northeastern United States, large metropolitan areas are so close together that they now form one continuous urban complex, extending from north of Boston to south of Washington, D.C. Geographer Jean Gottmann named this region Megalopolis, a Greek word meaning "great city"; others have called it the BosWash Corridor.

Other continuous urban complexes exist in the United States—the southern Great Lakes between Chicago and Milwaukee on the west and Pittsburgh on the east and southern California from Los Angeles to Tijuana. Among important examples in other MDCs are the German Ruhr (including the cities of Dortmund, Düsseldorf, and Essen), Randstad in the Netherlands (including the cities of Amsterdam, the Hague, and Rotterdam), and Japan's Tokaido (including the cities of Tokyo and Yokohama).

Within Megalopolis, the downtown areas of individual cities such as Baltimore, New York, and Philadelphia retain distinctive identities, and the urban areas are visibly separated from each other by open space used as parks, military bases, and dairy or truck farms. But at the periphery of the urban areas, the boundaries overlap. Washingtonians visit the Inner Harbor in downtown Baltimore, and Baltimoreans attend major-league hockey and basketball games in downtown Washington.

13.2.2 **MEGALOPOLIS**
Megalopolis extends more than 700 kilometers (440 miles) between Boston and Washington, D.C. The region contains one-fourth of the U.S. population on 2 percent of the country's total land area.

Boston

New York

Baltimore

Philadelphia

13.2.3 **CBDs OF THE FOUR LARGEST CITIES IN MEGALOPOLIS**

THE URBAN AREA

13.3 Fragmented Government

- Cities traditionally grew through annexation.
- Today, most urban areas have a large number of local governments.

Urban problems, such as traffic, solid-waste disposal, and the building of affordable housing, are regional in scale. But the fragmentation of local government in the United States makes it difficult to address regional problems.

ANNEXATION

As U.S. cities grew, they traditionally expanded through **annexation**, the process of legally adding land area to a city. Rules concerning annexation vary among states. Normally, land can be annexed into a city only if a majority of residents in the affected area vote in favor of doing so.

Peripheral residents generally desired annexation in the nineteenth century because the city offered better services, such as water supply, sewage disposal, trash pickup, paved streets, public transportation, and police and fire protection. Thus, as U.S. cities grew rapidly in the nineteenth century, the legal boundaries frequently changed to accommodate newly developed areas. For example, the city of Chicago expanded from 26 square kilometers (10 square miles) in 1837 to 492 square kilometers (190 square miles) in 1900.

Today, however, cities are less likely to annex peripheral land because the residents prefer to organize their own services rather than pay city taxes for them. As a result, today's cities are surrounded by a collection of suburban jurisdictions, whose residents prefer to remain legally independent of the large city. The number of local governments exceeds 1,400 in the New York area, 1,100 in the Chicago area, and 20,000 throughout the United States.

Originally, some of these peripheral jurisdictions were small, isolated towns that had a tradition of independent local government before being absorbed by urban growth. Others are newly created communities whose residents wish to live close to the large city but not legally be part of it.

Long Island, which extends for 150 kilometers (90 miles) east of New York City and is approximately 25 kilometers (15 miles) wide, contains nearly 800 local governments. The island includes two counties, two cities, 13 towns, 95 villages, 127 school districts, and more than 500 special districts (such as for garbage collection).

The multiplicity of local governments on Long Island leads to problems. When police or firefighters are summoned to the State University of New York at Old Westbury, two or three departments sometimes respond because the campus is in five districts. The boundary between the communities of Mineola and Garden City runs down the center of Old Country Road, a busy four-lane route. Mineola set a 40-mile-per-hour speed limit for the eastbound lanes, whereas Garden City set a 30-mile-per-hour speed limit for the westbound lanes.

13.3.1 **A DETROIT SUBURB: GROSSE POINTE CITY**
Grosse Pointe is 97 percent white, and household income is around twice the national average.

13.3.2 **LOCAL GOVERNMENTS IN THE DETROIT METROPOLITAN AREA.** The large number of local governments makes it difficult to plan at a regional scale to reduce income disparity and racial segregation.

METROPOLITAN GOVERNMENT

The large number of local government units has led to calls for a metropolitan government that could coordinate—if not replace—the numerous local governments in an urban area. Strong metropolitan-wide governments have been established in a few places in North America. Two kinds exist:

- **Federations.** Toronto has a metropolitan government created in 1953 through federation of 13 municipalities. Canada's other largest cities also have varying forms of regional government federations.
- **Consolidations.** Indianapolis and Miami are examples of U.S. urban areas that have consolidated city and county governments. Government functions that were once handled separately by The City of Indianapolis and Marion County are combined into a joint operation in the same office building. In Florida, the city of Miami and surrounding Dade County have combined some services, but the city boundaries have not been changed to match those of the county.

13.3.3 **THE CITY OF DETROIT** Detroit is 83 percent African American, and household income is around one-half the national average.

PATTERNS WITHIN URBAN AREAS

13.4 Models of Internal Structure

- Three models of urban structure have been developed.
- These models argue that cities grow in rings, wedges, and nodes.

Sociologists, economists, and geographers have developed three models to help explain where different types of people tend to live in an urban area—the concentric zone, sector, and multiple nuclei models.

The three models describing the internal social structure of cities were developed in Chicago, a city on a prairie. Except for Lake Michigan to the east, few physical features have interrupted the region's growth. Chicago includes a CBD known as the Loop, because elevated railway lines loop around it. Surrounding the Loop are residential suburbs to the south, west, and north. The three models were later applied to cities elsewhere in the United States and in other countries.

CONCENTRIC ZONE MODEL

According to the **concentric zone model**, created in 1923 by sociologist E. W. Burgess, a city grows outward from a central area in a series of five concentric rings, like the growth rings of a tree.

- The innermost zone is the CBD, where nonresidential activities are concentrated.
- A second ring, the zone in transition, contains industry and poorer-quality housing. Immigrants to the city first live in this zone in small dwelling units, frequently created by subdividing larger houses into apartments.
- The third ring, the zone of working-class homes, contains modest older houses occupied by stable, working-class families.
- The fourth zone has newer and more spacious houses for middle-class families.
- A commuters' zone beyond the continuous built-up area of the city is inhabited by people who work in the center but choose to live in dormitory towns for commuters.

13.4.1 **CONCENTRIC ZONE MODEL**
According to the model, a city grows in a series of rings that surround the CBD.

SECTOR MODEL

According to the **sector model**, developed in 1939 by land economist Homer Hoyt, the city develops in a series of sectors. As a city grows, activities expand outward in a wedge, or sector, from the center. Hoyt mapped the highest-rent areas for a number of U.S. cities at different times and showed that the highest social-class district usually remained in the same sector, although it moved farther out along that sector over time.

Once a district with high-class housing is established, the most expensive new housing is built on the outer edge of that district, farther out from the center. The best housing is therefore found in a corridor extending from downtown to the outer edge of the city. Industrial and retailing activities develop in other sectors, usually along good transportation lines.

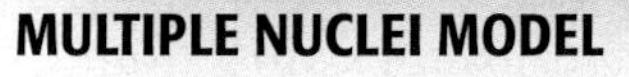

MULTIPLE NUCLEI MODEL

Geographers C. D. Harris and E. L. Ullman developed the **multiple nuclei model** in 1945. According to the multiple nuclei model, a city is a complex structure that includes more than one center around which activities revolve. Examples of these nodes include a port, neighborhood business center, university, airport, and park.

The multiple nuclei theory states that some activities are attracted to particular nodes, whereas others try to avoid them. For example, a university node may attract well-educated residents, pizzerias, and bookstores, whereas an airport may attract hotels and warehouses. On the other hand, incompatible land-use activities will avoid clustering in the same locations. Heavy industry and high-class housing, for example, rarely exist in the same neighborhood.

13.4.2 **SECTOR MODEL**
According to the model, a city grows in a series of wedges or corridors, which extend out from the CBD.

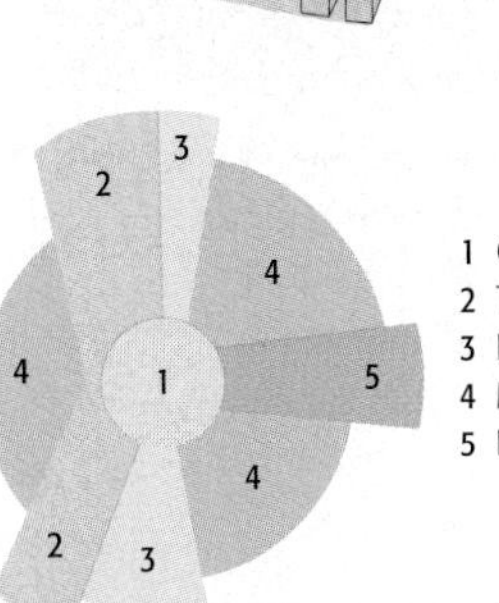

13.4.3 **MULTIPLE NUCLEI MODEL**
According to the model, a city consists of numerous nodes or centers, around which different types of people and activities cluster.

1 Central business district
2 Wholesale, light manufacturing
3 Low-class residential
4 Medium-class residential
5 High-class residential
6 Heavy manufacturing
7 Outlying business district
8 Residential suburb
9 Industrial suburb

PATTERNS WITHIN URBAN AREAS

13.5 Social Area Analysis

- The three models together explain where people live within cities.
- Census data can be used to map the distribution of social characteristics.

The three models help us understand where people with different social characteristics tend to live within an urban area. They can also help to explain why certain types of people tend to live in particular places.

Effective use of the models depends on the availability of data at the scale of individual neighborhoods. In the United States and many other countries, that information comes from a national census. Urban areas in the United States are divided into **census tracts** that contain approximately 5,000 residents and correspond, where possible, to neighborhood boundaries.

Every decade the U.S. Bureau of the Census publishes data summarizing the characteristics of the residents living in each tract. Examples of information the bureau publishes include the number of nonwhites, the median income of all families, and the percentage of adults who finished high school.

The spatial distribution of any of these social characteristics can be plotted on a map of the community's census tracts. Computers have become invaluable in this task because they permit rapid creation of maps and storage of voluminous data about each census tract. Social scientists can compare the distributions of characteristics and create an overall picture of where various types of people tend to live. This kind of study is known as social area analysis.

None of the three models taken individually completely explain why different types of people live in distinctive parts of the city. Critics point out that the models are too simple and fail to consider the variety of reasons that lead people to select particular residential locations. Because the three models are all based on conditions that existed in U.S. cities between the two world wars, critics also question their relevance to contemporary urban patterns in the United States or in other countries.

But if the models are combined rather than considered independently, they help geographers

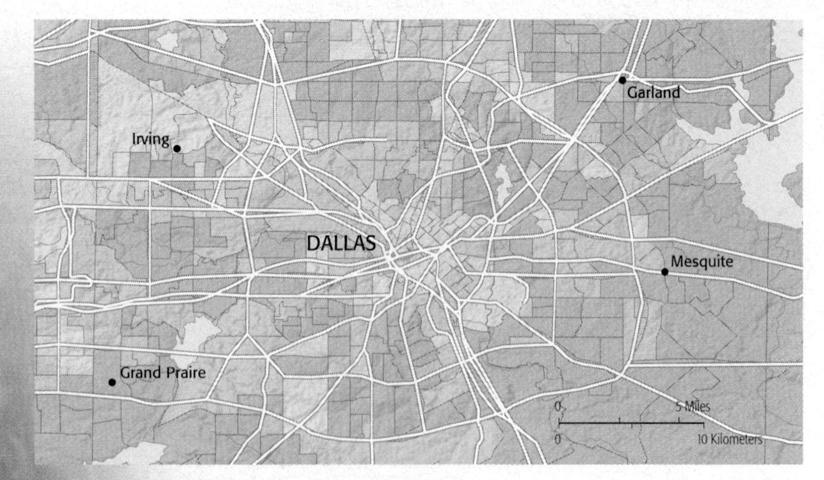

13.5.1 **EXAMPLE OF CONCENTRIC ZONE MODEL IN DALLAS**
The distribution of owner-occupants. The percentage of households that own their home is lower near the CBD and greater in the outer rings.

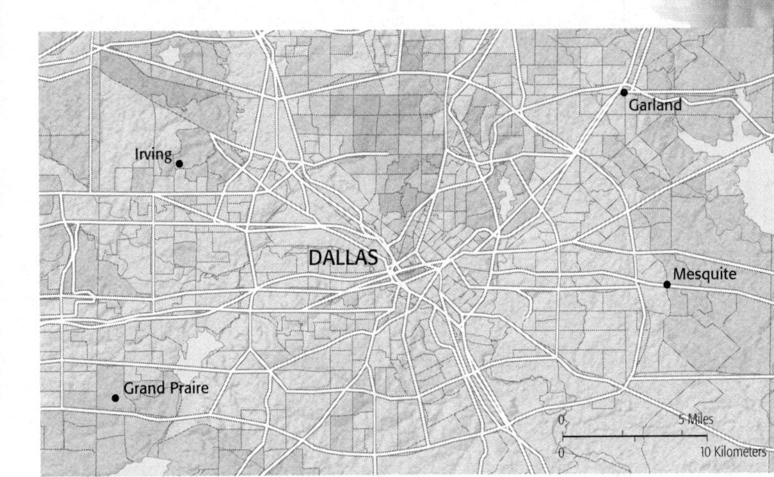

13.5.2 **EXAMPLE OF SECTOR MODEL IN DALLAS**
The distribution of high-income households. The median household income is the highest in a sector to the north.

Median household income 2000
- $75,000 and above
- $50,000 to $74,999
- below $50,000

explain where different types of people live in a city. People tend to reside in certain locations depending on their particular personal characteristics. This does not mean that everyone with the same characteristics must live in the same neighborhood, but the models say that most people prefer to live near others who have similar characteristics:

- Consider two families with the same income and ethnic background. One family owns its home, whereas the other rents. The concentric zone model suggests that the owner-occupant is much more likely to live in an outer ring and the renter in an inner ring.
- The sector theory suggests that given two families who own their homes, the family with the higher income will not live in the same sector of the city as the family with the lower income.
- The multiple nuclei theory suggests that people with the same ethnic or racial background are likely to live near each other.

Putting the three models together, we can identify, for example, the neighborhood in which a high-income, Asian American owner-occupant is most likely to live.

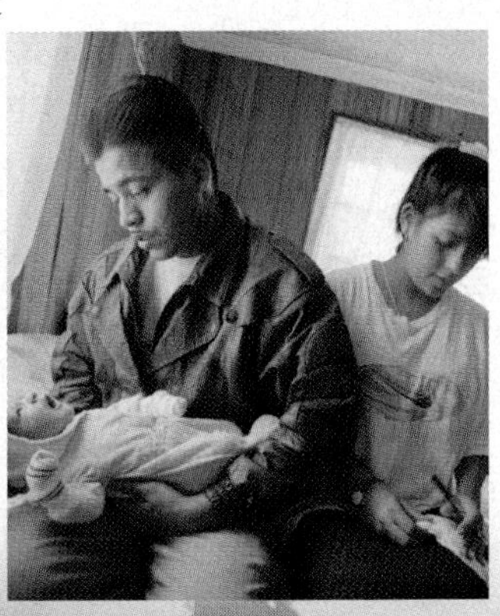

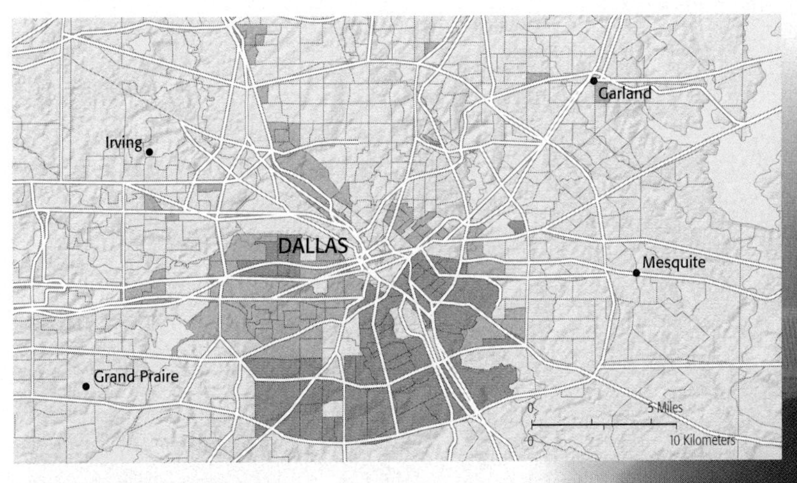

13.5.3 **EXAMPLE OF MULTIPLE NUCLEI MODEL IN DALLAS**
The distribution of minorities. The African American population is concentrated in a node to the south, and the Hispanic population in a node to the west.

- more than 60% African American
- more than 60% Hispanic
- other

PATTERNS WITHIN URBAN AREAS

13.6 Urban Patterns Outside the United States

- Downtowns are often very animated.
- Poor people are more likely to live in suburbs.

The three models may describe the spatial distribution of social classes in the United States, but American urban areas differ from those elsewhere in the world. These differences do not invalidate the models, but they do point out that social groups in other countries may not have the same reasons for selecting particular neighborhoods within their cities.

LESS DEVELOPED COUNTRIES

Three stages of development have influenced the structure of many cities in LDCs.

- **Precolonial cities.** Cities were often laid out surrounding a religious core, such as a mosque or a temple. For example, the Aztec city of Tenochtitlan, founded on the site of Mexico City, contained elaborate stone houses surrounding a massive Great Temple and a market center.
- **Colonial cities.** When Europeans gained control of Africa, Asia, and Latin America, they built new cities and demolished indigenous ones, including Tenochtitlan. The new European cities were laid out according to a standardized plan, with large public squares and wide streets.
- **Cities since independence.** Following independence, cities in LDCs have grown rapidly, because of a combination of a high natural increase rate and immigration from rural areas.

In LDCs, the poor are more likely to live in the suburbs, whereas the wealthy live near the center of cities, as well as in a sector extending from the center. Many of these poor suburban areas are **squatter settlements**. The United Nations estimated that 175 million people worldwide lived in squatter settlements in 2003.

Squatter settlements have few services because neither the city nor the residents can afford them. Latrines are usually designated by the settlement's leaders, and water is carried from a central well or dispensed from a truck. The settlements generally lack schools, paved roads, telephones, or sewers. Electricity service may be stolen by running a wire from the nearest power line. In the absence of bus service or available private cars, a resident may have to walk 2 hours to reach a place of employment.

Physical geography influences the location of squatter settlements. Mexico City's squatter settlements are located on either unstable hard-to-reach mountainsides or the equally unstable drained bed of Lake Texcoco.

13.6.1 SOCIAL AREAS IN MEXICO CITY
Higher-income householdss are more likely to live in the center of Mexico City, such as the Zona Rose (top). Lower-income households are more likely to live in suburbs, often in squatter settlements, such as below.

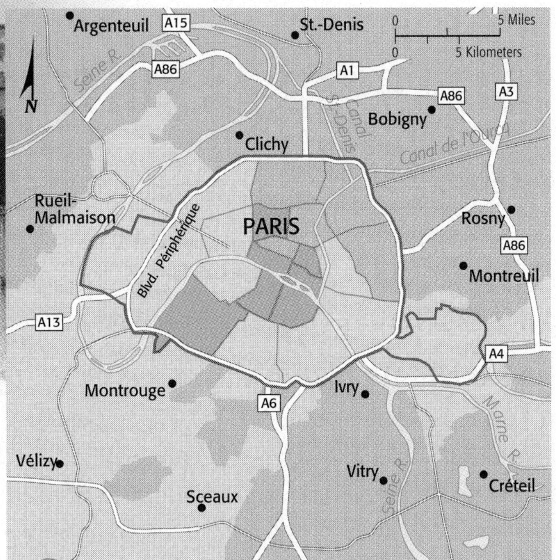

13.6.2 **SOCIAL AREAS IN PARIS, FRANCE**
Professionals with higher incomes are more likely to live in the center of Paris, such as the houses to the left. Factory workers with lower incomes and immigrants are more likely to live in suburbs, often in high-rise apartments, such as below.

- 25% or more top management
- 9% or more factory workers
- other areas

EUROPEAN CITIES

As in the United States, wealthier people in European cities cluster along a sector extending out from the CBD. In contrast to most U.S. cities, wealthy Europeans still live in the inner rings of the upper-class sector, not just in the suburbs. A central location provides proximity to the region's best shops, restaurants, cafés, and cultural facilities. Wealthy people are also attracted by the opportunity to occupy elegant residences in carefully restored, beautiful old buildings.

In the past, low-income people also lived in the center of European cities. Before the invention of electricity in the nineteenth century, social segregation was vertical: wealthier people lived on the first or second floors, whereas poorer people occupied the dark, dank basements, or they climbed many flights of stairs to reach the attics. Today, low-income people are less likely to live in European inner-city neighborhoods. Poor-quality housing has been renovated for wealthy people or demolished and replaced by offices or luxury apartment buildings.

People with lower incomes have been relegated to the outskirts of European cities. Vast suburbs containing dozens of high-rise apartment buildings house these people who were displaced from the inner city. European suburban residents face the prospect of long commutes by public transportation to reach jobs and other downtown amenities. Shops, schools, and other services are worse than in inner neighborhoods, and the suburbs are centers for crime, violence, and drug dealing. Because the housing is mostly in high-rise buildings, people lack large private yards. Many residents of these dreary suburbs are persons of color or recent immigrants from Africa or Asia who face discrimination and prejudice by Europeans.

CITY AND SUBURB

13.7 Suburbanization

- Suburbs sprawl outside American cities.
- Retailing as well as housing has grown in suburbs.

Most Americans live in suburbs. Public opinion polls in the United States and Western Europe show people's strong desire for suburban living. In most polls, more than 90 percent of respondents prefer the suburbs to the inner city.

SUBURBANIZATION OF RETAILING

Suburban residential growth has fostered change in traditional retailing patterns. Historically, urban residents bought food and other daily necessities at small neighborhood shops in the midst of housing areas and shopped in the CBD for other products. CBD sales have stagnated because suburban residents won't make the long journey there.

Instead, retailing has been increasingly concentrated in planned suburban shopping malls surrounded by generous parking lots. Malls have become centers for activities in suburban areas. Retired people go to malls for safe, vigorous walking exercise, or they sit on a bench to watch the passing scene. Teenagers arrive after school to meet their friends.

A shopping mall is built by a developer, who buys the land, builds the structures, and leases space to individual merchants. The key to a successful large shopping mall is the inclusion of one or more anchors. The anchors may be a supermarket and discount store in a smaller mall and several department stores in a larger mall. Most consumers go to a mall to shop at an anchor and, while there, patronize the smaller shops.

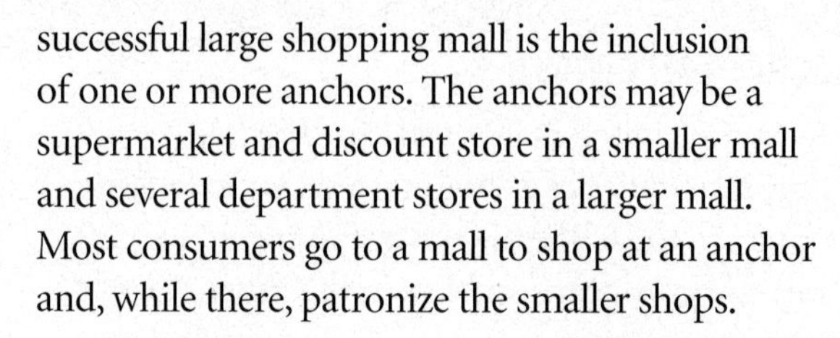

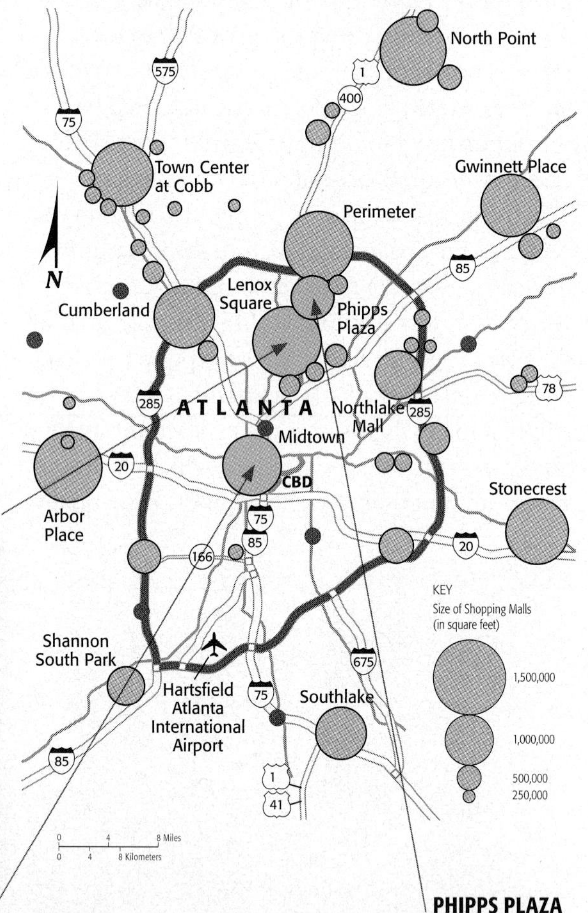

13.7.1 MAJOR RETAIL CENTERS IN ATLANTA
Most shopping malls in the Atlanta metropolitan area, as elsewhere in North America, are in the suburbs, especially near the beltway.

LENNOX SQUARE

UNDERGROUND ATLANTA

PHIPPS PLAZA

SUBURBAN SPRAWL

Suburban housing offers many attractions:

- A detached single-family dwelling rather than a row house or apartment.
- Private land surrounding the house.
- Space to park several cars at no cost.
- A greater opportunity for home ownership.
- A private retreat from the stress of urban living.

Families with children are especially attracted to suburbs, which offer more space for play and protection from the high crime rates and heavy traffic that characterize inner-city life. As incomes rose in the twentieth century, first in the United States and more recently in Western Europe, more families were able to afford to buy suburban homes.

U.S. suburbs are characterized by **sprawl**, which is the progressive spread of development over the landscape. When private developers select new housing sites, they seek cheap land that can be easily prepared for construction—land often not contiguous to the existing built-up area.

Land is not transformed immediately from farms to housing developments. Instead, developers buy farms for future construction of houses by individual builders. Developers frequently reject land adjacent to built-up areas in favor of detached isolated sites, depending on the price and physical attributes of the alternatives. The periphery of U.S. cities therefore looks like Swiss cheese, with pockets of development and gaps of open space.

The modern residential suburb is segregated in two ways. First, residents are separated from commercial and manufacturing activities that are confined to compact, distinct areas. Second, housing in a given suburban community is usually built for people of a single social class, with others excluded by virtue of the cost, size, or location of the housing.

SUBURBAN HOUSING, FLORIDA

CITY AND SUBURB

13.8 Inner-City Decline and Renewal

- Low-income residents concentrate in the inner-city neighborhoods in the U.S.
- Some U.S. inner-city neighborhoods have been gentrified.

Inner cities in the United States contain concentrations of low income people who face a variety of economic, social, and physical challenges very different from those faced by suburban residents.

INNER-CITY CHALLENGES

Inner-city residents are frequently referred to as a permanent **underclass** because they are trapped in an unending cycle of hardships:

- **Inadequate job skills**. Inner-city residents are increasingly unable to compete for jobs. They lack technical skills needed for most jobs because fewer than half complete high school.
- **Culture of poverty.** Unwed mothers give birth to two-thirds of the babies in U.S. inner-city neighborhoods, and 80 percent of children in the inner city live with only one parent. Because of inadequate child-care services, single mothers may be forced to choose between working to generate income and staying at home to take care of the children.
- **Crime.** Trapped in a hopeless environment, some inner-city residents turn to drugs. Although drug use is a problem in suburbs as well, rates of use have increased most rapidly in inner cities. Some drug users obtain money through criminal activities.
- **Homelessness.** Several million people are homeless in the United States. Most people are homeless because they cannot afford housing and have no regular income. Homelessness may have been sparked by family problems or job loss.
- **Poverty.** The concentration of low-income residents in inner-city neighborhoods of central cities has produced financial problems. These people require public services, but they can pay very little of the taxes to support them. Central cities face a growing gap between the cost of needed services in inner-city neighborhoods and the availability of funds to pay for them.
- **Deteriorated housing.** Inner-city housing is subdivided by absentee landlords into apartments for low-income families, a process known as **filtering**. Landlords stop maintaining houses when the rent they collect becomes less than the maintenance cost. In such a case, the building soon deteriorates and grows unfit for occupancy.

13.8.1 **INNER-CITY POVERTY, CINCINNATI, OHIO**
Homeless people camp under the I-75 bridge across the Ohio River.

13.8.2 **INNER-CITY GENTRIFICATION, BOSTON**
Faneuil Hall Marketplace, incorporating historic structures, including Quincy Market, attracts well-heeled residents and tourists to the center of Boston.

GENTRIFICATION

Gentrification is the process by which middle-class people move into deteriorated inner-city neighborhoods and renovate the housing. Most cities have at least one gentrified inner-city neighborhood. In a few cases, inner-city neighborhoods never deteriorated because the community's social elite maintained them as enclaves of expensive property. In most cases, inner-city neighborhoods have only recently been renovated by the city and by private investors.

Middle-class families are attracted to deteriorated inner-city housing for a number of reasons. First, houses may be larger, more substantially constructed, yet cheaper in the inner city than in the suburbs. Inner-city houses may also possess attractive architectural details such as ornate fireplaces, cornices, high ceilings, and wood trim.

Gentrified inner-city neighborhoods also attract middle-class individuals who work downtown. Inner-city living eliminates the strain of commuting on crowded freeways or public transit. Others seek proximity to theaters, bars, restaurants, and other cultural and recreational facilities located downtown. Renovated inner-city housing is especially appealing to single people and couples without school-age children.

Because renovating an old inner-city house can be nearly as expensive as buying a new one in the suburbs, cities encourage the process by providing low-cost loans and tax breaks. Public expenditures for renovation have been criticized as subsidies for the middle class at the expense of people with lower incomes, who are forced to move out of the gentrified neighborhoods because the rents in the area are suddenly too high for them.

CITY AND SUBURB

13.9 Urban Transportation

- Most trips in the United States are by private motor vehicle.
- Public transportation has made a modest comeback in some cities.

People do not travel aimlessly; their trips have a precise point of origin, destination, and purpose. More than half of all trips are work related. Shopping or other personal business and social journeys each account for approximately one-fourth of all trips. Sprawl makes people more dependent on transportation for access to work, shopping, and social activities.

MOTOR VEHICLES

Cars and trucks have permitted large-scale development of suburbs at greater distances from the center. Motor vehicle drivers have much greater flexibility in their choice of residence than was ever before possible.

The motor vehicle is an important user of land in the city. An average city allocates about one-fourth of its land to roads and parking lots. Valuable land is devoted to parking cars and trucks, although expensive underground and multistory parking structures can reduce the amount of ground-level space needed. Freeways cut a wide path through the heart of cities, and elaborate interchanges consume even more space.

Motor vehicles have costs beyond their purchase and operation: delays imposed on others, increased need for highway maintenance, construction of new highways, and pollution. The average American spends 36 hours per year sitting in traffic jams and uses 55 gallons of gasoline.

The U.S. government has encouraged the use of cars and trucks by paying 90 percent of the cost of limited-access high-speed interstate highways, which stretch for 74,000 kilometers (46,000 miles) across the country. The use of motor vehicles is also supported by policies that keep the price of fuel below the level found in Western Europe.

PUBLIC TRANSIT

Historically, the growth of suburbs was constrained by transportation problems. People lived in crowded cities because they had to be within walking distance of shops and places of employment.

The invention of the railroad in the nineteenth century enabled people to live in suburbs and work in the central city. Cities then built street railways (called trolleys, streetcars, or trams) and underground railways (subways) to accommodate commuters. Rail and trolley lines restricted suburban development to narrow ribbons within walking distance of the stations. The suburban explosion in the twentieth century relied on motor vehicles rather than railroads, especially in the United States.

In larger cities, public transportation is better suited than motor vehicles to moving large numbers of people, because each traveler takes up far less space. Public transportation is cheaper, less polluting, and more energy efficient than the automobile. It also is particularly suited to rapidly bringing a large number of people into a small area. Despite the obvious advantages of public transportation for commuting, only 5 percent of trips in U.S. cities are by public transit. Outside the big cities, public transportation is extremely rare or nonexistent.

Public transportation has been expanded in some U.S. cities to help reduce air pollution and conserve petroleum. New subway lines have been built in Atlanta, Baltimore, San Francisco, and Washington, and existing systems expanded in Boston, Chicago, and New York. The federal government has permitted Boston, New York, and other cities to use funds originally allocated for interstate highways to modernize rapid transit service instead. The trolley—now known by the more elegant term of fixed light-rail transit—is making a modest comeback in North America. California, the state that most symbolizes the automobile-oriented American culture, is the leader in construction of new fixed light-rail transit lines.

Despite modest recent successes, most public transportation systems are caught in a vicious circle because fares do not cover operating costs. As patronage declines and expenses rise, the fares are increased, which drives away passengers and leads to service reduction and still higher fares.

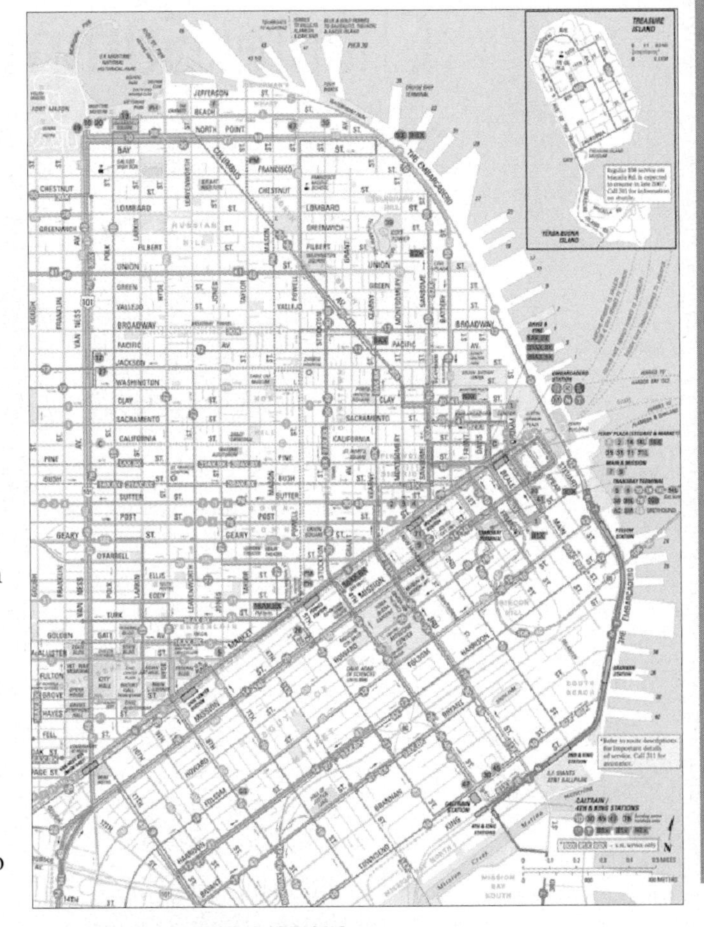

13.9.1 **TRANSPORTATION OPTIONS IN THE SAN FRANCISCO BAY AREA**
San Francisco has a dense network of public transit, including buses, subways, and cable cars. But most trips are made by car, including across the Golden Gate bridge (opposite page).

Chapter Summary

THE URBAN AREA

- The central business district (CBD) contains a large share of a city's business and public services.
- Urban areas have expanded beyond the legal boundaries of cities to encompass urbanized areas and metropolitan areas that are functionally tied to the cities.
- With suburban growth, most metropolitan areas have been fragmented into a large number of local governments.

PATTERNS WITHIN URBAN AREAS

- The concentric zone, sector, and multiple nuclei models describe where different types of people live within urban areas.
- The three models together help to understand that people live in different rings, sectors, and nodes depending on their stage in life, social status, and ethnicity.
- The distribution of people typical of U.S. cities is different in other countries.

CITY AND SUBURB

- Most Americans now live in suburbs that surround cities.
- Low-income inner-city residents face a variety of economic, social, and physical challenges.
- Tying together sprawling American urban areas is a dependency on motor vehicles.

YOKOHAMA, JAPAN, KANAGAWA HIGHWAY AND DAIKOKU PARKING AREA

Geographic Consequences of Change

What is the future for cities? As shown in this chapter, contradictory trends are at work simultaneously. Why does one inner-city neighborhood become a slum and another an upper-class district? Why does one city attract new shoppers and visitors while another languishes?

Inner-city residents may rarely venture out to suburbs. Lacking a motor vehicle, they have no access to most suburban locations. Lacking money, they do not shop in suburban malls or attend sporting events at suburban arenas. The spatial segregation of inner-city residents and suburbanites lies at the heart of the stark contrasts so immediately observed in any urban area.

Inner-city residential areas have physical problems stemming from the high percentage of older deteriorated housing, social problems stemming from the high percentage of low-income households, and economic problems stemming from a gap between demand for services and supply of local tax revenue.

Conversely, many people live in urban areas and never venture into inner-city neighborhoods or downtown. They live in suburbs, attend school in suburbs, work in suburbs, shop in suburbs, visit friends and family in suburbs, and attend movies and sports events in suburbs. Motor vehicles allow movement across urban areas without entering the center.

The suburban lifestyle as exemplified by the detached single-family house with surrounding yard attracts most people. Transportation improvements, most notably the railroad in the nineteenth century and the automobile in the twentieth century, have facilitated the sprawl of urban areas. Among the negative consequences of large-scale sprawl are segregation and inefficiency.

Several U.S. states have taken strong steps in the past few years to curb sprawl, reduce traffic congestion, and reverse inner-city decline. The goal is to produce a pattern of compact and contiguous development, while protecting rural land for agriculture, recreation, and wildlife protection. Legislation and regulations to limit suburban sprawl and preserve farmland has been called **smart growth.**

SHARE YOUR VOICE Student Essay

Carl Nicholas Kallof
Arizona State University

Where the Midwest Meets Mexico?

I'm a Phoenix native, born and raised in the Valley of the Sun. Not too long ago, Phoenix was declared one of the fastest growing cities in the United States. When I think back on my childhood, Phoenix didn't seem to even be on the map, and that was only the early 90s.

The Cohen Brothers made a movie called *Raising Arizona,* in which Tempe was more or less accurately portrayed as a trailer park. Now it is the crossroads of Mesa, downtown Phoenix, and Scottsdale, with smack in the center Arizona State University, a "New American University." Far from the Ivy League of the Northeast, the football schools of the Midwest, and the South, Arizona State has been positioning itself as a prestigious West Coast university like USC.

Of course, the city loves the growth. More growth means more money, especially when you consider that Arizona is now home to an exodus of Californians and Midwesterners. Why live in Los Angeles or Cleveland when you can buy a mansion in the desert for a measly $400,000?

But I do not love the growth. The more Californians and Midwesterners moving here means more traffic, higher prices, more demand for water, and more people.

The other rising demographic in Phoenix politics is that of the Mexicans, who speak Spanish. It will only be a matter of time before Arizona is a predominantly Spanish-speaking state, and yet everyday more from the Farm Belt flock to the suburbs of Gilbert and Glendale. Most of these people aren't going to bother to learn the Spanish language, something they consider a threat to their culture. It begs the question, "so why did you move here in the first place?"

I miss the days when Phoenix wasn't trying to be the next Los Angeles: when traffic wasn't as bad, when ASU's tuition wasn't as high, and when most of the residents from the Valley were actually born here.

Chapter **Resources**

FUTURE TRANSPORTATION SYSTEMS

The future health of urban areas depends on relieving traffic congestion.

GPS

Information about traffic congestion is now being transmitted through cell phones, pagers, e-mails, web sites, and in-car dashboard monitors. Reports can be tailored to individual journeys. To make these systems usable, vehicles manufactured since 2000 either are equipped with GPS or can have GPS capability added as systems become more widely available and less expensive to purchase and operate.

SMART HIGHWAYS

In North America, congestion is also being reduced through "smart" highways. Toronto and several California cities charge motorists higher tolls to drive on freeways during congested times. Attached to a vehicle is a transponder recording the time of day it is on the highway. A monthly bill sent to the vehicle's owner reflects the differential tolls.

CONGESTION CHARGING

Elsewhere, fees are being charged to drive in congested central areas. Singapore was the first city to do this in 1975. The most extensive effort is in London, which was started in 2003 and extended to a wider area 4 years later. Motorists wishing to drive into central London between 7 AM and 6 AM must pay a fee. Drivers pay through cell phone, text message, or web site, or in person at some shops. Cameras record the license plate of all vehicles entering central London. The license plates recorded by the cameras are matched with those that paid, and hefty fines are levied against violators. Studies showed that after imposing the congestion fee, the number of vehicles entering central London declined by around one-sixth and average speeds and air quality improved by around one-sixth.

FUTURE INTELLIGENT TRANSPORTATION SYSTEMS

Future intelligent transportation systems are likely to remove decisions from the drivers through hands-free driving. A motorist will drive to a freeway entrance, where the vehicle will be subjected to a thorough diagnostic (taking a half-second) to ensure that it has enough fuel and is in good operating condition.

A menu offers a choice of predetermined destinations, such as "home" or "office," or a destination can be programmed by hand. A release will send the vehicle accelerating automatically on the entrance ramp into the freeway. Sensors in the bumpers and fenders, attached to radar or GPS, alert vehicle systems to accelerate, brake, or steer as needed. Spacing between vehicles can be as little as 2 meters.

While the vehicle is automatically controlled, the "driver" swivels the seat to a workstation to make phone calls, check e-mail, surf the Internet, or write letters. Or the driver can read, watch television, or nap. When the vehicle nears the programmed freeway exit, a tone warns that the driver will have to take back control. The vehicle is halted on the exit ramp until the driver firmly presses the brake to release the "autodrive" system, much as cruise control is currently disengaged.

ON THE INTERNET

Data concerning any urban area can be found at the **U.S. Bureau of the Census** web site **www.census.gov.** The American Factfinder service provides information from the most recent census, as well as annual updates from the American Community Survey. Tables and maps can be generated for census tracts within urban areas as well as for entire urban areas. Access is also provided to data from earlier censuses.

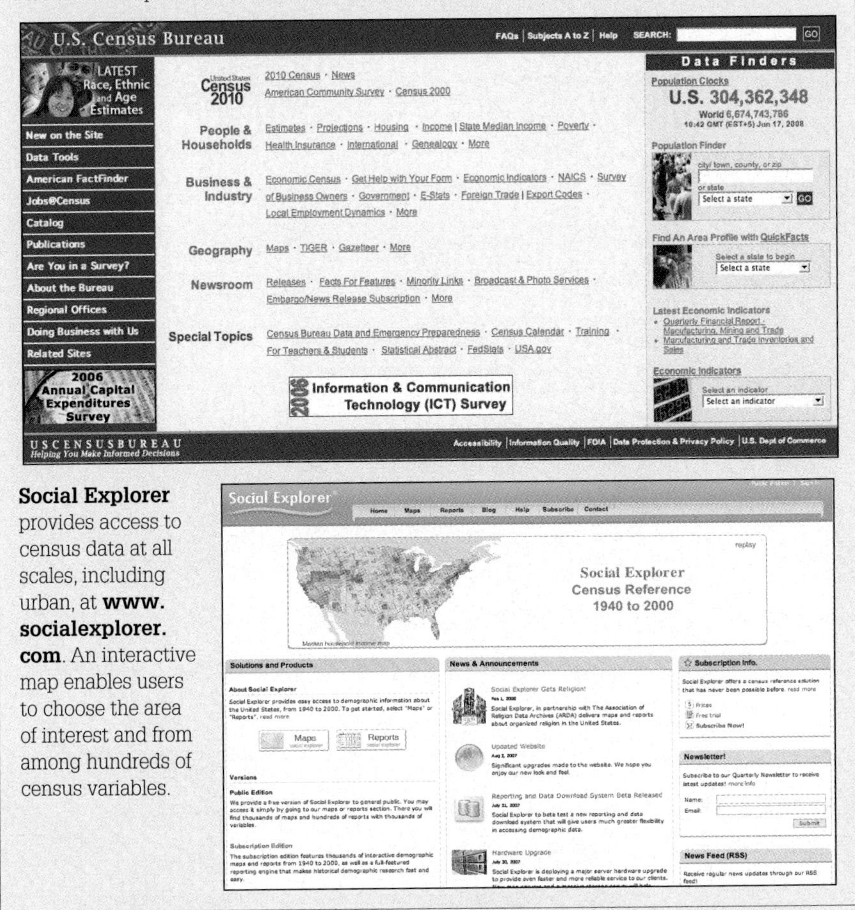

Social Explorer provides access to census data at all scales, including urban, at **www.socialexplorer.com**. An interactive map enables users to choose the area of interest and from among hundreds of census variables.

FURTHER READINGS

Berry, Brian J. L., and James O. Wheeler, eds. *Urban Geography in America, 1950–2000: Paradigms and Personalities.* New York: Routledge, 2005.

Ford, Larry R. "A New and Improved Model of Latin American City Structure." *Geographical Review* 86 (1996): 437–440.

Garreau, Joel. *Edge City: Life on the New Frontier.* New York: Doubleday, 1991.

Garvin, Alexander. *The American City: What Works, What Doesn't.* New York: McGraw-Hill, 1996.

Glaeser, E. L., and J. M. Shapiro. "Urban Growth in the 1990s: Is City Living Back?" *Journal of Regional Science* 43 (2003): 139–66.

Gottmann, Jean. *Megalopolis.* New York: Twentieth-Century Fund, 1961.

Harris, Chauncy D. "Diffusion of Urban Models: A Case Study." *Urban Geography* 19 (1998): 49–67.

———. "The Nature of Cities and Urban Geography in the Last Half Century." *Urban Geography* 18 (1997): 15–35.

Harris, Chauncy D., and Edward L. Ullman. "The Nature of Cities." *Annals of the American Academy of Political and Social Science* 143 (1945): 7–17.

Hoyt, Homer. *The Structure and Growth of Residential Neighborhoods.* Washington, DC: Federal Housing Administration, 1939.

Jacobs, Jane. *Death and Life of Great American Cities.* New York: Random House, 1961.

Knox, Paul L., and Steven Pinch. *Urban Social Geography: An Introduction*, 5th ed. New York: Wiley, 2006.

Levy, John. *Contemporary Urban Planning*, 7th ed. Upper Saddle River, NJ: Prentice Hall, 2005.

Mumford, Lewis. *The City in History.* New York: Harcourt, Brace, and World, 1961.

Park, Robert E., Ernest W. Burgess, and Roderick D. McKenzie, eds. *The City.* Chicago: University of Chicago Press, 1925.

KEY TERMS

Annexation
Legally adding land area to a city in the United States.

Census tract
An area delineated by the U.S. Bureau of the Census for which statistics are published; in urbanized areas, census tracts correspond roughly to neighborhoods.

Central business district (CBD)
The area of a city where retail and office activities are clustered.

Concentric zone model
A model of the internal structure of cities in which social groups are spatially arranged in a series of rings..

Filtering
A process of change in the use of a house, from single-family owner occupancy to abandonment.

Gentrification
A process of converting an urban neighborhood from a predominantly low-income renter-occupied area to a predominantly middle-class owner-occupied area.

Metropolitan statistical area (MSA)
In the United States, a central city of at least 50,000 population, the county within which the city is located, and adjacent counties meeting one of several tests indicating a functional connection to the central city.

Micropolitan statistical area
An urbanized area of between 10,000 and 50,000 inhabitants, the county in which it is found, and adjacent counties tied to the city.

Multiple nuclei model
A model of the internal structure of cities in which social groups are arranged around a collection of nodes of activities.

Sector model
A model of the internal structure of cities in which social groups are arranged around a series of sectors, or wedges, radiating out from the central business district (CBD).

Smart growth
Legislation and regulations to limit suburban sprawl and preserve farmland.

Sprawl
Development of new housing sites at relatively low density and at locations that are not contiguous to the existing built-up area.

Squatter settlement
An area within a city in a less developed country in which people illegally establish residences on land they do not own or rent and erect homemade structures.

Underclass
A group in society prevented from participating in the material benefits of a more developed society because of a variety of social and economic characteristics.

Urbanized area
In the United States, a central city plus its contiguous built-up suburbs.

Chapter 14 **RESOURCE ISSUES**

When you finish drinking a soda, do you pitch the can in the trash or place it in a recycling bin? When you leave a room, do you turn off the lights and computer, or leave them on? In winter, if you feel cold, do you put on a sweater, or do you turn up the thermostat?

People have always transformed Earth's land, water, and air for their benefit. But human actions in recent years have gone far beyond the impact of the past. With less than one-fourth of the world's population, MDCs consume most of the world's energy and generate most of its pollutants. Meanwhile, in LDCs 2 billion people live without clean water or sewers, and 1 billion live in cities with unsafe sulfur dioxide levels.

Plants and animals live in harmony with their environment, but people often do not. Geographers study the troubled relationship between human actions and the physical environment in which we live. From the perspective of human geographers, Earth offers a large menu of resources available for people to use. A **resource** is a substance in the environment that is useful to people, is economically and technologically feasible to access, and is socially acceptable to use. Resources include food, water, soil, plants, animals, and minerals.

The problem is that most resources are limited, and Earth has a tremendous number of consumers. Geographers observe two major misuses of resources:

- We deplete scarce resources, especially petroleum, natural gas, and coal, for energy production.
- We destroy resources through pollution of air, water, and soil.

These two misuses are the basic themes of this chapter.

SAN GORGONIO PASS WIND FARM, CALIFORNIA

THE KEY ISSUES IN THIS CHAPTER

DEPLETING RESOURCES

14.1 Nonrenewable Energy Resources

14.2 Energy Production and Reserves

14.3 Mineral Resources

POLLUTING RESOURCES

14.4 Air Pollution

14.5 Water Pollution

14.6 Land Pollution

CONSERVING AND REUSING RESOURCES

14.7 Renewable Resources

14.8 Recycling Resources

14.9 Sustainability

DEPLETING RESOURCES

14.1 Nonrenewable Energy Resources

- Most energy comes from the three fossil fuels.
- Energy is consumed primarily in businesses, homes, and transportation.

Five-sixths of the world's energy comes from three substances—oil, natural gas, and coal. In MDCs the remainder comes primarily from nuclear, solar, and **geothermal power**. Burning wood and **hydroelectric power** provide much of the remaining energy in LDCs.

THE THREE FOSSIL FUELS

Petroleum, natural gas, and coal are known as **fossil fuels**. A fossil fuel is the residue of plants and animals that were buried millions of years ago. As sediment accumulated over these remains, intense pressure and chemical reactions slowly converted them into the fossil fuels we use today. When we burn these substances today, we are releasing energy originally stored in plants and animals millions of years ago.

Historically, the most important energy source worldwide was **biomass fuel**, such as wood, plant material, and animal waste. Biomass fuel is burned directly or converted to charcoal, alcohol, or methane gas. Biomass remains the most important source of fuel in some LDCs, but during the past 200 years MDCs have converted to other energy sources.

Earth's energy resources can be divided between those that are renewable and those that are not:

- **Renewable energy** is replaced continually, or at least within a human life span. It has an essentially unlimited supply and is not depleted when used by people. Solar energy, hydroelectric, geothermal, **fusion**, and wind are examples.
- **Nonrenewable energy** forms so slowly that for practical purposes the supply is finite. The fossil fuels, as well as nuclear energy, are examples.

The world faces an energy problem in part because we are rapidly depleting the remaining supply of the three fossil fuels, especially petroleum. Once the present supply of fossil fuels is consumed, it is gone, and we must look to other resources for our energy. (Technically, fossil fuels are continually being formed, but the process takes millions of years, so humans must regard the current supply as essentially finite.)

14.1.1 **ENERGY CONSUMPTION**

Other Latin America 5%
Brazil 2%
Other Asia 7%
South Korea 2%
India 3%
United States 22%
China 14%
Canada 3%
Germany 3%
France 2%
United Kingdom 2%
Italy 2%
Other Western Europe 6%
Middle East 8%
Africa 2%
Japan 5%
Russia 7%
Other Eastern Europe 5%
MDCs
LDCs

SHARE OF WORLD CONSUMPTION

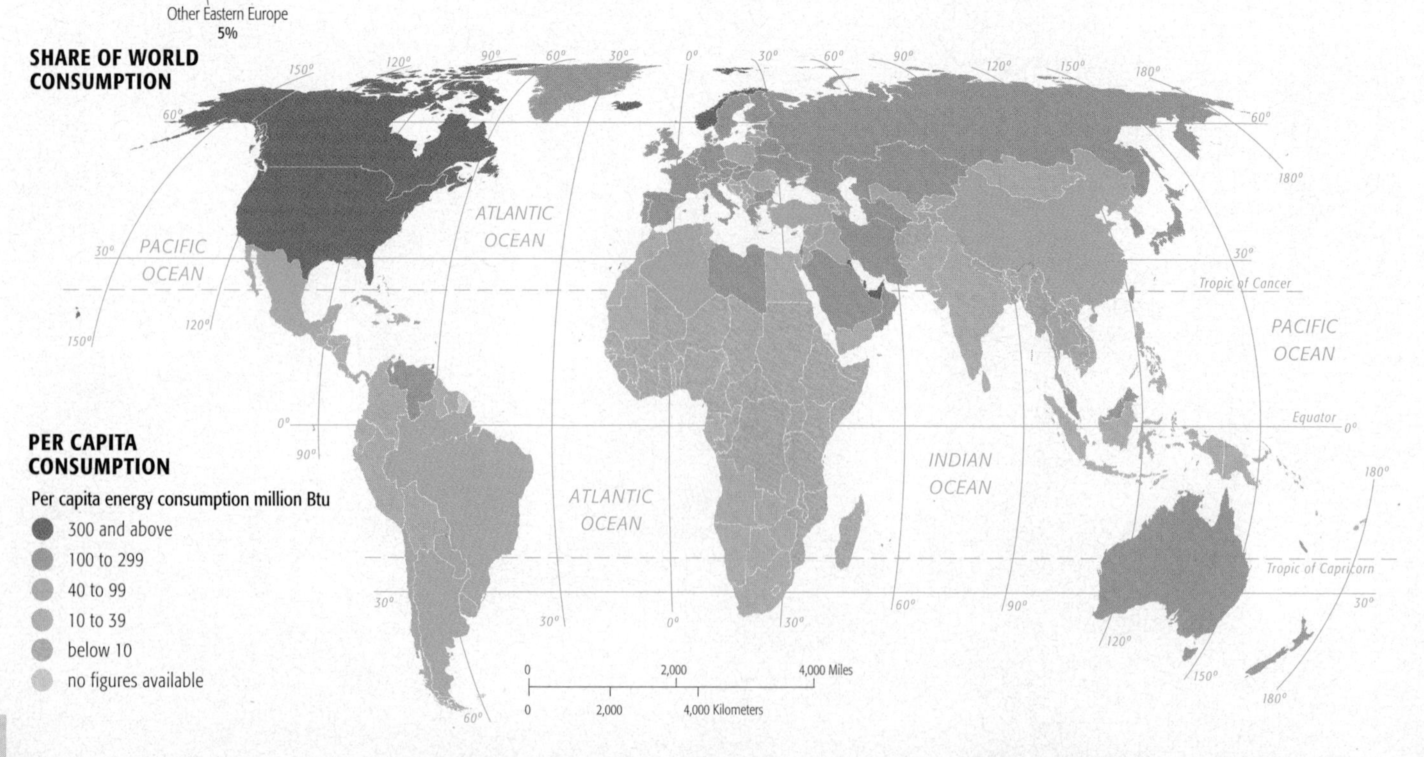

14.1.2 **MINING IN AN OPEN CAST COAL MINE, STOBSWOOD, NORTHUMBERLAND, ENGLAND**

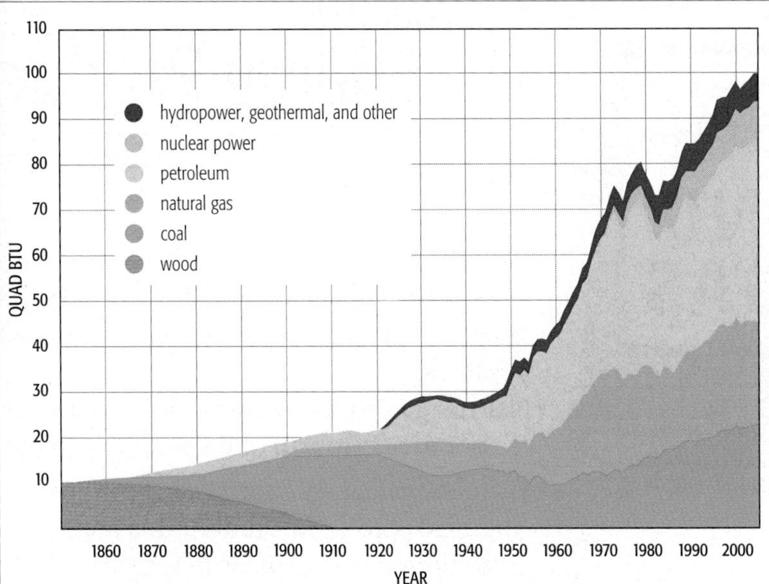

14.1.3 **U.S. ENERGY CONSUMPTION.**
Coal supplanted wood as the leading energy source in the United States in the late 1800s, as a consequence of the Industrial Revolution. Petroleum was first pumped in 1859, but it was not an important resource until the diffusion of motor vehicles in the twentieth century. Natural gas was originally burned off as a waste product of oil drilling but now heats millions of homes.

FOSSIL FUEL CONSUMPTION

The one-fourth of the world's population living in MDCs consume one-half of the world's energy. The United States alone, with just 5 percent of the world's population, accounts for 22 percent of world energy consumption. This high energy consumption by a modest percentage of the world's population supports a lifestyle rich in food, goods, services, comfort, education, and travel in MDCs.

As they promote development and cope with high population growth, LDCs are consuming much more energy. The share of world energy consumed by LDCs has increased from one-fourth in 1990 to nearly one-half in 2005 and is expected to increase much more. China, the world's second-largest consumer of energy—15 percent of world consumption in 2005—is expected to pass the United States around 2010. As a result of increased demand in LDCs, global consumption of petroleum is expected to increase by about 50 percent during the next two decades, whereas both coal and natural gas consumption are expected to double.

Energy is consumed in three principal places.

- **Businesses.** For U.S. businesses the main energy resource is coal, followed by natural gas and oil. Some businesses directly burn coal in their own furnaces. Others rely on electricity, mostly generated at coal-burning power plants.
- **Homes.** At home, energy is used primarily for the heating of living space and water. Natural gas is the most common source, followed by petroleum (heating oil and kerosene).
- **Transportation.** Almost all transportation systems operate on petroleum products, including automobiles, trucks, buses, airplanes, and most railroads. Only subways, streetcars, and some trains run on coal-generated electricity.

We can use other resources for heat, fuel, and manufacturing, but they are likely to be more expensive and less convenient to use than fossil fuels. And converting from fossil fuels will likely disrupt our daily lives and cause us hardship.

Because of dwindling supplies of fossil fuels, most of the buildings in which we live, work, and study will have to be heated another way. Cars, trucks, and buses will have to operate on some other energy source. The many plastic objects that we use (because they are made from petroleum) must be made with other materials.

14.1.4 **NATURAL GAS PIPELINE CONSTRUCTION IN THE AFRICAN DESERT (above) PUMPING PETROLEUM IN THE DESERT (below)**

DEPLETING RESOURCES

14.2 Energy Production and Reserves

- Remaining supplies of fossil fuels are not distributed uniformly.
- Petroleum reserves are especially limited and clustered in a handful of countries.

A FAMILIAR SIGN DURING THE 1970s.

Fossil fuels are distributed unevenly around the world.

- Coal. China is responsible for extracting 40 percent of the world's coal, and the United States 20 percent. The United States also has one-fourth of the world's proven coal reserves.
- **Natural gas.** Russia accounts for one-fourth of world production of natural gas and possesses one-fourth of the world's proven reserves.
- **Petroleum.** Five Middle Eastern countries have 60 percent of the world's oil reserves—about 20 percent in Saudi Arabia and 10 percent each in Iran, Iraq, Kuwait, and United Arab Emirates. Canada is now thought to have 14 percent of world petroleum reserves, located in extensive deposits of Alberta sands.

FOSSIL FUEL RESERVES

How much of the fossil fuel supply remains? The amount of energy remaining in deposits that have been discovered is called a **proven reserve.** Proven reserves for the fossil fuels are as follows:

FUEL	AMOUNT	YEARS REMAINING AT CURRENT USE
Petroleum	1.3 trillion barrels	50
Natural gas	175 trillion cubic meters	60
Coal	1 quadrillion metric tons	175

Unless substantial new proven reserves are found—or consumption decreases sharply—the world's petroleum and natural gas reserves will be depleted sometime in the twenty-first century. However, some deposits in the world have not yet been discovered. The energy in deposits that are undiscovered but thought to exist is a **potential reserve.** Estimates of potential reserves of petroleum and natural gas vary widely, from a few years to several hundred years.

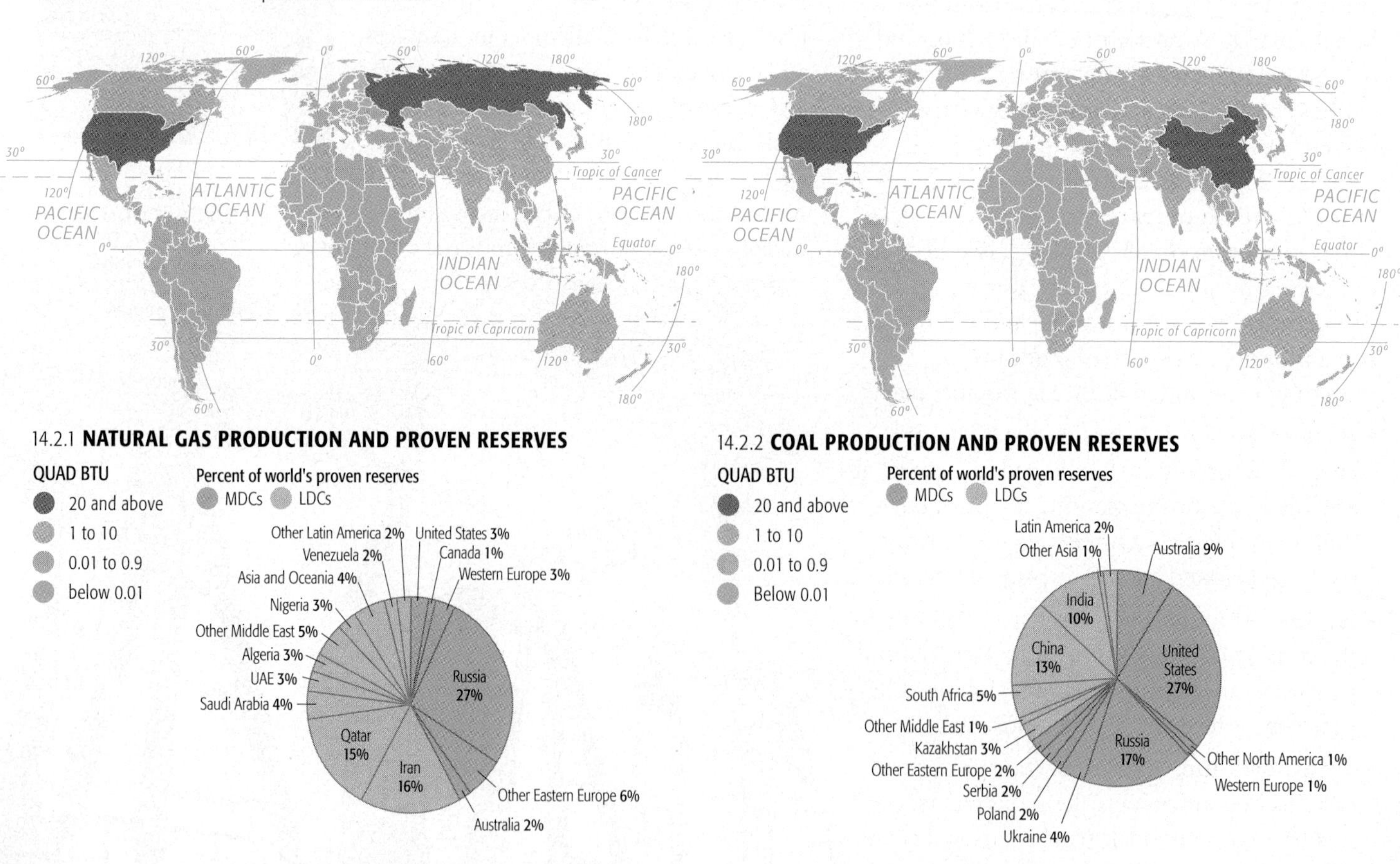

14.2.1 **NATURAL GAS PRODUCTION AND PROVEN RESERVES**

14.2.2 **COAL PRODUCTION AND PROVEN RESERVES**

PETROLEUM IMPORTS

Because MDCs consume more energy than they produce, they must import more fossil fuels, especially petroleum, from LDCs. The United States and Western Europe import more than half their petroleum, and Japan more than 90 percent.

The United States produced more petroleum than it consumed during the first half of the twentieth century. Beginning in the 1950s, the handful of large transnational companies then in control of international petroleum distribution determined that extracting domestic petroleum was more expensive than importing it from the Middle East.

As a result, U.S. petroleum imports increased from 14 percent of total consumption in 1954 to 58 percent in 2007. European countries and Japan have always depended on foreign petroleum because of limited domestic supplies. China changed from a net exporter to an importer of petroleum during the 1990s.

For importing countries, the major source of petroleum is the Middle East. Transnational companies based in MDCs originally exploited Middle Eastern petroleum fields and sold the petroleum at a low price to consumers in MDCs. At first, the transnationals set oil prices and paid the Middle Eastern governments only a small percentage of their oil profits.

Several Arab states in the Middle East, plus a handful of LDCs possessing substantial petroleum reserves, created the Organization of Petroleum Exporting Countries (OPEC) in 1960. OPEC's Arab members were angry at MDCs for supporting Israel during that nation's 1973 war with the Arab states of Egypt, Jordan, and Syria. In retaliation, Arab OPEC states refused to sell petroleum to the nations that had supported Israel.

During the winter of 1973–74, gasoline in the United States was rationed by license plate number (cars with licenses ending in an odd number could buy only on odd-numbered days). Production of steel, motor vehicles, and other energy-dependent industries plummeted in the United States in the wake of the boycott and never regained pre-boycott levels. LDCs were also hurt, because they depended on low-cost petroleum to spur industrial growth and to manufacture fertilizers.

OPEC lifted the boycott in 1974 and instead raised petroleum prices from $3 per barrel to more than $35 by 1981. Declining petroleum prices during the 1990s lulled consumers in MDCs to purchase gas-guzzling trucks and cheap airline tickets, but when petroleum shot up to more than $100 a barrel in the twenty-first century, consumers and manufacturers belatedly took notice.

The world will not literally "run out" of petroleum during the twenty-first century. However, at some point extracting the remaining petroleum reserves will prove so expensive and environmentally damaging that use of alternative energy sources will accelerate and dependency on petroleum will diminish.

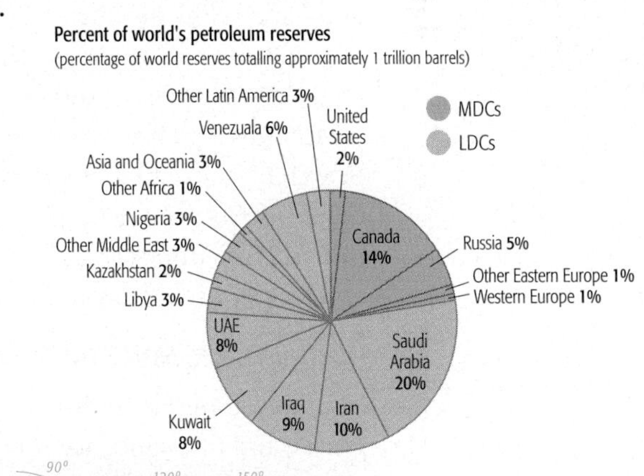

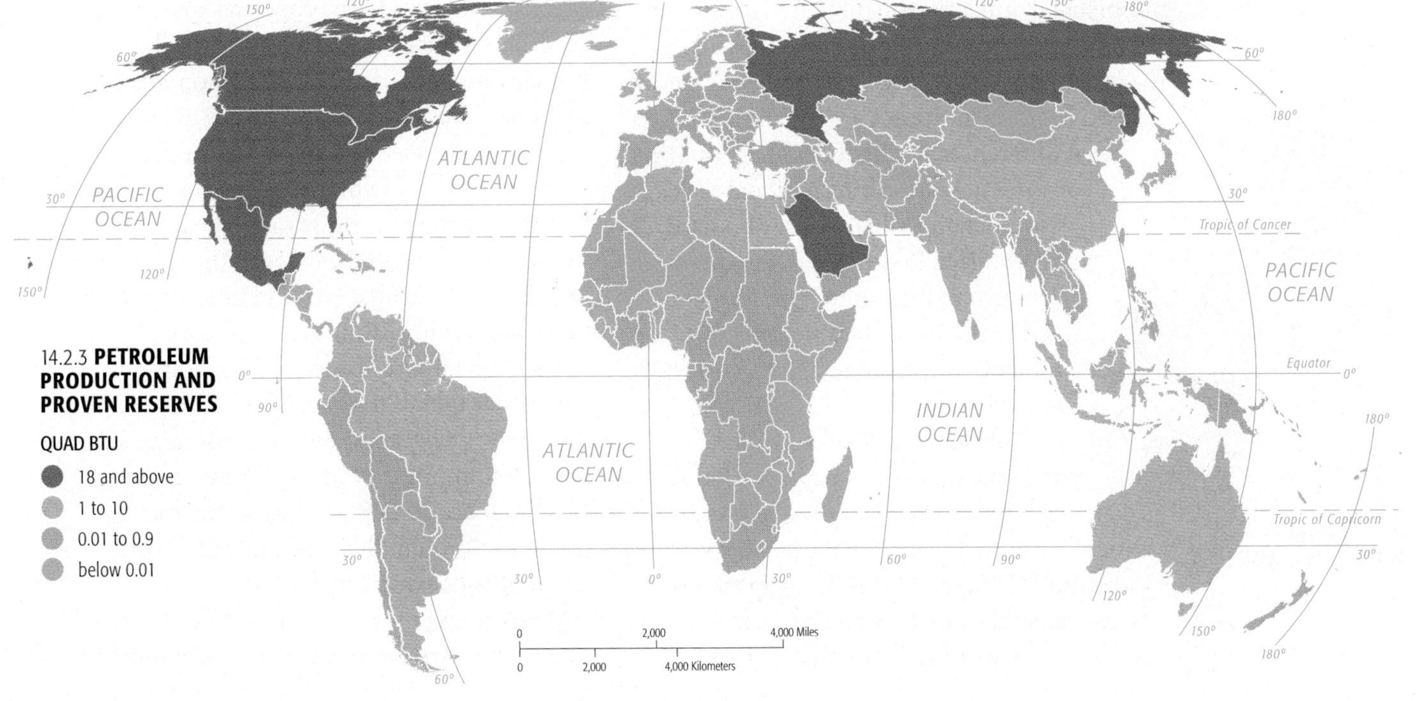

14.2.3 **PETROLEUM PRODUCTION AND PROVEN RESERVES**

DEPLETING RESOURCES

14.3 Mineral Resources

- Mineral resources are classified as nonmetallic, ferrous metallic, and nonferrous metallic.
- Some mineral resources are ubiquitous and abundant, whereas others are scarce and highly clustered.

Earth has 92 natural elements, but about 99 percent of the crust is composed of eight elements—oxygen, silicon, aluminum, iron, calcium, sodium, potassium, and magnesium. Oxygen alone accounts for nearly one-half of the crust and silicon more than one-fourth.

The eight most common elements combine with thousands of rare ones to form approximately 3,000 different minerals, all with their own properties of hardness, color, and density, as well spatial distribution. Each mineral is potentially a resource, if people find a use for it.

Mineral deposits are not uniformly distributed around the world. Most of the world's supply of particular minerals is found in a handful of countries. Countries such as Australia and China rank among leading producers of several minerals, whereas other countries have abundant supplies of only one mineral.

Minerals are either metallic or nonmetallic. In weight, more than 90 percent of the minerals that humans use are nonmetallic, but metallic minerals are important for economic activities and so carry relatively high value.

NONMETALLIC MINERAL RESOURCES

Building stones, including large stones, coarse gravel, and fine sand, account for 90 percent of nonmetallic mineral extraction. These minerals are fashioned into objects of daily use, such as roads and tools. The rocks and earthen materials used for these purposes are so common that differences in distribution are of little consequence at the international scale.

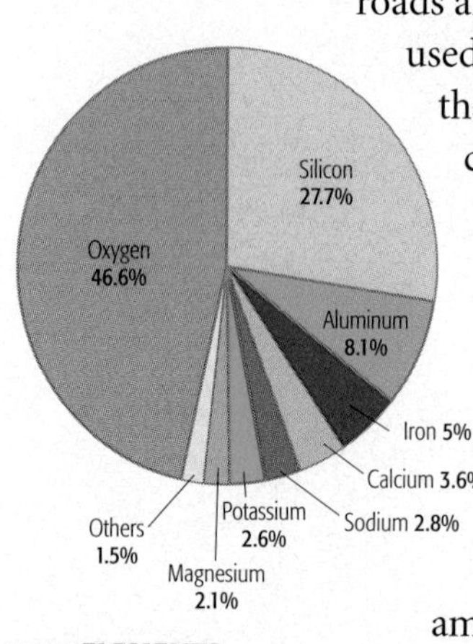

14.3.1 **ELEMENTS IN EARTH'S CRUST**

Nonmetallic minerals are also used for fertilizer. Because soils are often deficient in these minerals, farmers add them. Important nonmetallic mineral sources of fertilizers include the following.

- **Phosphorus.** Obtained from phosphate rock (rich in apatite), found among the marine sediments of old seabeds. Morocco possesses one-half of the world's reserves.
- **Potassium.** Obtained from the evaporation of saltwater. Principal sources include former Soviet Union countries, North America, and the Dead Sea.
- **Calcium.** Concentrated in subhumid soils such as the plains and prairies of the Western United States and Canada, as well as Russia's steppes.
- **Sulfur.** Used to make insecticides and herbicides, as well as fertilizers. North America produces one-fourth of the world total.

Nitrogen is also important in fertilizers. Though ubiquitous, it takes a lot of energy to capture nitrogen, so fossil fuel prices affect availability.

Another group of nonmetallic minerals, gemstones, are valued for their color and brilliance when cut and polished. Diamonds are also useful in manufacturing, because they are the strongest and hardest known material and have the highest thermal conductivity of any material at room temperature. Two-thirds of the world's diamonds are currently mined in Australia, Botswana, and Russia.

METALLIC MINERAL RESOURCES: FERROUS

Metallic minerals have properties that are especially valuable for fashioning machinery, vehicles, and other essential components of an industrialized society. They are to varying degrees malleable (able to be hammered into thin plates) and ductile (able to be drawn into fine wire) and are conductors of heat and electricity. Each metal possesses these qualities in different combinations and degrees and therefore has its distinctive set of uses.

Many metals are also capable of combining with other metals to form alloys with yet other distinctive properties. A mineral bearing a metal such as aluminum or iron is known as an ore. Nearly all ore contains at least some metallic mineral, although the concentration is often too low to justify extracting it.

Metals are known as ferrous or nonferrous. The term **ferrous** refers to iron ore and other

NONMETALLIC MINERALS	15% and above	5–15%
DIAMONDS	Botswana, Russia, Australia	Canada, Angola, Congo (Kinshasa), South Africa
NITROGEN	China	Russia, India, United States
PHOSPHORUS	United States, Morocco, China	Russia, Tunisia
SULFUR	United States	Canada, Russia, China, Japan
FERROUS METALS AND ALLOYS	**15% and above**	**5–15%**
IRON ORE	China, Brazil, Australia	India, Russia
CHROMIUM	South Africa	Kazakhstan, India
MAGNESIUM	China	Canada
MANGANESE	South Africa	Australia, Brazil, Gabon, China, Ukraine, India
MOLYBDENUM	United States, Chile, China	Canada, Peru
NICKEL	Russia	Australia, Canada, Indonesia, New Caledonia, Cuba
TIN	China, Indonesia	Peru, Bolivia
TITANIUM	Australia, South Africa, Canada	China, Norway, United States, India
TUNGSTEN	China	
NONFERROUS METALS	**15% and above**	**5–15%**
ALUMINUM (BAUXITE)	Australia	Brazil, China, Guinea, India, Jamaica
COPPER	Chile	United States, Indonesia, Peru, Australia
GOLD		South Africa, Australia, United States, China, Peru, Russia, Indonesia
LEAD	China, Australia	
PLATINUM	South Africa	United States, Peru
SILVER	Peru	Russia
ZINC	China	China, Mexico, Australia, Chile, Canada, Poland, United States

alloys used in the production of iron and steel.

- **Iron.** By far the world's most widely used ferrous metal, prized for its many assets: a good conductor of heat and electricity, magnetic and to be magnetized, and malleable into useful shapes. Humans began fashioning tools and weapons from iron 4,000 years ago.
- **Manganese.** Especially vital for making steel because it imparts toughness and carries off undesirable sulfur and oxygen. South Africa has more than one-half of the reserves, and Ukraine another one-fourth.
- **Chromium.** A principal component of stainless steel, because it helps keep a sharp cutting edge even at high temperatures. Chromium is extracted from chromite ore, one-half of which is mined in South Africa and one-third in Kazakhstan and India.
- **Titanium.** A lightweight, high-strength, corrosion-resistant metal used as white pigment in paint. Titanium is extracted primarily from the mineral ilmenite; Australia possesses one-third of the world's reserves.
- **Magnesium.** Relatively light yet strong, so used to produce lightweight, corrosion-resistant alloys, especially with aluminum to make beverage cans. China supplies three-fourths of the world's magnesium.
- **Molybdenum.** Imparts toughness and resilience to steel. Unlike the other rare metals discussed here, the United States plays a leading role, with one-half of world reserves.
- **Nickel.** Used primarily for stainless steel and high-temperature and electrical alloys. World reserves are only around 100 years at current rates of use.
- **Tin.** Valued for its corrosion-resistant properties, used for plating iron and steel. World reserves are estimated at only around 50 years and clustered in Australia.
- **Tungsten.** Makes very hard alloys with steel and is used to manufacture tungsten carbide for cutting tools. China is responsible for 90 percent of world production and one-half of world reserves.

METALLIC MINERAL RESOURCES: NONFERROUS

14.3.2 **ELEMENTS IN EARTH'S CRUST**

Nonferrous metals are utilized for products other than iron and steel.

- **Aluminum.** The most abundant nonferrous metal, lighter, stronger, and more resistant to corrosion than iron and steel; obtained primarily through extraction from bauxite ore.
- **Copper.** Valued for its high ductility, malleability, thermal and electrical conductivity, and resistance to corrosion, used primarily in electronics and constructing buildings.
- **Lead.** Very corrosion-resistant, dense, ductile, and malleable, used for thousands of years, first in building materials and pipes, then in ammunition, brass, glass, and crystal, and now primarily in motor-vehicle batteries.
- **Zinc.** Primarily a coating to protect iron and steel from corrosion and as an alloy to make brass.

World reserves of aluminum exceed 1,000 years at current rates of use, but reserves are extremely limited for the other three—less than 60 years for copper, 25 years for lead, and 45 years for zinc.

Nonferrous metals also include precious metals—silver, gold, and the platinum group. Silver and gold have been prized since ancient times for their beauty and durability. In addition to its use in jewelry, silver is a component of photographic film, gold is important in dentistry, and both are used in electronics. Platinum is used in motor vehicles for catalytic converters and fuel cells.

IRON ORE

LEAD ORE

POLLUTING RESOURCES

14.4 Air Pollution

- Air pollution occurs at global, regional, and local scales.
- Air pollution can cause global warming, damage lakes and vegetation, and harm animal health.

At ground level, Earth's average atmosphere is made up of about 78 percent nitrogen, 21 percent oxygen, and less than 1 percent argon. The remaining 0.04 percent includes several trace gases, some of which are critical. **Air pollution** is a concentration of trace substances at a greater level than occurs in average air. Air pollution concerns geographers at three scales—global, regional, and local.

GLOBAL-SCALE AIR POLLUTION

Two global-scale issues are global warming and ozone damage:

- **Global warming.** The average temperature of Earth's surface has increased by 1° Celsius (2° Fahrenheit) during the past century. Human actions, especially the burning of fossil fuels, may have caused this. Global warming of only a few degrees could melt the polar ice caps and raise the level of the oceans many meters.

  A concentration of trace gases in the atmosphere can block some of the heat leaving Earth's surface heading for space, thereby raising Earth's temperatures. When fossil fuels are burned, one of the trace gases, carbon dioxide, is discharged into the atmosphere.

  Plants and oceans absorb much of the discharges, but increased fossil-fuel burning during the past 200 years has caused the level of carbon dioxide in the atmosphere to rise by more than one-fourth, according to the UN Intergovernmental Panel on Climate Change. The increase in Earth's temperature, caused by carbon dioxide trapping some of the radiation emitted by the surface, is called the greenhouse effect.

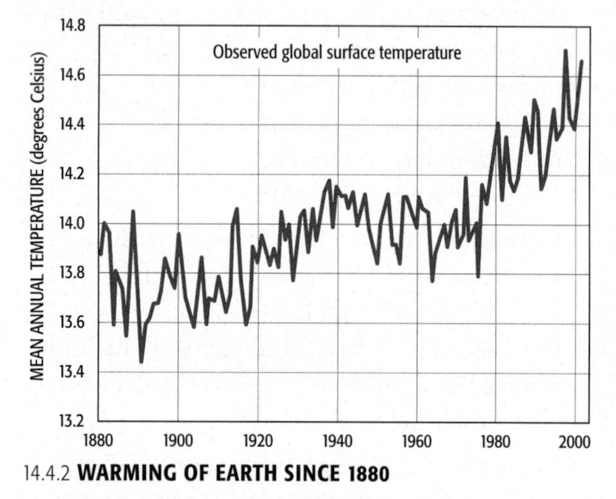

14.4.2 **WARMING OF EARTH SINCE 1880**

- **Global-scale ozone damage.** Earth's atmosphere has zones with distinct characteristics. The stratosphere—the zone between 15 and 50 kilometers (9 to 30 miles) above Earth's surface—contains a concentration of **ozone** gas. The ozone layer absorbs dangerous ultraviolet (UV) rays from the Sun. Were it not for the ozone in the stratosphere, UV rays would damage plants, cause skin cancer, and disrupt food chains.

  Earth's protective ozone layer is threatened by pollutants called **chlorofluorocarbons (CFCs).** CFCs such as *freon* were once widely used as coolants in refrigerators and air conditioners. When they leak from these appliances, the CFCs are carried into the stratosphere, where they break down Earth's protective layer of ozone gas.

14.4.1 **RECEDING OF NORTH POLAR ICE CAP**
Between 1979 and 2003 the polar ice cap has melted visibly.

REGIONAL-SCALE AIR POLLUTION

At the regional scale, air pollution may damage a region's vegetation and water supply through **acid deposition.** Sulfur oxides and nitrogen oxides, emitted by burning fossil fuels, enter the atmosphere, where they combine with oxygen and water. Tiny droplets of sulfuric acid and nitric acid form and return to Earth's surface as acid deposition. When dissolved in water, the acids may fall as **acid precipitation**—rain, snow, or fog. The acids can also be deposited in dust.

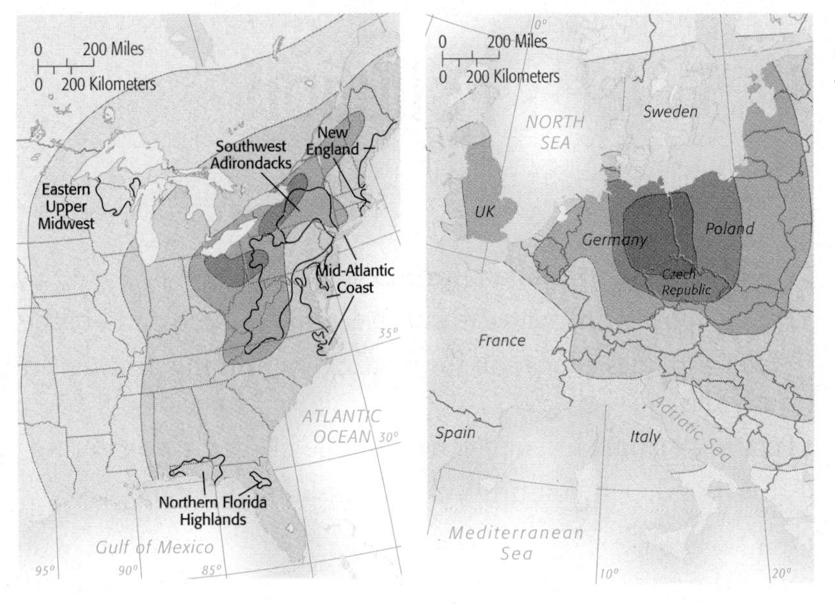

14.4.3 **ACID DEPOSITION IN NORTH AMERICA AND EUROPE**

Kilograms per hectare
- 60 and above
- 40 to 59
- 30 to 39
- 20 to 29
- 10 to 19
- below 10

Geographers are particularly interested in the effects of acid precipitation because the worst damage may not be experienced at the same location as the emission of the pollutants. Before they reach the surface, these acidic droplets might be carried hundreds of kilometers.

LOCAL-SCALE AIR POLLUTION

At the local scale, air pollution is especially severe in places where emission sources are concentrated, such as in urban areas. The air above urban areas may be especially polluted because a large number of factories, motor vehicles, and other polluters emit residuals in a concentrated area. Weather conditions may make it difficult for the emissions to dissipate.

Urban air pollution has three basic components:

- **Carbon monoxide.** Produced by improper combustion in power plants and vehicles.
- **Hydrocarbons.** Also produced by improper combustion, as well as from evaporation of paint solvents. Hydrocarbons and nitrogen oxides in the presence of sunlight form **photochemical smog**, which causes respiratory problems, stinging in the eyes, and an ugly haze over cities.
- **Particulates.** Dust and smoke particles, especially from factory smoke stacks and vehicle exhausts.

14.4.4 **SMOG OVER LOS ANGELES**

14.4.5 **ACID DEPOSITION IN NORTH CAROLINA**
Mount Mitchell State Park Fraser fir and red spruce evergreen trees killed by acid rain and balsam woolly aphids.

POLLUTING RESOURCES

14.5 Water Pollution

- Three water pollution sources are industries, sewers, and agriculture.
- The Aral Sea is one of the world's most extreme cases of water pollution.

Water serves many human purposes.

- People must drink water to survive, and they cook and bathe with water.
- Water provides a location for boating, swimming, fishing, and other recreation activities.
- People consume fish and other aquatic life and use water for agriculture.

These uses depend on fresh, clean, unpolluted water.

Clean water is not always available, because people also use water for purposes that pollute it. Pollution is widespread, because it is easy to dump waste into a river and let the water carry it downstream where it becomes someone else's problem. Water can decompose some waste without adversely impacting other activities, but the volume exceeds the capacity of many rivers and lakes to accommodate it.

THREE WATER POLLUTION SOURCES

Three main sources generate most water pollution:

- **Water-using industries**. Industries such as steel, chemicals, paper products, and food processing are major water polluters. Each requires a large amount of water in the manufacturing process and generates a lot of wastewater. Food processors, for example, wash pesticides and chemicals from fruit and vegetables. They also use water to remove skins, stems, and other parts. Water can also be polluted by industrial accidents, such as petroleum spills from ocean tankers and leaks from underground tanks at gasoline stations.
- **Municipal sewage.** In MDCs, sewers carry wastewater from sinks, bathtubs, and toilets to a municipal treatment plant, where most—but not all—of the pollutants are removed. The treated wastewater is then typically dumped back into a river or lake. In LDCs, sewer systems are rare, and wastewater usually drains untreated into rivers and lakes.
- **Agriculture.** Fertilizers and pesticides spread on fields to increase agricultural productivity are carried into rivers and lakes by the irrigation system or natural runoff. Expanded use of these products may help to avoid a global food crisis, yet they destroy aquatic life by polluting rivers.

These three sources of pollution can be divided into point sources and nonpoint sources. Point-source pollution enters a stream at a specific location, whereas nonpoint-source pollution comes from a large diffuse area. Manufacturers and municipal sewage systems tend to pollute through point sources, such as a pipe from a wastewater treatment plant. Farmers tend to pollute through nonpoint sources, such as by permitting fertilizer to wash from a field during a storm. Point-source pollutants are usually smaller in quantity and much easier to control. Nonpoint-sources usually pollute in greater quantities and are much harder to control.

14.5.1 **LEADING WATER POLLUTION SOURCES: MUNICIPAL SEWAGE (below), AGRICULTURE (right), AND INDUSTRY (far right).**

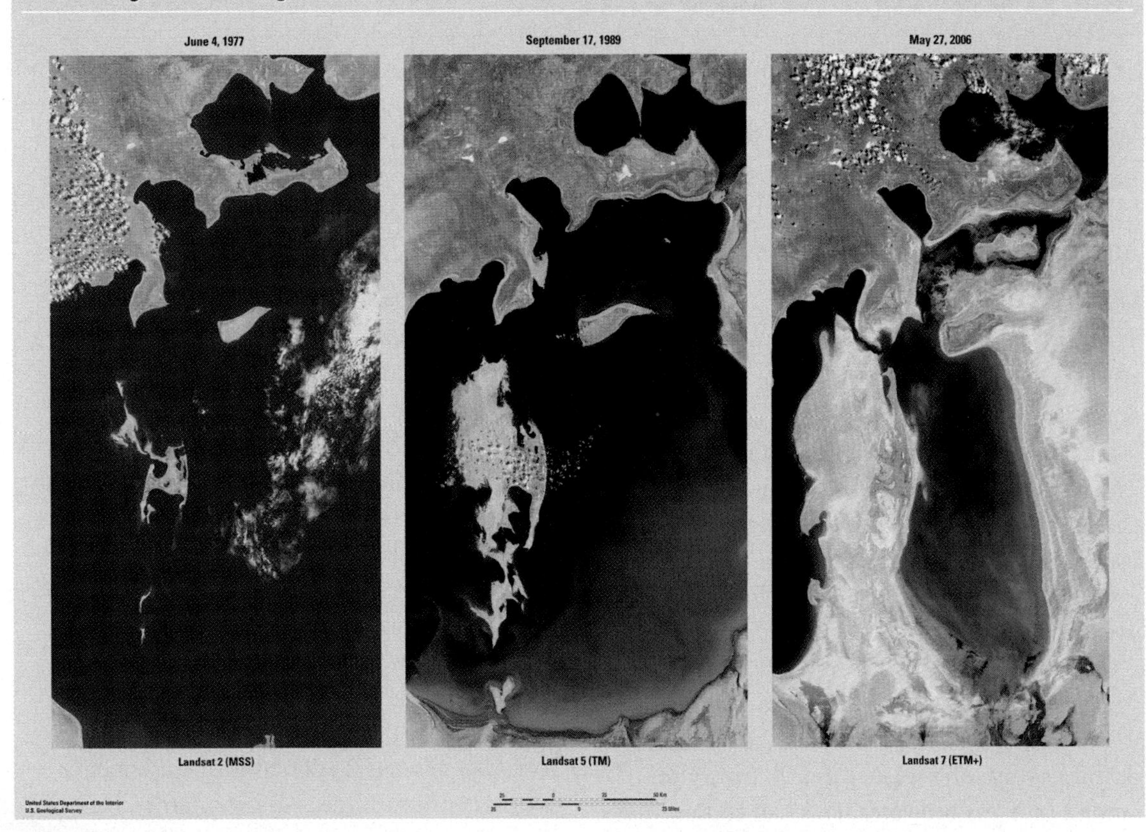

14.5.2 **THE ARAL SEA IN 1977 (left), 1989 (center), AND 2006 (right).**
The 1989 image shows an island forming in the sea. The 2006 image shows that the island has grown so large that it has divided the sea, and silting on the east side has turned the water area into a wasteland of salt.

THE DISAPPEARING ARAL SEA

One of the world's most extreme instances of water pollution is the Aral Sea (which is actually a lake) in the former Soviet Union, now divided between the countries of Kazakhstan and Uzbekistan. The world's fourth-largest lake in 1960, the Aral has been shrinking rapidly in area and volume and could disappear altogether by 2020. The Aral Sea declined from about 68,000 square kilometers in 1960 to 17,160 square kilometers in 2004 and approximately 7,000 in 2007.

The Aral Sea died because the Soviet Union diverted its tributary rivers, the Amu Dar'ya and the Syr Dar'ya, beginning in 1954, to irrigate cotton fields. Ironically, the cotton now is withering because winds pick up salt from the exposed lakebed and deposit it on the cotton fields.

Carp, sturgeon, and other fish species have disappeared, the last fish dying in 1983. Large ships lie aground in salt flats that were once the lake bed, near abandoned fishing villages that now lie tens of kilometers from the rapidly receding shore.

The destruction of the Aral Sea over several decades was little known and denied by the Soviet Union. Satellite imagery enabled geographers and other scientists to document without question the extent of the destruction of the Aral and to monitor precisely the speed of destruction.

The government of Kazakhstan announced a plan in 2003 to save the northern part of the Aral Sea. At the heart of the plan is construction of a dam to cut off the northern and southern portions. The purpose is to raise the water level and reduce the salinity in the north. The southern portion in Uzbekistan will disappear altogether in the years ahead.

POLLUTING RESOURCES

14.6 Land Pollution

- Solid waste is most often dumped in landfills.
- Paper is the most common solid waste.

When we consume a product, we also consume an unwanted by-product—a glass, metal, paper, or plastic box, wrapper, or container in which the product is packaged. About 2.1 kilograms (4.6 pounds) of solid waste per person is generated daily in the United States, including about 60 percent from residences and 40 percent from businesses.

LANDFILLS

The **sanitary landfill** is by far the most common strategy for disposal of solid waste in the United States: more than one-half of the country's waste is trucked to landfills and buried under soil. This strategy is the opposite of our disposal of gaseous and liquid wastes: we *disperse* air and water pollutants into the atmosphere, rivers, and eventually the ocean, but we *concentrate* solid waste in thousands of landfills.

Concentration would seem to eliminate solid-waste pollution, but it may only hide it—temporarily. Chemicals released by the decomposing solid waste can leak from the landfill into groundwater. This can contaminate water wells, soil, and nearby streams.

The number of landfills in the United States has declined by three-fourths since 1990. Thousands of small-town "dumps" have been closed and replaced by a small number of large regional ones. Better compaction methods, combined with expansion in the land area of some of the large regional ones, have resulted in expanded landfill capacity. At the same time, the two principal alternatives to disposing of solid waste in landfills—incineration and recycling—have both increased rapidly.

Some communities now pay to use landfills elsewhere. New Jersey and New York are two states that regularly try to dispose of their solid waste by transporting it out of state. New York City exports 25,000 tons of trash a day to other communities. Passaic County, New Jersey, hauls waste 400 kilometers (250 miles) west to Johnstown, Pennsylvania. San Francisco trucks solid waste to Altamont, California, 100 kilometers (60 miles) away.

14.6.1 **C & C LANDFILL, MARSHALL, MICHIGAN** Methane gas generated by decaying garbage in the landfill is being collected for use in generating electricity.

In LDCs, solid waste is carted off to landfills with much less stringent pollution controls than in MDCs.

Burning the trash reduces its bulk by about three-fourths, and the remaining ash demands far less landfill space. Incineration also provides energy—the incinerator's heat can boil water to produce steam heat or to operate a turbine that generates electricity. Given the shortage of space in landfills, the percentage of solid waste that is burned has increased rapidly during the past three decades, to one-sixth of solid waste.

HAZARDOUS WASTE

Solid waste, a mixture of many materials, may burn inefficiently. Burning releases some toxins into the air, and some remain in the ash. Thus solving one pollution problem may increase another.

Disposing of hazardous waste is especially difficult. Hazardous wastes include heavy metals (including mercury, cadmium, and zinc), PCB oils from electrical equipment, cyanides, strong solvents, acids, and caustics. These may have been unwanted by-products generated in manufacturing or discarded after use. If poisonous industrial residuals are not carefully placed in protective containers, the chemicals may leach into the soil and contaminate groundwater or escape into the atmosphere. Breathing air or consuming water contaminated with toxic wastes can cause cancer, mutations, chronic ailments, and even immediate death.

As toxic-waste disposal sites become increasingly hard to find, some European and North American firms have tried to transport their waste to West Africa, often unscrupulously. Some firms have signed contracts with West African countries, whereas others have found isolated locations to dump waste without official consent.

14.6.2 **MEXICO CITY'S GARBAGE PICKERS**
In Mexico City, thousands of people known as pepenadores, or garbage pickers, survive by going through rubbish. Older men make up the largest share of pepenadores. What they find is sold to companies for a small amount. In many cases, pepenadores actually live at the dump.

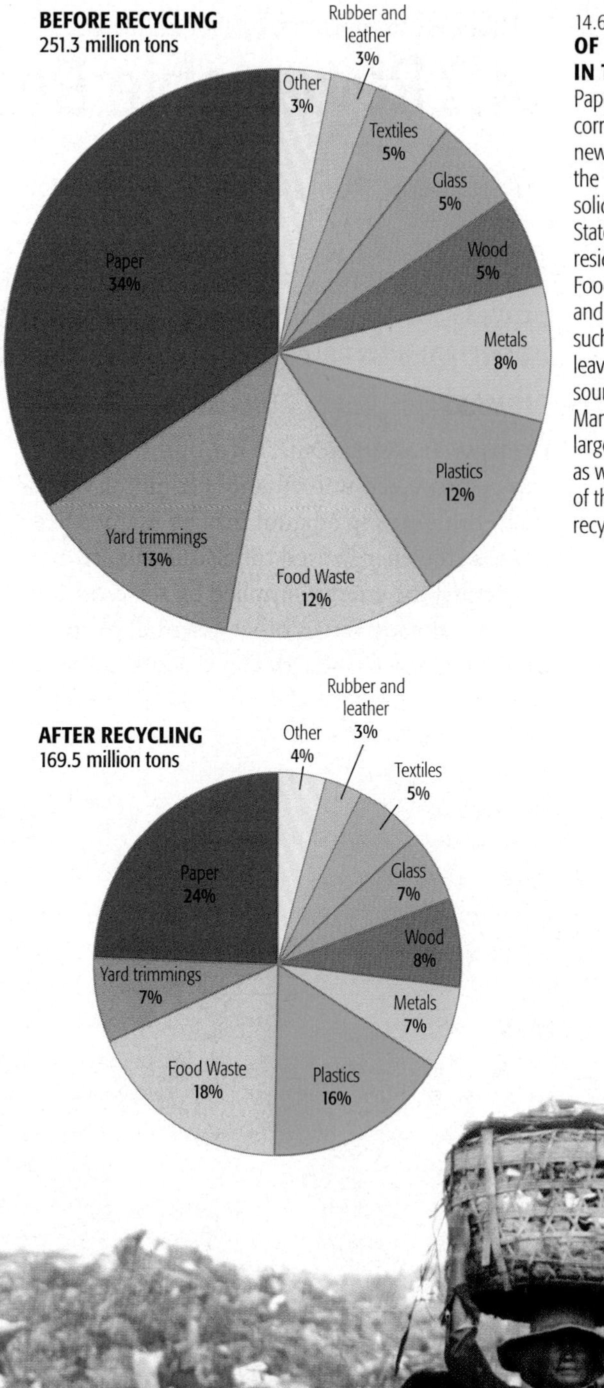

14.6.3 **SOURCES OF SOLID WASTE IN THE UNITED STATES**
Paper products, such as corrugated cardboard and newspapers, account for the largest percentage of solid waste in the United States, especially among residences and retailers. Food products, plastics, and debris from yards, such as grass clippings and leaves, are other important sources of solid waste. Manufacturers discard large quantities of metals, as well as paper. One-third of the solid waste is recycled.

CONSERVING AND REUSING RESOURCES

14.7 Renewable Resources

- Renewable energy sources provide alternatives to fossil fuels.
- Renewable energy sources include biomass, hydroelectric, wind, and solar power.

Renewable energy provides alternatives to fossil fuels. Biomass, water, wind, and the Sun provide sources of renewable energy. Nuclear power, though not renewable, has been an important alternative to fossil fuels in some places.

BIOMASS

Crops such as sugarcane, corn, and soybeans can be processed into ethanol for motor vehicle fuels. But burning ethanol may be inefficient because the energy used to produce the crops may equal the energy supplied by the crops. And biomass already serves other essential purposes, providing much of Earth's food, clothing, and shelter.

14.7.1 **ELECTRICITY FROM HYDROELECTRIC POWER**

Percent
- 75 and above
- 25 to 74
- 10 to 24
- 1 to 9
- below 1

PACIFIC OCEAN
ATLANTIC OCEAN
INDIAN OCEAN
Tropic of Cancer
Equator
Tropic of Capricorn

0 2,000 4,000 Miles
0 2,000 4,000 Kilometers

THREE GORGES DAM ON YANGTZE RIVER

HYDROELECTRIC

The world's second-most popular source of electricity, after coal, is hydroelectricity. Hydroelectric supplies about one-fourth of worldwide demand. Many LDCs depend on hydroelectric power for the vast majority of their electricity. Hydroelectric dams may flood formerly usable land, cause erosion, and upset ecosystems. The world's largest hydroelectric river dam, China's Three Gorges Dam, spanning the Yangtze River, has been especially criticized for its effects on local ecosystems.

SOLAR

The ultimate renewable resource is solar energy, supplied by the Sun. The Sun's energy is free and ubiquitous and utilizing it does not damage the environment or cause pollution. Solar energy is harnessed through either **passive systems** (such as heat-generating south-facing windows) or **active systems** (such as **photovoltaic cells**, which convert light energy to electrical energy).

NUCLEAR

Nuclear power supplies about one-sixth of the world's electricity. The big advantage of nuclear power is the large amount of energy released from a small amount of material. One kilogram of enriched nuclear fuel contains more than 2 million times the energy in 1 kilogram of coal. Waste from nuclear fuel is highly radioactive and lethal for thousands of years, and no one has yet devised permanent storage for it. Uranium, required for nuclear power, is a finite resource.

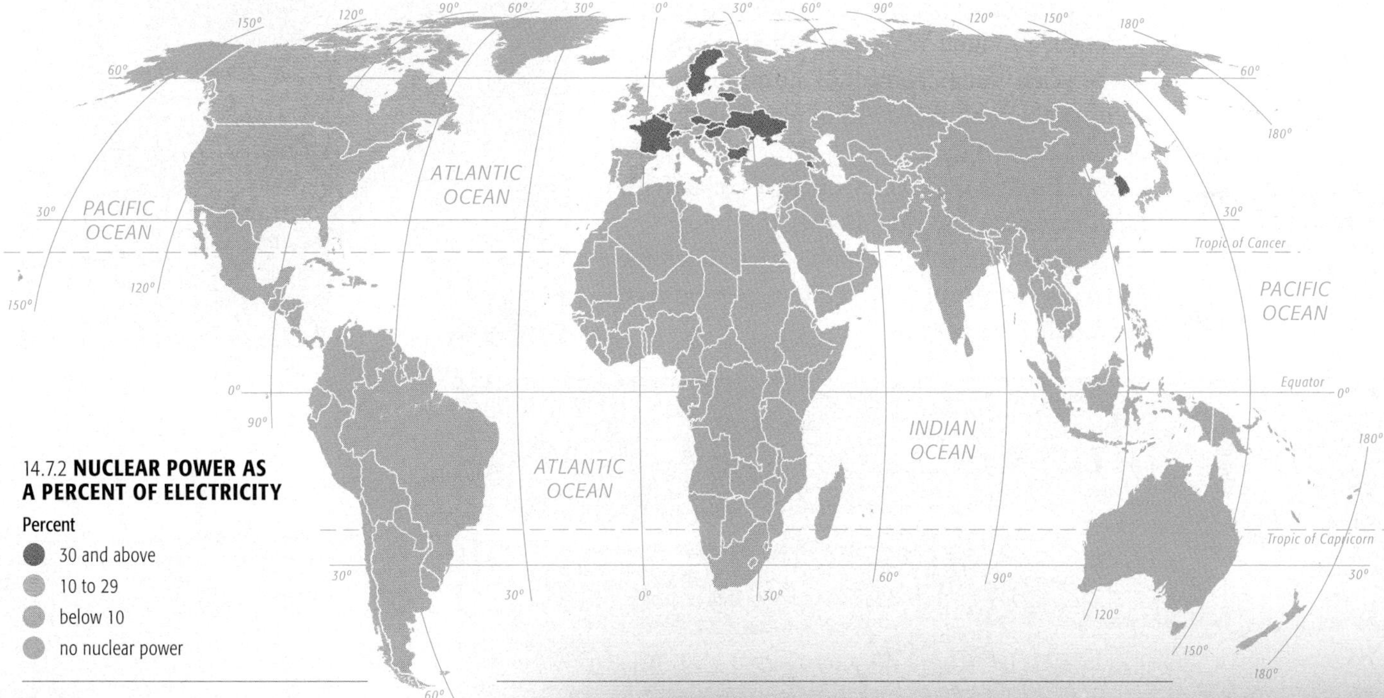

14.7.2 **NUCLEAR POWER AS A PERCENT OF ELECTRICITY**

Percent
- 30 and above
- 10 to 29
- below 10
- no nuclear power

WIND

Hundreds of wind "farms" consisting of dozens of wind turbines have been constructed across the United States. One-third of the country is considered windy enough to make wind power economically feasible. Like moving water, moving air can turn a turbine. Construction of wind turbines modifies the environment much less severely than construction of a dam across a river. However, some environmentalists oppose construction of wind turbines because they can be noisy and lethal for birds and bats. They also can be visually blighting when constructed on mountaintops or offshore of places of outstanding beauty.

CONSERVING AND REUSING RESOURCES

14.8 Recycling Resources

- Recyclables are collected in four ways.
- Some recycled material can be re-manufactured into new products.

Unwanted by-products are usually "thrown away," perhaps in a trash can. **Recycling** is the separation, collection, processing, marketing, and reuse of the unwanted material. Recycling increased in the United States from 7 percent of all solid waste in 1970 to 10 percent in 1980, 17 percent in 1990, and 33 percent in 2008.

As a result of recycling, about 82 million of the 251 million tons of solid waste generated in the United States in 2006 did not have to go to landfills and incinerators, compared to 34 million of 200 million tons generated in 1990.

The percentage of recovered materials varies widely by product: 52 percent of paper products are recycled and 45 percent of aluminum cans, compared to only 5 percent of plastic and 2 percent of food scraps.

Recycling involves two main series of activities. First, materials that would otherwise be thrown away are collected and sorted. Then the materials are manufactured into new products for which a market exists.

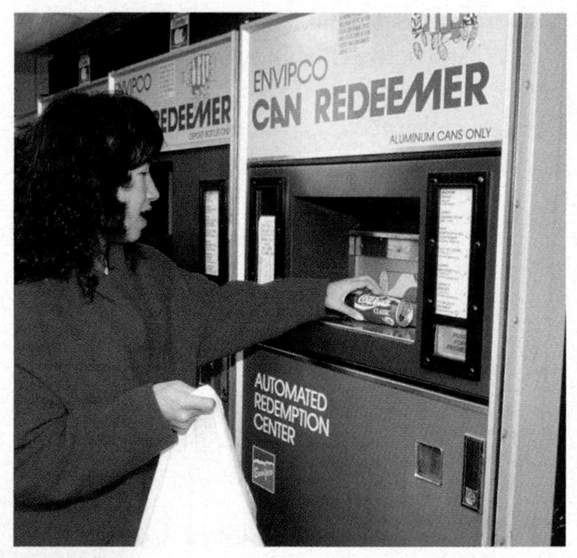

14.8.1 **FOUR METHODS OF RECYCLING**
Deposit programs (right), drop-off centers (below), curbside pick-up (below right), buy-back centers (far right).

PICK-UP AND PROCESSING

Recyclables are collected in four primary methods:

- **Deposit programs.** Return of glass and aluminum containers to retailers.
- **Drop-off centers.** Sites for individuals to leave recyclable materials, typically several large containers placed at a central location.
- **Curbside programs.** Recyclables placed at the curb in a separate container from the nonrecycled trash at a specified time each week.
- **Buy-back centers.** Commercial operations that pay consumers for recyclable materials, especially aluminum cans, but also sometimes plastic containers and glass bottles.

Regardless of the collection method, recyclables are sent to a materials recovery facility to be sorted and prepared into marketable commodities for manufacturing. Recyclables are bought and sold just like any other commodity: typical prices in recent years have been 30¢ per pound for plastic, $30 per ton for clear glass, and $90 per ton for corrugated paper. Prices for the materials change and fluctuate with the market.

RE-MANUFACTURING

Once cleaned and separated, the recyclables are ready to be manufactured into a marketable product. Four major manufacturing sectors accounted for more than half of the recycling activity—paper mills, steel mills, plastic converters, and iron and steel foundries. Common household items that contain recycled materials include newspapers and paper towels; aluminum, plastic, and glass soft drink containers; steel cans; and plastic laundry detergent bottles. Recycled materials are also used in such industrial applications as recovered glass in roadway asphalt ("glassphalt") or recovered plastic in carpeting, park benches, and pedestrian bridges.

- **Paper.** Most types can be recycled. Newspapers have been recycled profitably for decades, and recycling of other paper is growing, especially computer paper. The key to recycling is collecting large quantities of clean, well-sorted, uncontaminated, and dry paper.
- **Plastic.** Principal obstacle is that plastic types must not be mixed. Even a small amount of the wrong type of plastic can ruin the melt. The plastic industry has responded to this problem by developing a series of numbers marked inside triangles on the bottom of containers.
- **Glass.** Glass is 100 percent recyclable and can be used repeatedly with no loss in quality. Although unbroken clear glass is valuable, mixed color glass is nearly worthless, and broken glass is hard to sort.
- **Aluminum.** Scrap is readily accepted for recycling, although other metals are rarely accepted. The principal source of recycled aluminum is beverage containers.

14.8.2 **PAPER RECYCLING**
Forklift truck with clamping grip lifting a large bale of used paper at a paper recycling factory in Cuijk, The Netherlands.

CONSERVING AND REUSING RESOURCES

14.9 Sustainability

- Sustainable development utilizes resources at a rate that conserves them for the future.
- For geographers, biodiversity measures the number of species in an area.

Conservation is the sustainable use and management of natural resources such as wildlife, water, air, and Earth deposits to meet human needs, including food, medicine, and recreation.

- *Renewable resources* such as trees are conserved if they are consumed at a less rapid rate than they can be replaced.
- *Nonrenewable resources* such as fossil fuels are conserved if remaining reserves are maintained for future generations.

Conservation is compatible with development but only if natural resources are utilized in a careful rather than a wasteful manner.

Conservation differs from **preservation**, which is maintenance of resources in their present condition, with as little human impact as possible. The value of nature does not derive from human needs and interests, but from the fact that every plant and animal living on Earth has a right to exist and should be preserved regardless of the cost. The concept of preservation does not regard nature as a resource for human use.

14.9.1 **A LOGGING TRUCK BEING LOADED IN THE IDAHO MOUNTAINS**

SUSTAINABLE DEVELOPMENT

An increasingly important approach to careful utilization of resources is **sustainable development**, based on promotion of biodiversity. Sustainable development is "development that meets the needs of the present without compromising the ability of future generations to meet their own needs," according to the United Nations.

The concept of sustainable development is based on limiting the use of renewable resources to the level at which the environment can continue to supply them indefinitely. The amount of timber cut down in a forest or the number of fish removed from a body of water can be controlled at a level that does not reduce future supplies.

The UN's "sustainable development" was defined in the 1987 *Brundtland Report*, named for the chair of the World Commission on Environment and Development' Gro Harlem Brundtland, former Prime Minister of Norway. The *Brundtland Report* argued that sustainable development had to recognize the importance of economic growth while conserving natural resources. Environmental protection, economic growth, and social equity are linked because economic development aimed at reducing poverty can at the same time threaten the environment.

A rising level of economic development generates increased pollution, at least until a country reaches a GDP of about $5,000 per person, according to economists Gene Grossman and Alan Krueger. But the *Brundtland Report* was optimistic that environmental protection could be promoted at the same time as economic growth and social equity.

Some environmentally oriented critics have argued that it is too late to discuss sustainability. The World Wildlife Fund (WWF), for example, claims that the world surpassed its sustainable level around 1980. The WWF calculates that humans are currently using about 13 billion hectares of Earth's land area, including 3 billion hectares for cropland, 2 billion for forest, 7 billion for energy, and 1 billion for fishing, grazing, and built-up areas. However, according to the WWF, Earth has only 11.4 billion hectares of biologically productive land, so humans are already using all of the productive land and none is left for future growth.

BIODIVERSITY

Biological diversity, or **biodiversity** for short, refers to the variety of species across Earth as a whole or in a specific place. Sustainable development is promoted when biodiversity of a particular place or Earth as a whole is protected.

- For biologists, biodiversity refers particularly to the maintenance of genetic diversity within populations of plants and animals. Strategies to protect genetic diversity have been established on a global scale. Some endangered species have been protected by the Convention on International Trade in Endangered Species of Wild Fauna and Flora. Examples include the curtailing of logging, whaling, and taking of porpoises in tuna seines (nets).
- Geographers are especially concerned with biogeographic diversity, which is the number of species within a specific region or habitat. A community containing a large number of species is said to be species-rich, whereas an area with few species is species-poor. Strategies to protect biogeographic diversity vary among countries. Luxembourg protects 44 percent of its land and Ecuador 38 percent, whereas Cambodia, Iraq, and some former Soviet Union republics have no land under conservation.

Reduction of biodiversity through species extinction is especially troublesome in tropical forests. Three species per hour are extinguished in the tropics, and more than 5,000 species are considered in danger of extinction. Tropical forests occupy only 7 percent of Earth's land area but contain more than one-half of the world's species, including two-thirds of vascular plant species and one-third of avian species.

The principal cause of the high rate of extinction is cutting down forests. Rapid deforestation results from changing economic activities in the tropics, especially a decline in shifting cultivation. Tropical rain forests are disappearing at the rate of 10 to 20 million hectares (25 to 50 million acres) per year. The amount of Earth's surface allocated to tropical rain forests has been reduced to less than one-half of its original area during the past quarter century.

14.9.2 **BIODIVERSITY** The welcome sign at the main trailhead to the Monteverde "Cloud Forest" Reserve, Costa Rica.

Chapter **Summary**

DEPLETING RESOURCES

- Most energy comes from three nonrenewable fossil fuels—petroleum, natural gas, and coal.
- MDCs consume most of the fossil fuels and must import most of them from other countries.
- Petroleum reserves are especially scarce and located primarily in the Middle East.
- Nonmetallic and metallic mineral resources are also important for humans, and most of them are clustered in a handful of locations.

POLLUTING RESOURCES

- Air pollution occurs at global, regional, and local scales.
- Water is needed for human survival and is used to discharge waste, especially from industries, sewage, and agriculture.
- Solid waste is clustered in a small number of sanitary landfills.

CONSERVING AND REUSING RESOURCES

- Renewable energy sources include hydroelectric, wind, and solar power.
- Solid waste resources are recycled through pick-up, processing, and re-manufacturing.
- Sustainable development conserves resources for future generations and preserves biodiversity.

SAUDI ARABIA. RAS TANURA OIL REFINERY

Geographic Consequences of Change

Will depletion and destruction of Earth's resources lead to disaster? In 1968, a group of scientists known as the Club of Rome presented an influential argument for this position in a report titled *The Limits to Growth*. According to these scientists, many of whom were professors at the Massachusetts Institute of Technology, the combination of population growth, resource depletion, and unrestricted use of industrial technology will disrupt the world's ecology and economy and lead to mass starvation, widespread suffering, and destruction of the physical environment.

In a recent update, the authors argued that environmental destruction is proceeding at a more rapid rate than they had originally thought. If new sets of attitudes and policies toward environmental protection are not in place within 20 years, the environment will be permanently damaged, and people's standards of living will fall.

Most geographers recognize that unrestricted industrial and demographic growth will have negative consequences, but they do not believe that the dire predictions of *The Limits to Growth* are inevitable. Human actions have depleted some resources, but substitutes may be available. Although pollution degrades the physical environment, industrial growth can be compatible with environmental protection. Demand for food is increasing, but human actions are also expanding the capacity of Earth to provide food.

Nonetheless, future generations will pay the price if we continue to mismanage Earth's resources. Our shortsightedness could lead to shortages of energy to heat homes and operate motor vehicles. Our carelessness has already led to unsafe drinking water and toxic air in some places. Humans once believed Earth's resources to be infinite, or at least so vast that human actions could never harm or deplete them. But warnings from scientists, geographers, and governments are making it clear that resources are indeed limited.

SHARE YOUR VOICE Student Essay

Sitara Nayudu
UC Irvine

Breaking News: Environment Now Sexy

In our age of pop culture worship, people care more about celebrity babies and drama than environmental issues. Celebrities are stereotyped as exotic and exciting, while environmental advocates are stereotyped as hippy-like and close-minded.

As an environmental engineering major, I don't spend my time sporting hemp clothing and driving a natural gas car. Nor am I spray painting Hummers or chaining myself to trees. And when it comes down to class, I find environmental books to be quite tedious and boring.

However, I understand that environmental issues are real and immediate. The problem we face as a society is that real issues aren't presented as exciting and fashionable, but instead as severe and accusatory.

Deforestation, global warming and the contamination and depletion of natural resources are critical problems we are facing. As the human population continues to grow exponentially, Mother Earth is finding it harder and harder to sustain its inhabitants.

Let's take deforestation as an example: Currently forests cover 30% of the Earth's surface, but at the rate they are being cut down (swaths the size of Panama are gone every year), all the Earth's forests will disappear within the century. One of the biggest negative effects is the loss of habitat for millions of plant and animal species.

What most people don't realize is that a majority of our medications to treat cancer and other diseases come from the rainforests. And with only a fraction of the rainforests' many species discovered, what answers are still available out there for the illnesses and cancers that plague our people? Furthermore the world's forests help purify the air and water by sucking up tons of carbon dioxide and toxins daily.

Like deforestation, global warming and contamination/depletion of our natural resources are equally detrimental to our world's standard of living and well-being. What we need to do as a society is to stop pointing fingers and throwing out accusations and start presenting the issues as more appealing to the public. People are more likely to be involved in issues they find exciting and rewarding. Everyone wants to be a hero, so why not save the world?

Chapter Resources

THE CAR OF THE FUTURE

The most serious challenge to reducing pollution and conserving nonrenewable resources is reliance on petroleum as automotive fuel.

Consumers in MDCs will continue to rely on motor vehicles, while demand for vehicles is soaring in LDCs. So carmakers are scrambling to bring alternative fuel vehicles to the market. Among the alternatives are the following:

- **Hybrid vehicles.** Sales of hybrid vehicles have increased rapidly during the first decade of the twenty-first century, led by Toyota's success with the hybrid Prius. A gasoline engine powers the vehicle at high speeds, but at low speeds, when the gas engine is at its least efficient, an electric motor takes over. Energy wasted in coasting and braking is also captured as electricity and stored until needed.

- **Diesels.** Diesel engines burn fuel more efficiently, with greater compression at a higher temperature than conventional gas engines. Most new vehicles in Europe are diesel-powered, where they are valued for zippy acceleration on crowded roads, as well as for high gas mileage. Diesels have made limited inroads in the United States, because most diesel engines for cars have not met U.S. pollution-control regulations.

- **Biodiesels.** Biodiesel engines combine diesel fuel with approximately 5 percent biodiesel fuel, which is produced from vegetable oils or recycled restaurant grease.

- **Ethanol.** Ethanol is fuel made by distilling crops. Brazil has pioneered distilling sugar cane for fuel. In the United States, the use of corn has been encouraged, but this has proved controversial because the amount of fossil fuels needed to grow and distill the corn is comparable to—and possibly greater than—the amount saved in vehicle fuels. Furthermore, growing corn for ethanol diverts corn from the food chain, allegedly causing higher food prices in the United States and globally. More promising is ethanol distilled from cellulosic biomass, such as trees and grasses.

- **Plug-in electrics.** Batteries that recharge by plugging into an electric outlet supply the power for these vehicles. Principal limitation has been short range of batteries before they need recharging. To extend the range to that of a conventional gas engine, a plug-in electric car can have a small gas motor that recharges the batteries. A plug-in vehicle may save petroleum, but if the ultimate source of electricity is a coal or natural gas power plant, it is not conserving nonrenewable fossil fuels.

- **Hydrogen fuel cells.** Hydrogen forced through a PEM (polymer electrolyte membrane or proton exchange membrane) combines with oxygen from the air, producing an electric charge. The electricity powers an electric motor.

THE "GREENEST" ALTERNATIVE VARIES BY LOCATION

In places where most electricity is generated by natural gas, alternative fuels that require plugging into the electric grid may conserve petroleum at the expense of more rapid depletion of natural gas. Where most electricity is generated by coal, a plug-in may cause more air pollution. Plug-in vehicles may make more sense in states that depend on nuclear or hydroelectric power for electricity. States that depend on farm production may benefit from increased use of ethanol.

The United States as a whole gets 51 percent of electricity from coal, 20 percent from nuclear, 19 percent from natural gas, 6 percent from hydroelectric, and 4 percent from other sources of power.

Here are some variations among states.

PRINCIPAL SOURCE OF ELECTRICITY

COAL 90+%	NATURAL GAS 50+%	NUCLEAR 50+%	HYDROELECTRIC 50+%
West Virginia	Rhode Island	Vermont	Idaho
North Dakota	Nevada	Connecticut	Washington
Indiana	California	South Carolina	Oregon
Wyoming	Alaska	New Jersey	
Kentucky			

THREE LEADING SOURCES OF ELECTRICITY IN THE FIVE MOST POPULOUS STATES

California	55% natural gas	16% hydroelectric	15% nuclear
New York	34% nuclear	27% natural gas	16% coal
Texas	45% natural gas	39% coal	12% nuclear
Florida	42% natural gas	35% coal	17% nuclear
Illinois	52% coal	45% nuclear	

KEY TERMS

Acid deposition
Sulfur oxides and nitrogen oxides, emitted by burning fossil fuels, enter the atmosphere—where they combine with oxygen and water to form sulfuric acid and nitric acid—and return to Earth's surface.

Acid precipitation
Conversion of sulfur and oxygen oxides to acids that return to Earth as rain, snow, or fog.

Active solar energy systems
Solar energy system that collects energy through the use of mechanical or electrical devices like photovoltaic cells or flat-plate collectors.

Air pollution
Concentration of trace substances, such as carbon monoxide, sulfur dioxide, nitrogen oxides, hydrocarbons, and solid particulates, at a greater level than occurs in average air.

Biodiversity
The number of species within a specific habitat.

Biomass fuel
Fuel that derives from plant material and animal waste.

Chlorofluorocarbon (CFC)
A gas used as a solvent, a propellant in aerosols, a refrigerant, and in plastic foams and fire extinguishers.

Conservation
The sustainable use and management of a natural resource through consuming at a less rapid rate than the resource's replacement rate.

Ferrous
Metals, including iron ore, that are utilized in the production of iron and steel.

Fossil fuel
Energy source formed from the residue of plants and animals buried millions of years ago.

Fusion
Creation of energy by joining the nuclei of two hydrogen atoms to form helium.

Geothermal
Energy from steam or hot water produced from hot or molten underground rocks.

Greenhouse effect
Anticipated increase in Earth's temperature, caused by carbon dioxide (emitted by burning fossil fuels) trapping some of the radiation emitted by the surface.

Hydroelectric power
Power generated from moving water.

Nonferrous
Metals utilized to make products other than iron and steel.

Nonrenewable energy
A source of energy that is a finite supply capable of being exhausted.

Ozone
An unstable form of oxygen that absorbs ultraviolet radiation. The Ozone Layer is a zone between 15 and 50 kilometers (9 to 30 miles) above the Earth's surface.

Passive solar energy systems
Solar energy system that collects energy without the use of mechanical devices.

Photochemical smog
An atmospheric condition formed through a combination of weather conditions and pollution, especially from motor vehicle emissions.

Photovoltaic cell
Solar energy cells, usually made from silicon, that collect solar rays to generate electricity.

Potential reserve
The amount of energy in deposits not yet identified but thought to exist.

Preservation
Maintenance of a resource in its present condition, with as little human impact as possible.

Proven reserve
The amount of a resource remaining in discovered deposits.

Recycling
The separation, collection, processing, marketing, and reuse of unwanted material.

Renewable energy
A resource that has a theoretically unlimited supply and is not depleted when used by humans.

Resource
A substance in the environment that is useful to people, is economically and technologically feasible to access, and is socially acceptable to use.

Sanitary landfill
A place to deposit solid waste, where a layer of earth is bulldozed over garbage each day to reduce emissions of gases and odors from the decaying trash, to minimize fires, and to discourage vermin.

Sustainable development
The level of development that can be maintained in a country without depleting resources to the extent that future generations will be unable to achieve a comparable level of development.

FURTHER READINGS

Aubrecht, Gordon J. II. *Energy: Physical, Environmental, and Social Impact.* Upper Saddle River, NJ: Pearson Prentice Hall, 3d ed., 2006.

Barr, Stewart, "Strategies for Sustainability: Citizens and Responsible Environmental Behaviour." *Area* 35 (2003): 227–40.

Brown, Lester R., et al. *State of the World.* New York and London: W. W. Norton & Co., annually since 1984.

Buttimer, Anne, ed. *Sustainable Landscapes and Lifeways: Scale and Appropriateness.* Cork, Ireland: Cork University Press, 2001.

Commoner, Barry. *Making Peace with the Planet.* New York: New Press, 1990.

Cutter, Susan L., and William H. Renwick. *Exploitation, Conservation, Preservation: A Geographic Perspective on Natural Resource Use*, 4th ed. Danvers, MA: John Wiley, 2004.

Ehrlich, Paul R., and Anne H. Ehrlich. *Betrayal of Science and Reason: How Anti-Environmental Rhetoric Threatens Our Future.* Washington: Island Press, 1998.

Gerdes, Louise I., ed. *Pollution: Opposing Viewpoints.* Detroit: Greenhaven Press, 2006.

Kidd, J. S., and Renee A. Kidd. *Air Pollution: Problems and Solutions.* New York: Chelsea House, 2006.

Leone, Daniel A., ed. *Is the World Heading Toward an Energy Crisis?* Farmington Hills, MI: Greenhaven Press, 2006.

Livingston, James V., ed. *Focus on Water Pollution Research.* New York: Nova Science Publishers, 2006.

Mallon, Karl, ed. *Renewable Energy Policy and Politics: A Handbook for Decision-making.* London and Sterling, VA: Earthscan, 2006.

Meadows, Donnela H., Jorgen Randers, and Dennis L. Meadows. *Limits to Growth: The 30-Year Update.* White River Junction, VT: Chelsea Green Publishing, 2004.

ON THE INTERNET

Statistics on production, consumption, and reserves of different energy sources in the United States and worldwide can be found on the web site of the U.S. Department of Energy's Energy Information Administration **(www.eia.doe.gov)**. The U.S. Environmental Protection Agency provides information organized by various types of pollution on its website **www.epa.gov.**

GLOSSARY

Acid deposition Sulfur oxides and nitrogen oxides, emitted by burning fossil fuels, enter the atmosphere—where they combine with oxygen and water to form sulfuric acid and nitric acid—and return to Earth's surface.

Acid precipitation Conversion of sulfur and oxygen oxides to acids that return to Earth as acidic rain, snow, or fog.

Active solar energy systems Solar energy system that collects and stores energy through the use of mechanical or electrical devices such as photovoltaic cells or flat-plate collectors.

Agribusiness Commercial agricultural concerns often owned by larger corporations and characterized by the integration of different steps into the larger food-processing industry.

Agricultural density The ratio of the number of farmers to the total amount of land suitable for agriculture.

Agriculture The deliberate effort to modify a portion of Earth's surface through the cultivation of crops and the raising of livestock for sustenance or economic gain.

Air pollution Concentration of trace substances, such as carbon monoxide, sulfur dioxide, nitrogen oxides, hydrocarbons, and solid particulates, at a greater level than occurs in average air.

Animism Belief that objects, such as plants and stones, or natural events, such as thunderstorms and earthquakes, have a discrete spirit and conscious life.

Annexation Legally adding land area to a city in the United States.

Apartheid Laws (no longer in effect) in South Africa that physically separated different races into different geographic areas.

Arable land Land suitable for agriculture

Arithmetic density The total number of people divided by the total land area.

Balance of power Condition of roughly equal strength between opposing countries or alliances of countries.

Balkanization Process by which a state breaks down through conflicts among its ethnicities.

Balkanized A descriptive term to indicate a small geographic area that cannot successfully be organized into one or more stable states because it is inhabited by many ethnicities with complex, long-standing antagonisms toward each other.

Basic industries Industries that sell their products or services primarily to consumers outside the settlement.

Biodiversity The number of species within a specific habitat.

Biomass fuel Fuel that derives from plant material and animal waste.

Boundary Invisible line that marks the extent of a state's territory.

Branch (religion) A large and fundamental division within a religion.

Break-of-bulk point A location where transfer is possible from one mode of transportation to another.

Bulk-gaining industry An industry in which the final product weighs more or comprises a greater volume than the inputs.

Bulk-reducing industry An industry in which the final product weighs less or comprises a lower volume than the inputs.

Business services Services that primarily meet the needs of other businesses, including professional, financial, and transportation.

Census tract An area delineated by the U.S. Bureau of the Census for which statistics are published; in urbanized areas, census tracts correspond roughly to neighborhoods.

Central business district (CBD) The area of a city where retail and office activities are clustered.

Central place A market center for the exchange of services by people attracted from the surrounding area.

Central place theory A theory that explains the distribution of services wherein people from surrounding areas are attracted to a market center for the exchange of services.

Cereal grain A grass yielding grain for food.

Chain migration Migration of people to a specific location because relatives or members of the same nationality previously moved there.

Chlorofluorocarbon (CFC) A gas used as a solvent, a propellant in aerosols, a refrigerant, and in plastic foams and fire extinguishers.

City-state A sovereign state comprising a city and its immediate hinterland.

Clustered rural settlement A rural settlement in which the houses and farm buildings of each family are situated close to each other and fields surround the settlement.

Colonialism Attempt by one country to establish settlements and to impose its political, economic, and cultural principles in another territory.

Colony A territory that is legally tied to a sovereign state rather than completely independent.

Commercial agriculture Agriculture undertaken primarily to generate products for sale off the farm.

Compact state A state in which the distance from the center to any boundary does not vary significantly.

Concentration The spread of something over a given area.

Concentric zone model A model of the internal structure of cities in which social groups are spatially arranged in a series of rings.

Conservation The sustainable use and management of a natural resource through consumption at a less rapid rate than the resource's replacement rate.

Consumer services Businesses that provide services primarily to individual consumers, including retail services and education, health, and leisure services.

Contagious diffusion The rapid, widespread diffusion of a feature or trend throughout a population.

Cosmogony A set of religious beliefs concerning the origin of the universe.

Cottage industry Manufacturing based in homes rather than in a factory, commonly found prior to the Industrial Revolution.

Counterurbanization Net migration from urban to rural areas in more developed countries.

Crop rotation The practice of rotating the use of different fields from crop to crop each year to avoid exhausting the soil.

Crude birth rate (CBR) The total number of live births in a year for every 1,000 people alive in the society.

Crude death rate (CDR) The total number of deaths in a year for every 1,000 people alive in the society.

Cultural landscape Fashioning of a natural landscape by a cultural group.

Culture The body of customary beliefs, social forms, and material traits that together constitute a group of people's distinct tradition.

Custom The frequent repetition of an act, to the extent that it becomes characteristic of the group of people performing the act.

Demographic transition The process of change in a society's population from a condition of high crude birth and death rates and low rate of natural increase to a condition of low crude birth and death rates, low rate of natural increase, and a higher total population.

Denglish Combination of German and English.

Denomination A division of a branch religion that unites a number of local congregations in a single legal and administrative body.

Density The frequency with which something exists within a given unit of area.

Dependency ratio The number of people under the age of 15 and over age 64 compared to the number of people active in the labor force.

Development A process of improvement in the material conditions of people through diffusion of knowledge and technology.

Dialect A regional variety of a language distinguished by vocabulary, spelling, and pronunciation.

Diffusion The process of spread of a feature or trend from one place to another over time.

Dispersed rural settlement A rural settlement pattern characterized by isolated farms rather than clustered villages.

Distance decay The diminishing importance and eventual disappearance of a phenomenon with increasing distance from its origin.

Distribution The arrangement of something across Earth's surface.

Doubling time The number of years needed to double a population, assuming a constant rate of natural increase.

Economic base A community's collection of basic industries.

Elongated state A state with a long, narrow shape.

Emigration Migration from a location.

Environmental determinism A nineteenth- and early twentieth-century approach to the study of geography that argued that the general laws sought by human geographers could be found in the physical sciences. Geography was therefore the study of how the physical environment caused human activities.

Epidemiologic transition Distinctive causes of death in each stage of the demographic transition.

Epidemiology Branch of medical science concerned with the incidence, distribution, and control of diseases that affect large numbers of people.

Ethnic cleansing Process in which a more powerful ethnic group forcibly removes a less powerful one in order to create an ethnically homogeneous region.

Ethnic religion A religion with a relatively concentrated spatial distribution whose principles are likely to be based on the physical characteristics of the particular location in which its adherents are concentrated.

Ethnicity Identity with a group of people that share distinct physical and mental traits as a product of common heredity and cultural traditions.

Expansion diffusion The spread of a feature or trend among people from one area to another in a snowballing process.

Extinct language A language that was once used by people in daily activities but is no longer used.

Fair trade Alternative in international trade that emphasizes small businesses, worker-owned and democratically run cooperatives. It requires employers to pay workers fair wages, permit union organizing, and comply with minimum environmental and safety standards.

Federal state An internal organization of a state that allocates most powers to the units of local government.

Ferrous metal Metals, including iron ore, that are used in the production of iron and steel.

Filtering A process of change in the use of a house from single-family owner occupancy to abandonment.

Floodplain The area subject to flooding according to historical trends.

Folk culture Culture traditionally practiced by a small, homogeneous, rural group living in relative isolation from other groups.

Forced migration Permanent movement usually compelled by cultural factors.

Fordist production Form of mass production in which each worker is assigned one specific task to perform repeatedly.

Foreign direct investment Investment made by a foreign company in the economy of another country.

Formal region (or uniform or homogeneous region) An area in which everyone shares one or more distinctive characteristics.

Fossil fuel Energy source formed from the residue of plants and animals buried millions of years ago.

Fragmented state A country that includes several discontinuous pieces of territory.

Franglais A term used by the French for English words that have entered the French language; a combination of français and anglais, the French words for *French* and *English*, respectively.

Functional region (or nodal region) An area organized around a node or focal point.

Fundamentalism Literal interpretation and strict adherence to basic principles of a religion; a religious branch, denomination, or sect.

Fusion Creation of energy by joining the nuclei of two hydrogen atoms to form helium.

Gender Empowerment Measure (GEM) Compares the ability of women and men to participate in economic and political decision making.

Gender-related Development Index (GDI) Compares the level of development of women with that of both sexes.

Gentrification A process of converting an urban neighborhood from a predominantly low-income renter-occupied area to a predominantly middle-class owner-occupied area.

Geographic information system (GIS) A computer system that stores, organizes, analyzes, and displays geographic data.

Geothermal Energy from steam or hot water produced from hot or molten underground rocks.

Gerrymandering Process of redrawing legislative boundaries for the purpose of benefiting the party in power.

Global Positioning System (GPS) A system that determines the precise position of something on Earth through a series of satellites, tracking stations, and receivers.

Globalization Actions or processes that involve the entire world and result in making something worldwide in scope.

Grain Seed of a cereal grass.

Gravity model A model that holds that the potential use of a service at a particular location is directly related to the number of people in a location and inversely related to the distance people must travel to reach the service.

Green revolution Rapid diffusion of new agricultural technology, especially new high-yield seeds and fertilizers.

Greenhouse effect Anticipated increase in Earth's temperature, caused by carbon dioxide (emitted by burning fossil fuels) trapping some of the radiation emitted by the surface.

Greenwich Mean Time (GMT) The time in that time zone encompassing the prime meridian, or 0° longitude.

Gross domestic product (GDP) The value of the total output of goods and services produced in a country in a given time period, normally 1 year.

Guest workers Workers who migrate to the more developed countries of Northern and Western Europe, usually from Southern and Eastern Europe or from North Africa, in search of higher-paying jobs.

Habit A repetitive act performed by a particular individual.

Hearth The region from which innovative ideas originate.

Hierarchical diffusion The spread of a feature or trend from one key person or node of authority or power to other persons or places.

Human Development Index (HDI) Constructed by the United Nations as an indicator of the level of development for each country when combining income, literacy, education, and life expectancy.

Hydroelectric power Power generated from moving water.

Ideograms The system of writing used in China and other East Asian countries in which each symbol represents an idea or a concept rather than a specific sound, as is the case with letters in English.

Immigration Migration to a new location.

Industrial Revolution A series of improvements in industrial technology during the eighteenth century that transformed the process of manufacturing goods.

Infant mortality rate (IMR) The total number of deaths in a year among infants under 1 year old for every 1,000 live births in a society.

Intensive subsistence agriculture A form of subsistence agriculture in which farmers must expend a relatively large amount of effort to produce the maximum feasible yield from a parcel of land.

Internal migration Permanent movement within a particular country.

International migration Permanent movement from one country to another.

Interregional migration Permanent movement from one region of a country to another within the same country.

Intraregional migration Permanent movement within one region of a country.

Isolated language A language that is unrelated to any other languages and therefore not attached to any language family.

Just-in-time delivery Shipment of parts and materials to arrive at a factory at the time they are needed.

Labor-intensive industry An industry for which labor costs comprise a high percentage of total expenses.

Landlocked state A country that does not have a direct outlet to the sea.

Language A system of communication through the use of speech; a collection of sounds understood by a group of people to have the same meaning.

Language branch A collection of languages related through a common language family that existed several thousand years ago.

Language family A collection of languages related to each other through a common ancestor before recorded history.

Language group A collection of languages within a branch that share a common origin in the relatively recent past and display relatively few differences in grammar and vocabulary.

Latitude The numbering system used to indicate the location of parallels drawn on a globe and measuring distance north and south of the equator (0°).

Life expectancy The average number of years an individual can be expected to live given current social, economic, and medical conditions. Life expectancy at birth is the average number of years a newborn infant can expect to live.

Lingua franca A language mutually understood and commonly used in trade by people who have different native languages.

Literacy rate The percentage of a country's people who can read and write.

Literary tradition A language that is written as well as spoken.

Location The position of anything on Earth's surface.

Longitude The numbering system used to indicate the location of meridians drawn on a globe and measuring distance east and west of the prime meridian (0°).

Map A two-dimensional, or flat, representation of Earth's surface or a portion of it.

Maquiladora Factories built by U.S. companies in Mexico near the U.S. border to take advantage of much lower labor costs in Mexico.

Market area (or hinterland) The area surrounding a central place from which people are attracted to use the place's goods and services.

Meridian An arc drawn on a map between the North and South poles.

Micropolitan statistical area An urbanized area of between 10,000 and 50,000 inhabitants, the county in which it is found, and adjacent counties tied to the city.

Migration Form of relocation diffusion involving a permanent move to a new location.

Migration transition Change in the migration pattern in a society that results from industrialization, population growth, and other social and economic changes that also produce the demographic transition.

Missionary An individual who helps to diffuse a universalizing religion.

Mobility All types of movement from one location to another.

Monotheism The doctrine or belief of the existence of only one god.

Multi-ethnic state State that contains more than one ethnicity.

Multinational state State that contains two or more ethnic groups with traditions of self-determination that agree to coexist peacefully by recognizing each other as distinct nationalities.

Multiple nuclei model A model of the internal structure of cities in which social groups are arranged around a collection of nodes of activities.

Nationalism Loyalty and devotion to a particular nationality.

Nationality Identity with a group of people that share legal attachment and personal allegiance to a particular place as a result of being born there.

Nation-state A state whose territory corresponds to that occupied by a particular ethnicity that has been transformed into a nationality.

Natural increase rate (NIR) The percentage growth of a population in a year, computed as the crude birth rate minus the crude death rate.

Net migration The difference between the level of immigration and the level of emigration.

New international division of labor Transfer of some types of jobs, especially those requiring low-paid, less-skilled workers, from more developed to less developed countries.

Nonbasic industries Industries that sell their products primarily to consumers in the community.

Nonferrous metal Metals used to make products other than iron and steel.

Nonrenewable energy A source of energy that is in finite supply and capable of being exhausted.

Official language The language adopted for use by the government for the conduct of business and publication of documents.

Outsourcing A decision by a corporation to turn over much of the responsibility for production to independent suppliers.

Overpopulation The number of people in an area exceeds the capacity of the environment to support life at a decent standard of living.

Ozone An unstable form of oxygen that absorbs ultraviolet radiation. The Ozone Layer is a zone between 15 and 50 kilometers (9 to 30 miles) above Earth's surface.

Pandemic Disease that occurs over a wide geographic area and affects a very high proportion of the population.

Parallel A circle drawn around the globe parallel to the equator and at right angles to the meridians.

Passive solar energy systems Solar energy system that collects energy without the use of mechanical devices.

Pastoral nomadism A form of subsistence agriculture based on herding domesticated animals.

Pattern The geometric or regular arrangement of something in a study area.

Perforated state A country that completely surrounds another one.

Photochemical smog An atmospheric condition formed through a combination of weather conditions and pollution, especially derived from motor vehicle emissions.

Photovoltaic cell Solar energy cells, usually made from silicon, that collect solar rays to generate electricity.

Physiological density The number of people per unit of area of arable land, which is land suitable for agriculture.

Pilgrimage A journey to a place considered sacred for religious purposes.

Place A specific point on Earth distinguished by a particular character.

Plantation A large farm in tropical and subtropical climates that specializes in the production of one or two crops for sale, usually to a more developed country.

Polder Land created by the Dutch for use in agriculture, housing or businesses by draining water from an area.

Polytheism Belief in or worship of more than one god.

Popular culture Culture found in a large, heterogeneous society that shares certain habits despite differences in other personal characteristics.

Possibilism The theory that the physical environment may set limits on human actions, but people have the ability to adjust to the physical environment and choose a course of action from many alternatives.

Post-Fordist production Adoption by companies of work rules such as the allocation of workers to teams and consensus problem solving.

Potential reserve The amount of energy in deposits not yet identified but thought to exist.

Preservation Maintenance of a resource in its present condition with as little human impact as possible.

Primary sector The portion of the economy concerned with the direct extraction of materials from Earth's surface, generally through agriculture, although sometimes by mining, fishing, and forestry.

Primate city rule A pattern of settlements in a country, such that the largest settlement has more than twice as many people as the second-ranking settlement.

Primate city The largest settlement in a country, if it has more than twice as many people as the second-ranking settlement.

Prime meridian The meridian, designated as 0° longitude, that passes through the Royal Observatory at Greenwich, England.

Productivity The value of a particular product compared to the amount of labor needed to make it.

Projection The system used to transfer locations from Earth's surface to a flat map.

Prorupted state An otherwise compact country with a large projecting extension.

Proven reserve The amount of a resource remaining in discovered deposits.

Public services Services offered by the government to provide security and protection for citizens and businesses.

Pull factor Factor that induces people to move into a new location.

Push factor Factor that induces people to leave old residences.

Quotas (migration) Laws that place maximum limits on the number of people who can immigrate to a country each year.

Race Identity with a group of people descended from a common ancestor.

Racism Belief that race is the primary determinant of human traits and capacities and that racial differences produce an inherent superiority of a particular race.

Racist A person who subscribes to the beliefs of racism.

Ranching A form of commercial agriculture in which livestock graze over an extensive area.

Range (of a service) The maximum distance people are willing to travel to use a service.

Rank-size rule Pattern of settlements in a country, such that the *n*th largest settlement is $1/n$ the population of the largest settlement.

Recycling The separation, collection, processing, marketing, and reuse of unwanted material.

Refugees People who are forced to migrate from their home country and cannot return for fear of persecution because of their race, religion, nationality, membership in a social group, or political opinion.

Region An unique area of Earth with its combination of features.

Relocation diffusion The spread of a feature or trend through bodily movement of people from one place to another.

Remote sensing The acquisition of data about Earth's surface from a satellite orbiting the planet or other long-distance methods.

Renewable energy A resource that has a theoretically unlimited supply and is not depleted when used by humans.

Resource A substance in the environment that is useful to people, is economically and technologically feasible to access, and is socially acceptable to use.

Ridge tillage System of planting crops on ridge tops in order to reduce farm production costs and promote greater soil conservation.

Sanitary landfill A place to deposit solid waste where a layer of earth is bulldozed over garbage each day to reduce emissions of gases and odors from the decaying trash, to minimize fires, and to discourage vermin.

Scale Generally, the relationship between the portion of Earth being studied and Earth as a whole, specifically the relationship between the size of an object on a map and the size of the actual feature on Earth's surface.

Secondary sector The portion of the economy concerned with manufacturing useful products through processing, transforming, and assembling raw materials.

Sect A relatively small group that has broken away from an established denomination.

Sector model A model of the internal structure of cities in which social groups are arranged around a series of sectors, or wedges, radiating out from the central business district (CBD).

Seed agriculture Reproduction of plants through annual introduction of seeds, which are the result of sexual fertilization.

Self-determination Concept that ethnicities have the right to govern themselves.

Sharecropper A person who works fields rented from a landowner and pays the rent and loans by turning over to the landowner a share of the crops.

Shifting cultivation A form of subsistence agriculture in which people shift activity from one field to another; each field is used for crops for a relatively few years and left fallow for a relatively long period.

Site factors Location characteristics related to the costs of production at the plant, such as land, labor, and capital.

Site The physical character of a place.

Situation factors Location characteristics related to the transportation of materials into and from a factory.

Situation The location of a place relative to other places.

Slash-and-burn agriculture Another name for shifting cultivation, so named because fields are cleared by slashing the vegetation and burning the debris.

Smart growth Legislation and regulations to limit suburban sprawl and preserve farmland.

Sovereignty Ability of a state to govern its territory free from control of its internal affairs by other states.

Space The physical gap or interval between two objects.

Space-time compression The reduction in the time it takes to diffuse something to a distant place, as a result of improved communications and transportation systems.

Spanglish Combination of Spanish and English.

Spatial interaction The movement of physical processes, human activities, and ideas within and among regions.

Sprawl Development of new housing sites at relatively low density and at locations that are not contiguous to the existing built-up area.

Squatter settlement An area within a city in a less developed country in which people illegally establish residences by erecting homemade structures on land they do not own or rent.

State An area organized into a political unit and ruled by an established government with control over its internal and foreign affairs.

Stimulus diffusion The spread of an underlying principle, even though a specific characteristic is rejected.

Structural adjustment program Economic policies imposed on LDCs by international agencies to create conditions encouraging international trade, such as reducing government spending, controlling inflation, selling publicly owned utilities to private corporations, and eliminating governmental waste and corruption.

Subsistence agriculture Agriculture designed primarily to provide food for direct consumption by the farmer and the farmer's family.

Sustainable agriculture Farming methods that preserve long-term productivity of land and minimize pollution, typically by rotating soil-restoring crops with cash crops and reducing inputs of fertilizer and pesticides.

Sustainable development The level of development that can be maintained in a country without depleting resources to the extent that future generations will be unable to achieve a comparable level of development.

Swidden A patch of land cleared for planting through slashing and burning.

Taboo A restriction on behavior imposed by social custom.

Tertiary sector The portion of the economy concerned with transportation, communications, and utilities, sometimes extended to the provision of all goods and services to people in exchange for payment.

Threshold The minimum number of people needed to support a service.

Toponym The name given to a portion of Earth's surface.

Total fertility rate (TFR) The average number of children a woman will have throughout her childbearing years.

Transnational corporation A company that conducts research, operates factories, and sells products in many countries, not just where its headquarters or shareholders are located.

Triangular slave trade A practice, primarily during the eighteenth century, in which European ships transported slaves from Africa to Caribbean islands, molasses from the Caribbean to Europe, and trade goods from Europe to Africa.

Truck farming Commercial gardening and fruit farming; term derived from *truck*, a Middle English word meaning bartering or the exchange of commodities.

Underclass A group in society prevented from participating in the material benefits of a more developed society because of a variety of social and economic characteristics.

Undocumented immigrants People who enter a country without proper documents.

Unitary state An internal organization of a state that places most power in the hands of central government officials.

Universalizing religion A religion that attempts to appeal to all people, not just those living in a particular location.

Urbanization An increase in the percentage and in the number of people living in urban settlements.

Urbanized area In the United States, a central city plus its contiguous, built-up suburbs.

Value added The gross value of the product minus the costs of raw materials and energy to produce it.

Vegetative planting Reproduction of plants by direct cloning from existing plants.

Vernacular region (or perceptual region) An area that people believe exists as part of their cultural identity.

Wet rice Rice planted on dry land in a nursery and then moved to a deliberately flooded field to promote growth.

PHOTO CREDITS

DK would like to thank Louise Thomas for help with the image research, Holly Jackman for design assistance, and Eloise Musgrove for editorial assistance.

Note: Photos that are identified in the book as numbered figures appear first within each chapter listing, each prefaced with the 3-digit figure number, followed by the credit. Photos that are *not* numbered in the text are grouped secondly within each chapter listing, with the chapter number first, followed by the text page number, a letter identifier as needed, and finally the credit.

Frontmatter FM.v Getty Images FM.vii Mark Ralston/AFP/Getty Images FM.vii.left Frans Lemmens/Alamy Images FM.vii.mid Mark Henley/Panos Pictures FM.vii.right CORBIS FM.viii-ix ©Jim Zukerman/Alamy FM.viii.left Macduff Everton/CORBIS FM.viii.mid James Burger/Alamy/Alamy Images FM.viii.right Adam Woolfitt/Robert Harding Travel/Photolibrary.com FM.ix.left Steve Dunwell/Getty Images FM.ix.mid Neville Elder/Corbis Sygma FM.ix.right Boris Heger/Peter Arnold FM.x-xi © Jon Arnold Images Ltd/Alamy FM.x.left Natalie Behring/Getty Images FM.x.mid Stephen Wilkes/Stone/Getty Images FM.x.right NASA FM.xi.left Barry Lewis/CORBIS FM.xi.right A. T. Willett/Alamy

Chapter 1 1.1.1a James Mellaart 1.1.1c Alan Hills The British Museum 1.1.2 Library of Congress 1.1.3 Mappa Mundi, c.1290 (vellum), 13th c,Richard of Haldingham (Richard de Bello)(fl.c.1260-1305)/The Bridgeman Art Library 1.1.4 The Art Archive/Bodleian Library Oxford 1.1.5 Library of Congress 1.1.6 National Maritime Museum Picture Library, London, England 1.1.7 British Library Board. Picture#080802 1.1.8 The Map House of London/Getty Images 1.1.9 Courtesy of David Rumsey Map Collection, www.davidrumsey.com 1.1.10 Image Makers/Getty Images 1.2.1a Cameron Davidson/Alamy Images 1.2.2a RiMaPics/Alamy Images 1.3.9a Satellite imagery provided by GlobeXplorer.com 1.3.9b Satellite imagery provided by GlobeXplorer.com 1.3.9c Satellite imagery provided by GlobeXplorer.com 1.3.9d Satellite imagery provided by GlobeXplorer.com 1.4.3 Google Maps © Google, and are used with permission 1.5.1 Satellite imagery provided by GlobeXplorer.com 1.5.1b aerialarchives.com/Alamy Images 1.5.2a Wendy Chen/Getty Images 1.5.3 Alan Curtis/Alamy Images 1.2A Frans Lemmens/Alamy Images 1.3A Mary Evans Picture Library/The Image Works 1.3B Alan Curtis/Alamy Images 1.3C Kevin Foy/Alamy Images 1.4A Mary Evans Picture Library/The Image Works 1.6A Stefano Bianchetti/CORBIS 1.6B Bettmann/CORBIS 1.6C Courtesy of University of Leipzig 1.6D Collection #PA46M139#1/University of Kentucky 1.6E Beinecke Rare Book and Manuscript Library, Yale University 1.17 David Crausby/Alamy Images 1.17A Kevin Foy/Alamy Images 1.17B vario images GmbH & Co.KG/Alamy Images 1.17C Robert Harding Picture Library Ltd/Alamy Images 1.17D Andrew Woodley/Alamy Images 1.17E Alison Wright/Alamy Images 1.19A Getty Images 1.22A Mary Evans Picture Library/The Image Works 1.22B Alan Curtis/Alamy Images 1.22C Kevin Foy/Alamy Images 1.22D From "One Planet Many People—Atlas of Our Changing Environment, United Nations Programme, 2005 1.23A James Rubenstein

Chapter 2 2.2.2 Ed Kashi/CORBIS 2.5.1 Fredrik Naumann/Panos Pictures 2.7.2 Qilai Shen/Panos Pictures 2.8.1 Wissam Al-Okaili/AFP/Getty Images 2.8.3 Romeo Rancoco/Reuters/Corbis 2.9.3 Mark Burnett/Alamy Images 2.26A Mark Henley/Panos Pictures 2.27A Staffan Widstand/Corbis Edge 2.27B ER Productions/CORBIS 2.27C Maher Attar/Photolibrary.com 2.28A Will & Deni McIntyre/Stone/Getty Images 2.29C Oldrich Karasek/Alamy Images 2.31A Staffan Widstand/Corbis Edge 2.32A Alamy Images 2.33A ER Productions/CORBIS 2.33B Mike Kemp/Alamy Images 2.34A Joel Stettenheim/CORBIS 2.37A Michel Setboun/CORBIS 2.37B David R. Frazier Photolibrary/Alamy Images 2.37C Holger Leue/Alamy Images 2.38A The Granger Collection 2.38B Maher Attar/Photolibrary.com 2.39A Carl & Ann Purcell/CORBIS 2.40A Owen Franken/CORBIS 2.40B Mark Henley/Panos Pictures 2.44A Gideon Mendel/CORBIS 2.46A Staffan Widstand/Corbis 2.46B ER Productions/CORBIS 2.46C Maher Attar/Photolibrary.com 2.46D Chad Ehlers/Photolibrary.com 2.47A James Rubenstein 2.48A Kapoor Baldev/Sygma/Corbis/Sygma 2.48B Chris Stowers/Panos Pictures

Chapter 3 3.1.1 Alison Wright/CORBIS 3.1.2 Sebastian Forsyth/Alamy Images 3.1.3 G.M.B. Akash/Panos Pictures 3.2.1 Danita Delimont/Alamy Images 3.2.2 Jim West/Alamy Images 3.2.3 Jim Sugar/CORBIS 3.2.4 Jim West/Alamy Images 3.3.1 Carlos Barria/Reuters/Corbis 3.3.3 Mark Ralston/AFP/Getty Images

3.4.1 Mark Henley/Panos Pictures 3.4.2 Tomas van Houtryve/Panos Pictures 3.4.3 Bettmann/CORBIS 3.4.4 Regis Bossu/Sygma/Corbis 3.4.5 Henry Westheim Photgraphy/Alamy 3.5.1 The Granger Collection, New York 3.5.2 Library of Congress 3.5.3 David Turnley/CORBIS 3.6.2 The Travel Library/Rex USA 3.6.3 Ambient Images/Alamy 3.7.1 Paul Smith/Panos Pictures 3.7.2 Rick D'Elia/Corbis 3.7.3 Getty Images 3.7.4 Getty Images 3.8.1 Archive Holdings/Getty Images 3.8.3 Vario Images GmbH & Co. KG/Alamy 3.9.2 Eric Nathan/Alamy Images 3.9.3 Getty Images 3.9.4 Oote Boe Photography/Alamy 3.9.5 Gerald French/ CORBIS 3.50A CORBIS 3.51A Library of Congress 3.51B Getty Images 3.51C Bettmann/CORBIS 3.52A Alison Wright/CORBIS 3.52A1 David H. Wells/CORBIS 3.52B The Travel Library/Rex USA 3.52C Bettmann/CORBIS 3.57A Rafael Marchante/Reuters/Corbis 3.62A Library of Congress 3.64A Christopher J. Morris/CORBIS 3.65A Getty Images 3.66A Bettmann/CORBIS 3.67A Hulton Archive/Getty Images 3.70A Alison Wright/CORBIS 3.70B The Travel Library/Rex USA 3.70C Bettmann/CORBIS 3.70D Library of Congress 3.71 James Rubenstein 3.72A PoodlesRock/CORBIS 3.72B Reuters/Corbis 3.73B Stephanie Maze/CORBIS

Chapter 4 4.1.1 Cathrine Wessel/CORBIS 4.1.2 Bettmann/CORBIS 4.2.1 Courtesy Microsoft Corporation, Andrew Fiore and Marc Smith, 2006 4.74A Macduff Everton/CORBIS 4.75A Choice/Alamy Images 4.75B Bettmann/CORBIS 4.75C James Leynse/CORBIS 4.77A Mark Wilson/Getty Images 4.77B Sylvain Grandadam/Stone/Getty Images 4.78A Lou Linwei/Alamy Images 4.78B Chris Willson/Alamy Images 4.79A Caro/Alamy 4.79B Choice/Alamy Images 4.80A Paul M. Thompson/Alamy Images 4.80B Alistair Berg/Alamy Images 4.80C Richard Wareham Fotografie/Alamy Images 4.81A Michael S. Yamashita/CORBIS 4.81B Joe Fox/Alamy Images 4.81C Robert Cianflone/Allsport Concepts/Getty Images 4.82A Bettmann/ CORBIS 4.83A G.E. Kidder Smith/CORBIS 4.83B Photo Collection Alexander Alland, Sr./CORBIS 4.83C Richard Johnson/Alamy Images 4.83D Bettmann/CORBIS 4.84A WILDLIFE GmbH/Alamy Images 4.84B Robert Holmes/CORBIS 4.85A Cephas Picture Library/Alamy 4.85B Tim Graham/Alamy Images 4.85C Scott Kemper/Alamy Images 4.86A Andrea Matone/Alamy Images 4.86B Storm Stanley/Robert Harding World Imagery/Corbis 4.86C David Turnley/CORBIS 4.86D Reuters/Corbis 4.87A Roger Cracknell/Alamy Images 4.87B David Turnley/CORBIS 4.87C Charles O. Cecil/Alamy Images 4.87D Charles O. Cecil/Alamy Images 4.87E Sarah Hadley/Alamy Images 4.89A Qilai Shen/Panos Pictures 4.90A James Leynse/CORBIS 4.90B Huw Jones/Alamy/Alamy Images 4.90C Henry Westheim/Alamy Images 4.91A Howard Davies/ CORBIS 4.92A A ROOM WITH VIEWS/Alamy Images 4.92B Michael Snell/Alamy Images 4.93A Nic Bothma/epa/Corbis 4.93B Tony Roberts/CORBIS 4.94A Choice/Alamy Images 4.94B Bettmann/CORBIS 4.94C James Leynse/CORBIS 4.94D Petr Svarc/Alamy Images 4.95A James Rubenstein 4.96A Paul Chesley/Stone/Getty Images 4.97A Guiziou Franck/hemis. fr/Alamy Images 4.97B James Pomerantz/Corbis

Chapter 5 5.4.3 North Wind Picture Archives/Alamy Images 5.5.3 Drive Images/Alamy Images 5.7.3a swissworld.org 5.8.1 Rosemary Behan/Alamy Images 5.8.2 Jeff Morgan tourism and leisure/Alamy Images 5.98A James Burger/Alamy/Alamy Images 5.99B EIGHTFISH/Alamy Images 5.99C Rosemary Behan/ Alamy Images 5.102A David J. & Janice L. Frent Collection/CORBIS 5.102B Stephane Frances/Sygma/ Corbis/Sygma 5.103A Paul Souders/Corbis 5.104A Per KarlssonBKWine.com/Alamy Images 5.110A SCPhotos/Alamy 5.110B EIGHTFISH/Alamy Images 5.112A Erwin Gavic, www.autosnelwegen.net 5.113A William Campbell/Sygma/Corbis 5.114A Richard T. Nowitz/CORBIS 5.115A David Robertson/Alamy Images 5.116A J.A. Kraulis/Masterfile Stock Image Library 5.117A Rudi Von Briel/PhotoEdit . 5.118B EIGHTFISH/Alamy Images 5.118C Rosemary Behan/Alamy Images 5.118D Natalie Behring/Panos Pictures 5.119A James Rubenstein

Chapter 6 6.1.1 David Samuel Robbins/CORBIS 6.1.2 Alistair Duncan/Dorling Kindersley 6.1.3 JTB Photo/Photolibrary.com 6.1.4 ArkReligion.com/Alamy Images 6.2.1 Peter Barritt/Alamy Images 6.2.2 Alan Novelli/Alamy Images 6.2.3 luminous/Alamy/Alamy Images 6.2.4 Jim Zuckerman/Alamy/Alamy Images 6.4.2 Scala/Art Resource, NY 6.4.3 Brian A. Vikander/CORBIS 6.4.4 Iain Lowson/Alamy Images 6.5.1 Todd Bigelow/Getty Images 6.5.2 BRIAN HARRIS/Alamy Images 6.6.1 Courtesy of the Library of Congress 6.6.2 ArkReligion.com/Alamy/Alamy Images 6.6.4 Dinodia Images/Alamy 6.7.1 Dan Porges/Peter Arnold . 6.7.2 Keren Su/CORBIS 6.7.3 Carol Beckwith/Angela Fisher/Getty Images 6.7.4 Raghu Rai/Magnum Photos 6.8.3 Peter Macdiarmid/Getty Images 6.9.2 Israel images/Alamy Images 6.122A Adam Woolfitt/Robert Harding Travel/Photolibrary.com 6.123A Jim Zuckerman/Alamy/Alamy Images 6.123B ArkReligion.com/ Alamy/Alamy Images 6.123C Peter Macdiarmid/Getty Images, .Getty News 6.126A David Lyons/Alamy Images 6.126B FALKENSTEINFOTO/Alamy Images 6.127A Fabian von Poser/Alamy Images 6.127A1 Peter M. Wilson/Alamy/Alamy Images 6.128A Joel Wintermantle/Alamy Images 6.129A Michael Ventura/ Alamy Images 6.135A Linda Whitwam Dorling Kindersley 6.135A1 Rolf Bruderer/CORBIS 6.138A Giles

Peress/Magnum Photos,. 6.142A China Tourism Press/Getty Images,. 6.142B Jim Zuckerman/Alamy/Alamy Images 6.142C ArkReligion.com/Alamy/Alamy Images 6.142D Peter Macdiarmid/Getty Images, .Getty News 6.143A James Rubenstein 6.144A Reinhard Krause/Reuters/Corbis 6.144B Shaul Schwarz/Corbis/Sygma

Chapter 7 7.2.1 Collection of The New-York Historical Society, #46093 7.3.2 Bettmann/CORBIS 7.4.6 Stapleton Collection/CORBIS 7.9.4A Otto Lang/CORBIS 7.9.4B Nigel Chandler/Sygma/CORBIS 7.9.4C Matthias Schrader/dpa/Landov Media 7.146A Steve Dunwell/Getty Images 7.147A Collection of The New-York Historical Society, #46093 7.147B David Turnley/CORBIS 7.147C David Turnley/CORBIS 7.149A David Butow/CORBIS SABA 7.150A North Wind Picture Archives/Alamy 7.156A Peter Turnley/CORBIS 7.157A David Turnley/CORBIS 7.161A Peter Turnley/CORBIS 7.162A David Turnley/CORBIS 7.163A Bernard Bisson/Sygma/Corbis 7.163C Peter Turnley/CORBIS 7.166A Collection of The New-York Historical Society, #46093 7.166B David Turnley/CORBIS 7.166C David Turnley/CORBIS 7.166D Jim West/Alamy Images 7.166E Andre Jenny/Alamy Images 7.167A James Rubenstein 7.168A United States Department of Defense

Chapter 8 8.1.2 Jeremy Woodhouse/Alamy Images 8.2.2 Guy Vanderelst/Getty Images 8.2.3 Medioimages/Photodisc/Getty Images 8.2.4 David Lees/CORBIS 8.6.1 Bernard Bisson/Sygma/Corbis 8.7.1 Reproduced from James Parton, Caricature and Other Comic Art. New York: Harper Brothers, 1877, p. 316. Courtesy of the Library of Congress 8.8.3 UPI/Corbis/Bettmann 8.8.4L US State Department 8.8.4R US State Department 8.9.1 Sean Adair/Reuters/Corbis 8.9.2 Mike Stewart/Corbis/Sygma 8.9.3 Sion Touhig/CORBIS 8.9.4 Guillermo Navarro/CORBIS 8.170A Neville Elder/Corbis Sygma 8.171A Guy Vanderelst/Getty Images 8.171B Jonathan Blair/CORBIS 8.171C Sion Touhig/CORBIS 8.172A Mike King/Alamy Images 8.176A John Dakers; Eye Ubiquitous/CORBIS 8.179A Derek Hudson/Sygma/Corbis 8.180A Martin Harvey/Alamy Images 8.181A Andrew Caballero Reynolds/Alamy Images 8.181B Jonathan Blair/CORBIS 8.190A Guy Vanderelst/Getty Images 8.190B Jonathan Blair/CORBIS 8.190C Sion Touhig/CORBIS 8.190D Frischling Steven/CORBIS SYGMA 8.191 James Rubenstein 8.192A Reuters/CORBIS

Chapter 9 9.1.1 Getty Images 9.1.2 Alice Attie/epa/CORBIS 9.1.3 Karen Kasmauski/CORBIS 9.1.6 thislifeAfrica/Alamy Images 9.2.7 Stock Connection Distribution/Alamy 9.3.2 G.M.B. Akash/Panos Pictures 9.3.7 vario images GmbH & Co.KG/Alamy 9.4.3 Danita Delimont/Alamy 9.5.2 Images&Stories/Alamy Images 9.5.6 Aubrey Wade/Panos Pictures 9.6.2 Juan Vrijdag/Panos Pictures 9.7.1 PeerPoint/Alamy Images 9.7.2 EIGHTFISH/Alamy Images 9.8.1 Getty Images 9.8.3 Philippe Lissac/Panos Pictures 9.9.1 Ron Giling/Peter Arnold 9.194A Boris Heger/Peter Arnold 9.195A Giacomo Pirozzi/Panos Pictures 9.195B Peter Barker/Panos Pictures 9.195C Ron Giling/Peter Arnold 9.196A Mikkel Ostergaard/Panos Pictures 9.197A Tom & Dee Ann McCarthy/CORBIS 9.198A Helen King/CORBIS 9.198B Giacomo Pirozzi/Panos Pictures 9.199A Giacomo Pirozzi/Panos Pictures 9.202A Corbis Premium RF/Alamy 9.202B Giacomo Pirozzi/Panos Pictures 9.204A Caroline Penn/Panos Pictures 9.206A Dinodia Images/Alamy 9.209A Nick Cobbing/Rex USA 9.209B Chung Sung-Jun/Getty Images 9.211A Philippe Lissac/Panos Pictures 9.212A Stuart Kelly/Alamy 9.212B Ron Giling/Peter Arnold,. 9.213A Sue Cunningham Photographic/Alamy 9.213B David Marsden/Photolibrary.com 9.214A Giacomo Pirozzi/Panos Pictures 9.214B Peter Barker/Panos Pictures 9.215A James Rubenstein 9.216A Qilai Shen/Panos Pictures

Chapter 10 10.3.1 Pegaz/Alamy 10.3.2 Microsoft Virtual Earth. Getmapping plc. 10.3.3 AfriPics.com/Alamy 10.5.1 Joao Luiz Bulcao/CORBIS 10.5.2 Andy Aitchison/CORBIS 10.5.4 Keren Su/CORBIS 10.5.5 Rob Walls/Alamy 10.7.2 Dung Vo Trung/CORBIS 10.7.3 Peter Johnson/CORBIS 10.8.2.b Justin Kase zsixz/Alamy 10.8.2a Trip/Alamy Images 10.8.2c Nigel Cattlin/Alamy 10.8.2d DWP Imaging/Alamy 10.8.2e Stock Connection Distribution/Alamy 10.9.1 Findlay/Alamy Images 10.218A Natalie Behring/Getty Images 10.219A Gary Dublanko/Alamy 10.219B imagebroker/Alamy 10.219C Dung Vo Trung/CORBIS 10.221A Gary Dublanko/Alamy 10.222A Aerial Archives/Alamy 10.226A Paulo Fridman/CORBIS 10.230A imagebroker/Alamy 10.231A inga spence/Alamy 10.234A Malcolm Case-Green/Alamy 10.238A Gary Dublanko/Alamy 10.238B imagebroker/Alamy 10.238C Dung Vo Trung/CORBIS 10.238D Paolo Negri/Getty Images 10.239A James Rubenstein 10.240A Eye Candy Images/Alamy Images 10.240B Howard Burditt/Reuters/Corbis

Chapter 11 11.1.1 Hulton Archive/Getty Images 11.1.2 Dave King/Dorling Kindersley, Courtesy of The Science Museum, London 11.1.3 The Granger Collection, New York 11.3.2 Ludo Kuipers/CORBIS 11.3.3 Peter Endig/epa/CORBIS 11.3.4 Ed Kashi/CORBIS 11.4.1 Bettmann/CORBIS 11.4.2 Ve Streano/CORBIS

11.4.5 Ewasko/Stone/Getty Images 11.6.1 Oliver Berg/dpa/CORBIS 11.6.2 Glowimages/Getty Images 11.6.3 Dan Lamont/CORBIS 11.6.4 Johner Images/Getty ImagesJohner Images 11.6.5 Alan Schein/zefa/Corbis 11.7.1 Yann Layma/Getty Images,. 11.7.2 Hideo Kurihara/Alamy Images 11.7.3 Panoramic Images/Getty Images 11.9.1 Daniel Leclair/Reuters/Corbis 11.9.4 Richard T. Nowitz/CORBIS 11.242A Stephen Wilkes/Stone/Getty Images 11.243A Hulton Archive/Getty Images 11.243B Ewasko/Stone/Getty Images 11.243C Hideo Kurihara/Alamy Images 11.245A Mary Evans Picture Library/The Image Works 11.248A David Kennedy/epa/CORBIS 11.252A Kevin Fleming/CORBIS 11.258A Lowell Georgia/CORBIS 11.259A Michael S. Yamashita/CORBIS 11.259B Jerry Arcieri/Corbis/CORBIS 11.262A Hulton Archive/Getty Images 11.262B Ewasko/Stone/Getty Images 11.262C Hideo Kurihara/Alamy Images 11.262D Christian Charisius/Reuters/Corbis 11.263A James Rubenstein 11.264A Hulton-Deutsch Collection/CORBIS

Chapter 12 12.2.1 Roger Wood/CORBIS 12.2.2 Robert Harding Picture Library Ltd/Alamy 12.2.3 Historical Picture Archive/CORBIS 12.2.4 Bettmann/CORBIS 12.5.3a Nic Cleave Photography/Alamy Images 12.5.3b Bobbi Lane/Beateworks/CORBIS 12.5.3c Robert Landau/CORBIS 12.6.1 Robert Harding World Imagery/CORBIS 12.7.3a Gwendolyn Mambo/Alamy Images 12.7.3b API/Alamy Images 12.7.3c James Leynse/CORBIS 12.7.3d Enigma/Alamy Images 12.7.3e Yadid Levy/Alamy Images 12.8.3 Ben Pipe/Alamy Images 12.266A NASA 12.267A David R. Frazier Photolibrary/Alamy Images 12.267B Atlantide Phototravel/CORBIS 12.267C Gwendolyn Mambo/Alamy Images 12.268A Art Kowalsky/Alamy Images 12.268B Visions of America, LLC/Alamy 12.268C dbimages/Alamy 12.269A Jan Tadeusz/Alamy Images 12.269B Nic Cleave Photography/Alamy Images 12.269C Wolfgang Kaehler/Alamy Images 12.269D travelbild.com/Alamy Images 12.269E Charles Stirling (Travel)/Alamy Images 12.269F Jon Arnold Images Ltd/Alamy Images 12.273A Bill Wymar/Alamy Images 12.274A Peter Ravallo/Alamy/Alamy Images 12.279A Atlantide Phototravel/CORBIS 12.283A Ron Chapple Stock/Alamy Images 12.284A The Bon-Ton Stores,. 12.285A Courtesy United Dairy Farmers,. 12.286A David R. Frazier Photolibrary, ./Alamy Images 12.286B Atlantide Phototravel/CORBIS 12.286C Gwendolyn Mambo/Alamy Images 12.286D David R. Frazier Photolibrary, ./Alamy Images 12.286E James Rubenstein 12.287A James Rubenstein 12.288A Kevin Schafer/Alamy Images 12.288B Eric Nathan/Alamy Images

Chapter 13 13.290A Barry Lewis/CORBIS 13.3.1 Dennis Cox/Creative Eye/MIRA.com 13.3.3 Jeff Haynes/AFP/Getty Images 13.6.1 Peter Adams Photography/Alamy 13.7.1 Dennis MacDonald/Alamy Images 13.8.1 Tom Uhlman/Alamy Images 13.8.2 David Davis Photoproductions/Alamy Images 13.291A SuperStock/Alamy Images 13.291B Mike Powell/Lifesize/Getty Images 13.291C Lester Lefkowitz/Stone/Getty Images 13.292A Panoramic Images/Getty Images 13.293A Najlah Feanny/CORBIS SABA 13.293B Jeff Greenberg/Alamy Images 13.293C Kim Karpeles/Alamy Images 13.295A David Noble Photography/Alamay 13.295B Ambient Images ./Alamy 13.295C Ball Miwako/Alamy Images 13.295D SuperStock/Alamy Images 13.298A Visions of America, LLC/ALamy 13.298B imagebroker/Alamy 13.299A Mike Powell/Lifesize/Getty Images 13.299B Ron Levine/Getty Images 13.300A Joel Sartore/GettyNational Geographic Society 13.300B Corbis 13.301A Image Source/Getty Images 13.301B Visions of America, LLC/Alamy 13.301C Janet Jarman/CORBIS 13.301D Jeffery Allan Salter/Corbis-SABA Press Photos 13.302A Mark Edwards/Peter Arnold 13.302B Terrazas Glavan Monica-Unep/Alamy Images 13.303A Michael Cogliantry/Getty Images 13.303B Michel Setboun/Digital/CORBIS 13.304A Andre Jenny/Alamy Images 13.304B Nik Wheeler/CORBIS 13.304D Andre Jenny/Alamy Images 13.305A Lester Lefkowitz/Stone/Getty Images 13.306A Tom Uhlman/Alamy 13.306B Getty Images 13.307A Derrick Alderman/Alamy Images 13.307B Jason Lindsey/Alamy Images 13.308A Martin Roemers/Panos Pictures 13.309A Robert Clay/Alamy Images 13.309B Getty Images 13.309C Phil Schermeister/CORBIS 13.310A SuperStock/Alamy Images 13.310B Mike Powell/Lifesize/Getty Images 13.310C Lester Lefkowitz/Getty Images 13.310D Flashfilm/Getty Images 13.311A James Rubenstein 13.312A imac/Alamy Images 13.312B LH Images/Alamy Images 13.312C MShieldsPhotos/Alamy Images

Chapter 14 14.1.4 B. Constantino/Alamy Images 14.2.3 Anthony Hope/Alamy Images 14.4.2left NASA 14.4.2right NASA 14.4.4 David R. Frazier Photolibrary/Alamy 14.4.5 Tina Manley/North America/Alamy 14.5.1 Chris Howes/Getty Images 14.5.2 Image courtesy of USGS 14.6.1 Jim West/Alamy Images 14.6.2 KLJ Photographic/Alamy 14.8.1a SCPhotos/Alamy 14.8.1b David R. Frazier Photolibrary/Alamy 14.8.1c Ball Miwako/Alamy 14.8.1d InsideOutPix/Alamy 14.8.2 Lourens Smak/Alamy 14.9.1 David R. Frazier Photolibrary/Alamy Images 14.9.2 Martin Shields/Alamy Images 14.314A A. T. Willett/Alamy 14.315A Paul Fleet/Alamy Images 14.315B Chris Howes/Getty Images 14.315C SCPhotos/Alamy 14.317A Paul

Fleet/Alamy Images 14.318A H. Armstrong Roberts/ClassicStock/CORBIS 14.321A Dave King Dorling Kindersley, Courtesy of The Science Museum, London 14.321B Colin Keates Dorling Kindersley, Courtesy of the Natural History Museum, London 14.324A Neil McAllister/Alamy Images 14.325A Anthony Vizard; Eye Ubiquitous/CORBIS 14.328A China Images/Alamy 14.328B vario images GmbH & Co.KG/Alamy 14.329A Peter Donaldson/Alamy Images 14.329B Jack Hobhouse/Alamy Images 14.334A Paul Fleet/Alamy Images 14.334B Chris Howes/Getty Images 14.334C SCPhotos/Alamy 14.334D Tom Hanley/Alamy Images 14.335A James Rubenstein

INDEX

C

J

K

S

W

Y

Z

With ***Contemporary Human Geography***, renowned author James M. Rubenstein has partnered with Pearson Prentice Hall and the information architects at Dorling Kindersley to provide an exciting new approach to the study of Human Geography that is inviting, informative, and challenging.

Praise for *Contemporary Human Geography*

"This is geography as it should be. The text provides superb graphics that can only encourage students to grasp the breadth of geographic expression on the landscape."

CLAUDIA LOWE, FULLERTON COLLEGE

"Most texts are written for other professors, not for students. This one is different. Very different."

OWEN DWYER, INDIANA UNIVERSITY–PURDUE UNIVERSITY INDIANAPOLIS

"Students are more likely to do the reading, and more likely to digest this material than they are in a conventional textbook."

ED CARR, UNIVERSITY OF SOUTH CAROLINA

"I think that previous texts have included way too much detail and are very difficult to cover in one semester. This type of text will help excite students about geography as a discipline, and move beyond the idea that geography is about memorization and location."

WENDY SHAW, SOUTHERN ILLINOIS UNIVERSITY

"As a springboard for the study of human geography, this text is a refreshing new approach. It excites students to further inquiry and investigation."

DANIEL VARA, FITCHBURG STATE COLLEGE

ABOUT THE COVER

The Crown Fountain in Chicago's Millennium Park displays faces of a broad spectrum of local citizens onto the fountain towers. The City of Chicago decided to allocate prime downtown real estate to this bold, contemporary, intera ... for the cover of this innovative textbook.

Prentice Hall
is an imprint of
PEARSON
DK
www.mygeos...
www.pearsonhighered.com